BIRKHÄUSER

Progress in Mathematics
Volume 263

Stevo Todorcevic

Walks on Ordinals and Their Characteristics

Birkhäuser
Basel · Boston · Berlin

Stevo Todorcevic
Université Paris VII – C.N.R.S.
UMR 7056
2, Place Jussieu – Case 7012
75251 Paris Cedex 05
France
e-mail: stevo@math.jussieu.fr

Department of Mathematics
University of Toronto
Toronto M5S 2E4
Canada
e-mail: stevo@math.toronto.edu

and

Mathematical Institute, SANU
Kneza Mihaila 35
11000 Belgrad
Serbia
e-mail: stevo@mi.sanu.ac.yu

2000 Mathematics Subject Classification 03E10, 03E75, 05D10, 06A07, 46B03, 54D65, 54A25

Library of Congress Control Number: 2007933914

Bibliographic information published by Die Deutsche Bibliothek. Die Deutsche Bibliothek lists this publication in the Deutsche Nationalbibliografie; detailed bibliographic data is available in the Internet at http://dnb.ddb.de

ISBN 978-3-7643-8528-6 Birkhäuser Verlag AG, Basel · Boston · Berlin

Basel · Boston · Berlin
P.O. Box 133, CH-4010 Basel, Switzerland
Part of Springer Science+Business Media
Printed on acid-free paper produced from chlorine-free pulp. TCF∞
Printed in Germany

ISBN 978-3-7643-8528-6

e-ISBN 978-3-7643-8529-3

9 8 7 6 5 4 3 2 1

www.birkhauser.ch

Contents

Chapter 1

Introduction

1.1 Walks and the metric theory of ordinals

This book is devoted to a particular recursive method of constructing mathematical structures that live on a given ordinal θ, using a single transformation $\xi \mapsto C_\xi$ which assigns to every ordinal $\xi < \theta$ a set C_ξ of smaller ordinals that is closed and unbounded in the set of ordinals $< \xi$. The transfinite sequence

$$C_\xi \ (\xi < \theta)$$

which we call a 'C-sequence' and on which we base our recursive constructions may have a number of 'coherence properties' and we shall give a detailed study of them and the way they influence these constructions. Here, 'coherence' usually means that the C_ξ's are chosen in some canonical way, beyond the already mentioned and natural requirement that C_ξ is closed and unbounded in ξ for all ξ. For example, choosing a canonical 'fundamental sequence' of sets $C_\xi \subseteq \xi$ for $\xi < \varepsilon_0$, relying on the specific properties of the Cantor normal form for ordinals below the first ordinal satisfying the equation $x = \omega^x$, is a basis for a number of important results in proof theory. In set theory, one is interested in longer sequences as well and usually has a different perspective in applications, so one is naturally led to use some other tools besides the Cantor normal form. It turns out that the sets C_ξ can not only be used as 'ladders' for climbing up in recursive constructions but also as tools for 'walking' from an ordinal β to a smaller one α,

$$\beta = \beta_0 > \beta_1 > \cdots > \beta_{n-1} > \beta_n = \alpha,$$

where the 'step' $\beta_i \to \beta_{i+1}$ is defined by letting β_{i+1} be the minimal point of C_{β_i} that is bigger than or equal to α. This notion of a 'walk' and the corresponding 'characteristics' and 'distance functions' constitute the main body of study in this book. We show that the resulting 'metric theory of ordinals' is a theory of considerable intrinsic interest which provides not only a unified approach to a number of classical problems in set theory but is also easily applicable to other areas of mathematics. For example, highly applicable characteristics of the walk are defined on the basis of the corresponding 'traces'. The most natural trace of the walk is its 'upper trace' defined simply to be the set

$$\mathrm{Tr}(\alpha, \beta) = \{\beta_0 > \beta_1 > \cdots > \beta_{n-1} > \beta_n\}$$

of places visited along the way, which is of course most naturally enumerated in decreasing order. Another important trace of the walk is its 'lower trace,' the set

$$\Lambda(\alpha, \beta) = \{\lambda_0 \leq \lambda_1 \leq \cdots \leq \lambda_{n-2} \leq \lambda_{n-1}\},$$

where $\lambda_i = \max(\bigcup_{j=0}^{i} C_{\beta_j} \cap \alpha)$ for $i < n$. The traces are usually used in defining various binary operations on ordinals $< \theta$, the most prominent of which is the 'square-bracket operation' that gives us a way to transfer the quantifier 'for every unbounded set' to the quantifier 'for every closed and unbounded set'. It is perhaps not surprising that this reduction of quantifiers has proven to be quite useful in constructions of mathematical structures on θ where one needs to have some grip on substructures of cardinality θ.

From the metric theory of ordinals based on analysis of walks, one also learns that the triangle inequality of an ultrametric

$$\varrho(\alpha, \gamma) \leq \max\{\varrho(\alpha, \beta), \varrho(\beta, \gamma)\}$$

has three versions, depending on the natural ordering between the ordinals α, β and γ. The three versions of the inequality are in fact of a quite different character and occur in quite different places and constructions in set theory. For example, the most frequent occurrence is the case $\alpha < \beta < \gamma$, when the triangle inequality becomes something that one can call 'transitivity' of ϱ. Considerably more subtle is the case $\alpha < \gamma < \beta$ of this inequality[1]. It is this case of the inequality that captures most of the coherence properties found in this article. It is also an inequality that has proven to be quite useful in applications.

A large portion of the book is organized as a discussion of four basic characteristics of the walk $\rho, \rho_0, \rho_1, \rho_2$ and ρ_3. The reader may choose to follow the analysis of any of these functions in various contexts. The characteristic $\rho_0(\alpha, \beta)$ codes the entire walk $\beta = \beta_0 > \beta_1 > \cdots > \beta_{n-1} > \beta_n = \alpha$ by simply listing the positions of β_{i+1} in the set C_{β_i} for $i < n$. While this looks simple-minded, the resulting mapping ρ_0 is a rather remarkable object. For example, in the realm of the space ω_1 of countable ordinals, it gives us a canonical example of a special Aronszajn tree of increasing sequences of rationals which has the additional remarkable property that, when ordered lexicographically, its cartesian square can be covered by countably many chains. In other words, the single characteristic ρ_0 of walks on countable ordinals gives two critical structures, one in the class of

[1] It appears that the third case $\beta < \alpha < \gamma$ of this inequality is rarely a reasonable assumption to be made in this context.

so-called Lipschitz trees and the other in the class of linear orderings. For higher cardinals θ, analysis of ρ_0 leads us to some interesting finitary characterizations of hyper inaccessible cardinals. This is given in some detail in Chapter 6 of this book.

The characteristic $\rho_1(\alpha, \beta)$ loses a considerable amount of information about the walk as it records only the maximal order type among the sets

$$\{C_{\beta_0} \cap \alpha, C_{\beta_1} \cap \alpha, \ldots, C_{\beta_{n-1}} \cap \alpha\}.$$

Nevertheless it gives us the first example of what we call a 'coherent mapping'. The class of coherent mappings and trees in the case $\theta = \omega_1$ exhibits an unexpected structure that we study in great detail in Chapter 4 of the book. The fine structure in the class of 'coherent trees' is based on the metric notion of a 'Lipschitz mapping' between trees. The profusion of such mappings between coherent trees eventually leads us to the so-called 'Lipschitz Map Conjecture' that has proven crucial for the final resolution of the basis problem for uncountable linear orderings and that is presented in the same chapter. For higher cardinals θ the characteristic ρ_1 and its local versions offer a rich source of so-called 'unbounded functions' that have some applications.

The characteristic $\rho_2(\alpha, \beta)$ simply counts the number of steps of the walk from β to α. While this also looks rather simple minded, the remarkable properties of the corresponding function ρ_2 become especially apparent on higher cardinals θ. Important properties of this characteristic are its coherence and its unboundedness. The coherence property of ρ_2 requires the corresponding C-sequence C_ξ $(\xi < \theta)$ to be 'coherent' in the sense that $C_\alpha = C_\beta \cap \alpha$ whenever α is a limit point of C_β. On the other hand, the unboundedness of ρ_2 translates into a requirement that the corresponding C-sequence C_ξ $(\xi < \theta)$ be 'nontrivial'[2], a condition that eventually leads us to a simple and natural characterization of weakly compact cardinals that we choose to reproduce in some detail in Chapter 6.

Finally, the characteristic $\rho_3(\alpha, \beta)$ attaches one of the digits 0 or 1 to the walk according to the behavior of the last step $\beta_{n-1} \to \beta_n = \alpha$. The full analysis of this characteristic is currently available only in the reals of the space ω_1, where ρ_3 becomes a rather canonical example of a sequence-coherent mapping with values in $\{0, 1\}$ and with properties reminiscent of those appearing in the well-known notion of a Hausdorff gap in the quotient algebra $\mathcal{P}(\omega)/\text{fin}$ (another critical object that shows up in many problems about this quotient structure).

The true 'metric theory of ordinals' comes only with development of the characteristic $\rho(\alpha, \beta)$ of the walk that takes advantage of the so-called 'full lower trace' of the walk. The depth of this characteristic is apparent even in the space ω_1 of countable ordinals, but its full power comes at higher cardinals θ and especially at θ that are successors of singular cardinals. The full analysis of the characteristic ρ requires C_ξ $(\xi < \theta)$ to be a so-called 'square sequence' or in other words requires

[2]We say that C_ξ $(\xi < \theta)$ is *nontrivial* if there is no closed and unbounded set $C \subseteq \theta$ such that. for all limit points α of C, there is $\beta \geq \alpha$ such that $C \cap \alpha \subseteq C_\beta$.

the most widely known coherence condition on this sequence, which says that if α is a limit point of C_β, then $C_\alpha = C_\beta \cap \alpha$. It is not surprising that this characteristic has the largest number of applications, many of which are reproduced in this book. We have already mentioned that its development in the case $\theta = \omega_1$ was the initial impulse for development of the so-called metric theory of countable ordinals that has already a rich spectrum of applications. At higher cardinals θ the characteristic ρ can be used in facilitating set-theoretic forcing constructions of rather special objects and we shall reproduce some of these constructions in Chapter 7 of this book. While, as said above, the full development of ρ requires C_ξ $(\xi < \theta)$ to be a 'square sequence', the function ρ itself holds considerable information about the notion of 'square sequences.' For example in Chapter 7, we use ρ to turn an arbitrary square sequence C_ξ $(\xi < \theta)$ into a non-special one by expressing the usual order relation among ordinals $< \theta$ as an increasing union of tree-orderings that come from square sequences on θ themselves. This again leads us to some applications that we choose to reproduce in detail at the end of Chapter 7.

We have already mentioned that one of the important outcomes of our study of walks on ordinals is the 'square-bracket operation', a transformation which to every pair $\alpha < \beta$ of ordinals $< \theta$ assigns an ordinal $[\alpha\beta]$ belonging to the upper trace $\mathrm{Tr}(\alpha, \beta)$ of the walk from β to α. We have also mentioned that the choice of $[\alpha\beta]$ has to be rather careful in order to reduce an arbitrary unbounded subset A of θ to the corresponding set of values

$$\{[\alpha\beta] : \alpha, \beta \in A \text{ and } \alpha < \beta\}$$

that contains a closed and unbounded subset of θ relative to some fixed stationary set $\Gamma \subseteq \theta$, which the C-sequence C_ξ $(\xi < \theta)$ avoids[3]. We present several variations on the way $[\alpha\beta]$ is chosen, each of which works best in some particular context. The common feature of these definitions of $[\alpha\beta]$ is that they are all based on the oscillation mapping

$$\mathrm{osc} : \mathcal{P}(\theta)^2 \to \mathrm{Card}$$

defined by

$$\mathrm{osc}(x, y) = |x \setminus (\sup(x \cap y) + 1)/ \sim |,$$

where $\sim$ is the equivalence relation on $x \setminus (\sup(x \cap y) + 1)$ defined by letting $\alpha \sim \beta$ if and only if the closed interval determined by the ordinals α and β contains no point from y. In other words, $\mathrm{osc}(x, y)$ is simply the number of convex pieces that the set $x \setminus (\sup(x \cap y) + 1)$ is split into by the set y. The original theory of the oscillation mapping osc has been developed in the realm of partial functions from θ into θ. In other words, there is a well-developed theory of the oscillation mapping

$$\mathrm{osc}(s, t) = |\{\xi : s(\xi) \leq t(\xi) \text{ but } s(\xi^+) > t(\xi^+)\}|,^4$$

[3]C_ξ $(\xi < \theta)$ *avoids* Γ if $C_\alpha \cap \Gamma = \emptyset$ for all limit ordinals $\alpha < \theta$.

[4]Here, ξ^+ is the immediate successor of ξ in the common domain of s and t.

(see, for example, [111]), but the general theory works equally well and it will be in part reproduced here in Chapters 8 and 9. The common feature of all results of such a theory is identification of the notion of 'unbounded', in either of the two contexts, in such a way that the typical oscillation result would say that the set of values $\mathrm{osc}(x, y)$ the oscillation mapping takes when x and y run inside two 'unbounded' sets is in some sense rich. In our context of defining the square-bracket operation, the sets x and y in $\mathrm{osc}(x, y)$ are members of our C-sequence C_ξ ($\xi < \theta$) on which we base the notion of walk, and the notion of 'unbounded' becomes the familiar notion of nontriviality of C_ξ ($\xi < \theta$). This makes the square-bracket operation $[\alpha\beta]$ well defined in a wide variety of contexts and therefore quite applicable.

Judging from the applications found so far, it appears that in order to make a particular variation of the oscillation mapping or the square-bracket operation useful, one needs to be be able to give a quite precise estimate of its behavior, not only on unbounded subsets of θ but also on families of θ pairwise-disjoint finite subsets of θ. It is for this reason that definite results about any particular variation of osc and $[\alpha\beta]$ presented in this book will typically be about families A of pairwise-disjoint finite subsets of θ. Given such a family A of cardinality θ, by going to a subfamily, we may assume that elements of A have some fixed finite cardinality n. So, given a in A one can view it enumerated increasingly as $a(0), \ldots, a(n-1)$. All variations of the square-bracket operation that we present will have the property that the set of ordinals $\xi < \theta$ that can be represented as

$$\xi = [a(0)b(0)] = [a(1)b(1)] = \cdots = [a(n-1)b(n-1)],$$

for some $a < b$ in A, contains a closed and unbounded set relative to some fixed stationary set $\Gamma \subseteq \theta$ which the C-sequence C_α ($\alpha < \theta$) avoids. Many applications however require that we know the values $[a(i)b(j)]$ when $i \neq j < n$ and when a and b run through A. It turns out that modulo taking a 'projection' $[\![\cdot\cdot]\!]$ of $[\cdot\cdot]$, (or in other words, modulo composing $[\cdot\cdot]$ with a map from θ into θ) for many of the square-bracket operations that we define in this book, the particular set

$$\{[\![a(i)b(j)]\!] : i, j < n \text{ and } i \neq j\}$$

of values will be independent of the choice of $a \neq b$ in A. This turns out to be crucial in several applications of $[\cdot\cdot]$ presented in this book. Naturally, one would also like to know whether one can define a variation on $[\cdot\cdot]$ where we would have freedom of getting arbitrary values of the form $[a(i)b(j)]$ independently of whether $i = j$ or not. It turns out that this is indeed possible for some choices of θ, though the corresponding definitions are necessarily less general as they do not apply in the case $\theta = \omega_1$ since otherwise one would be able to prove that the countable chain condition is not productive without appealing to additional axioms of set theory.[5] In case $\theta = \omega_1$, we do have symmetric binary operations with some degree

[5]Recall that the countable chain condition is a productive property under MA_{ω_1} (see [36], 41E).

of freedom in that direction (see Chapter 2), but the exact breaking point between this and what requires additional axioms of set theory has yet to be determined.

This book will also present some higher-dimensional characteristics of the walk, though in that context the full theory is yet to be developed. For example, in Chapter 10, we consider the characteristic $\tau(\alpha,\beta,\gamma)$ which to any given three ordinals $\alpha<\beta<\gamma<\theta$ assigns the place where the walk from γ to α branches from the walk from γ to β. It turns out that when C_ξ $(\xi<\theta^+)$ is a square sequence, the characteristic τ can be used to 'step-up' objects living on θ to objects on its successor θ^+. For example, one application of this characteristic is found in the proof that Chang's Conjecture[6] is equivalent to a 3-dimensional Ramsey-theoretic statement saying that, for every coloring of $[\omega_2]^3$ with ω_1 colors, there is an uncountable set $B\subseteq\omega_2$ which misses at least one of the colors. The 3-dimensional characteristic $\chi(\alpha,\beta,\gamma)$ that simply measures the length of the common parts of the walks $\gamma\to\alpha$ and $\gamma\to\beta$ can be used for detecting when a subset Γ of θ admits a rich restriction of the 3-dimensional version of the oscillation mapping,

$$\mathrm{osc}:[\theta^+]^3\longrightarrow\omega,$$

defined on the basis of its 2-dimensional version as follows:

$$\mathrm{osc}(\alpha,\beta,\gamma)=\mathrm{osc}(C_{\beta_s}\setminus\alpha, C_{\gamma_t}\setminus\alpha),$$

where $s=\rho_0(\alpha,\beta)\restriction\chi(\alpha,\beta,\gamma)$ and $t=\rho_0(\alpha,\gamma)\restriction\chi(\alpha,\beta,\gamma)$. Here β_s is the member β_i of the trace of the walk $\beta=\beta_0>\cdots>\beta_l=\alpha$ whose code is the sequence s, i.e., $\rho_0(\beta_i,\beta)=s$, and similarly γ_t is the term γ_j of the walk $\gamma=\gamma_0>\cdots>\gamma_k=\alpha$ whose code is the sequence t. In Chapter 10 we show that our analysis of the 3-dimensional version of the oscillation mapping leads naturally to a square-bracket operation in that dimension, though the full analogy is yet to be completed as one still needs to determine the behavior of this operation on families of pairwise-disjoint finite subsets of θ^+ (which at the moment seems elusive).

The final section of Chapter 10 is concerned with generalizing the basic notion of walks to the context of sets of ordinals rather than ordinals themselves, with the goal of obtaining two-cardinal versions of the square-bracket operation. For example, we show that for every pair of infinite cardinals $\kappa<\lambda$ with κ regular, there is a mapping $c:[[\lambda]^\kappa]^2\to\lambda$ such that, for every cofinal subset U of $[\lambda]^\kappa$ and every $\xi\in\lambda$, there exist $x\subset y$ in U such that $c(x,y)=\xi$. Fuller analogues of the square-bracket operation are however obtained only in certain cases. For example, assuming that there is a stationary subset S of $[\lambda]^\omega$ which is equinumerous with a locally countable[7] subset of $[\lambda]^\omega$, one can define a square-bracket operation

$$[\cdot\cdot]_S:[[\lambda]^\omega]^2\to[\lambda]^\omega$$

[6] Recall that Chang's conjecture is the model-theoretic statement claiming that every model of a countable signature that has the form $(\omega_2,\omega_1,<,\dots)$ has an uncountable elementary submodel M such that $M\cap\omega_1$ is countable.

[7] A subset $K\subseteq[\lambda]^\omega$ is locally countable, if the set $\{x\in K: x\subseteq a\}$ is countable for every $a\in[\lambda]^\omega$.

with the property that, for every cofinal subset U of $[\lambda]^\omega$, the set of all $s \in S$ that are not of the form $[xy]_S$ for some $x \subset y$ in U is not stationary in $[\lambda]^\omega$. Since under the same assumption about the stationary set S, this set can be partitioned into $|S|$ pairwise-disjoint stationary subsets, one gets the following: For every stationary subset S of $[\lambda]^\omega$ that is equinumerous with a locally countable subset of $[\lambda]^\omega$, there is a projection

$$[\![..]\!]_S : [[\lambda]^\omega]^2 \to S$$

of the square-bracket operation $[..]_S$ with the property that, for every cofinal set $U \subseteq [\lambda]^\omega$ and every $z \in S$, there exist $x \subset y$ in U such that $[\![xy]\!]_S = z$.

An interesting phenomenon that one realizes while analysing walks on ordinals is the special role of the first uncountable ordinal ω_1 in this theory. Any natural coherency requirement on the sets C_ξ ($\xi < \theta$) that one finds in this theory is satisfiable in the case $\theta = \omega_1$. How natural the notion of walk in this context is can be seen from the fact that basically all of its characteristics lead us in one way or the other to some 'critical' structure that shows up in various rough classifications of mathematical structures. For example, any of the characteristics ρ, ρ_0, ρ_1, ρ_2 and ρ_3 of the walk that we study here lead us to the canonical linear ordering appearing on the list of five linear orderings that forms a basis for the class of all uncountable linear orderings. The first uncountable cardinal is the only cardinal on which the theory can be carried out without relying on additional axioms of set theory. The first uncountable cardinal is also the place where the theory has its deepest applications as well as its most important open problems. This special role can perhaps be best explained by the fact that many set-theoretical problems, especially those coming from other fields of mathematics, are usually concerned only about the duality between the countable and the uncountable rather than some intricate relationship between two or more uncountable cardinalities. [8] For example, consider the classical problem coming from topology asking if the hereditary separability of a given regular space X is equivalent to its dual requirement that every collection of open subsets of X has a countable subcollection with the same union.[9] It turns out that this problem has a reformulation in terms of the behavior of mappings of the form $c : [\omega_1]^2 \longrightarrow 2$ on uncountable families A of pairwise-disjoint finite subsets of ω_1. In other words, a mapping that has a certain complex behavior on uncountable families of pairwise disjoint finite sheafs would lead to an example of a regular space in which one of the implications fails. For example, one can produce a hereditary separable non-Lindelöf space assuming there is $c : [\omega_1]^2 \longrightarrow 2$ with the property that, for every uncountable family A of pairwise-disjoint finite subsets of ω_1, all of some fixed size n for every position

[8]This is of course not to say that an intricate relationship between two or more uncountable cardinalities may not be a profitable detour in the course of solving such a problem. In fact, this is one of the reasons for our attempt to develop the metric theory of ordinals without restricting ourselves to the realm of countable ordinals.

[9]In other words, every subspace of X is Lindelöf.

$i_0 < n$ and every requirement $h : n \longrightarrow 2$, there exist $a < b$ in A such that

$$c(a(i_0), b(j)) = h(j) \text{ for all } j < n.$$

On the other hand, there is a regular hereditary Lindelöf non-separable space if there is $c : [\omega_1]^2 \longrightarrow 2$ with the dual property that, for every uncountable family A of pairwise-disjoint finite subsets of ω_1, all of some fixed size n for every position $j_0 < n$ and every requirement $h : n \longrightarrow 2$, there exist $a < b$ in A such that

$$c(a(i), b(j_0)) = h(i) \text{ for all } i < n.$$

It turns out that one cannot produce c having the first property without appealing to additional axioms of set theory, but that a variation of the oscillation mapping that we reproduce in Chapter 2 of this book will give us a c with the second property and therefore a regular hereditary Lindelöf non-separable space.

One can find this kind of application of the oscillation mapping or the square-bracket operation on ω_1 in other areas of mathematics as well. For example consider the problem of finding a large subspace on which a given homogeneous polynomial $P : X \longrightarrow \mathbb{C}$ is zero, where X is some Banach space over the field $\mathbb{C}$ of complex numbers.[10] It is known that if X is infinite-dimensional, then there will always be an infinite-dimensional subspace Y of X on which P is zero, but can one find larger Y assuming X is not separable? More precisely, if X is not separable, can one find a non-separable subspace Y of X on which P vanishes? Note that a counterexample would be a polynomial $P : X \longrightarrow \mathbb{C}$ which takes all the complex values on any non-separable subspace of X. So it is natural to try appealing to the square-bracket operation for constructing such a polynomial P. It turns out that this is indeed possible and we shall reproduce it in Chapter 5 of this book for the space $X = \ell_1(\omega_1)$. In order to apply the square-bracket operation we need to take one of its convenient projections $[\![\cdot\cdot]\!]_{\mathcal{G}_0}$ with only countably many values, which one may assume to form a countable dense subset D of the unit disc of the complex plane. Then the polynomial $P : \ell_1(\omega_1) \longrightarrow \mathbb{C}$ is defined as

$$P(x) = \sum [\![\alpha\beta]\!]_{\mathcal{G}_0} x_\alpha x_\beta.$$

In order to establish that P takes all the values from $\mathbb{C}$ on any closed non-separable subspace of $\ell_1(\omega_1)$, one uses the following property of the projection $[\![\cdot\cdot]\!]_{\mathcal{G}_0}$: For every uncountable family A of pairwise-disjoint finite subsets of ω_1, all of the same size n, there is uncountable set $B \subseteq A$, an equivalence relation $\sim$ on $n = \{0, 1, \ldots, n-1\}$, and a single mapping $h : n \times n \rightarrow D$ such that

$$[\![a(i)b(j)]\!]_{\mathcal{G}_0} = h(i,j) \text{ for all } a < b \text{ in } B \text{ and } i \nsim j.$$

[10]Recall that a homogeneous polynomial of degree n on a Banach space X is a mapping $P : X \longrightarrow \mathbb{C}$ of the form $P(x) = \Phi(x, x, \ldots, x)$, where $\Phi : X^n \longrightarrow \mathbb{C}$ is a bounded symmetric n-linear form on X.

Moreover, for every uncountable $C \subseteq B$ and for every $g : n \times n \longrightarrow D$ there exist $a < b$ in C such that

$$[\![a(i)b(j)]\!]_{\mathcal{G}_0} = g(i,j) \text{ for all } i,j < n \text{ with } i \sim j.$$

This particular application makes it clear how useful it is to know behavior of the square-bracket operation not only on uncountable subsets of ω_1 but also on arbitrary uncountable families A of pairwise-disjoint subsets of ω_1.

Minimal walks in general and the metric theory of ordinals in particular are applicable to problems that are not necessarily problems about uncountable structures but rather about the behavior of countable substructures of a fixed large structure. We demonstrate this in Chapter 3 where we present some applications of the metric theory of countable ordinals to classical problems from infinite-dimensional geometry, such as, for example, the distortion problem or the unconditional basic sequence problem. Given a function

$$\varrho : [\eta]^2 \longrightarrow \omega$$

that is locally finite[11] and that satisfies the two ultrametric inequalities,[12] one can use it to define *special functionals* on the vector space $c_{00}(\eta)$ of all finitely supported maps from the ordinal η into the reals. Let e_α $\alpha < \eta$ be the standard Hamel basis of $c_{00}(\eta)$. Let e^*_α $\alpha < \eta$ be the corresponding sequence of biorthogonal functionals. We say that a sequence $(E_i)_{i<n}$ of finite subsets of ω_1 is *special* if $E_i < E_j$[13] for $i < j < n$, and if

$$|E_0| = 1 \text{ and } |E_j| = \sigma_\varrho(E_0, E_1, \ldots, E_{j-1}) \text{ for } 0 < j < n.$$

Here $\sigma_\varrho(E_0, E_1, \ldots, E_{j-1})$ is the integer that codes in some natural way the finite metric structure induced by ϱ on the union of these sets. To a finite set $E \subseteq \eta$, we associate the vector and functional on $c_{00}(\eta)$,

$$x_E = \frac{1}{|E|^{1/2}} \sum_{\alpha \in E} e_\alpha \quad \text{and} \quad \phi_E = \frac{1}{|E|^{1/2}} \sum_{\alpha \in E} e^*_\alpha.$$

Given a special sequence $(E_i)_{i<n}$ of finite subsets of ω_1, the corresponding *special functional* is defined by $\sum_{i<n} \phi_{E_i}$. Special functionals induce a norm on $c_{00}(\eta)$, and if we let X_η denote the completion of $c_{00}(\eta)$ under this norm, the sequence e_α $(\alpha < \eta)$ will be a weakly null sequence in X_η with no infinite unconditional subsequence. More precisely, we show in Chapter 3 that, if $(E_i)_{i<\omega}$ is any infinite special sequence of finite subsets of η and, if for $i < \omega$ we let $v_i = x_{E_i}$ as defined above, then for every $n < \omega$ and every sequence $(a_i)_{i \le n} \subseteq [-1, +1]$ of scalars,

$$\max_{0 \le k \le n} \Big| \sum_{i=0}^{k} a_i \Big| \le \Big\| \sum_{i=0}^{n} a_i v_i \Big\| \le (3 + \varepsilon) \max_{0 \le k \le n} \Big| \sum_{i=0}^{k} a_i \Big| .$$

[11] I.e., the set $\{\xi < \alpha : \varrho(\xi, \alpha) \le n\}$ is finite for every $\alpha < \eta$ and $n < \omega$.
[12] I.e., $\varrho(\alpha, \gamma) \le \max\{\varrho(\alpha, \beta), \varrho(\beta, \gamma)\}$ and $\varrho(\alpha, \beta) \le \max\{\varrho(\alpha, \gamma), \varrho(\beta, \gamma)\}$ for all $\alpha < \beta < \gamma$.
[13] We let $E < F$ if every ordinal from E is smaller than every ordinal from F.

We have already noted that the largest ordinal η supporting such a ϱ-function is the first uncountable ordinal ω_1. It follows therefore that one can have an uncountable weakly-null sequence with no *infinite* unconditional subsequence. It follows that the function ϱ can be used to control *all* countably infinite subsets of the long weakly-null sequence e_α $(\alpha < \omega_1)$. With a considerable amount of additional work one can even build a reflexive Banach space $\mathcal{X}_{\omega_1}$ with a Schauder basis of length ω_1 with no *infinite* unconditional basic sequence. Extending these ideas further one can use ϱ to build another Banach space $\mathcal{X}^0_{\omega_1}$ with a Schauder basis of length ω_1, which on one hand keeps the distortion and all the conditional structure of the space $\mathcal{X}_{\omega_1}$ at the level of its nonseparable subspaces, but on the other hand, every infinite-dimensional closed subspace of $\mathcal{X}^0_{\omega_1}$ contains an isomorphic copy of the space c_0, or in other words separable subspaces of $\mathcal{X}^0_{\omega_1}$ do not hold any conditional structure nor do they allow their norms to be distorted.

1.2 Summary of results

Many chapters of the book can be read independently from each other. Once the basic definition of the walk is understood, the reader may choose to follow the development of a particular characteristic in various contexts. The reader inclined towards applications of the methods of this book may start by reading about a particular application and go back towards background material that is needed. In this section, we include a short summary that might help the reader in finding specific results presented in this book.

The first section of Chapter 2 is the one to be first read as it presents the notion of minimal walk along a given C-sequence and the corresponding characteristic ρ_0 that codes it. The resulting tree $T(\rho_0)$ is a tree of height ω_1 that has all of its levels countable and which admits a natural strictly increasing map into the rationals. This could be the shortest construction of such a tree found in the literature and surely is the most canonical one. This could be seen for example on the basis of the fact that if we order $T(\rho_0)$ lexicographically we obtain an uncountable linearly ordered set whose cartesian square can be covered by countably many chains. The proof of this fact depends on the development of the full lower trace of the minimal walk and the reader is advised to skip this on the first reading and go instead to the second section of Chapter 2 which presents the characteristic ρ_1 and the corresponding tree $T(\rho_1)$ for which this fact is easier to establish. The characteristic ρ_1 has its own interesting application presented in the same section. For example, we show that the tree $T(\rho_1)$ naturally leads us to an example of a homogeneous non metrizable compactum that can be represented as a weakly compact subset of some Banach space. As another application of ρ_1 presented in the same section is a functor that transfers a given graph G on the vertex set ω_1 to a graph G^* on the same vertex set such that if G^* has an uncountable clique then the vertex set ω_1 can be covered by countably many sets that are cliques of G. Moreover, we show that G^* satisfies the countable chain condition provided G

satisfies a slightly stronger version of this condition saying that for every uncountable family $\mathcal{F}$ of pairwise disjoint finite subsets of ω_1, we can find $a \neq b$ in $\mathcal{F}$ such that $\{\alpha, \beta\}$ is an edge of G for every $\alpha \in a$ and $\beta \in b$. The point of this is that essentially all known ccc graphs on ω_1 do satisfy this stronger form of the countable chain condition and therefore this in particular shows that many of the standard consequences of MA_{ω_1} are Ramsey-theoretic in nature. Recall that it is still not known if MA_{ω_1} is in fact equivalent to the statement that every ccc graph on the vertex set ω_1 contains an uncountable clique. We finish this section with a proof that unlike $T(\rho_0)$ the tree $T(\rho_1)$ does not always admit a strictly increasing map into the rationals. This amounts to measuring how much information about the minimal walk is lost when one passes from the full code $\rho_0(\alpha, \beta)$ to the maximal weight $\rho_1(\alpha, \beta)$ of the walk.

The third section of Chapter 2 presents an important theme that is going to be developed in later parts of the book, the theme of the oscillation mapping. More precisely, in Section 2.3, we present two oscillation mappings osc_0 and osc_1 corresponding to the upper and lower trace of the minimal walk, respectively. Of specific applications of these two oscillation mappings, we present a rather absolute decomposition of ω_1 into infinitely many pairwise disjoint stationary subsets, and an example of a regular hereditarily Lindelöf topological space that is not separable. In Section 2.3 we introduce two new characteristics ρ_2 and ρ_3. The characteristic ρ_2 while quite simple minded in that it counts only the number of steps of the minimal walk its full power will become apparent only in later parts of the book when dealing with walks on larger cardinals. In fact, as far as we know, ρ_2 could be the first nontrivial two-place mapping from a large cardinal θ into ω in the sense that it takes arbitrarily high value from ω on any product of two unbounded subsets of θ. Recall that large cardinals θ typically have the Ramsey-theoretic properties saying that maps $f : [\theta]^2 \longrightarrow \omega$ are constant on $[\Gamma]^2$ for large subsets $\Gamma \subseteq \theta$. The mapping ρ_3 while considerably more subtle makes sense only in the realm of countable ordinals. To a given walk from a countable ordinal β down to a smaller ordinal α, the characteristic ρ_3 assigns one of the digits 0 or 1 according to what happens on the last step of the walk. Since it is a coherent mapping it shares many properties with the characteristic ρ_1 though we shall show that, unlike $T(\rho_1)$, if we order lexicographically $T(\rho_3)$, we are not always guaranteed that the cartesian square of the corresponding uncountable linear ordering can be covered by countably many chains. On the other hand, it is true that every uncountable subset X of $T(\rho_3)$ contains an uncountable subset Y which when ordered lexicographically has the property that its cartesian square can be covered by countably many chains. This again amounts to measuring how much of the information is lost about the minimal walks by passing from the characteristic $\rho_1(\alpha, \beta)$ to the characteristic $\rho_3(\alpha, \beta)$. However, while ρ_3 looses much of the information about the minimal walk, it still gives us a rather interesting and canonical object on ω_1 that is very much reminiscent of the classical notion of a Hausdorff gap in the quotient algebra $\mathcal{P}(\omega)/\mathrm{fin}$.

In Chapter 3 we develop the metric theory of countable ordinals concentrated around the two ultrametric triangle inequalities

$$d(\alpha,\gamma) \le \max\{d(\alpha,\beta), d(\beta,\gamma)\} \text{ and } d(\alpha,\beta) \le \max\{d(\alpha,\gamma), d(\beta,\gamma)\},$$

whenever $\alpha < \beta < \gamma$. We have already mentioned two applications of this theory found in this chapter, a reflexive Banach space $\mathcal{X}_{\omega_1}$ with a Schauder basis of length ω_1 with no infinite unconditional basic sequence, and another Banach space $\mathcal{X}^0_{\omega_1}$ with a Schauder basis of length ω_1 that is, on one hand saturated with copies of the space c_0 but on the other hand its norm admits an arbitrarily high distortion relative to the class of nonseparable subspaces of $\mathcal{X}^0_{\omega_1}$. Chapter 3 presents also the characteristic ρ of the minimal walk which besides the two mentioned triangle inequalities has many other interesting properties that show their full power on larger cardinals than ω_1. In Chapter 3 we do present two important objects that are naturally derived from ρ, the Cohen name for a Souslin tree and a particularly canonical example of a Hausdorff gap in $\mathcal{P}(\omega)/\mathrm{fin}$. The Souslin tree has domain ω_1 and the property that no infinite ground model set can be its chain or antichain. In Section 3.4 we study the general theory of functions $\varrho : [\omega_1]^2 \longrightarrow \omega$ that satisfy the two ultrametric triangle inequalities and that are locally finite in the sets that the set $\{\xi < \alpha : \varrho(\xi,\alpha) \le n\}$ is finite for every $\alpha < \omega_1$ and every $n < \omega$. We show that there is a vast variety of such mappings including the universal one whose construction we reproduce in the same section.

In Chapter 4 we develop a metric theory of trees that lies behind the properties of the tree $T(\rho_1)$. The basic notion here is that of a Lipschitz map between trees, a level preserving map with the property that $\Delta(g(x), g(y)) \ge \Delta(x,y)$ for all x and y in its domain. This leads us to the notion of a Lipschitz tree, an uncountable tree T with the property that every level preserving map from an uncountable subset of T into T has a Lipschitz restriction on some uncountable subset of its domain. The main purpose of Chapter 4 is to study the class $\mathcal{C}$ of Lipschitz trees as a structure equipped with the quasi ordering $S \le T$ defined to hold whenever there is a Lipschitz map from the tree S into the tree T, or equivalently whenever there is a strictly increasing map from S into T. We also examine the relationship between the class $\mathcal{C}$ and the larger class $\mathcal{A}$ of Aronszajn trees. Most of the theory is developed without appeal to additional axioms of set theory though in the same places we have used either MA_{ω_1} or the Proper Forcing Axiom. It turns out that Lipschitz trees share many of the properties of the tree $T(\rho_1)$ such as for example the property that when ordered lexicographically the cartesian square of the corresponding uncountable linear ordering can be covered by countably many chains. It turns out that all trees $T(\rho_0)$, $T(\rho_1)$, $T(\rho_2)$, $T(\rho_3)$, and $T(\rho)$ associated to various characteristics of walks in ω_1 are Lipschitz. It turns out also that the class $\mathcal{C}$ is totally ordered under $\le$. In fact the chain $(\mathcal{C}, \le)$ is discrete in the sense that every T in $\mathcal{C}$ admits a naturally defined shift $T^{(1)}$ that forms an immediate successor of T in $\mathcal{C}$. In fact, the shift $T^{(1)}$ is also an immediate successor of T even in the bigger class $\mathcal{A}$ and this is essentially equivalent to Shelah's Conjecture saying that an uncountable ordering either contains an uncountable well ordered or conversely well ordered subset, an uncountable separable ordered subset, or an uncountable subset whose cartesian square can be covered by countably many chains. It turns out that the chain $\mathcal{C}$ is not well ordered and this in particular solves an old problem of R. Laver asking if the class $\mathcal{A}$ is well quasi-ordered under a stronger ordering than $\leq$. Chapter 4 contains many other structural results about the classes $\mathcal{C}$ and $\mathcal{A}$. For example, we show that while $\mathcal{A}$ is not totally ordered under $\leq$, the chain $\mathcal{C}$ is both cofinal and coinitial in $(\mathcal{A}, \leq)$.

In Chapter 5 we introduce and study the square-bracket operation $[\alpha\beta]$ for pairs α and β of countable ordinals. The ordinal $[\alpha\beta]$ is taken from the upper trace of the minimal walk from β to α. In later parts of the book we shall see several variation of this choice of $[\alpha\beta]$ but the basic idea is always the same and it is based on the oscillation mapping osc_0 of upper traces exposed above in Section 2.3. In Chapter 5 itself we present two other variations on the square-bracket operation, one based on special Aronszajn tree and the other on the tree of all finite binary sequences. What the square-bracket operation adds to the space ω_1 of countable ordinals is a rigidity which can formally be expressed by the fact that there is a sentence of $L(Q^2)$ which has only rigid models. This is presented in Example 5.1.10 which solves a problem of Ebbinghaus and Flum who showed that every model of a sentence of $L(Q)$ has a nontrivial automorphism. In Section 5.3 we present tree geometrical application of the square-bracket operation. The first one is an example of a projective geometry of points and hyperplanes in $\mathbb{R}^n$. The second one is the example of a complex polynomial on $\ell(\omega_1)$ with no nonseparable null subspace already mentioned above. The third application of $[\cdot\cdot]$ is an example of a reflexive Banach space $\mathcal{X}$ with a Schauder basis of length ω_1 in which every operator T can be written as $\lambda I + S$, where S is an operator with separable range. While in some sense this example is subsumed by the example of the Banach space $\mathcal{X}_{\omega_1}$ from Chapter 3 we have included it as its construction uses a quite different set of ideas. In Section 5.4 we give a formal explanation of how the square-bracket operation reduces the quantification over uncountable subsets of ω_1 to that over closed and unbounded subsets of ω_1. Answering a question of W.H. Woodin, we show that there is a natural functor based on $[\cdot\cdot]$ which to every subset Γ on ω_1 associates a graph $K_\Gamma \subseteq [\omega_1]^2$ such that, modulo PFA or Woodin's axiom $(*)$, a subset Γ of ω_1 contains a closed and unbounded set if and only if K_Γ contains $[X]^2$ for some uncountable $X \subseteq \omega_1$.

In Chapter 6 we develop the characteristic ρ_0 of the minimal walk in the general context. Recall that in the context of walks on ω_1 the tree $T(\rho_0)$ that corresponds to the characteristics ρ_0 is *special* in the sense that it can be decomposed into countably many antichains. In Chapter 6 we show that a strongly inaccessible cardinal θ is not Mahlo precisely when one is able to find a C-sequence in θ for which the corresponding tree $T(\rho_0)$ *special* in the sense that it admits a regressive mapping that is not constant on any subset of $T(\rho_0)$ that cannot be covered by less than θ antichains. The idea is then used in providing a unified approach towards

Ramsey-theoretic characterizations of n-Mahlo cardinals in terms of the existence of min-homogeneous sets relative to regressive maps of the form $f : [\Gamma]^{n+2} \longrightarrow \theta$ or $f : [\Gamma]^{n+3} \longrightarrow \theta$. In the same chapter we develop the general theory of the characteristics ρ_1 and ρ_2. In particular, in Section 6.2 we develop the local version $\rho_1^\kappa : [\theta]^2 \longrightarrow \kappa$ of ρ_1 for an arbitrary regular cardinal $\kappa < \theta$. The main interest in this variation of ρ_1 is that under some very mild assumptions about θ one obtains mappings from $[\theta]^2$ into κ that have strong unboundedness properties. Similar unboundedness property of the characteristic $\rho_2 : [\theta]^2 \longrightarrow \omega$ is equivalent to the non triviality property of the C-sequence C_α $(\alpha < \theta)$ saying that there is no closed and unbounded subset C of θ such that for every $\alpha < \theta$ there is $\beta \geq \alpha$ such that $C \cap \alpha \subseteq C_\beta$. This gives a vast variety of cardinals θ for which one can find a C-sequence C_α $(\alpha < \theta)$ such that the corresponding function $\rho_2 : [\theta]^2 \longrightarrow \omega$ is strongly unbounded. This can be seen from the characterization of weakly compact cardinals given in Section 6.3 which says that a strongly inaccessible cardinal θ is weakly compact if and only if for every C-sequence C_α $(\alpha < \theta)$ there is a closed and unbounded set $C \subseteq \theta$ such that for all $\alpha < \theta$ there is $\beta \geq \alpha$ such that $C \cap \alpha = C_\beta \cap \alpha$. We finish Chapter 6 with a particular topological application of the unboundedness property of ρ_2

In Chapter 7 we study walks based on C-sequences C_α $(\alpha < \theta)$ that satisfy the coherence property saying that $C_\alpha = C_\beta \cap \alpha$ whenever α is a limit point of C_β. We call such C-sequences *square sequences*. It is this condition on the C-sequence which permits that the full theory of walks on ω_1 be lifted to the level θ that supports it. For example, we show that the full lower trace $\mathrm{F}(\alpha, \beta)$ of walks along square sequence keeps all its properties from the context of countable ordinals. As an application we show that the characteristic ρ_2 in this context has an interesting coherence property saying that $\sup_{\xi<\alpha} |\rho_2(\xi, \alpha) - \rho_2(\xi, \beta)| < \infty$ for all $\alpha < \beta < \theta$. This has an interesting topological interpretation saying that the square of the sequential fan with θ edges has tightness equal to θ. Another interpretation of the same fact is the statement that the P-ideal dichotomy is incompatible with the existence of a nontrivial square sequence on any cardinal θ that is larger than ω_1. The main reason of imposing the coherence property on a given C-sequence is however motivated by attempts to develop the theory of the characteristic ρ in the general context. This is done in Section 7.2 where we also deal with local versions ρ^κ of this characteristic. A typical application of this new theory is Theorem 7.2.14 saying that if a regular uncountable cardinal $\theta \neq \omega_1$ carries a nontrivial square sequence then it also carries one for this the corresponding tree $(\theta, <^2)$ is not special. Complementing a well-known result of Laver and Shelah, this shows that the existence of a weakly compact cardinal is the exact consistence strength of the statement that all Aronszajn trees on ω_2 are special. Special square sequence is something that is more closely related to Jensen's notion of square sequences on successor cardinals and we develop the corresponding theory of walks in the following sections of Chapter 7. Particularly interesting is the case of successors of regular cardinals which we consider in Section 7.4 where we develop the theory of the corresponding characteristic ρ and the derived set-mapping $D : [\kappa^+]^2 \longrightarrow [\kappa^+]^{<\kappa}$ defined by

$$D\{\alpha, \beta\} = \{\xi \leq \min\{\alpha, \beta\} : \rho\{\xi, \alpha\} \leq \rho\{\alpha, \beta\}\}.$$

In Section 7.5 we use the set-mapping D to construct interesting forcing notions. For example, we show that if a regular cardinal κ is λ-inaccessible then there is a λ-closed κ-cc forcing notion that introduces a Souslin tree on κ^+ as well as a forcing notion that introduces a Kurepa family on κ. So, in particular, we show that if $\Box_{\omega_1}$ holds then there is a property K forcing notion that introduces a Souslin tree on ω_2. In the same section we reproduce a proof that under $\Box_{\omega_1}$ there is a property K forcing notion that introduces a locally compact scattered topology on ω_2 all of whose Cantor-Bendixson ranks are countable. We deal with successors of singular cardinals κ in Section 7.5. The main applications of the characteristic ρ in this context are in producing Jensen matrices $\mathrm{J}_{\alpha n}$ $(\alpha < \kappa^+, n < \omega)$ of subsets of κ^+. As one application of Jensen matrices we present a construction of a cofinal Kurepa family of countable subsets of κ^+. Cofinal Kurepa families are quite useful objects. We show this by using them in constructing Bernstein decompositions of arbitrary Hausdorff spaces, or in constructing coherent families of finite-to-one maps indexed by countable subsets of κ^+.

In Chapter 8 we develop the general theory of the oscillation mapping of traces and the corresponding square-bracket operation. In fact we give three variations on the basic idea behind the definition of the square-bracket operation with a particular emphasis on the properties of the corresponding projections. Projections are something that is typically needed in applications and they do tend to impose accessibility conditions on the cardinal θ. For example, we show that if, for example, θ is the first inaccessible cardinal then there is a projection $o : [\theta]^2 \longrightarrow \omega$ of the oscillation mapping which takes all the values from ω on any set of the form $[\Gamma]^2$ for Γ unbounded subset of θ. However if one wants such a projection of the oscillation mapping that could be useful in constructing interesting complex polynomials on $\ell_1(\theta)$ or interesting bilinear mappings one needs to impose the condition that θ is not bigger than the continuum. It turns out that the basic definition of the square-bracket operation from ω_1 lifts without difficulties when ω_1 is replaced by a cardinal θ that is a successor of a regular cardinal or more generally that admits a non reflecting stationary subset. A typical application of this result is the statement that for every regular cardinal κ there is a 2-nilpotent group G of cardinality κ^+ such that every abelian subgroup of G has cardinality at most κ. To obtain a projection $[\![\cdot\cdot]\!]$ of $[\cdot\cdot]$ that would have the *generic* behaviour[14] one needs to vary the original idea using the assumption that the cardinal θ is in some sense large. We present two such variations with some topological applications of the corresponding projections. For example, we show that for every regular uncountable cardinal κ there is a topological group G of cellularity κ whose cartesian

[14]In other words, that would have the property that for every family A of θ pairwise disjoint finite subsets of θ all of some fixed size n and for every mapping $h : n \times n \longrightarrow \theta$ there exist $a < b$ in A such that $[\![a(i)b(j)]\!] = h(i, j)$ of all $i, j < n$.

square has cellularity $> \kappa$. In fact, this result is true for arbitrary uncountable cardinal κ that is not necessarily regular. This is proved in a similar fashion using the colorings defined on successors of singular cardinals exposed in Chapter 9 of this book.

In the first section of Chapter 9, we use the characteristic ρ in producing partial square sequence that can sometimes be used in places of full square sequences. In the second section of Chapter 9, we show that for a regular uncountable cardinal κ, there is a structure of the form $(\kappa^+, \kappa, R_n)_{n<\omega}$ with no elementary substructure B of size κ with $B \cap \kappa$ bounded in κ just in case there is a *strongly unbounded* function $f : [\kappa^+]^2 \longrightarrow \kappa$ satisfying the two ultrametric triangle inequalities mentioned above. This gives us a useful reformulation of the well-known model-theoretic statement known as Chang's Conjecture which is simply the statement that every structure of the form $(\omega_2, \omega_1, R_n)_{n<\omega}$ has an uncountable elementary substructure B such that $B \cap \omega_1$ is countable. In Section 9.3 we use this reformulation to show that Chang's Conjecture is also equivalent to the statement that every ccc poset forces that every mapping of the form $g : \omega_2 \times \omega_2 \longrightarrow \omega$ must be constant on some product of two infinite subsets of ω_2. This is then used in Section 9.4 in analyzing possible Ramsey-theoretic reformulations of the Continuum Hypothesis.

In Section 10.1 we develop a general stepping-up procedure that lifts structures living on a given cardinal θ to similarly behaved structure living on its successor θ^+. We manage this by just assuming the existence of a square sequence on θ^+ though we are able to step-up some complex combinatorial statements like these of Hajnal and Komjath where originally one would have expected that the higher-gap morasses are needed. As an application, we show that using $\square_{\omega_1}$ one can construct a reflexive Banach space X with a Schauder basis of length ω_2 with the property that every bounded linear operator $T : X \longrightarrow X$ can be written as a sum of an operator with a separable range and a diagonal operator with only countably many changes of eigenvalues. In Section 10.2 we show that Chang's Conjecture is equivalent to the purely Ramsey-theoretic statement saying that for every $f : [\omega_2]^2 \longrightarrow \omega_1$ there is an uncountable set $\Gamma \subseteq \omega_2$ such that $f''[\Gamma]^2 \neq \omega_1$. In Section 10.3 we introduce the 3-dimensional oscillation mapping and use it in producing a coloring $c : [\omega_2]^3 \longrightarrow \omega$ that takes all of its values from ω on the symmetric cube of any uncountable subset of ω_2.

The last section of Chapter 10 is an attempt to start the corresponding theory of two cardinal walks, an analogous theory that would work in the context of structures of the form $\mathcal{P}_\kappa(\lambda) = \{x \subseteq \lambda : |x| < \kappa\}$ in place of ordinals and cardinals. In fact we were able to define a notion of walk in this context that is sufficiently rich for giving us the analogue of the square-bracket operation in this context. For example, we show that for every pair $\kappa < \lambda$ of cardinals with κ regular there is a projection $[\![\cdot\cdot]\!] : [[\lambda]^\kappa]^2 \longrightarrow \lambda$ of the square-bracket operation in the context of $[\lambda]^\kappa$ (or more formally, in the context of $\mathcal{P}_{\kappa^+}(\lambda)$) such that for every cofinal subset U of $[\lambda]^\kappa$ and every $\xi \in \lambda$ there exist $x \subset y$ in U such that $[\![xy]\!] = \xi$. A particularly interesting case of this result is when $\kappa = \omega$ which allows a further elaboration. For example, we show that if there is a stationary subset of $[\lambda]^\omega$ that is equinumerous with a locally countable subset of $[\lambda]^\omega$ then there is a variation of the square-bracket operation $[\cdot\cdot]_S : [[\lambda]^\omega]^2 \longrightarrow [\lambda]^\omega$ with properties quite analogous to those of the square-bracket operation of ω_1 (which is really the case $\kappa = \omega$ and $\lambda = \omega_1$) in the sense that for every cofinal subset U of $[\lambda]^\omega$ the set of all $s \in S$ that are not of the form $[xy]_S$ for $x \subset y$ in U is not stationary in $[\lambda]^\omega$. One of the points of this variation of the square bracket operation is that the set S can be split into $|S|$ pairwise disjoint stationary sets, and since quite frequently S has cardinality θ that is bigger than λ, we can define projections $[\![\cdot\cdot]\!]_S : [[\lambda]^\kappa]^2 \longrightarrow \theta$ that take more than λ colors (more precisely, $\theta = |S|$ colors) on any cofinal subset U of $[\lambda]^\omega$.

1.3 Prerequisites and notation

We have tried to keep the prerequisites needed for mastering the material of this book to a minimum. Though no specific training in set theory is necessary, the reader should be familiar with the notion of ordinal and recursive definitions and inductive proofs on them. This all can be found in most of the introductory texts on the subject such as, for example, the newer ones [51], [64] and [55], or the older text [62] which has particularly detailed expositions of the recursive definitions over ordinals. By comparing, the reader will notice that we are using standard notation in essentially complete agreement with these textbooks. In these sources the reader will find all the operations and properties of ordinals that we will use, but if more complete treatment is needed the reader may also consult one of the specialized texts like [7]. We shall also look at ordinals with their natural topology induced by order. It is in this context that we refer to 'closed' subsets of a particular ordinal θ. The notions of 'unbounded' and 'cofinality' in this context refer of course to the natural ordering of θ. The combinatorially inclined readers will notice that we are adopting the Erdős–Rado notation for the symmetric powers $[S]^\theta$, the collections of all subsets of the set S of cardinality θ. Of special interest are of course the finite symmetric powers $[S]^k$ and the mappings defined on them because of the clear connection that this book has with Ramsey theory (see [40], [33] and [133]). In fact most of our characteristics of walks are mappings with domains equal to symmetric squares $[\theta]^2$ of some ordinals θ. A pair $\{\alpha, \beta\} \in [\theta]^2$ is usually assumed to be written such that $\alpha < \beta$. In other words, sometimes it is convenient to identify the symmetric square $[\theta]^2$ with the set

$$\{(\alpha, \beta) \in \theta^2 : \alpha < \beta\}$$

of ordered pairs. In this way a given characteristic $f : [\theta]^2 \longrightarrow \eta$ of the walk can be identified with the sequence $f_\beta : \beta \longrightarrow \eta$ $(\beta < \theta)$ of fiber mappings defined by

$$f_\beta(\alpha) = f(\alpha, \beta).$$

Also we can more clearly express when the given characteristic f is 'coherent' in some way by referring to the coherence between the corresponding fiber mappings. Moreover, this identification allows us to use the notation $f(\alpha, \beta)$ instead of the more cumbersome notation $f(\{\alpha, \beta\})$. Basically all our characteristics

$$f : [\theta]^2 \longrightarrow \eta$$

of walks are defined recursively in the sense that $f(\alpha, \beta)$ is defined on the basis of the values $f(\alpha', \beta')$ on lexicographically smaller pairs (α, β). All the recursive definitions require us to specify the boundary values $f(\alpha, \alpha)$ which are typically taken to be constant values such us 0 or $\emptyset$ depending on the context. It is for this reason that sometimes we implicitly assume that the diagonal $\{(\alpha, \alpha) : \alpha < \theta\}$ is a part of the domain of f.

1.4 Acknowledgements

We are grateful to a large number of people who have read versions of the manuscript and suggested corrections and improvements or helped us with preparation of the final manuscript. Of these we would like in particular to mention Maxim Burke, Christine Härtl, Akihiro Kanamori, Bernhard König, Piotr Koszmider, Carlos Martinez, Justin Moore and Luis Pereira. Special thanks are due to Jordi Lopez-Abad for helping us in designing the diagrams and figures of the book.

Chapter 2

Walks on Countable Ordinals

2.1 Walks on countable ordinals and their basic characteristics

The space ω_1 of countable ordinals is by far the most interesting space considered in this book. There are many mathematical problems whose combinatorial essence can be reformulated as problems about ω_1, which is in some sense the smallest uncountable structure. What we mean by 'structure' is ω_1 together with a system C_α $(\alpha < \omega_1)$ of fundamental sequences, i.e., a system with the following two properties:

(a) $C_{\alpha+1} = \{\alpha\}$,

(b) C_α is an unbounded subset of α of order-type ω, whenever α is a countable limit ordinal > 0.

Despite its simplicity, this structure can be used to derive virtually all known other structures that have been defined so far on ω_1. There is a natural recursive way of picking up the fundamental sequences C_α, a recursion that refers to the Cantor normal form which works well for, say, ordinals $< \varepsilon_0$[1]. For longer fundamental sequences one typically relies on some other principles of recursive definitions and one typically works with fundamental sequences with as few extra properties as possible. We shall see that the following assumption is what is frequently needed and will therefore be implicitly assumed whenever necessary:

(c) if α is a limit ordinal, then C_α does not contain limit ordinals.

---

[1]One is tempted to believe that the recursion can be stretched all the way up to ω_1 and this is probably the way P.S. Alexandroff discovered the phenomenon that regressive mappings on ω_1 must be constant on uncountable sets (see [2] and [3, Appendix]).

Definition 2.1.1. A *step* from a countable ordinal β towards a smaller ordinal α is the minimal point of C_β that is $\geq \alpha$. The cardinality of the set $C_\beta \cap \alpha$, or better to say the order-type of this set, is the *weight* of the step.

Definition 2.1.2. A *walk* (or a *minimal walk*) from a countable ordinal β to a smaller ordinal α is the sequence $\beta = \beta_0 > \beta_1 > \cdots > \beta_n = \alpha$ such that for each $i < n$, the ordinal β_{i+1} is the step from β_i towards α.

Analysis of this notion leads to several two-place functions on ω_1 that have a rich structure and many applications. Let us expose some of these functions.

Definition 2.1.3. The *full code* of the walk is the function $\rho_0 : [\omega_1]^2 \longrightarrow \omega^{<\omega}$, defined recursively by

$$\rho_0(\alpha, \beta) = \langle |C_\beta \cap \alpha| \rangle^\frown \rho_0(\alpha, \min(C_\beta \setminus \alpha)),$$

with the boundary value[2] $\rho_0(\alpha, \alpha) = \emptyset$ where the symbol $^\frown$ refers to the sequence obtained by concatenating the one-term sequence $\langle |C_\beta \cap \alpha| \rangle$ with the already known finite sequence $\rho_0(\alpha, \min(C_\beta \setminus \alpha)$ of integers.

Definition 2.1.4. The *upper trace* of the walk from β to α is the finite set $\mathrm{Tr}(\alpha, \beta)$ of ordinals between α and β visited during the minimal walk from β to α. Note the following recursive definition of this trace,

$$\mathrm{Tr}(\alpha, \beta) = \{\beta\} \cup \mathrm{Tr}(\alpha, \min(C_\beta \setminus \alpha))$$

where the boundary value is $\mathrm{Tr}(\alpha, \alpha) = \{\alpha\}$. Note also the following expression of the upper trace in terms of the full code,

$$\mathrm{Tr}(\alpha, \beta) = \{\xi : \rho_0(\xi, \beta) \sqsubseteq \rho_0(\alpha, \beta)\},$$

where $\sqsubseteq$ denotes the ordering of being an initial segment among finite sequences of ordinals. One also has the recursive definition of this trace.

Definition 2.1.5. The *lower trace* of the walk from β to α is the set

$$\mathrm{L}(\alpha, \beta) = \{\lambda(\xi, \beta) : \xi \in \mathrm{Tr}(\alpha, \beta) \text{ and } \xi \neq \beta\},$$

where

$$\lambda(\xi, \beta) = \max\{\max(C_\eta \cap \xi) : \eta \in \mathrm{Tr}(\xi, \beta) \text{ and } \eta \neq \xi\}.$$

Note the following recursive definition of this trace,

$$\mathrm{L}(\alpha, \beta) = (\mathrm{L}(\alpha, \min(C_\beta \setminus \alpha)) \setminus \max(C_\beta \cap \alpha)) \cup \{\max(C_\beta \cap \alpha)\}^3$$

with the boundary value $\mathrm{L}(\alpha, \alpha) = \emptyset$.

[2]Note that while ρ_0 operates on the set $[\omega_1]^2$ of unordered pairs of countable ordinals, the formal recursive definition of $\rho_0(\alpha, \beta)$ does involve the term $\rho_0(\alpha, \alpha)$ and so we have to give it as a *boundary value* $\rho_0(\alpha, \alpha) = \emptyset$. The boundary value is always specified when such a function is recursively defined and will typically be either 0 or the empty sequence $\emptyset$. This convention will be used even when a function e with domain such as $[\omega_1]^2$ is given to us beforehand and we consider its diagonal value $e(\alpha, \alpha)$.

[3]If $C_\beta \cap \alpha = \emptyset$, we let $\mathrm{L}(\alpha, \beta) = \mathrm{L}(\alpha, \min(C_\beta \setminus \alpha))$.

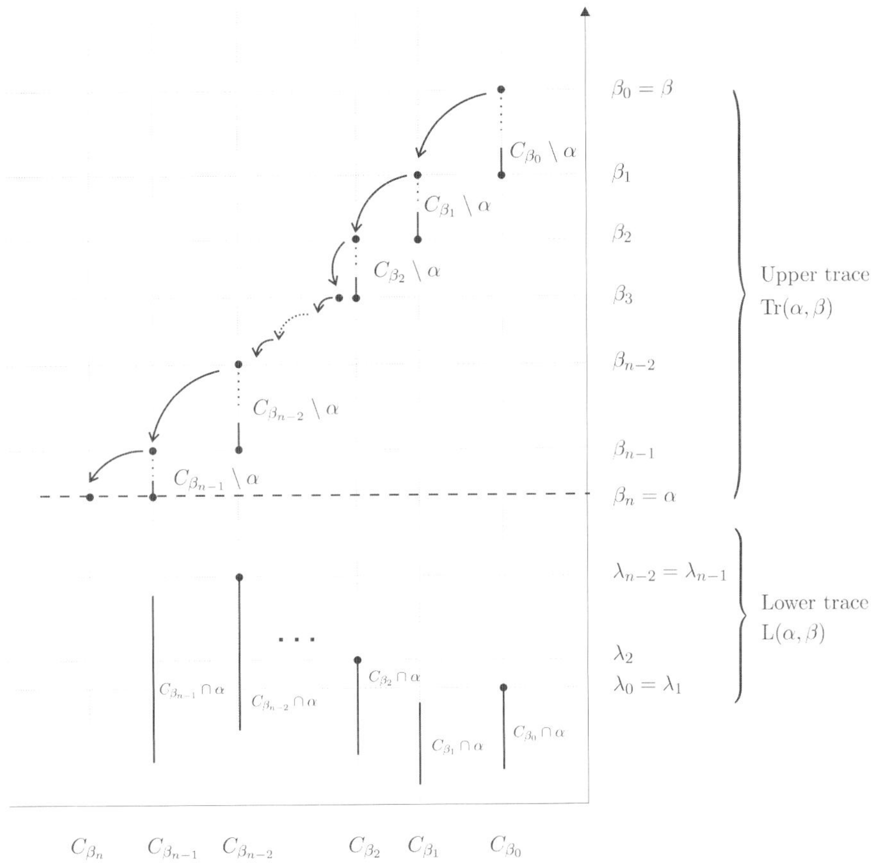

Figure 2.1: The walk and its traces.

Thus, if $\beta = \beta_0 > \beta_1 > \cdots > \beta_n = \alpha$ is the minimal walk from β to α (i.e., $\beta_0 = \beta$, $\beta_n = \alpha$, and $\beta_{i+1} = \min(C_{\beta_i} \setminus \alpha)$ for $i < n$), then

$$\mathrm{Tr}(\alpha, \beta) = \{\beta_i : 0 \leq i \leq n\} \text{ and } \mathrm{L}(\alpha, \beta) = \{\lambda_i : 0 \leq i < n\},$$

where $\lambda_i = \max(\bigcup_{j=0}^{i} C_{\beta_j} \cap \alpha)$. The following immediate facts about these notions will be frequently and often implicitly used.

Lemma 2.1.6. *For* $\alpha \leq \beta \leq \gamma$,

(a) $\alpha > \mathrm{L}(\beta, \gamma)$ *implies that* $\rho_0(\alpha, \gamma) = \rho_0(\beta, \gamma)^\frown \rho_0(\alpha, \beta)$ *and therefore that* $\mathrm{Tr}(\alpha, \gamma) = \mathrm{Tr}(\beta, \gamma) \cup \mathrm{Tr}(\alpha, \beta)$,

(b) $\mathrm{L}(\alpha, \beta) > \mathrm{L}(\beta, \gamma)$ *implies that* $\mathrm{L}(\alpha, \gamma) = \mathrm{L}(\beta, \gamma) \cup \mathrm{L}(\alpha, \beta)$.[4] □

[4]For two sets of ordinals F and G, by $F < G$ we denote the fact that every ordinal from F is smaller than every ordinal from G. When $G = \{\alpha\}$, we write $F < \alpha$ rather than $F < \{\alpha\}$.

Definition 2.1.7. The *full lower trace* of the minimal walk is the function $\mathrm{F} : [\omega_1]^2 \longrightarrow [\omega_1]^{<\omega}$, defined recursively by

$$\mathrm{F}(\alpha, \beta) = \mathrm{F}(\alpha, \min(C_\beta \setminus \alpha)) \cup \bigcup_{\xi \in C_\beta \cap \alpha} \mathrm{F}(\xi, \alpha),$$

with the boundary value $\mathrm{F}(\alpha, \alpha) = \{\alpha\}$ for all α.

Lemma 2.1.8. *For all* $\alpha \leq \beta \leq \gamma < \omega_1$,

(a) $\mathrm{F}(\alpha, \gamma) \subseteq \mathrm{F}(\alpha, \beta) \cup \mathrm{F}(\beta, \gamma)$,
(b) $\mathrm{F}(\alpha, \beta) \subseteq \mathrm{F}(\alpha, \gamma) \cup \mathrm{F}(\beta, \gamma)$.

Proof. The proof is by induction on γ. Let

$$\gamma_1 = \min(C_\gamma \setminus \alpha).$$

Towards proving (a), we first show that the part $\mathrm{F}(\alpha, \gamma_1)$ of $\mathrm{F}(\alpha, \gamma)$ is included in $\mathrm{F}(\alpha, \beta) \cup \mathrm{F}(\beta, \gamma)$. Consider first the case $\gamma_1 < \beta$. By the inductive hypothesis,

$$\mathrm{F}(\alpha, \gamma_1) \subseteq \mathrm{F}(\alpha, \beta) \cup \mathrm{F}(\gamma_1, \beta) \subseteq \mathrm{F}(\alpha, \beta) \cup \mathrm{F}(\beta, \gamma),$$

since $\gamma_1 \in C_\gamma \cap \beta$, and therefore, $\mathrm{F}(\gamma_1, \beta) \subseteq \mathrm{F}(\beta, \gamma)$. If $\gamma_1 \geq \beta$, then $\gamma_1 = \min(C_\gamma \setminus \beta)$ and again $\mathrm{F}(\beta, \gamma_1) \subseteq \mathrm{F}(\beta, \gamma)$. By the inductive hypothesis,

$$\mathrm{F}(\alpha, \gamma_1) \subseteq \mathrm{F}(\alpha, \beta) \cup \mathrm{F}(\beta, \gamma_1) \subseteq \mathrm{F}(\alpha, \beta) \cup \mathrm{F}(\beta, \gamma),$$

so we are done in this case as well. Let us now consider the factor $\mathrm{F}(\xi, \alpha)$ of $\mathrm{F}(\alpha, \gamma)$, where $\xi \in C_\gamma \cap \alpha$. Using the inductive hypothesis we get

$$\mathrm{F}(\xi, \alpha) \subseteq \mathrm{F}(\alpha, \beta) \cup \mathrm{F}(\xi, \beta) \subseteq \mathrm{F}(\alpha, \beta) \cup \mathrm{F}(\beta, \gamma),$$

since $\xi \in C_\gamma \cap \beta$ and $\mathrm{F}(\xi, \beta)$ is by definition included in $\mathrm{F}(\beta, \gamma)$.

To prove (b) consider first the case $\gamma_1 < \beta$. By the inductive hypothesis,

$$\mathrm{F}(\alpha, \beta) \subseteq \mathrm{F}(\alpha, \gamma_1) \cup \mathrm{F}(\gamma_1, \beta).$$

Since $\gamma_1 \in C_\gamma \cap \beta$, the set $\mathrm{F}(\gamma_1, \beta)$ is included in $F(\beta, \gamma)$. Similarly by the definition of $\mathrm{F}(\alpha, \gamma)$, the set $\mathrm{F}(\alpha, \gamma_1)$ is included in $\mathrm{F}(\alpha, \gamma)$.

Suppose now that $\gamma_1 \geq \beta$. Note that in this case $\gamma_1 = \min(C_\gamma \cap \beta)$, so $\mathrm{F}(\beta, \gamma_1)$ is included in $\mathrm{F}(\beta, \gamma)$. Using the inductive hypothesis for $\alpha \leq \beta \leq \gamma_1$ we have

$$\mathrm{F}(\alpha, \beta) \subseteq \mathrm{F}(\alpha, \gamma_1) \cup \mathrm{F}(\beta, \gamma_1) \subseteq \mathrm{F}(\alpha, \gamma) \cup \mathrm{F}(\beta, \gamma). \qquad \square$$

Lemma 2.1.9. *For all* $\alpha \leq \beta \leq \gamma$,

(a) $\rho_0(\alpha, \beta) = \rho_0(\min(\mathrm{F}(\beta, \gamma) \setminus \alpha), \beta)^\frown \rho_0(\alpha, \min(\mathrm{F}(\beta, \gamma) \setminus \alpha))$,
(b) $\rho_0(\alpha, \gamma) = \rho_0(\min(\mathrm{F}(\beta, \gamma) \setminus \alpha), \gamma)^\frown \rho_0(\alpha, \min(\mathrm{F}(\beta, \gamma) \setminus \alpha))$.

Proof. The proof is again by induction on γ. Let $\alpha_1 = \min(\mathrm{F}(\beta,\gamma)\setminus\alpha)$ and $\gamma_1 = \min(C_\gamma \setminus \alpha)$. Let ξ_1 be the minimal $\xi \in (C_\gamma \cap \beta)\cup\{\min(C_\gamma\setminus\beta)\}$ such that $\alpha_1 \in \mathrm{F}(\xi,\beta)$ (or $\mathrm{F}(\beta,\xi)$).

Applying the inductive hypothesis to $\alpha \le \xi_1 \le \beta$ (or $\alpha \le \beta \le \xi_1$) and noting that $\alpha_1 = \min(\mathrm{F}(\xi_1,\beta)\setminus\alpha)$ (or $\alpha_1 = \min(\mathrm{F}(\beta,\xi_1)\setminus\alpha)$), we get

$$\rho_0(\alpha,\beta) = \rho_0(\alpha_1,\beta)^\frown\rho_0(\alpha,\alpha_1), \tag{2.1.1}$$

the part (a) of Lemma 2.1.9. Note that depending on whether $\alpha \le \gamma_1 \le \beta$ or $\alpha \le \beta \le \gamma_1$, we have that $\mathrm{F}(\gamma_1,\beta) \subseteq \mathrm{F}(\beta,\gamma)$ or $\mathrm{F}(\beta,\gamma_1)\subseteq \mathrm{F}(\beta,\gamma)$ because in the latter case γ_1 is also equal to $\min(C_\gamma\setminus\beta)$. Define

$$\alpha_2 := \min(\mathrm{F}(\gamma_1,\beta)\setminus\alpha) \text{ or } \alpha_2 := \min(\mathrm{F}(\beta,\gamma_1)\setminus\alpha),$$

depending on whether $\alpha \le \gamma_1 \le \beta$ or $\alpha \le \beta \le \gamma_1$, respectively. Note that in any of the cases of the definition of α_2, we have that $\alpha_2 \ge \alpha_1$. Applying the inductive hypothesis to $\alpha \le \gamma_1 \le \beta$ (or $\alpha \le \beta \le \gamma_1$) we get:

$$\rho_0(\alpha,\beta) = \rho_0(\alpha_2,\beta)^\frown\rho_0(\alpha,\alpha_2), \tag{2.1.2}$$

$$\rho_0(\alpha,\gamma_1) = \rho_0(\alpha_2,\gamma_1)^\frown\rho_0(\alpha,\alpha_2). \tag{2.1.3}$$

Comparing (2.1.1) and (2.1.2) we see that $\rho_0(\alpha,\alpha_1)$ is a tail of $\rho_0(\alpha,\alpha_2)$, so the equation (2.1.3) can be rewritten as

$$\rho_0(\alpha,\gamma_1) = \rho_0(\alpha_2,\gamma_1)^\frown\rho_0(\alpha_1,\alpha_2)^\frown\rho_0(\alpha,\alpha_1) = \rho_0(\alpha_1,\gamma_1)^\frown\rho_0(\alpha,\alpha_1). \tag{2.1.4}$$

Finally since $\rho_0(\alpha,\gamma) = \langle|C_\gamma\cap\alpha|\rangle^\frown\rho_0(\alpha,\gamma_1) = \rho_0(\gamma_1,\gamma)^\frown\rho_0(\alpha,\gamma_1)$ holds, the equation (2.1.4) gives us the desired conclusion (b) of Lemma 2.1.9. □

Definition 2.1.10. Recall that *right lexicographical ordering* on $\omega^{<\omega}$ denoted by $<_r$ refers to the total ordering defined by letting $s <_r t$ if s is an end-extension of t, or if $s(i) < t(i)$ for $i = \min\{j : s(j) \ne t(j)\}$. Using the identification between a countable ordinal α and the α-sequence $(\rho_0)_\alpha$ of elements of $\omega^{<\omega}$, the ordering $<_r$ induces the following ordering $<_{\rho_0}$ on ω_1:

$$\alpha <_{\rho_0} \beta \quad \text{iff} \quad \rho_0(\xi,\alpha) <_r \rho_0(\xi,\beta), \tag{2.1.5}$$

where $\xi = \Delta(\alpha,\beta) = \min\{\eta \le \min\{\alpha,\beta\} : \rho_0(\eta,\alpha) \ne \rho_0(\eta,\beta)\}$.[5]

Lemma 2.1.11. *The cartesian square of the total ordering $<_{\rho_0}$ of ω_1 is the union of countably many chains.*

[5]Note that when the ordinal ξ is equal to $\min\{\alpha,\beta\}$ then $\max\{\alpha,\beta\} <_{\rho_0} \min\{\alpha,\beta\}$ which is a consequence of the fact that the empty sequence is the $<_r$-maximal element of $\omega^{<\omega}$.

Proof. It suffices to decompose the set of all pairs (α, β) where $\alpha < \beta$. To each such pair we associate a hereditarily finite set $p(\alpha, \beta)$ which codes the finite structure obtained from $\mathrm{F}(\alpha, \beta) \cup \{\beta\}$ by adding relations that describe the way ρ_0 acts on it. To show that this parametrization works, suppose we are given two pairs (α, β) and (γ, δ) such that

$$p(\alpha, \beta) = p(\gamma, \delta) \text{ and } \alpha <_{\rho_0} \gamma.$$

We must show that $\beta \leq_{\rho_0} \delta$. Let

$$\xi_{\alpha\beta} = \min(\mathrm{F}(\alpha, \beta) \setminus \Delta(\alpha, \gamma)), \text{ and}$$
$$\xi_{\gamma\delta} = \min(\mathrm{F}(\gamma, \delta) \setminus \Delta(\alpha, \gamma)).$$

Note that $\mathrm{F}(\alpha, \beta) \cap \Delta(\alpha, \gamma) = \mathrm{F}(\gamma, \delta) \cap \Delta(\alpha, \gamma)$ so $\xi_{\alpha\beta}$ and $\xi_{\gamma\delta}$ correspond to each other in the isomorphism of the (α, β) and (γ, δ) structures. It follows that:

$$\rho_0(\xi_{\alpha\beta}, \alpha) = \rho_0(\xi_{\gamma\delta}, \gamma)(= t_{\alpha\gamma}), \tag{2.1.6}$$

$$\rho_0(\xi_{\alpha\beta}, \beta) = \rho_0(\xi_{\gamma\delta}, \delta)(= t_{\beta\delta}). \tag{2.1.7}$$

Applying Lemma 2.1.9, we get:

$$\rho_0(\Delta(\alpha, \gamma), \alpha) = t_{\alpha\gamma}{}^\frown\rho_0(\Delta(\alpha, \gamma), \xi_{\alpha\beta}), \tag{2.1.8}$$

$$\rho_0(\Delta(\alpha, \gamma), \gamma) = t_{\alpha\gamma}{}^\frown\rho_0(\Delta(\alpha, \gamma), \xi_{\gamma\delta}). \tag{2.1.9}$$

It follows that $\rho_0(\Delta(\alpha, \gamma), \xi_{\alpha\beta}) \neq \rho_0(\Delta(\alpha, \gamma), \xi_{\gamma\delta})$. Applying Lemma 2.1.9 for β and δ and the ordinal $\Delta(\alpha, \gamma)$ we get:

$$\rho_0(\Delta(\alpha, \gamma), \beta) = t_{\beta\delta}{}^\frown\rho_0(\Delta(\alpha, \gamma), \xi_{\alpha\beta}), \tag{2.1.10}$$

$$\rho_0(\Delta(\alpha, \gamma), \delta) = t_{\beta\delta}{}^\frown\rho_0(\Delta(\alpha, \gamma), \xi_{\gamma\delta}). \tag{2.1.11}$$

It follows that $\rho_0(\Delta(\alpha, \gamma), \beta) \neq \rho_0(\Delta(\alpha, \gamma), \delta)$. This shows $\Delta(\alpha, \gamma) \geq \Delta(\beta, \delta)$. A symmetrical argument shows the other inequality $\Delta(\beta, \delta) \geq \Delta(\alpha, \gamma)$. It follows that

$$\Delta(\alpha, \gamma) = \Delta(\beta, \delta)(= \bar{\xi}).$$

Our assumption is that $\rho_0(\bar{\xi}, \alpha) <_r \rho_0(\bar{\xi}, \gamma)$ and since these two sequences have $t_{\alpha\gamma}$ as common initial part, this reduces to

$$\rho_0(\bar{\xi}, \xi_{\alpha\beta}) <_r \rho_0(\bar{\xi}, \xi_{\gamma\delta}). \tag{2.1.12}$$

On the other hand $t_{\beta\delta}$ is a common initial part of $\rho_0(\bar{\xi}, \beta)$ and $\rho_0(\bar{\xi}, \delta)$, so their lexicographical relationship depends on their tails which by (2.1.10) and (2.1.11) are equal to $\rho_0(\bar{\xi}, \xi_{\alpha\beta})$ and $\rho_0(\bar{\xi}, \xi_{\gamma\delta})$ respectively. Referring to (2.1.12) we conclude that indeed $\rho_0(\bar{\xi}, \beta) <_r \rho_0(\bar{\xi}, \delta)$, i.e., $\beta <_{\rho_0} \delta$. □

The following results show that the ordering $C(\rho_0) = (\omega_1, <_{\rho_0})$ is a canonical object in the class of uncountable linear orderings.

Theorem 2.1.12. *Under* MA_{ω_1}, *the ordering* $C(\rho_0)$ *is a* minimal *uncountable linear ordering in the sense that every linear ordering that embeds into* $C(\rho_0)$ *but is not equivalent*[6] *to it must be countable.*

Proof. Let Ω be a given uncountable subset of ω_1. We need to construct a mapping $f : \omega_1 \to \Omega$ that is strictly increasing relative to the ordering $<_{\rho_0}$. Choose a sufficiently fast closed and unbounded subset D of ω_1. For $\xi < \omega_1$, let $\delta(\xi)$ denote the maximal element of D that is smaller than or equal to ξ and let $\delta^+(\xi)$ denote the minimal member of D strictly above ξ. Let $\mathcal{P}$ be the collection of all finite partial mappings p from ω_1 into Ω such that:

(1) $\alpha <_{\rho_0} \beta$ in $\mathrm{dom}(p)$ implies $p(\alpha) <_{\rho_0} p(\beta)$,
(2) $\delta(p(\alpha)) = \delta^+(\alpha)$ for all $\alpha \in \mathrm{dom}(p)$,
(3) $\delta(\Delta(\alpha, \beta)) = \delta(\Delta(p(\alpha), p(\beta)))$ for all $\alpha \neq \beta \in \mathrm{dom}(p)$,
(4) p preserves *splitting patterns* in the sense that $\Delta(\alpha, \beta) > \Delta(\beta, \gamma)$ if and only if $\Delta(p(\alpha), p(\beta)) > \Delta(p(\beta), p(\gamma))$ for every triple $\alpha, \beta, \gamma \in \mathrm{dom}(p)$.

We consider $\mathcal{P}$ as a partially ordered set under the inclusion ordering and we claim that $\mathcal{P}$ satisfies the countable chain condition. So let p_ξ $(\xi < \omega_1)$ be a given sequence of elements of $\mathcal{P}$. Going to a subsequence, we assume that p_ξ $(\xi < \omega_1)$ form an increasing Δ-system[7] of functions with some root function r. The best way to view p_ξ's is as finite partial functions from the tree[8]

$$T(\rho_0) = \{(\rho_0)_\beta \restriction \alpha : \alpha \leq \beta < \omega_1\},$$

identifying a countable ordinal α with the function $\rho_{0\alpha} : \alpha \to \omega$. So each p_ξ generates two finite $\wedge$-closed subtrees of $T(\rho_0)$, one with its domain and the other with its range. Going to a subsequence of p_ξ $(\xi < \omega_1)$, we may further assume that for every $\xi < \eta < \omega_1$, the pair of subtrees associated with p_ξ is isomorphic to the pair of subtrees associated with p_η as witnessed by the natural order isomorphism between the two finite sets of ordinals $\mathrm{dom}(p_\xi)$ and $\mathrm{dom}(p_\eta)$ that fixes $\mathrm{dom}(r)$. Note that by Lemma 2.1.11 not only the square of $C(\rho_0)$ but also all of its finite cartesian powers can be covered by countably many chains. So identifying p_ξ's with the naturally chosen elements of $C(\rho_0)^{2k}$, where k is the common cardinality of $\mathrm{dom}(p_\xi) \setminus \mathrm{dom}(r)$, and going to an uncountable subsequence of p_ξ $(\xi < \omega_1)$, we may assume that it forms a chain in $C(\rho_0)^{2k}$ which means that $p_\xi \cup p_\eta$ satisfies (1) and (2) for all $\xi < \eta < \omega_1$. Referring to the proof of Lemma 2.1.11, we can find $\xi < \eta < \omega_1$ (and in fact an uncountable subsequence of such ordinals) such that for a single ordinal δ,

$$\Delta(\alpha_\xi(i), \alpha_\eta(i)) = \Delta(p_\xi(\alpha_\xi(i)), p_\eta(\alpha_\eta(i))) = \delta$$

---

[6] Recall that for two linear orderings K and L, one writes $K \leq L$ if there is a strictly increasing mapping from K into L. Recall also that two linear orderings K and L are *equivalent*, in notation $K \equiv L$, if $K \leq L$ and $L \leq K$.

[7] Here, we use the well-known Δ-*System Lemma* (see, for example, [64, II.1.5]).

[8] We consider $T(\rho_0)$, as well as any other tree whose members are transfinite sequences of objects, ordered by the natural initial-segment relation $\sqsubseteq$. By $\sqsubset$, we denote the strict version of $\sqsubseteq$.

for all $i = 1, \ldots, k$, where $\alpha_\xi(1), \ldots, \alpha_\xi(k)$ and $\alpha_\eta(1), \ldots, \alpha_\eta(k)$ are increasing enumerations of $\mathrm{dom}(p_\xi) \setminus \mathrm{dom}(r)$ and $\mathrm{dom}(p_\eta) \setminus \mathrm{dom}(r)$, respectively. It should be clear that this means that $p_\xi \cup p_\eta$ satisfies (3) and (4) as well.

If we choose the closed unbounded set D to be the trace on ω_1 of some continuous $\in$-chain of countable elementary submodels of $(H(\omega_1), \in)$, one easily checks that for all $\alpha < \omega_1$, the set

$$\mathcal{D}_\alpha = \{p \in \mathcal{P} : \alpha \in \mathrm{dom}(p)\}$$

is a dense subset of $\mathcal{P}$. So an application of MA_{ω_1} gives us a mapping $f : \omega_1 \to \Omega$ that is strictly increasing relative to the ordering $<_{\rho_0}$. □

Corollary 2.1.13. *Under* MA_{ω_1}*, every uncountable linear ordering L whose cartesian square is the union of countably many chains contains an isomorphic copy of the ordering $C(\rho_0)$ or its reverse $C(\rho_0)^*$.*

Proof. Choose an arbitrary one-to-one mapping $f : C(\rho_0) \to L$. Let $\mathcal{P}$ be the set of all finite subsets of $C(\rho_0)$ on which f is strictly increasing. We consider $\mathcal{P}$ as a partially ordered set ordered by the inclusion. If $\mathcal{P}$ satisfies the countable chain condition, then an application of MA_{ω_1} would give us an uncountable subset K of $C(\rho_0)$ on which f is strictly increasing. It follows that $K \leq L$. Since by Theorem 2.1.12, $C(\rho_0) \leq K$, this gives us the conclusion of the corollary. So we are left with analysing the case when $\mathcal{P}$ fails to satisfy the countable chain condition. So let $\mathcal{X}$ be a given uncountable subset of $\mathcal{P}$ consisting of pairwise incompatible members of $\mathcal{P}$. Applying the Δ-system lemma and noting that the root can't contribute to the incompatibilities of members of $\mathcal{P}$, subtracting it from each member of $\mathcal{P}$, we may assume that $\mathcal{X}$ consists of pairwise-disjoint finite subsets of $C(\rho_0)$ all of some fixed size $k \geq 1$. We show that in this case there is an uncountable subset K on which f is decreasing. Recall that $C(\rho_0)$ can naturally be viewed as a subset of the lexicographically ordered tree $T(\rho_0)$. Similarly, the ordering L can be identified with a subset of a lexicographically ordered Aronszajn tree T. So considering the $\wedge$-closures of elements of $\mathcal{X}$ as well as their f-images, we perform a Δ-system argument on them, obtaining an uncountable subfamily $\mathcal{X}_0$ of $\mathcal{X}$ where the corresponding two families of trees form Δ-systems with roots R and S, respectively. Moreover, as in the previous proof, we assume that the corresponding structures are isomorphic via natural isomorphisms. Note now that since for two different elements p and q of $\mathcal{X}_0$ the mapping f is not increasing on the union $p \cup q$, there must be $\alpha \in p$ and $\beta \in q$ occupying the same position in the increasing enumerations of p and q, respectively such that

$$\alpha <_{\rho_0} \beta \text{ iff } f(\alpha) >_{\rho_0} f(\beta). \tag{2.1.13}$$

As above, we may identify a $p \in \mathcal{X}_0$ with the k-tuple $\langle p(1), \ldots, p(k)\rangle$ of elements of $C(\rho_0)$ that enumerates p increasingly according to the ordering of $C(\rho_0)$. Similarly, we identify the image $f''p$ with the k-tuple $\langle f(p(1)), \ldots, f(p(k))\rangle$

of elements of L. So going to an uncountable subfamily of $\mathcal{X}_0$, we may assume that $\{\langle p(1),\ldots,p(k)\rangle : p \in \mathcal{X}_0\}$ is a chain in $C(\rho_0)^k$ and that $\{\langle f(p(1)),\ldots,f(p(k))\rangle : p \in \mathcal{X}_0\}$ is a chain in L^k. Using 2.1.13 it follows now that f is decreasing on $K = \{p(1) : p \in \mathcal{X}_0\}$. Combining this with Theorem 2.1.12, we get the desired conclusion $C(\rho_0)^* \leq K^* \leq L$ of the corollary. □

Remark 2.1.14. It follows that, up to the equivalence, the ordering $C(\rho_0)$ and its reverse $C(\rho_0)^*$ will appear in any basis[9] of the class of uncountable linear orderings as long as the axiom used in finding the basis includes the weak forms of MA_{ω_1} used in the proofs of Theorem 2.1.12 and Corollary 2.1.13. In fact, the combination of the results of [10] and [77] shows that under the Proper Forcing Axiom,[10] the list

$$\omega_1,\ \omega_1^*,\ B,\ C(\rho_0),\ C(\rho_0)^* \tag{2.1.14}$$

form a basis for the class of all uncountable linear orderings, where B is any set of reals of cardinality $\aleph_1$. We shall return to this subject matter in Section 4.4 below.

Notation 2.1.15. *Well-ordered sets of rationals.* The set $\omega^{<\omega}$ ordered by the right lexicographical ordering $<_r$ is a particular copy of the rationals of the interval $(0, 1]$ which we are going to denote by $\mathbb{Q}_r$ or simply by $\mathbb{Q}$. The next lemma shows that for a fixed α, the map $\xi \mapsto \rho_0(\xi,\alpha)$ is a strictly increasing function from α into $\mathbb{Q}_r$. We let $(\rho_0)_\alpha$, or simply $\rho_{0\alpha}$, denote this function and we will identify it with its range, i.e., view it as a member of the tree $\sigma\mathbb{Q}_r$ of all well-ordered subsets of $\mathbb{Q}_r$, ordered by end-extension.

Lemma 2.1.16. $\rho_0(\alpha,\gamma) <_r \rho_0(\beta,\gamma)$ *whenever* $\alpha < \beta < \gamma$. □

It follows that the sequence $\rho_{0\alpha} : \alpha \to \mathbb{Q}_r$ $(\alpha < \omega_1)$ of mappings that one associates to ρ_0 defines the downwards closed subtree

$$T(\rho_0) = \{(\rho_0)_\beta \restriction \alpha : \alpha \leq \beta < \omega_1\}$$

of $\sigma\mathbb{Q}_r$. Note that by Lemma 2.1.9, for a fixed α and arbitrary $\beta \geq \alpha$, the restriction $(\rho_0)_\beta \restriction \alpha$ is determined by the way $(\rho_0)_\beta$ acts on the full lower trace $\mathrm{F}(\alpha,\beta)$, a finite subset of α. Hence all levels of $T(\rho_0)$ are countable, and therefore $T(\rho_0)$ is a particular example of an *Aronszajn tree*, a tree of height ω_1 with all levels and chains countable.

Theorem 2.1.17. *The subtree* $T(\rho_0)$ *of* $\sigma\mathbb{Q}_r$ *admits a strictly increasing map into the rationals.*[11] □

This is easily deducible from the following fact about ρ_0 that is of independent interest.

[9]Recall that a class $\mathcal{B}$ of linear orderings is a *basis* for a class $\mathcal{X}$ of linear orderings if for every $L \in \mathcal{X}$ there is $K \in \mathcal{B}$ such that $K \leq L$. For example, Corollary 2.1.13 shows that under MA_{ω_1}, the class of all uncountable linear orderings whose cartesian squares can be covered by countably many chains has a two-element basis $C(\rho_0)$, $C(\rho_0)^*$.
[10]The statement of this axiom and some of its applications can be found for example in [13].
[11]Trees of this kind are also known under the name *special Aronszajn trees*

Lemma 2.1.18. *$\{\xi < \beta : \rho_0(\xi, \beta) = \rho_0(\xi, \gamma)\}$ is a closed subset of β whenever $\beta < \gamma$.*

Proof. Let $\alpha < \beta$ be a given accumulation point of this set and let $\beta = \beta_0 > \cdots > \beta_n = \alpha$ and $\gamma = \gamma_0 > \cdots > \gamma_m = \alpha$ be the traces of the walks $\beta \to \alpha$ and $\gamma \to \alpha$, respectively. Being an accumulation point of the set $\{\xi < \beta : \rho_0(\xi, \beta) = \rho_0(\xi, \gamma)\}$, the ordinal α is in particular a limit ordinal, so we can find an ordinal $\xi < \alpha$ such that $\rho_0(\xi, \beta) = \rho_0(\xi, \gamma)$ and

$$\xi > \max(C_{\beta_i} \cap \alpha) \text{ and } \xi > \max(C_{\gamma_j} \cap \alpha) \text{ for all } i < n \text{ and } j < m. \qquad (2.1.15)$$

Note that this in particular means that the walk $\alpha \to \xi$ is a common tail of the walks $\beta \to \xi$ and $\gamma \to \xi$. Subtracting $\rho_0(\xi, \alpha)$ from $\rho_0(\xi, \beta)$ we get $\rho_0(\alpha, \beta)$ and subtracting $\rho_0(\xi, \alpha)$ from $\rho_0(\xi, \gamma)$ we get $\rho_0(\alpha, \gamma)$. It follows that $\rho_0(\alpha, \beta) = \rho_0(\alpha, \gamma)$. □

It follows that $T(\rho_0)$ does not branch at limit levels. Now note the general fact that any downward closed subtree of $\sigma\mathbb{Q}$ which is finitely branching at limit nodes admits a strictly increasing mapping into $\mathbb{Q}$, so the tree $T(\rho_0)$ is a special subtree of $\sigma\mathbb{Q}$.

Definition 2.1.19. Identifying the power set of $\mathbb{Q}$ with the particular copy $2^{\mathbb{Q}}$ of the Cantor set, define for every countable ordinal α,

$$G_\alpha = \{x \in 2^{\mathbb{Q}} : x \text{ end-extends no } \rho_{0\beta} \restriction \alpha \text{ for } \beta \geq \alpha\}.$$

Lemma 2.1.20. *G_α $(\alpha < \omega_1)$ is an increasing sequence of proper G_δ-subsets of the Cantor set whose union is equal to the Cantor set.* □

Lemma 2.1.21. *The set $X = \{\rho_{0\beta} : \beta < \omega_1\}$ considered as a subset of the Cantor set $2^{\mathbb{Q}}$ has universal measure zero.*

Proof. Let μ be a given nonatomic Borel measure on $2^{\mathbb{Q}}$. For $t \in T(\rho_0)$, set

$$P_t = \{x \in 2^{\mathbb{Q}} : x \text{ end-extends } t\}.$$

Note that each P_t is a perfect subset of $2^{\mathbb{Q}}$ and therefore is μ-measurable. Let

$$S = \{t \in T(\rho_0) : \mu(P_t) > 0\}.$$

Then S is a downward closed subtree of $\sigma\mathbb{Q}$ with no uncountable antichains. By an old result of Kurepa (see [105]), no Souslin tree admits a strictly increasing map into the reals (as for example $\sigma\mathbb{Q}$ does). It follows that S must be countable and so we are done. □

2.2 The coherence of maximal weights

Definition 2.2.1. The *maximal weight* of the walk is the characteristic

$$\rho_1 : [\omega_1]^2 \longrightarrow \omega$$

defined recursively by

$$\rho_1(\alpha, \beta) = \max\{|C_\beta \cap \alpha|, \rho_1(\alpha, \min(C_\beta \setminus \alpha))\},$$

with the boundary value $\rho_1(\alpha, \alpha) = 0$ for all α. Thus $\rho_1(\alpha, \beta)$ is simply the maximal integer appearing in the sequence $\rho_0(\alpha, \beta)$ defined above.

Lemma 2.2.2. *For all $\alpha < \beta < \omega_1$ and $n < \omega$,*

(a) *$\{\xi \leq \alpha : \rho_1(\xi, \alpha) \leq n\}$ is finite,*
(b) *$\{\xi \leq \alpha : \rho_1(\xi, \alpha) \neq \rho_1(\xi, \beta)\}$ is finite.*

Proof. The proof is by induction. To prove (a) it suffices to show that for every $n < \omega$ and every $A \subseteq \alpha$ of order-type ω there is $\xi \in A$ such that $\rho_1(\xi, \alpha) > n$. Let $\eta = \sup(A)$. If $\eta = \alpha$ one chooses arbitrary $\xi \in A$ with the property that $|C_\alpha \cap \xi| > n$, so let us consider the case $\eta < \alpha$. Let $\alpha_1 = \min(C_\alpha \setminus \eta)$. By the inductive hypothesis there is $\xi \in A$ such that:

$$\xi > \max(C_\alpha \cap \eta), \tag{2.2.1}$$

$$\rho_1(\xi, \alpha_1) > n. \tag{2.2.2}$$

Note that $\rho_0(\xi, \alpha) = \langle |C_\alpha \cap \eta| \rangle^\frown \rho_0(\xi, \alpha_1)$, and therefore

$$\rho_1(\xi, \alpha) \geq \rho_1(\xi, \alpha_1) > n.$$

To prove (b) we show by induction that for every $A \subseteq \alpha$ of order-type ω there exists $\xi \in A$ such that $\rho_1(\xi, \alpha) = \rho_1(\xi, \beta)$. Let $\eta = \sup(A)$ and let $\beta_1 = \min(C_\beta \setminus \eta)$. Let $n = |C_\beta \cap \eta|$ and let

$$B = \{\xi \in A : \xi > \max(C_\beta \cap \eta) \text{ and } \rho_1(\xi, \beta_1) > n\}.$$

Then B is infinite, so by the induction hypothesis we can find $\xi \in B$ such that $\rho_1(\xi, \alpha) = \rho_1(\xi, \beta_1)$. Then

$$\rho_1(\xi, \beta) = \max\{n, \rho_1(\xi, \beta_1)\} = \rho_1(\xi, \beta_1),$$

so we are done. □

Definition 2.2.3. Consider the following linear ordering $<_{\rho_1}$ on ω_1 that is now based on the mapping ρ_1 rather than ρ_0:

$$\alpha <_{\rho_1} \beta \quad \text{iff} \quad \rho_1(\xi, \alpha) <_r \rho_1(\xi, \beta), \tag{2.2.3}$$

where $\xi = \Delta(\alpha, \beta) = \min\{\eta < \min\{\alpha, \beta\} : \rho_1(\eta, \alpha) \neq \rho_1(\eta, \beta)\}$.[12]

[12] When $\rho_1(\xi, \alpha) = \rho_1(\xi, \beta)$ for all $\xi < \min\{\alpha, \beta\}$ then it is understood that $\min\{\alpha, \beta\} <_{\rho_1} \max\{\alpha, \beta\}$.

Lemma 2.2.4. *The cartesian square of the total ordering $<_{\rho_1}$ of ω_1 is the union of countably many chains.*

Proof. It suffices to decompose $\{\langle\alpha,\beta\rangle : \alpha<\beta<\omega_1\}$ into countably many chains. For $\alpha<\beta<\omega_1$, let

$$D_{\alpha\beta} = \{\xi : \xi = \alpha \text{ or } \xi < \alpha \text{ and } \rho_1(\xi,\alpha) \neq \rho(\xi,\beta)\},$$

and let $n_{\alpha\beta}$ be the maximal value $\rho_1(\xi,\alpha)$ or $\rho_1(\xi,\beta)$ for ξ in $D_{\alpha\beta}$. Let

$$F_{\alpha\beta} = \{\xi : \xi = \alpha \text{ or } \xi < \alpha \text{ and } \rho_1(\xi,\alpha) \leq n_{\alpha,\beta} \text{ or } \rho_1(\xi,\beta) \leq n_{\alpha,\beta}\}.$$

Note that $D_{\alpha\beta} \subseteq F_{\alpha\beta}$. The lemma is proved once we show that, for some $\alpha<\beta$ and $\gamma<\delta$, we have $\alpha <_{\rho_1} \gamma$ and

$$n_{\alpha\beta} = n_{\gamma\delta} = n,\ \rho_{1\alpha} \restriction F_{\alpha\beta} \cong \rho_{1\gamma} \restriction F_{\gamma\delta}{}^{13},\ \rho_{1\beta} \restriction F_{\alpha\beta} \cong \rho_{1\delta} \restriction F_{\gamma\delta}{}^{14}, \tag{2.2.4}$$

and then prove that $\beta <_{\rho_1} \delta$. For this we first observe that $\Delta(\alpha,\gamma) = \Delta(\beta,\delta)$ is an easy consequence of 2.2.4. Let ξ denote the common value of the function Δ.

If $\xi \in F_{\alpha\beta} \setminus F_{\gamma\delta}$, then

$$\rho_1(\xi,\gamma) = \rho_1(\xi,\delta) > n \geq \rho_1(\xi,\beta)$$

which means that $\beta <_{\rho_1} \delta$. If on the other hand $\xi \in F_{\gamma\delta} \setminus F_{\alpha\beta}$, then

$$\rho_1(\xi,\alpha) > n \geq \rho_1(\xi,\gamma)$$

contradicting our assumption $\alpha <_{\rho_1} \gamma$. If $\xi \notin F_{\alpha\beta} \cup F_{\gamma\delta}$, then

$$\rho_1(\xi,\beta) = \rho_1(\xi,\alpha) < \rho_1(\xi,\gamma) = \rho_1(\xi,\delta)$$

which gives us the desired inequality $\beta <_{\rho_1} \delta$. □

Let $C(\rho_1)$ denote the linearly ordered set $(\omega_1, <_{\rho_1})$. Then working as in the case of the ordering $<_{\rho_0}$, we have the following result showing that $C(\rho_1)$ also serves as a basis member for the class of uncountable linear orderings.

Theorem 2.2.5. *Assuming* MA_{ω_1}, *the ordering $C(\rho_1)$ and its reverse $C(\rho_1)^*$ serve as a two-element basis for the class of all uncountable linear orderings whose cartesian squares can be covered by countably many chains.* □

We now note the following unboundedness property of ρ_1 which is an immediate consequence of Lemma 2.2.2 and which will be elaborated on in later sections of this book.

[13] Here, $\rho_{1\alpha} : \alpha \to \omega$ are defined from ρ_1 in the usual way, $\rho_{1\alpha}(\xi) = \rho_1(\xi,\alpha)$.

[14] Here the isomorphism relation $\cong$ between two finite partial functions p and q from ω_1 into ω refers to the fact that there is an order-preserving bijection between their domains transfering one function to the other.

Lemma 2.2.6. *For every pair $\mathcal{A}$ and $\mathcal{B}$ of uncountable families of pairwise-disjoint finite subsets of ω_1 and every positive integer n, there exist uncountable subfamilies $\mathcal{A}_0 \subseteq \mathcal{A}$ and $\mathcal{B}_0 \subseteq \mathcal{B}$ such that, for every pair $a \in \mathcal{A}_0$ and $b \in \mathcal{B}_0$ such that $a < b$,*[15] *we have $\rho_1(\alpha, \beta) > n$ for all $\alpha \in a$ and $\beta \in b$.* □

Corollary 2.2.7. *The tree*

$$T^*(\rho_1) = \{t : \alpha \longrightarrow \omega : \ \alpha < \omega_1, t =^* \rho_{1\alpha}\}^{16}$$

is a coherent, homogeneous, Aronszajn tree that admits a strictly increasing map into the reals. □

Definition 2.2.8. Consider the following extension of $T^*(\rho_1)$:

$$\widetilde{T}(\rho_1) = \{t : \alpha \longrightarrow \omega : \alpha < \omega_1 \text{ and } t \restriction \xi \in T^*(\rho_1) \text{ for all } \xi < \alpha\}.$$

If we order $\widetilde{T}(\rho_1)$ by the right lexicographical ordering $<_r$ we get a complete linearly ordered set. It is not continuous as it contains jumps of the form

$$[t^\frown\langle m\rangle, t^\frown\langle m+1\rangle^\frown\vec{0}],$$

where $t \in T^*(\rho_1)$ and $m < \omega$. Removing the right-hand points from all the jumps we get a linearly ordered continuum which we denote by $\widetilde{A}(\rho_1)$.

Lemma 2.2.9. *$\widetilde{A}(\rho_1)$ is a homogeneous nonreversible ordered continuum that can be represented as the union of an increasing ω_1-sequence of Cantor sets.*

Proof. For a countable limit ordinal $\delta > 0$, set

$$\widetilde{A}_\delta(\rho_1) = \{t \in \widetilde{A}(\rho_1) : l(t) \leq \delta\}.$$

Then each $\widetilde{A}_\delta(\rho_1)$ is a closed subset of $\widetilde{A}(\rho_1)$ with the level $T^*_\delta(\rho_1)$ being a countable order-dense subset. One easily concludes from this that $\widetilde{A}_\delta(\rho_1)$ (δ limit $< \omega_1$) is the required decomposition of $\widetilde{A}(\rho_1)$.

To show that $\widetilde{A}(\rho_1)$ is homogeneous, consider two pairs $x_0 < x_1$ and $y_0 < y_1$ of non-endpoints of $\widetilde{A}(\rho_1)$. Choose a countable limit ordinal δ strictly above the lengths of x_0, x_1, y_0 and y_1. So we have a Cantor set $\widetilde{A}_\delta(\rho_1)$, an order-dense subset $T^*_\delta(\rho_1)$ of $\widetilde{A}_\delta(\rho_1)$, and two pairs of points $x_0 < x_1$ and $y_0 < y_1$ in

$$\widetilde{A}_\delta(\rho_1) \setminus T^*_\delta(\rho_1).$$

By Cantor's theorem there is an order isomorphism $\sigma : \widetilde{A}_\delta(\rho_1) \longrightarrow \widetilde{A}_\delta(\rho_1)$ such that:

$$\sigma'' T^*_\delta(\rho_1) = T^*_\delta(\rho_1), \tag{2.2.5}$$

$$\sigma(x_i) = y_i \text{ for } i < 2. \tag{2.2.6}$$

[15] For sets of ordinals a and b we write $a < b$ whenever $\alpha < \beta$ for all $\alpha \in a$ and $\beta \in b$.

[16] Here, $=^*$ denotes the fact that the functions agree on all but finitely many arguments.

Extend σ to the rest of $\widetilde{A}(\rho_1)$ by the formula

$$\sigma(t) = \sigma(t \restriction \delta)^\frown t \restriction [\delta, l(t)).$$

It is easily checked that the function σ is the required order-isomorphism.

Let us now prove that there is no order-reversing bijection

$$\pi : \widetilde{A}(\rho_1) \longrightarrow \widetilde{A}(\rho_1).$$

Since $T^*(\rho_1)$ with the lexicographical ordering does not have a countable order-dense subset, the image $\pi''T^*(\rho_1)$ must have sequences of length bigger than any given countable ordinal. So for every limit ordinal δ we can fix t_δ in $T^*(\rho_1)$ of length $\geq \delta$ such that $\pi(t_\delta)$ also has length $\geq \delta$. Let

$$f(\delta) = \max\{\xi < \delta : t_\delta(\xi) \neq \pi(t_\delta)(\xi)\} + 1.$$

By the Pressing Down Lemma[17], find an uncountable set Γ of countable limit ordinals, a countable ordinal $\bar{\xi}$ and $s, t \in T^*_{\bar{\xi}}(\rho_1)$ such that for all $\delta \in \Gamma$:

$$f(\delta) = \bar{\xi}, \tag{2.2.7}$$

$$t_\delta \restriction \bar{\xi} = s \text{ and } \pi(t_\delta) \restriction \bar{\xi} = t. \tag{2.2.8}$$

Moreover we may assume that

$$\{t_\delta \restriction \delta : \delta \in \Gamma\} \text{ and } \{\pi(t_\delta) \restriction \delta : \delta \in \Gamma\}$$

are antichains of the tree $T^*(\rho_1)$. Pick $\gamma \neq \delta$ in Γ and suppose for definiteness that $t_\gamma <_l t_\delta$. Note that the ordinal $\Delta(t_\gamma, t_\delta)$ where this relation is decided is above $\bar{\xi}$ and is smaller than both γ and δ. Similarly the ordinal $\Delta(\pi(t_\gamma), \pi(t_\delta))$ where the lexicographical ordering between $\pi(t_\gamma)$ and $\pi(t_\delta)$ is decided also belongs to the interval

$$(\bar{\xi},\ \min\{\gamma, \delta\}).$$

It follows that $\Delta(\pi(t_\gamma), \pi(t_\delta)) = \Delta(t_\gamma, t_\delta)(= \xi)$ and that

$$\pi(t_\gamma)(\xi) = t_\gamma(\xi) < t_\delta(\xi) = \pi(t_\delta(\xi)),$$

contradicting the fact that π is order-reversing. □

Definition 2.2.10. The set $\widetilde{T}(\rho_1)$ has another natural structure, a topology generated by the family of sets of the form

$$\widetilde{V}_t = \{u \in \widetilde{T}(\rho_1) : t \sqsubseteq u\},$$

for t a node of $T^*(\rho_1)$ of successor length as a clopen subbase. Let $T^0(\rho_1)$ denote the set of all nodes of $T^*(\rho_1)$ of successor length. Then $\widetilde{T}(\rho_1)$ can be regarded as

[17] This is a well-known fact saying that regressive mappings defined on large subsets of ω_1 (or any other uncountable cardinal) are constant on large subsets of ω_1 (see, for example, [64, II.6.15]).

the set of all downward closed chains of the tree $T^0(\rho_1)$ and the topology on $\widetilde{T}(\rho_1)$ is simply the topology one obtains from identifying the power set of $T^0(\rho_1)$ with the cube

$$\{0,1\}^{T^0(\rho_1)}$$

with its Tychonoff topology[18]. $\widetilde{T}(\rho_1)$ being a closed subset of the cube is compact. In fact $\widetilde{T}(\rho_1)$ has some very strong topological properties such as the property that closed subsets of $\widetilde{T}(\rho_1)$ are its retracts.

Lemma 2.2.11. *$\widetilde{T}(\rho_1)$ is a homogeneous Eberlein compactum*[19]*.*

Proof. The proof that $\widetilde{T}(\rho_1)$ is homogeneous is quite similar to the corresponding part of the proof of Lemma 2.2.9. To see that $\widetilde{T}(\rho_1)$ is an Eberlein compactum, i.e., that the function space $\mathcal{C}(\widetilde{T}(\rho_1))$ is weakly compactly generated, let $\{X_n\}$ be a countable antichain decomposition of $T^0(\rho_1)$ and consider the set

$$K = \{2^{-n}\chi_{\tilde{V}_t} : n < \omega, t \in X_n\} \cup \{\chi_\emptyset\}.$$

Note that K is a weakly compact subset of $\mathcal{C}(\widetilde{T}(\rho_1))$ which separates the points of $\widetilde{T}(\rho_1)$. □

In the next application the coherent sequence $\rho_{1\alpha} : \alpha \longrightarrow \omega$ $(\alpha < \omega_1)$ of finite-to-one maps needs to be turned into a coherent sequence of maps that are actually one-to-one. One way to achieve this is via the following formula:

$$\bar{\rho}_1(\alpha,\beta) = 2^{\rho_1(\alpha,\beta)} \cdot (2 \cdot |\{\xi \leq \alpha : \rho_1(\xi,\beta) = \rho_1(\alpha,\beta)\}| + 1).$$

Lemma 2.2.12.

(a) *$\bar{\rho}_1(\alpha,\gamma) \neq \bar{\rho}_1(\beta,\gamma)$ for all $\alpha < \beta < \gamma < \omega_1$.*
(b) *$\{\xi \leq \alpha : \bar{\rho}_1(\xi,\alpha) \neq \bar{\rho}_1(\xi,\beta)\}$ is finite for all $\alpha < \beta < \omega_1$.*

Proof. Suppose $\alpha \leq \beta < \gamma$ and that $\bar{\rho}_1(\alpha,\gamma) = \bar{\rho}_1(\beta,\gamma)$. Then $\rho_1(\alpha,\gamma) = \rho_1(\beta,\gamma) = n$ and

$$|\{\xi \leq \alpha : \rho_1(\xi,\gamma) = n\}| = |\{\xi \leq \beta : \rho_1(\xi,\gamma) = n\}|.$$

Since the set on the left-hand side is an initial segment of the set of the right-hand side, the sets must in fact be equal. It follows that $\alpha = \beta$. This proves (a).

To prove (b), by Lemma 2.2.2 the set

$$D_{\alpha\beta} = \{\xi \leq \alpha : \rho_1(\xi,\alpha) \neq \rho_1(\xi,\beta)\}$$

---

[18] This is done by identifying a subset V of $T^0(\rho_1)$ with its characteristic function $\chi_V : T^0(\rho_1) \longrightarrow 2$.

[19] Recall that a compactum X is *Eberlein* if its function space $\mathcal{C}(X)$ can be generated by a subset which is compact in the weak topology.

is finite. Let

$$m = \max\{\rho(\xi,\eta) : \xi \in D_{\alpha\beta}, \eta \in \{\alpha,\beta\}\} + 1.$$

By Lemma 2.2.2, the proof of (b) is finished if we show that for $\xi < \alpha$, if $\rho_1(\xi,\alpha) > m$ and $\rho_1(\xi,\beta) > m$, then $\bar{\rho_1}(\xi,\alpha) = \bar{\rho_1}(\xi,\beta)$. From the definition of m, we must have that $j = \rho_1(\xi,\alpha) = \rho_1(\xi,\beta) > m$, and that

$$\{\eta \leq \xi : \rho(\eta,\alpha) = i\} = \{\eta \leq \xi : \rho(\eta,\alpha) = i\}.$$

It follows that if j denotes the cardinality of this set, then

$$\bar{\rho_1}(\xi,\alpha) = 2^i(2j+1) = \bar{\rho_1}(\xi,\beta).$$

This completes the proof. □

Define $\bar{\rho}_{1\alpha}$ from $\bar{\rho}_1$ just as $\rho_{1\alpha}$ was defined from ρ_1; then the $\bar{\rho}_{1\alpha}$'s are one-to-one and coherent. From ρ_1 one also has a natural sequence r_α $(\alpha < \omega_1)$ of elements of ω^ω defined as

$$r_\alpha(n) = |\{\xi \leq \alpha : \rho_1(\xi,\alpha) \leq n\}|.$$

Note that r_β eventually strictly dominates r_α whenever $\alpha < \beta$.

Definition 2.2.13. The sequences $e_\alpha = \bar{\rho}_{1\alpha}$ $(\alpha < \omega_1)$ and r_α $(\alpha < \omega_1)$ can be used in describing a functor

$$G \longmapsto G^*,$$

which to every graph G on ω_1 associates another graph G^* on ω_1 as follows:

$$\{\alpha,\beta\} \in G^* \text{ iff } \{e_\alpha^{-1}(l), e_\beta^{-1}(l)\} \in G \tag{2.2.9}$$

for all $l < \Delta(r_\alpha, r_\beta)$ for which these preimages are both defined and different[20].

Lemma 2.2.14. *Suppose that every uncountable family $\mathcal{F}$ of pairwise-disjoint finite subsets of ω_1 contains two sets A and B such that $A \otimes B \subseteq G$[21]. Then the same is true about G^* provided the uncountable family $\mathcal{F}$ consists of finite cliques[22] of G^*.*

Proof. Let $\mathcal{F}$ be a given uncountable family of pairwise-disjoint finite cliques of G^*. We may assume that all members of $\mathcal{F}$ are of some fixed size k.

Consider a countable limit ordinal $\delta > 0$ and an A in $\mathcal{F}$ with all its elements above δ. Let $n = n(A,\delta)$ be the minimal integer such that for all $\alpha < \beta$ in $A \cup \{\delta\}$:

$$\Delta(r_\alpha, r_\beta) < n,\text{[23]} \tag{2.2.10}$$

[20] As it will be clear from the proof of the following lemma, the functor $G \longrightarrow G^*$ can equally be based on any other coherent sequence $e_\alpha : \alpha \longrightarrow \omega$ $(\alpha < \omega_1)$ of one-to-one mappings and any other sequence r_α $(\alpha < \omega_1)$ of pairwise distinct reals.

[21] Here, $A \otimes B = \{\{\alpha,\beta\} : \alpha \in A, \beta \in B, \alpha \neq \beta\}$.

[22] A *clique* of G^* is a subset C of ω_1 with the property that $[C]^2 \subseteq G^*$.

[23] Recall that for $s \neq t \in \omega^\omega$, we let $\Delta(s,t) = \min\{k : s(k) \neq t(k)\}$.

$$e(\xi,\alpha) \neq e(\xi,\beta) \text{ for some } \xi \leq \alpha \text{ implies } e(\xi,\alpha), e(\xi,\beta) \leq n. \tag{2.2.11}$$

Let

$$H(\delta, A) = \{\xi \leq \omega_1 : e(\xi,\alpha) \leq n(\delta, A) \text{ for some } \alpha \geq \xi \text{ from } A \cup \{\delta\}\}.$$

Taking the transitive collapse $\bar{H}(\delta, A)$ of $H(\delta, A)$, the sequence

$$e_\alpha \upharpoonright H(\delta, A)\ (\alpha \in A)$$

collapses to a k-sequence $\bar{s}(\delta, A)$ of mappings with integer domains. Let $\bar{r}(\delta, A)$ denote the k-sequence

$$r_\alpha \upharpoonright (n(\delta, A) + 1)\ (\alpha \in A)$$

enumerated in increasing order. Hence every $A \in \mathcal{F}$ above δ generates a quadruple

$$(H(\delta, A) \cap \delta, \bar{r}(\delta, A), \bar{s}(\delta, A), n(\delta, A))$$

of parameters. Since the set of quadruples is countable, by the assumption on the graph G we can find two distinct members A_δ and B_δ of $\mathcal{F}$ above δ that generate the same quadruple of parameters denoted by

$$(H_\delta, \bar{r}_\delta, \bar{s}_\delta, n_\delta).$$

Moreover, we choose A_δ and B_δ to satisfy the following isomorphism condition.

For every $l \leq n_\delta$, $i < k$, if α is the ith member of A_δ, β the ith member of B_δ and if $e_\alpha^{-1}(l)$ and $e_\beta^{-1}(l)$ are both defined, then they are G-connected to the same elements of $H_\delta \cap \delta$. (2.2.12)

Let m_δ be the minimal integer $> n_\delta$ such that

$$\Delta(r_\alpha, r_\beta) < m_\delta \text{ for all } \alpha \in A_\delta \text{ and } \beta \in B_\delta. \tag{2.2.13}$$

Let

$$I_\delta = \{\xi : e(\xi,\alpha) \leq m_\delta \text{ for some } \alpha \geq \xi \text{ in } A_\delta \cup B_\delta\}.$$

Let $\bar{p}_\delta$ be the k-sequence $r_\alpha \upharpoonright m_\delta (\alpha \in A_\delta)$ enumerated in increasing order and similarly let $\bar{q}_\delta$ be the k-sequence $r_\beta \upharpoonright m_\delta(\beta \in B_\delta)$ enumerated increasingly. Let $\bar{t}_\delta$ and $\bar{u}_\delta$ be the transitive collapse of $e_\alpha \upharpoonright I_\delta$ $(\alpha \in A_\delta)$ and $e_\beta \upharpoonright I_\delta(\beta \in B_\delta)$, respectively. By the Pressing Down Lemma there is an unbounded $\Gamma \subseteq \omega_1$ and tuples

$$(H, \bar{r}, \bar{s}, n) \text{ and } (I, \bar{p}, \bar{q}, \bar{t}, \bar{u}, m)$$

such that for all $\delta \in \Gamma$:

$$(H_\delta, \bar{r}_\delta, \bar{s}_\delta, n_\delta) = (H, \bar{r}, \bar{s}, n), \tag{2.2.14}$$

$$(I_\delta \cap \delta, \bar{p}_\delta, \bar{q}_\delta, \bar{t}_\delta, \bar{u}_\delta, m_\delta) = (I, \bar{p}, \bar{q}, \bar{t}, \bar{u}, m). \tag{2.2.15}$$

Moreover we assume the following analogue of (2.2.12) where $k_1 = |I_\gamma \setminus \gamma|$ for some (equivalently all) $\gamma \in \Gamma$:

> For every $\gamma, \delta \in \Gamma$ and $i < k_1$, if η is the ith member of I_γ and if ξ is the ith member of I_δ in their increasing enumerations, then η and ξ are G-connected to the same elements of the root I, and if η happens to be the i^*th member of A_γ (or B_γ) for some $i^* < k$, then ξ must be the i^*th member of A_δ (B_δ resp.) and vice versa. (2.2.16)

By our assumption about the graph G there exist $\gamma < \delta$ in Γ such that:

$$\max(I_\gamma) < \delta, \tag{2.2.17}$$

$$(I_\gamma \setminus \gamma) \otimes (I_\delta \setminus \delta) \subseteq G. \tag{2.2.18}$$

The proof of Lemma 2.2.14 is finished once we show that

$$A_\gamma \otimes B_\delta \subseteq G^*. \tag{2.2.19}$$

Let $i, j < k$ be given and let α be the ith member of A_γ and let β be the jth member of B_δ.

Case 1: $i \neq j$. Pick an $l < \Delta(r_\alpha, r_\beta)$. By (2.2.10), $l < n$. Assume that $e_\alpha^{-1}(l)$ and $e_\beta^{-1}(l)$ are both defined.

Subcase 1.1: $e_\alpha^{-1}(l) < \gamma$ and $e_\beta^{-1}(l) < \delta$. By the first choice of parameters, $e_\alpha^{-1}(l)$ and $e_\beta^{-1}(l)$ are members of the set H which is an initial part of $H(\gamma, A_\gamma)$ and $H(\delta, B_\delta)$. Therefore, we have that the behavior of e_α and e_β on H is encoded by the ith and jth term of the sequence $\bar{s}$, respectively. In particular, we have

$$e_\beta^{-1}(l) = e_{\beta''}^{-1}(l), \tag{2.2.20}$$

where β'' =the jth member of A_γ. By our assumption that $[A_\gamma]^2 \subseteq G^*$ we infer that $\{\alpha, \beta''\} \in G^*$. Referring to the definition (2.2.9) we conclude that $e_\alpha^{-1}(l)$ and $e_\beta^{-1}(l)$ must be G-connected if they are different.

Subcase 1.2: $e_\alpha^{-1}(l) < \gamma$ and $e_\beta^{-1}(l) \in I_\delta \setminus \delta$. Let

$$\beta' = \text{the } j\text{th member of } B_\gamma.$$

By the choice of parametrization, the position of $e_{\beta'}^{-1}(l)$ in I_γ is the same as the position of $e_\beta^{-1}(l)$ in I_δ so by (2.2.16) their relationship to the point $e_\alpha^{-1}(l)$ of the root I is the same. Similarly, by the choice of the first set of parameters, letting β'' be the jth member of A_γ, the position of $e_{\beta'}^{-1}(l)$ in $H(\gamma, B_\gamma)$ and $e_{\beta''}^{-1}(l)$ in $H(\gamma, A_\gamma)$ is the same, so from (2.2.12) we conclude that their relationship with $e_\alpha^{-1}(l)$ is the same. However, we have checked in the previous subcase that $e_\alpha^{-1}(l)$ and $e_{\beta''}^{-1}(l)$ are G-connected in case they are different.

Subcase 1.3: $e_\alpha^{-1}(l) \in I_\gamma \setminus \gamma$ and $e_\beta^{-1}(l) < \delta$. This is essentially symmetric to the previous subcase.

Subcase 1.4: $e_\alpha^{-1}(l) \in I_\gamma \setminus \gamma$ and $e_\beta^{-1}(l) \in I_\delta \setminus \delta$. The fact that $\{e_\alpha^{-1}(l), e_\beta^{-1}(l)\} \in G$ in this case follows from (2.2.18).

Case 2: $i = j$. Consider an $l < \Delta(r_\alpha, r_\beta)$. Note that now we have $l < m$ (see (2.2.13)).

Subcase 2.1: $e_\alpha^{-1}(l) < \gamma$ and $e_\beta^{-1}(l) < \delta$. Then $e_\alpha^{-1}(l)$ and $e_\beta^{-1}(l)$ are elements of I which is an initial part of both I_γ and I_δ. Therefore the jth ($= i$th) term of $\bar{u}$ encodes both $e_\beta \restriction I$ and $e_{\beta'} \restriction I$ (see the definition of β' above). It follows that

$$e_\beta^{-1}(l) = e_{\beta'}^{-1}(l).$$

If $l \le n$ then $e_{\beta'}^{-1}(l)$ belongs to H and since the jth($= i$th) term of $\bar{s}$ encodes both $e_\alpha \restriction H$ and $e_\beta \restriction H$, it follows that

$$e_\alpha^{-1}(l) = e_{\beta'}^{-1}(l).$$

If $l > n$ then $e_\alpha^{-1}(l)$ and $e_{\beta'}^{-1}(l)$ are not members of H so from (2.2.11) and the definition of

$$H(\gamma, A_\gamma) \cap \gamma = H = H(\gamma, B_\gamma) \cap \gamma$$

we conclude that

$$e_\gamma(e_\alpha^{-1}(l)) = e_\alpha(e_\alpha^{-1}(l)) = l = e_{\beta'}(e_{\beta'}^{-1}(l)) = e_\gamma(e_{\beta'}^{-1}(l)).$$

Since e_γ is one-to-one we conclude that

$$e_\alpha^{-1}(l) = e_{\beta'}^{-1}(l) = e_\beta^{-1}(l).$$

Subcase 2.2: $e_\alpha^{-1}(l) < \gamma$ and $e_\beta^{-1}(l) \in I_\delta \setminus \delta$. Recall that β' is the ith ($= j$th) member of B_γ, so by the first choice of parameters the relative position of $e_{\beta'}^{-1}(l)$ in $H(\gamma, B_\gamma)$ must be the same as the relative position of $e_\alpha^{-1}(l)$ in $H(\gamma, A_\gamma)$, i.e., it must belong to the root H. Note that if j^* is the position of β in I_δ, then j^* must also be the position of β' in I_γ. It follows that the j^*th term of $\bar{u}$ encodes both

$$e_\beta \restriction I_\delta \text{ and } e_{\beta'} \restriction I_\gamma = e_{\beta'} \restriction I.$$

So it must be that $e_\beta^{-1}(l)$ belongs to the root I, a contradiction. So this subcase never occurs.

Subcase 2.3: $e_\alpha^{-1}(l) \in I_\gamma \setminus \gamma$ and $e_\beta^{-1}(l) < \delta$. This is symmetric to the previous subcase, so it also never occurs.

Subcase 2.4: $e_\alpha^{-1}(l) \in I_\gamma \setminus \gamma$ and $e_\beta^{-1}(l) \in I_\delta \setminus \delta$. Then $\{e_\alpha^{-1}(l), e_\beta^{-1}(l)\} \in G$ follows from (2.2.18).

This completes the proof of Lemma 2.2.14. □

Lemma 2.2.15. *If there is uncountable $\Gamma \subseteq \omega_1$ such that $[\Gamma]^2 \subseteq G^*$, then ω_1 can be decomposed into countably many sets Σ such that $[\Sigma]^2 \subseteq G$.*

Proof. Fix an uncountable $\Gamma \subseteq \omega_1$ such that $[\Gamma]^2 \subseteq G^*$. For a finite binary sequence s of length equal to some $l+1$, set

$$\Gamma_s = \{\xi < \omega_1 : e(\xi, \alpha) = l \text{ for some } \alpha \text{ in } \Gamma \text{ with } s \subseteq r_\alpha\}.$$

Then the sets Γ_s cover ω_1 and $[\Gamma_s]^2 \subseteq G$ for all s. □

Remark 2.2.16. Let G be the comparability graph of some Souslin tree T. Then for every uncountable family $\mathcal{F}$ of pairwise-disjoint cliques of G (finite chains of T) there exist $A \neq B$ in $\mathcal{F}$ such that $A \cup B$ is a clique of G (a chain of T). However, it is not hard to see that G^* fails to have this property (i.e., the conclusion of Lemma 2.2.14). This shows that some assumption on the graph G in Lemma 2.2.14 is necessary. There are indeed many graphs that satisfy the hypothesis of Lemma 2.2.14. Many examples appear when one is trying to apply Martin's axiom to some Ramsey-theoretic problems. Note that the conclusion of Lemma 2.2.14 is simply saying that the poset of all finite cliques of G^* is ccc, while its hypothesis is a bit stronger than the fact that the poset of all finite cliques of G is ccc in all of its finite powers. Applying Lemma 2.2.15 to the case when G is the incomparability graph of some Aronszajn tree, we see that the statement saying that all Aronszajn trees are special is a purely Ramsey-theoretic statement in the same way Souslin's hypothesis is.

We finish this section by showing that the assumption (c) on a given C-sequence C_α $(\alpha < \omega_1)$ on which we base our walks, and the function ρ_1, is not sufficient to give us the stronger conclusion that the tree $T(\rho_1)$ can be decomposed into countably many antichains (see Corollary 2.2.7). To describe an example, we need the following notation, given that we have fixed one C-sequence C_α $(\alpha < \omega_1)$. Let $C_\alpha(0) = 0$ and for $0 < n < \omega$, let $C_\alpha(n)$ denote the nth element of C_α according to its increasing enumeration with the convention that $C_{\alpha+1}(n) = \alpha$ for all $n > 0$. We assume that the C-sequence is chosen so that for a limit ordinal $\alpha > 0$ and a positive integer n, the ordinal $C_{\alpha+1}(n)$ is at least $n+1$ steps away from the closest limit ordinal below it. For a set D of countable ordinals, let D^0 denote the set of successor ordinals in D. We also fix, for each countable ordinal α a one-to-one function e_α whose domain is the set α^0 of all successor ordinals $< \alpha$ and range included in ω, which cohere in the sense that $e_\alpha(\xi) = e_\beta(\xi)$ for all but finitely many successor ordinals $\xi \in \alpha^0 \cap \beta^0$. For each $r \in ([\omega]^{<\omega})^\omega$, we associate another C-sequence C^r_α $(\alpha < \omega_1)$ by letting $C^r_\alpha = C_\alpha \cap [\xi^r_\alpha, \alpha) \cup \bigcup_{n \in \omega} D^r_\alpha(n)$, where

$$D^r_\alpha(n) = \{\xi \in [C_\alpha(n), C_\alpha(n+1))^0 : e_\alpha(\xi) \in r(n)\},$$

and where $\xi^r_\alpha = \sup(\bigcup_{n \in \omega} D^r_\alpha(n))$. (This definition really applies only when α is a limit ordinal; for successor ordinals we put $C^r_{\alpha+1} = \{\alpha\}$.) This gives us a C-sequence C^r_α $(\alpha < \omega_1)$ and we can consider the corresponding $\rho^r_1 : [\omega_1]^2 \longrightarrow \omega$

as above. This function will have the properties stated in Lemma 2.2.2, so the corresponding tree

$$T(\rho_1^r) = \{(\rho_1^r)_\beta \restriction \alpha : \alpha \leq \beta < \omega_1\}$$

is a coherent tree of finite-to-one mappings that admits a strictly increasing function into the real line. The following fact shows that $T(\rho_1^r)$ will typically not be decomposable into countably many antichains.

Lemma 2.2.17. *If r is a Cohen real, then $T(\rho_1^r)$ has no stationary antichain.*

Proof. The family $[\omega]^{<\omega}$ of finite subsets of ω equipped with the discrete topology and its power $([\omega]^{<\omega})^\omega$ with the corresponding product topology and the given Cohen real r is to meet all dense G_δ subsets of this space that one can explicitly define. Thus in particular, $\xi_\alpha^r = \alpha$, and therefore, $C_\alpha^r = \bigcup_{n\in\omega} D_\alpha^r(n)$ for every limit ordinal α. Since the tree $T(\rho_1^r)$ is coherent, to establish the conclusion of the lemma, it suffices to show that for every stationary $\Gamma \subseteq \omega_1$ there exists $\gamma, \delta \in \Gamma$ such that $(\rho_1^r)_\gamma \subset (\rho_1^r)_\delta$. This in turn amounts to showing that for every stationary $\Gamma \subseteq \omega_1$, the set

$$U_\Gamma = \{x \in ([\omega]^{<\omega})^\omega : (\exists \gamma, \delta \in \Gamma)\ (\rho_1^x)_\gamma \sqsubset (\rho_1^x)_\delta\}$$

is a dense-open subset of $([\omega]^{<\omega})^\omega$. So, given a finite partial function p from ω into $[\omega]^{<\omega}$, it is sufficient to find a finite extension q of p such that the basic open subset of $([\omega]^{<\omega})^\omega$ determined by q is included in U_Γ. Let n be the minimal integer that is bigger than all integers appearing in the domain of p or any set of the form $p(j)$ for $j \in \mathrm{dom}(p)$. For $\gamma \in \Gamma$, set

$$F_n(\gamma) = \{\xi \in \gamma^0 : e_\gamma(\xi) \leq n\}.$$

Applying the Pressing Down Lemma, we obtain a finite set $F \subseteq \omega_1$ and a stationary set $\Delta \subseteq \Gamma$ such that $F_n(\gamma) = F$ for all $\gamma \in \Delta$ and such that, if $\alpha = \max(F)+1$, then $e_\gamma \restriction \alpha = e_\delta \restriction \alpha$ for all $\gamma, \delta \in \Delta$. A similar application of the Pressing Down Lemma will give us an integer $m > n$ and two ordinals $\gamma < \delta$ in Δ such that

(1) $C_\gamma(j) = C_\delta(j)$ for all $j \leq m$,
(2) $C_\delta(m+1) > \gamma + 1$, and
(3) $e_\gamma \restriction (C_\gamma(m)+1)^0 = e_\delta \restriction (C_\gamma(m)+1)^0$.

Extend the partial function p to a partial function q with domain $\{0, 1, \ldots, m\}$ such that:

(4) $q(m) = \{e_\delta(\gamma+1)\}$, and
(5) $q(j) = \emptyset$ for any $j < m$ not belonging to $\mathrm{dom}(p)$.

Choose any $x \in ([\omega]^{<\omega})^\omega$ extending the partial mapping q. Then from the choices of the objects n, Δ, F, m, γ, δ and q, we have that

(6) $\gamma + 1 \in C_\delta^x$, and
(7) $C_\delta^x \cap \gamma$ is an initial segment of C_γ^x.

It follows that, given a $\xi < \gamma$, the walk from δ to ξ along the C-sequence C^x_β ($\beta < \omega_1$) either leads to the same finite string of the corresponding weights as the walk from γ to ξ, or else it starts with two steps to $\gamma + 1$ and γ, and then follows the walk from γ to ξ. Since $\rho^x_1(\xi, \delta)$ and $\rho^x_1(\xi, \gamma)$ are by definition maximums of these two strings of weights, we conclude that $\rho^x_1(\xi, \delta) \geq \rho^x_1(\xi, \gamma)$. On the other hand, note that by (7), in the second case, the weight $|C^x_\delta \cap \xi|$ of the first step from δ to ξ is less than or equal to the weight $|C^x_\gamma \cap \xi|$ of the first step from γ to ξ. It follows that we have also the other inequality $\rho^x_1(\xi, \delta) \leq \rho^x_1(\xi, \gamma)$. Hence we have shown that $\rho^x_1(\beta, \gamma) = \rho^x_1(\beta, \delta)$ for all $\beta < \gamma$, or in other words, that $(\rho^x_1)_\gamma \sqsubset (\rho^x_1)_\delta$. This finishes the proof. □

Question 2.2.18. What is the condition one needs to put on a given C-sequence C_α ($\alpha < \omega_1$) in order to guarantee that the corresponding tree $T(\rho_1)$ can be decomposed into countably many antichains?

2.3 Oscillations of traces

In this section we shall consider the version of the oscillation mapping

$$\text{osc} : ([\omega_1]^{<\omega})^2 \longrightarrow \text{Card}$$

defined by

$$\text{osc}(x, y) = |x/E(x, y)|,$$

where $E(x, y)$ is the equivalence relation on x defined by letting $\alpha E(x, y)\beta$ if and only if the closed interval determined by α and β contains no point from $y \setminus x$. So, $\text{osc}(x, y)$ is simply the number of convex pieces into which the set x is split by the set $y \setminus x$. The resulting 'oscillation theory' reproduced here in part has shown to be a quite useful and robust coding technique. In this section, we give some applications to this effect, and in Sections 8.1 and 9.4 below we shall extend the oscillation theory to a more general context.

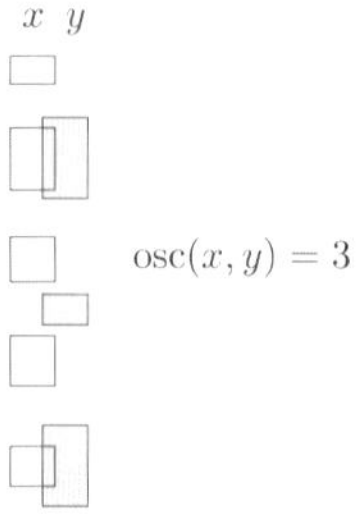

Figure 2.2: The oscillation mapping osc.

Definition 2.3.1. For $\alpha < \beta < \omega_1$, set

$$\mathrm{osc}_0(\alpha, \beta) = \mathrm{osc}(\mathrm{Tr}(\Delta_0(\alpha, \beta), \alpha), \quad \mathrm{Tr}(\Delta_0(\alpha, \beta), \beta)),$$

where $\Delta_0(\alpha, \beta) = \min\{\xi \leq \alpha : \rho_0(\xi, \alpha) \neq \rho_0(\xi, \beta)\} - 1$[24].

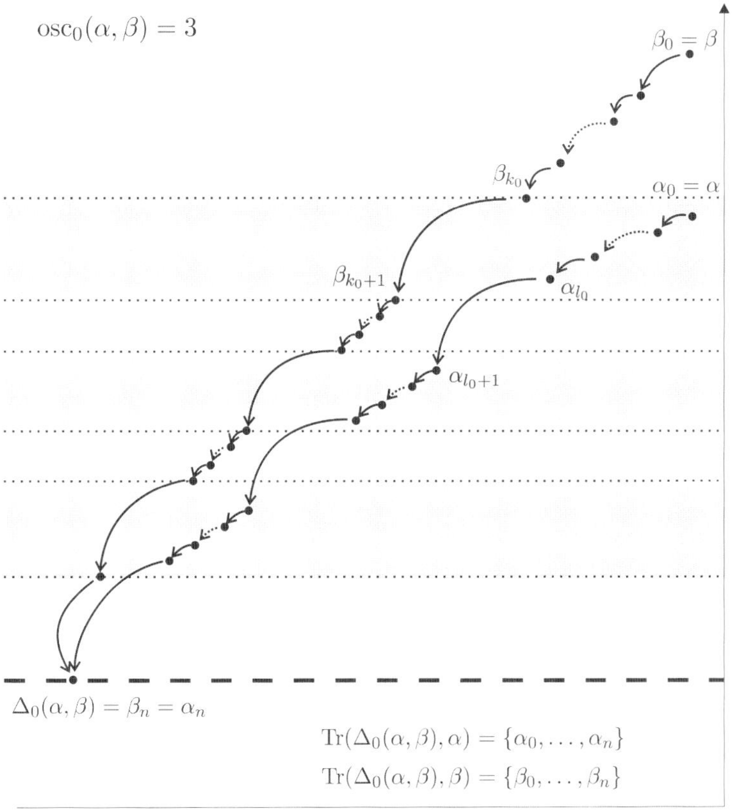

Figure 2.3: The oscillation mapping osc_0.

The following result shows that the upper traces do realize all possible oscillations in any uncountable subset of ω_1.

Lemma 2.3.2. *For every uncountable subset Γ of ω_1 and every positive integer n, there is an integer l such that for all $k < n$ there exist $\alpha, \beta \in \Gamma$ such that $\mathrm{osc}_0(\alpha, \beta) = l + k$.*

Proof. Fix a sequence $\mathcal{M}_i$ $(i \leq n)$ of continuous $\in$-chains of length ω_1 of countable elementary submodels of $(H_{\omega_2}, \in)$ containing all the relevant objects such that $\mathcal{M}_i$

[24] Note that by Lemma 2.1.18 this is well defined.

is an element of the minimal model of $\mathcal{M}_{i+1}$ for all $i < n$. Let M_n be the minimal model of $\mathcal{M}_n$ and pick a $\beta \in \Gamma$ above $\delta_n = M_n \cap \omega_1$. Since $\mathcal{M}_{n-1} \in M_n$ we can find $M_{n-1} \in \mathcal{M}_{n-1} \cap M_n$ such that $\delta_{n-1} = M_{n-1} \cap \omega_1 > \mathrm{L}(\delta_n, \beta)$. Similarly, we can find $M_{n-2} \in \mathcal{M}_{n-1} \cap M_{n-1}$ such that $\delta_{n-2} = M_{n-2} \cap \omega_1 > \mathrm{L}(\delta_{n-1}, \delta_n)$, and so on. This will give us an $\in$-chain M_i $(i \leq n)$ of countable elementary submodels of $(H_{\omega_2}, \in)$ containing all relevant objects such that if we let $\delta_{n+1} = \beta$ and $\delta_i = M_i \cap \omega_1$ for $i \leq n$, then $\delta_i > \mathrm{L}(\delta_{i+1}, \delta_{i+2})$ for all $i < n$. Pick an ordinal $\xi \in C_{\delta_0}$ above $\mathrm{L}(\delta_0, \beta)$. Then (see Lemma 2.1.6 above), $\delta_i \in \mathrm{Tr}(\xi, \beta)$ for all $i \leq n$ and in fact $\delta_0 \to \xi$ is the last step of the walk from β to ξ. Let $t = \rho_0(\xi, \beta)$, $t_i = \rho_0(\delta_i, \beta)$ for $i \leq n$. Then $t_n \sqsubset t_{n-1} \sqsubset \cdots \sqsubset t_0 \sqsubset t$ and $t = t_0{}^\frown |C_{\delta_0} \cap \xi|$. Let

$$\Gamma_0 = \{\gamma \in \Gamma : t = \rho_0(\xi, \gamma) \text{ and } \rho_{0\gamma} \restriction \xi = \rho_{0\beta} \restriction \xi\}.$$

Then $\beta \in \Gamma_0$, and since $\rho_{0\beta} \restriction \xi \in M$, we have that $\Gamma_0 \in M_0$. For $\Sigma \subseteq \Gamma_0$, let

$$\mathcal{T}_\Sigma = \{x \in [\omega_1]^{\leq |t|} : x \sqsubseteq \mathrm{Tr}(\xi, \alpha) \text{ for some } \alpha \in \Sigma\}$$

considered as a tree ordered by the relation $\sqsubseteq$. Call a subset Σ of Γ_0 *large* if $\mathcal{T}_\Sigma$ contains a subtree $\mathcal{T}$ whose root is $\{\xi\}$ and in which a node x of cardinality $< |t|$ must either have exactly one immediate successor or there is $i \leq n$ and uncountably many $\gamma < \omega_1$ such that $x \cup \{\gamma\} \in \mathcal{T}$ and $\rho_0(\gamma, \alpha) = t_i$ for all $\alpha \in \Sigma$ such that $x \cup \{\gamma\} \sqsubseteq \mathrm{Tr}(\xi, \alpha)$. A simple elementarity argument using models $M_0 \in M_1 \in \cdots \in M_n$ shows that Γ_0 itself is large. Note also that if one decomposes a large subset of Γ_0 into countably many subsets, at least one of them must be large. So we can find an ordinal ξ_0 in the interval $[\xi, \delta_0)$ and a large $\Sigma \in M_0$ such that $\xi_0 = \Delta_0(\alpha, \beta)$ for all $\alpha \in \Sigma$. Let $s = \rho_0(\xi_0, \beta)$. Since the walk from β to ξ_0 must pass through all the δ_i $(i \leq n)$ we know that $s \sqsupseteq t_0 \sqsupset t_1 \sqsupset \cdots \sqsupset t_n$. Choose a subtree $\mathcal{T} \in M_0$ of $\mathcal{T}_\Sigma$ witnessing that Σ is large. Pick an immediate successor $\{\xi, \zeta\}$ of $\{\xi\}$ such that $\zeta > \xi_0$. Then (see Lemma 2.1.6) for every $\alpha \in \Sigma$,

$$t_0 = \rho_0(\zeta, \alpha) \text{ and } \mathrm{Tr}(\xi_0, \alpha) = \mathrm{Tr}(\xi_0, \zeta) \cup \mathrm{Tr}(\zeta, \alpha).$$

Choose in M_0 a branching node $x_1 \sqsupset \{\xi, \zeta\}$ whose immediate successors $x_1 \cup \{\gamma\} \in \mathcal{T}$ have the property that $\rho_0(\gamma, \alpha) = t_1$ for all $\alpha \in \Sigma$ such that $x \cup \{\gamma\} \sqsubseteq \mathrm{Tr}(\xi, \alpha)$. Let

$$l = \mathrm{osc}(\mathrm{Tr}(\xi_0, \zeta) \cup x_1, \mathrm{Tr}(\xi_0, \beta)).$$

Choose an arbitrary non-negative integer $k < n$. Working in M_1, we choose an immediate successor $x_1 \cup \{\gamma_1\} \in \mathcal{T}$ of x_1 such that $\gamma_1 > \delta_1 \cap \mathrm{Tr}(\xi_0, \beta)$ and a branching node $x_2 \sqsupset x_1 \cup \{\gamma_1\}$ of $\mathcal{T}$ corresponding to t_2. Then working in M_2, we choose an immediate successor $x_2 \cup \{\gamma_2\} \in \mathcal{T}$ of x_2 such that $\gamma_2 > \delta_2 \cap \mathrm{Tr}(\xi_0, \beta)$ and a branching node $x_3 \sqsupset x_2 \cup \{\gamma_2\}$ of $\mathcal{T}$ corresponding to t_3, and so on. After k steps we arrive at an immediate successor $x_k \cup \{\gamma_k\} \in \mathcal{T} \cap M_k$ of x_k such that $\gamma_k > \delta_k \cap \mathrm{Tr}(\xi_0, \beta)$. Working in M_k we choose $\alpha \in \Sigma$ such that $x_k \cup \{\gamma_k\} \sqsubseteq \mathrm{Tr}(\xi, \alpha)$. Let $x = \mathrm{Tr}(\xi_0, \alpha)$, let $y = \mathrm{Tr}(\xi_0, \beta)$, and let $z = \mathrm{Tr}(\xi_0, \zeta) \cup x_1$. Then

$$x/E(x, y) = z/E(z, y) \cup \{x_2 \setminus x_1, x_3 \setminus x_2, \ldots, x \setminus x_k\}.$$

It follows that $\operatorname{osc}(\alpha, \beta) = l + k$. Since k was an arbitrary non-negative integer $\leq n$, this finishes the proof. □

It should be clear that the idea of the previous proof with very minor modifications also works in proving the following multidimensional version of Lemma 2.3.2.

Lemma 2.3.3. *For every uncountable family $\mathcal{G}$ of pairwise-disjoint finite subsets of ω_1 all of some fixed size m, and every positive integer n, there exist integers l_i $(i < m)$ and $a, b \in \mathcal{G}$ such that* $\operatorname{osc}_0(a(i), b(i)) = l_i + k$ *for all $i < m$.*[25] □

Consider the following projection of the oscillation mapping osc_0,

$$\operatorname{osc}_0^*(\alpha, \beta) = \max\{m \in \omega : \ 2^m | \operatorname{osc}_0(\alpha, \beta)\}.$$

One would need to choose a more involved projection in order to take full advantage of the multi-dimensional version of Lemma 2.3.2 but already this projection is giving us interesting results.

Corollary 2.3.4. *For every uncountable $\Gamma \subseteq \omega_1$ and every positive integer k there exist $\alpha, \beta \in \Gamma$ such that* $\operatorname{osc}_0^*(\alpha, \beta) = k$. □

Corollary 2.3.5. *Every inner model M of set theory which correctly computes ω_1 contains a partition of ω_1 into infinitely many pairwise-disjoint subsets that are stationary in the universe V of all sets.*

Proof. Our assumption about M means in particular that the C-sequence C_ξ $(\xi < \omega_1)$ can be chosen to be an element of M. It follows that the oscillation mapping osc_0^* is also an element of M. It follows that for each $\alpha < \omega_1$, the sequence of sets

$$S_{\alpha n} = \{\beta > \alpha : \operatorname{osc}_0^*(\alpha, \beta) = n\} \ (n < \omega)$$

belongs to M. So it suffices to show that there must be $\alpha < \omega_1$ such that for every $n < \omega$ the set $S_{\alpha n}$ is stationary in V. This is an immediate consequence of Corollary 2.3.4. □

Remark 2.3.6. In [67], Larson showed that Corollary 2.3.5 cannot be extended to partitions of ω_1 into uncountably many pairwise-disjoint stationary sets. He also showed that under the Proper Forcing Axiom, for every mapping $c : [\omega_1]^2 \to \omega_1$ there is a stationary set $S \subseteq \omega_1$ such that for all $\alpha < \omega_1$,

$$\{c(\alpha, \beta) : \beta \in S \setminus (\alpha + 1)\} \neq \omega_1.$$

Note that by Corollary 2.3.4 this cannot be extended to mappings c with countable ranges.

[25] Here $a(i)$ denotes the ith member of a according to the increasing enumeration of a.

The original oscillation theory has been developed in the realm of functions with domain and ranges included on ω, i.e., as a theory of the oscillation mapping

$$\mathrm{osc} : (\omega^{\leq\omega})^2 \to \omega + 1$$

that counts the number of times two such functions change in dominating each other. Some of this will be reproduced in Section 9.4 below. In a quite similar manner, one can also count oscillations between two finite partial functions x and y from ω_1 into ω as follows,

$$\mathrm{osc}(x, y) = |\{\xi \in \mathrm{dom}(x) : x(\xi) \leq y(\xi) \text{ but } x(\xi^+) > y(\xi^+)\}|,$$

where ξ^+ denotes the minimal ordinal in $\mathrm{dom}(x)$ above ξ if there is one.

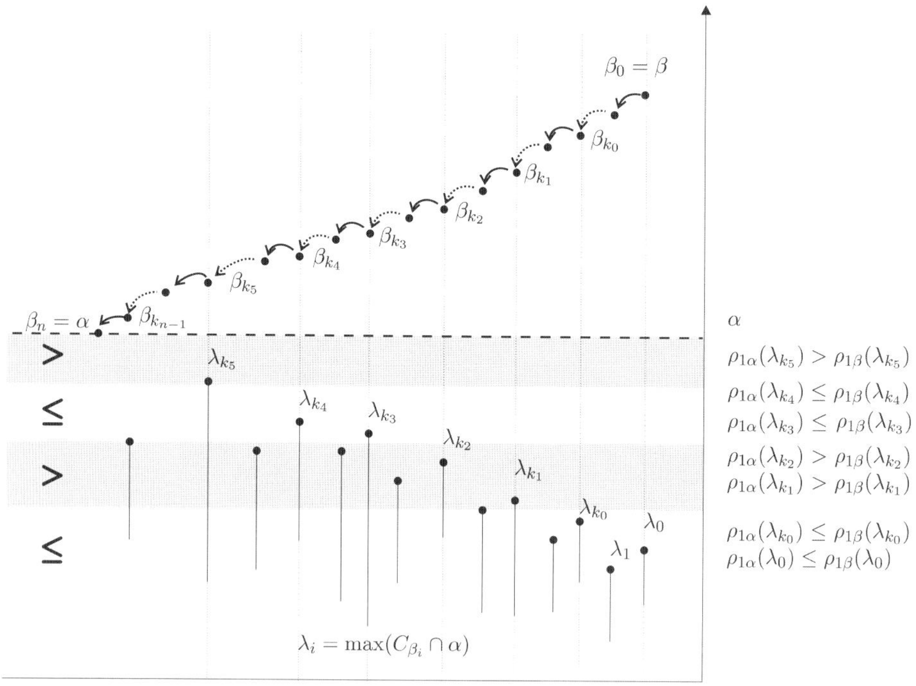

Figure 2.4: The oscillation mapping osc_1.

Definition 2.3.7. For $\alpha < \beta < \omega_1$, set (see Figure 2.4),

$$\mathrm{osc}_1(\alpha, \beta) = \mathrm{osc}(\rho_{1\alpha} \restriction \mathrm{L}(\alpha, \beta), \quad \rho_{1\beta} \restriction \mathrm{L}(\alpha, \beta)).$$

We have the following analogue of Lemma 2.3.2 which is now true even in its rectangular form.

Lemma 2.3.8. *For every pair* Γ *and* Σ *of uncountable subsets of* ω_1 *and every positive integer* n*, there is an integer* l *such that for all* $k < n$ *there exist* $\alpha \in \Gamma$ *and* $\beta \in \Sigma$ *such that* $\mathrm{osc}_1(\alpha, \beta) = l + k$.

Proof. Fix a continuous $\in$-chain $\mathcal{N}$ of length ω_1 of countable elementary submodels of $(H_{\omega_2}, \in)$ containing all the relevant objects and fix also a countable elementary submodel M of $(H_{\omega_2}, \in)$ that contains $\mathcal{N}$ as an element. Let $\delta = M \cap \omega_1$. Fix $\alpha_0 \in \Gamma \setminus \delta$ and $\beta_0 \in \Sigma \setminus \delta$. Let $L_0 = \mathrm{L}(\delta, \beta_0)$. By exchanging the steps in the following construction, we may assume that $\rho_1(\xi, \alpha_0) > \rho_1(\xi, \beta_0)$ for $\xi = \max(L_0)$. Choose $\xi_0 \in C_\delta$ such that $\xi_0 > L_0$ and such that the three mappings $\rho_{1\alpha_0}$, $\rho_{1\beta_0}$, and $\rho_{1\delta}$ agree on the interval $[\xi_0, \delta)$. By elementarity of M there will be $\delta_0^+ > \delta$, $\alpha_0^+ \in \Gamma \setminus \delta_0^+$, and $\beta_0^+ \in \Sigma \setminus \delta_0^+$ realizing the same type as δ, α_0, and β over the parameters L_0 and ξ_0. Let $L_1^+ = \mathrm{L}(\delta, \delta_0^+)$. Note that $\xi_0 \leq \min(L_1^+)$. It follows that, in particular, the mappings $\rho_{1\alpha_0^+}$ and $\rho_{1\beta_0^+}$ agree on L_1^+. Choose an $N \in \mathcal{N}$ belonging to M such that $N \cap \omega_1 > L_1^+$ and choose $\xi_1 \in C_\delta \setminus N$. Then we can find $\delta_1 > \delta$ and $\alpha_1^+ \in \Gamma \setminus \delta_1$ and $\beta_1 \in \Sigma \setminus \delta_1$ realizing the same type as δ, α_0^+, and β_0^+ over the objects accumulated so far, L_0, ξ_0, L_1^+, N, and ξ_1. Let $L_1^{++} = \mathrm{L}(\delta, \delta_1)$ and let t_1^+ be the restriction of $\rho_{1\alpha_1^+}$ on $\max(L_1^+) + 1$. Then the set of $\alpha \in \Gamma$ whose $\rho_{1\alpha}$ end-extends t_1^+ belongs to the submodel N and is uncountable, since clearly it contains the ordinal α_1^+ which does not belong to N. So by the elementarity of N and Lemma 2.2.6 there is $\alpha_1 \in \Gamma \setminus \delta$ such that $\rho_{1\alpha_1}$ end-extends t_1^+ and

$$\rho_1(\xi, \alpha_1) > \rho_1(\xi, \beta_1) \text{ for all } \xi \in L_1^{++}. \tag{2.3.1}$$

Let t_1 be the restriction of $\rho_{1\alpha_1}$ on $\max(L_1^{++})$. Then $t_1 \in M$ and every $\alpha \in \Gamma$ whose $\rho_{1\alpha}$ end-extends t_1 satisfies (2.3.1) and

$$\rho_1(\xi, \alpha) = \rho_1(\xi, \beta_1) \text{ for all } \xi \in L_1^+. \tag{2.3.2}$$

Note also that $\mathrm{L}(\delta, \beta_1) = L_0 \cup L_1^+ \cup L_1^{++}$. Using similar reasoning we can find $t_2 \in M \cap T(\rho_1)$ end-extending t_1, an $\alpha_2 \in \Gamma \setminus \delta$ whose $\rho_{1\alpha_2}$ end-extends t_2, finite sets $\delta > \max(L_2^{++}) > \max(L_2^+) > \max(L_1^{++})$, and $\beta_2 \in \Sigma \setminus \delta$ such that $\mathrm{L}(\delta, \beta_2) = L_0 \cup L_1^+ \cup L_1^{++} \cup L_2^+ \cup L_2^{++}$ and such that the analogues of (2.3.1) and (2.3.2) hold for all $\alpha \in \Gamma$ whose $\rho_{1\alpha}$ end-extends t_2, and so on. This procedure gives us a sequence $L_0 < L_1^+ < L_1^{++} < \cdots < L_n^+ < L_n^{++}$ of finite subsets of δ, an increasing sequence $t_1 \sqsubset t_2 \sqsubset \cdots \sqsubset t_n$ of nodes of $T(\rho_0) \cap M$, and a sequence β_k $(1 \leq k \leq n)$ such that for all $1 \leq k \leq n$, and all $\alpha \in \Gamma$ whose $\rho_{1\alpha}$ end-extends t_k,

$$\mathrm{L}(\delta, \beta_k) = L_0 \cup L_1^+ \cup L_1^{++} \cup \cdots \cup L_k^+ \cup L_k^{++}, \tag{2.3.3}$$

$$\rho_1(\xi, \alpha) > \rho_1(\xi, \beta_k) \text{ for all } \xi \in L_k^{++}, \tag{2.3.4}$$

$$\rho_1(\xi, \alpha) = \rho_1(\xi, \beta_k) \text{ for all } \xi \in L_k^+. \tag{2.3.5}$$

Choose $\alpha \in \Gamma \cap M$ whose $\rho_{1\alpha}$ end-extends t_n such that the set $L_n = \mathrm{L}(\alpha, \delta)$ lies above L_n^{++} and such that all the $\rho_{1\beta_k}$'s agree on L_n. Note that by Lemma 2.1.6 for each $1 \leq k \leq n$,

$$\mathrm{L}(\alpha, \beta_k) = L_0 \cup L_1^+ \cup L_1^{++} \cup \cdots \cup L_k^+ \cup L_k^{++} \cup L_n. \tag{2.3.6}$$

Let $l = \mathrm{osc}(\rho_{1\alpha} \restriction (L_0 \cup L_n),\ \ \rho_{1\beta} \restriction (L_0 \cup L_n))$ where β is equal to one of the β_k's. Note that l does not depend on which β_k we choose and that according to (2.3.4), (2.3.5) and (2.3.6), for each $1 \leq k \leq n$, we have that $\mathrm{osc}_1(\alpha, \beta_k) = l + k$, as required. □

Note that in the above proof the set Γ can be replaced by any uncountable family of pairwise-disjoint sets giving us the following slightly more general conclusion.

Lemma 2.3.9. *For every uncountable family $\mathcal{G}$ of pairwise-disjoint finite subsets of ω_1 all of some fixed size m, every uncountable subset Σ of ω_1, and every positive integer n, there exist integers l_i $(i < m)$ and $a \in \mathcal{G}$ such that for all $k < n$ there exist $\beta \in \Sigma$ such that $\mathrm{osc}_1(a(i), \beta) = l_i + k$ for all $i < m$.* □

Having in mind an application we define a projection of osc_1 as follows. For a real x, let $[x]$ denote the greatest integer not bigger than x. Choose a mapping $* : \mathbb{R} \to \omega$ such that $*(x) = n$ if $x - [x] \in \mathrm{I}_n$, where $\mathrm{I}_n = (\frac{1}{n+3}, \frac{1}{n+2}]$. For x in $\mathbb{R}$ we use the notation x^* for $*(x)$. Choose a one-to-one sequence x_ξ $(\xi < \omega_1)$ of elements of the first half of the unit interval $\mathrm{I} = [0,1]$ which together with the point 1 form a set that is linearly independent over the field of rational numbers. Finally, set

$$\mathrm{osc}_1^*(\alpha, \beta) = (\mathrm{osc}_1(\alpha, \beta) \cdot x_\alpha)^*.$$

Then Lemma 2.3.9 turns into the following statement about osc_1^*.

Lemma 2.3.10. *For every uncountable family $\mathcal{G}$ of pairwise-disjoint finite subsets of ω_1 all of some fixed size m, every uncountable subset Σ of ω_1, and every mapping $h : m \to \omega$, there exist $a \in \mathcal{G}$ and $\beta \in \Sigma$ such that $\mathrm{osc}_1^*(a(i), \beta) = h(i)$ for all $i < m$.*

Proof. Let ε be half of the length of the shortest interval of the form $\mathrm{I}_{h(i)}$ $(i < m)$. Note that for $a \in \mathcal{G}$, the points $1, x_{a(0)}, \ldots, x_{a(m-1)}$ are rationally independent, so a standard fact from number theory (see for example [46]; XXIII) gives us that there is an integer n_a such that for every $y \in \mathrm{I}^m$ there exist $k < n_a$ such that $|k \cdot x_{a(i)} - y_i| < \varepsilon \pmod 1$ for all $i < m$. Going to an uncountable subfamily of $\mathcal{G}$, we may assume that there is n such that $n_a = n$ for all $a \in \mathcal{G}$. By Lemma 2.3.9 there exists $a \in \mathcal{G}$ and l_i $(i < m)$ such that for every $k < n$ there is $\beta_k > a$ such that $\mathrm{osc}_1(a(i), \beta_k) = l_i + k$ for all $i < m$. For $i < m$, let $y_i = l_i \cdot x_{a(i)}$ and let z_i be the middle-point of the interval $\mathrm{I}_{h(i)}$. Find $k < n$ such that for all $i < m$,

$$|k \cdot x_{a(i)} - (z_i - y_i)| < \varepsilon \pmod 1.$$

Then $\mathrm{osc}_1^*(a(i), \beta_k) = h(i)$ for all $i < m$, as required. □

Corollary 2.3.11. *There is a regular hereditarily Lindelöf space which is not separable.*

Proof. For $\beta < \omega_1$, let $f_\beta \in \{0,1\}^{\omega_1}$ be defined by letting $f_\beta(\beta) = 1$, $f_\beta(\gamma) = 0$ for $\gamma > \beta$, and $f_\beta(\alpha) = \min\{1, \mathrm{osc}_1^*(\alpha,\beta)\}$ for $\alpha < \beta$. Consider $F = \{f_\beta : \beta < \omega_1\}$ as a subspace of $\{0,1\}^{\omega_1}$. Clearly F is not separable. That F is hereditarily Lindelöf follows easily from Lemma 2.3.10. □

Remark 2.3.12. The projection osc_0^* of the oscillation mapping osc_0 appears in [109] as the historically first such map with more than four colors that takes all of its values on every symmetric square of an uncountable subset of ω_1. The variations osc_1 and osc_1^* are on the other hand very recent and are due to J.T. Moore [78] who made them in order to obtain the conclusion of Corollary 2.3.11. Concerning Corollary 2.3.11 we note that the dual implication behaves quite differently since, assuming the Proper Forcing Axiom all, hereditarily separable regular spaces are hereditarily Lindelöf (see [111]).

2.4 The number of steps and the last step functions

In this section we show that a very natural characteristic associated to the minimal walks between countable ordinals lead to functions that have coherence and nontriviality properties very much reminiscent of the Hausdorff gap phenomenon that will be a subject of our study in Section 3.1 below.

Definition 2.4.1. The *number of steps* of the minimal walk is the two-place function $\rho_2 : [\omega_1]^2 \longrightarrow \omega$ defined recursively by

$$\rho_2(\alpha,\beta) = \rho_2(\alpha, \min(C_\beta \setminus \alpha)) + 1,$$

with the boundary condition $\rho_2(\gamma,\gamma) = 0$ for all γ.

This is an interesting mapping which is particularly useful on higher cardinalities and especially in situations where the more informative mappings ρ_0, ρ_1 and ρ lack their usual coherence properties. Later on we shall devote a whole section to ρ_2 but here we list only few of its basic properties. We start with the coherence property that this function enjoys.

Lemma 2.4.2. $\sup\{|\rho_2(\xi,\alpha) - \rho_2(\xi,\beta)| : \xi < \alpha\} < \infty$ *for all* $\alpha < \beta < \omega_1$.

Proof. Suppose the conclusion of the lemma fails for some $\alpha < \beta < \omega_1$. Then for every $k < \omega$, we can find $\xi_k < \alpha$ such that $|\rho_2(\xi_k,\alpha) - \rho_2(\xi_k,\beta)| > k$. We may assume that the sequence of ξ_k's is strictly increasing and let $\delta = \sup_k \xi_k$. Then δ is a limit ordinal $\leq \alpha$, so the lower traces of walks from α to δ and β to δ have a common upper bound $\gamma < \delta$. Then by Lemma 2.1.6, for every ordinal $\xi \in [\gamma,\delta)$, we have that

$$\rho_0(\xi,\alpha) = \rho_0(\alpha,\delta)^\frown\rho_0(\delta,\xi) \text{ and } \rho_0(\xi,\beta) = \rho_0(\beta,\delta)^\frown\rho_0(\delta,\xi). \tag{2.4.1}$$

It follows that for every $\xi \in [\gamma, \delta)$,

$$|\rho_2(\xi, \alpha) - \rho_2(\xi, \beta)| \leq |\rho_2(\delta, \alpha) - \rho_2(\delta, \beta)|, \tag{2.4.2}$$

and so in particular, $\xi_k \notin [\gamma, \delta)$ for all k such that $k > |\rho_2(\delta, \alpha) - \rho_2(\delta, \beta)|$, a contradiction. □

We mention also the following unboundedness property of this function which introduces another theme to be explored fully in later sections of this book.

Lemma 2.4.3. *For every uncountable family A of pairwise-disjoint finite subsets of ω_1, all of some fixed size n, and for every integer k, there exist an uncountable subfamily B of A such that for all $a < b$ in B, we have $\rho_2(a(i), b(j)) \geq k$ for all $i, j < n$.*[26]

Proof. The proof is by induction on k. So suppose that our given family A already satisfies that $\rho_2(a(i), b(j)) \geq k$ for all $i, j < n$ and all $a < b$ in A. We shall find uncountable $B \subseteq A$ such that $\rho_2(a(i), b(j)) \geq k + 1$ for all $i, j < n$ and all $a < b$ in B. To this end, for each limit ordinal $\delta < \omega_1$, we fix $b_\delta > \delta$ in A. Then there is $\eta_\delta < \delta$ such that

$$\rho_0(\xi, \beta) = \rho_0(\beta, \delta)^\frown \rho_0(\delta, \xi) \text{ for all } \xi \in [\eta_\delta, \delta) \text{ and } \beta \in b_\delta. \tag{2.4.3}$$

Find a stationary set $\Gamma \subseteq \omega_1$ and $\gamma < \omega_1$ such that $\eta_\delta = \eta$ for all $\delta \in \Gamma$. Choose a stationary subset Ξ of Γ such that

$$\gamma < b_\gamma < \delta < b_\delta \text{ for all } \gamma < \delta \text{ in } \Xi. \tag{2.4.4}$$

Form the family $A^* = \{\{\gamma\} \cup b_\gamma : \gamma \in \Xi\}$ of pairwise-disjoint sets of size $n + 1$. By the inductive hypothesis there is uncountable $B^* \subseteq A^*$ such that $\rho_2(a(i), b(j)) \geq k$ for all $i, j < n + 1$ and all $a < b$ in B^*. Then the uncountable subfamily

$$B = \{b \setminus \{\min(b)\} : b \in B^*\}$$

of A has the property that $\rho_2(a(i), b(j)) \geq k + 1$ for all $i, j < n$ and all $a < b$ in B. □

The unboundedness property of ρ_2 given by Lemma 2.4.3 shows, in particular, that while the fiber maps $\rho_2(\cdot, \alpha)$ $(\alpha < \omega_1)$ are at a finite distance from each other, there is no global map with domain ω_1 that has a finite distance from all of these fibers.

Corollary 2.4.4. *For every $g : \omega_1 \longrightarrow \omega$ there is $\alpha < \omega_1$ such that*

$$\sup\{|\rho_2(\xi, \alpha) - g(\xi)| : \xi < \alpha\} = \infty.$$

[26] Here, $a < b$ signifies the fact $\alpha < \beta$ whenever $\alpha \in a$ and $\beta \in b$, while $a(i)$ denotes the ith element of a relative to its increasing enumeration $\{a(0), a(1), \ldots, a(n-1)\}$.

Proof. Otherwise we can find an integer k and an unbounded set $\Gamma \subseteq \omega_1$ such that $\sup\{|\rho_2(\xi,\alpha) - g(\xi))| : \xi < \alpha\} \leq k$ for all $\alpha \in \Gamma$. Find unbounded $\Xi \subseteq \omega_1$ and an integer l such that g takes the constant value l on Ξ. Forming a family of pairs by taking an element from Ξ and the other from Γ and applying Lemma 2.4.3, we can find uncountable sets $\Xi_0 \subseteq \Xi$ and $\Gamma_0 \subseteq \Gamma$ such that

$$\rho_2(\xi,\alpha) > k + l + 1 \text{ for all } \xi \in \Xi_0, \alpha \in \Gamma_0 \text{ such that } \xi < \alpha. \tag{2.4.5}$$

Consider an $\alpha \in \Gamma_0$ that is above $\xi_0 = \min \Xi_0$. Then

$$k \geq |\rho_2(\xi_0,\alpha) - g(\xi_0))| > k + l + 1 - l = k + 1, \tag{2.4.6}$$

a contradiction. □

The main object of study in this section, however, is another natural characteristic of the minimal walk. This new mapping ρ_3 not only has strong coherence properties but its full analysis will lead us naturally to some finer requirements that one can put on the given C-sequences on which we base our walks.

Definition 2.4.5. The *last step function* is the characteristic $\rho_3 : [\omega_1]^2 \longrightarrow 2$ of the walk defined by letting

$$\rho_3(\alpha,\beta) = 1 \text{ iff } \rho_0(\alpha,\beta)(\rho_2(\alpha,\beta) - 1) = \rho_1(\alpha,\beta).$$

In other words, we let $\rho_3(\alpha,\beta) = 1$ just in case the last step of the walk $\beta \to \alpha$ comes with the maximal weight.

Lemma 2.4.6. *$\{\xi < \alpha : \rho_3(\xi,\alpha) \neq \rho_3(\xi,\beta)\}$ is finite for all $\alpha < \beta < \omega_1$.*

Proof. It suffices to show that for every infinite $\Gamma \subseteq \alpha$ there exists $\xi \in \Gamma$ such that $\rho_3(\xi,\alpha) = \rho_3(\xi,\beta)$. Shrinking Γ we may assume that for some fixed $\bar{\alpha} \in F(\alpha,\beta)$ and all $\xi \in \Gamma$:

$$\bar{\alpha} = \min(F(\alpha,\beta) \setminus \xi), \tag{2.4.7}$$

$$\rho_1(\xi,\alpha) = \rho_1(\xi,\beta), \tag{2.4.8}$$

$$\rho_1(\xi,\alpha) > \rho_1(\bar{\alpha},\alpha), \tag{2.4.9}$$

$$\rho_1(\xi,\beta) > \rho_1(\bar{\alpha},\beta). \tag{2.4.10}$$

It follows (see 2.1.9) that for every $\xi \in \Gamma$:

$$\rho_0(\xi,\alpha) = \rho_0(\bar{\alpha},\alpha)^\frown \rho_0(\xi,\bar{\alpha}), \tag{2.4.11}$$

$$\rho_0(\xi,\beta) = \rho_0(\bar{\alpha},\beta)^\frown \rho_0(\xi,\bar{\alpha}). \tag{2.4.12}$$

So for any $\xi \in \Gamma$, $\rho_3(\xi,\alpha) = 1$ iff the last term of $\rho_0(\xi,\bar{\alpha})$ is its maximal term iff $\rho_3(\xi,\beta) = 1$. □

The sequence $(\rho_3)_\alpha : \alpha \longrightarrow 2$ $(\alpha < \omega_1)$[27] is therefore coherent in the sense that $(\rho_3)_\alpha =^* (\rho_3)_\beta \restriction \alpha$ whenever $\alpha < \beta$. We need to show that the sequence is not trivial, i.e., that it cannot be uniformized by a single total map from ω_1 into 2. In other words, we need to show that ρ_3 still contains enough information about the C-sequence C_α $(\alpha < \omega_1)$ from which it is defined. For this it will be convenient to assume that C_α $(\alpha < \omega_1)$ satisfies the following natural condition:

(d) If $\alpha = \lambda + \omega$ for some limit ordinal λ, then $C_\alpha = \{\lambda + n : m < n < \omega\}$ for some non-negative integer m; if α is a limit of limit ordinals and if ξ occupies the nth place in the increasing enumeration of C_α, then $\xi = \lambda + m$ for some limit ordinal λ and integer $m > n$.

Definition 2.4.7. Let Λ denote the set of all countable limit ordinals and for an integer $n \in \omega$, let $\Lambda + n = \{\lambda + n : \lambda \in \Lambda\}$.

Lemma 2.4.8. *$\rho_3(\lambda + n, \beta) = 1$ for all but finitely many n with $\lambda + n < \beta$.*

Proof. Clearly we may assume that $\alpha = \lambda + \omega \leq \beta$. Then there is $n_0 < \omega$ such that for every $n \geq n_0$ the walk $\beta \to \lambda + n$ passes through α. By (d) we know that $C_\alpha = \{\lambda + n : m < n < \omega\}$ for some non-negative integer m, so in any such walk $\beta \to \lambda + n$ there is only one step from α to $\lambda + n$. So choosing $n > n_0, m, \rho_1(\alpha, \beta)$ we will ensure that the last step of $\beta \to \lambda + n$ comes with the maximal weight, i.e., $\rho_3(\lambda + n, \beta) = 1$. □

Lemma 2.4.9. *For all $\beta < \omega_1$, $n < \omega$, the set $\{\lambda \in \Lambda : \lambda + n < \beta$ and $\rho_3(\lambda + n, \beta) = 1\}$ is finite.*

Proof. Given an infinite subset Γ of $(\Lambda + n) \cap \beta$ we need to find a $\lambda + n \in \Gamma$ such that $\rho_3(\lambda + n, \beta) = 0$. Shrinking Γ if necessary assume that

$$\rho_1(\lambda + n, \beta) > n + 2$$

for all $\lambda + n \in \Gamma$. So if $\rho_3(\lambda + n, \beta) = 1$ for some $\lambda + n \in \Gamma$, then the last step of $\beta \to \lambda + n$ would have to be of weight $> n + 2$ which is impossible by our assumption (d) about C_α $(\alpha < \omega_1)$. □

The meaning of these properties of ρ_3 is perhaps easier to comprehend if we reformulate them in a way that resembles the original formulation of the existence of Hausdorff gaps. It is not surprising that this sort of variation on the classical Hausdorff gap phenomenon has appeared first in a topological study that analyses how the space of *subuniform*[28] ultrafilters on ω_1 is embedded into the space of all ultrafilters on ω_1.

---

[27] Recall the way one always defines the fiber functions from a two-variable function applied to the context of ρ_3: $(\rho_3)_\alpha(\xi) = \rho_3(\xi, \alpha)$.

[28] An ultrafilter on ω_1 is subuniform if it is nonprincipal and if it concentrates on a countable subset of ω_1. The space $\mathrm{SU}(\omega_1)$ of subuniform ultrafilters on ω_1 is an open subspace of the Čech–Stone remainder of the discrete space on ω_1. Lemma 2.4.10 says that there is a continuous $\{0, 1\}$-valued function on $\mathrm{SU}(\omega_1)$ which does not extend to a continuous function defined on the whole Čech–Stone remainder (see [131] and [23]).

Lemma 2.4.10. *Let $B_\alpha = \{\xi < \alpha : \rho_3(\xi, \alpha) = 1\}$ for $\alpha < \omega_1$. Then:*

1. $B_\alpha =^* B_\beta \cap \alpha$ *for* $\alpha < \beta$,
2. $(\Lambda + n) \cap B_\beta$ *is finite for all* $n < \omega$ *and* $\beta < \omega_1$,
3. $\{\lambda + n : n < \omega\} \subseteq^* B_\beta$ *whenever* $\lambda + \omega \leq \beta$. □

In particular, there is is no uncountable $\Gamma \subseteq \omega_1$ such that $\Gamma \cap \beta \subseteq^* B_\beta$ for all $\beta < \omega_1$, and so the tree

$$T(\rho_3) = \{(\rho_3)_\beta \restriction \alpha : \alpha \leq \beta < \omega_1\}$$

contains no uncountable chains. On the other hand, the P-ideal[29] $\mathfrak{I}$ generated by B_β ($\beta < \omega_1$) is large as it contains all intervals of the form $[\lambda, \lambda + \omega)$. The following general dichotomy about P-ideals shows that here indeed we have quite a canonical example of a P-ideal on ω_1.

Definition 2.4.11. The P-ideal dichotomy. For every P-ideal $\mathfrak{I}$ of countable subsets of some set S either:

1. there is uncountable $X \subseteq S$ such that $[X]^\omega \subseteq \mathfrak{I}$, or
2. S can be decomposed into countably many sets orthogonal to $\mathfrak{I}$.

Remark 2.4.12. It is known that the P-ideal dichotomy is a consequence of the Proper Forcing Axiom and moreover that it does not contradict the Continuum Hypothesis (see [122]). This is an interesting dichotomy which will be used in this article for testing various notions of coherence as we encounter them. In fact, this was the original and still most important reason for isolating this dichotomy from the rest of the consequences of PFA (see[111], Chapter 8).

Let us now turn our attention to the lexicographical ordering given by ρ_3, or more precisely, the lexicographical ordering among the fibers $(\rho_3)_\alpha$ of ρ_3 induced by the global lexicographical ordering of the coherent tree $T(\rho_3)$.

Definition 2.4.13. Consider the linear ordering $<_{\rho_3}$ on the set of countable limit defined as follows:

$$\alpha <_{\rho_3} \beta \quad \text{iff} \quad \rho_3(\xi, \alpha) <_r \rho_3(\xi, \beta), \tag{2.4.13}$$

where $\xi = \Delta(\alpha, \beta) = \min\{\eta < \min\{\alpha, \beta\} : \rho_3(\eta, \alpha) \neq \rho_3(\eta, \beta)\}$.[30]

It turns out that chain decompositions of the cartesian square of this lexicographical ordering is more closely tied to the assumption that the whole tree $T(\rho_3)$ is coverable by countably many more antichains than in the case of the ordering $<_{\rho_1}$ and the corresponding tree $T(\rho_1)$. Recall that in Lemma 2.2.17 we

[29]Recall that an ideal $\mathfrak{I}$ of subsets of some set S is a *P-ideal* if for every sequence A_n ($n < \omega$) of elements of $\mathfrak{I}$ there is B in $\mathfrak{I}$ such that $A_n \setminus B$ is finite for all $n < \omega$. A set X is *orthogonal* to $\mathfrak{I}$ if $X \cap A$ is finite for all A in $\mathfrak{I}$.

[30]If $\rho_3(\xi, \alpha) = \rho_3(\xi, \beta)$ for all $\xi < \min\{\alpha, \beta\}$, then it is understood that $\min\{\alpha, \beta\} <_{\rho_3} \max\{\alpha, \beta\}$.

have shown that there could be choices of the C-sequence C_α $(\alpha < \omega_1)$ for which the corresponding tree $T(\rho_1)$ is not special, and in fact contains no stationary antichain, though the cartesian square of the ordering $C(\rho_1) = (\omega_1, <_{\rho_1})$ is always decomposable into countably many chains. A similar generic choice of the C-sequence (satisfying the condition (d) above) will give us that the corresponding tree $T(\rho_3)$ is also not special, or more precisely, that it contains no stationary antichain. Moreover, a close examination of the proof shows that if ρ_3 is based on the generic C-sequence, then the cartesian square of $(\omega_1, <_{\rho_3})$ cannot be decomposed into countably many chains. This leads us to the following natural restriction.

Definition 2.4.14. A set Σ of countable ordinals is *special* if the corresponding subset $\{(\rho_3)_\alpha : \alpha \in \Sigma\}$ of the tree $T(\rho_3)$ can be decomposed into countably many antichains.

Note that since the restriction of the tree $T(\rho_3)$ to the set Λ of countable limit ordinals admits a strictly increasing real-valued function,[31] every uncountable subset of ω_1 contains an uncountable special subset.

Lemma 2.4.15. *If Σ is a special subset of ω_1, then the cartesian square of the total ordering $<_{\rho_3}$ on Σ is the union of countably many chains.*

Proof. Note that similarly to the tree $T(\rho_1)$, the restriction $T(\rho_3) \restriction \Lambda$ of $T(\rho_3)$ to the set Λ of countable limit ordinals admits a strictly increasing real-valued function. It follows that if Λ^0 denotes a set of countable limit ordinals of the form $\alpha + \omega$, then there is $a : T(\rho_3) \restriction \Lambda^0 \to \omega$ such that $a(s) \neq a(t)$ for every pair of distinct comparable nodes of $T(\rho_3) \restriction \Lambda^0$. We also assume that a is one-to-one on levels of $T(\rho_3) \restriction \Lambda^0$.

Consider a pair α and β of ordinals from Σ such that $\alpha < \beta$. Let λ_α be the maximal limit ordinal $\leq \alpha$, and let

$$D_{\alpha\beta} = \{\lambda \in \Lambda : \lambda = \lambda_\alpha \text{ or } \rho_3(\xi, \alpha) \neq \rho_3(\xi, \beta) \text{ for some } \xi \in [\lambda, \lambda + \omega)\}.$$

Then $D_{\alpha\beta}$ is a finite set of countable ordinals. Let

$$D^+_{\alpha\beta} = \bigcup \{[\lambda, \lambda + \omega) : \lambda \in D_{\alpha\beta}\}.$$

Let $p_{\alpha\beta}$ be the isomorphism type of the structure on $D^+_{\alpha\beta}$ whose relations code the restrictions of the fibers $(\rho_3)_\alpha$ and $(\rho_3)_\beta$ on this set, as well as the two mappings $\lambda \mapsto a((\rho_3)_\alpha \restriction \lambda + \omega)$ and $\lambda \mapsto a((\rho_3)_\beta \restriction \lambda + \omega)$ defined on the subset $D_{\alpha\beta}$ of $D^+_{\alpha\beta}$.

Since there exist only countably many isomorphism types of these structures and since the set Σ is special, the conclusion of the lemma would follow once we establish the following: Suppose that we have two pairs $\alpha < \beta$ and $\gamma < \delta$ of ordinals from Σ such that $\alpha <_{\rho_3} \beta$, that $p_{\alpha\beta} = p_{\gamma\delta}$, and that the pair of nodes

[31] Note that if the whole tree $T(\rho_3)$ admits a strictly increasing real-valued function, then it is in fact decomposable into countably many antichains.

$(\rho_3)_\alpha$ and $(\rho_3)_\gamma$ and the pair of nodes $(\rho_3)_\beta$ and $(\rho_3)_\delta$ are incomparable in $T(\rho_3)$. We shall show that $\beta <_{\rho_3} \delta$.

Let $\xi = \Delta(\alpha, \gamma)$. Note that ξ is not a limit ordinal and that $\xi < \min\{\alpha, \gamma\}$. Since the mappings $\lambda \mapsto a((\rho_3)_\alpha \restriction \lambda + \omega)$ and $\lambda \mapsto a((\rho_3)_\gamma \restriction \lambda + \omega)$ and the mappings $\lambda \mapsto a((\rho_3)_\beta \restriction \lambda + \omega)$ and $\lambda \mapsto a((\rho_3)_\delta \restriction \lambda + \omega)$ are isomorphic, we conclude that the set $D = D_{\alpha\beta} \cap \xi$ is an initial segment of both sets $D_{\alpha\beta}$ and $D_{\gamma\delta}$. Let λ_ξ be the maximal limit ordinal $\leq \xi$. Using the fact that the restriction of the fibers $(\rho_3)_\alpha$ and $(\rho_3)_\gamma$ on $D^+_{\alpha\beta}$ and $D^+_{\gamma\delta}$, respectively, are isomorphic, we conclude that $\lambda + \omega \leq \lambda_\xi < \xi$ for $\lambda = \max D$. Then in particular, $\lambda_\xi \notin D_{\alpha\beta} \cup D_{\alpha\beta}$, so the fibers $(\rho_3)_\alpha$ and $(\rho_3)_\beta$ and the fibers $(\rho_3)_\gamma$ and $(\rho_3)_\delta$ have identical restrictions on the interval $[\lambda_\xi, \lambda_\xi + \omega)$. Let $D^+ = \bigcup\{[\lambda, \lambda + \omega) : \lambda \in D\}$. Then we get the equality

$$(\rho_3)_\beta \restriction D^+ = (\rho_3)_\delta \restriction D^+$$

as an immediate consequence of the fact that $(\rho_3)_\beta \restriction D^+_{\alpha\beta}$ is isomorphic to $(\rho_3)_\beta \restriction D^+_{\alpha\beta}$ and the fact that D is an initial segment of the sets $D_{\alpha\beta}$ and $D_{\gamma\delta}$. Since $D^+_{\alpha\beta} \cap \lambda_\xi = D^+ = D^+_{\gamma\delta} \cap \lambda_\xi$, we conclude that

$$(\rho_3)_\beta \restriction \lambda_\xi \setminus D^+ = (\rho_3)_\alpha \restriction \lambda_\xi \setminus D^+ = (\rho_3)_\gamma \restriction \lambda_\xi \setminus D^+ = (\rho_3)_\delta \restriction \lambda_\xi \setminus D^+.$$

It follows that $(\rho_3)_\beta \restriction \lambda_\xi = (\rho_3)_\delta \restriction \lambda_\xi$. Combining this with the facts that the fibers $(\rho_3)_\alpha$ and $(\rho_3)_\beta$ and the fibers $(\rho_3)_\gamma$ and $(\rho_3)_\delta$ have identical restrictions on the interval $[\lambda_\xi, \lambda_\xi + \omega)$, we conclude that $\xi = \Delta(\beta, \delta)$ and that

$$\rho_3(\xi, \beta) = \rho(\xi, \alpha) < \rho_3(\xi, \gamma) = \rho_3(\xi, \delta).$$

This gives us the desired conclusion $\beta <_{\rho_3} \delta$ that finishes the proof. □

Let $C(\rho_3)$ denote the linearly ordered set $(\omega_1, <_{\rho_3})$. Then as in the case of $C(\rho_0)$ and $C(\rho_1)$, we have the following result proved along similar lines of reasoning and showing that $C(\rho_3)$ is yet another realization of the canonical minimal ordering on ω_1 whose cartesian square is the union of countably many chains.

Theorem 2.4.16. *Assuming* MA_{ω_1}, *the ordering* $C(\rho_3)$ *and its reverse* $C(\rho_3)^*$ *serve as a two-element basis for the class of all uncountable linear orderings whose cartesian squares can be covered by countably many chains.* □

Chapter 3

Metric Theory of Countable Ordinals

3.1 Triangle inequalities

In this section we study a characteristic of the minimal walk that satisfies certain triangle inequalities reminiscent of those found in an ultra-metric space. Some applications of the corresponding metric-like theory of ω_1 will appear already in this section and some of them will later on get separate treatments.

Definition 3.1.1. Define $\rho : [\omega_1]^2 \longrightarrow \omega$ recursively by,

$$\rho(\alpha, \beta) = \max\{|C_\beta \cap \alpha|, \rho(\alpha, \min(C_\beta \setminus \alpha)), \rho(\xi, \alpha) : \quad \xi \in C_\beta \cap \alpha\}, \tag{3.1.1}$$

with the boundary condition $\rho(\alpha, \alpha) = 0$ for all α.

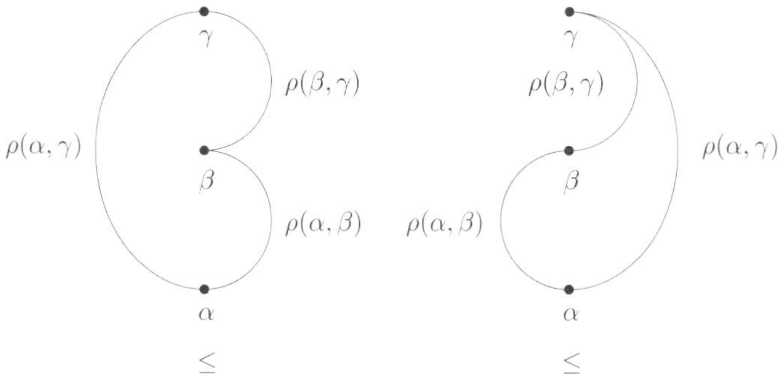

Figure 3.1: The two triangle inequalities of ρ.

The following lemma lists three basic properties of the ρ-function. As we go on, we will discover several other properties of this function that will be closely tied with properties of the C-sequence on which we base our walks.

Lemma 3.1.2. *For all* $\alpha < \beta < \gamma < \omega_1$ *and* $n < \omega$,

(a) $\{\xi \leq \alpha : \rho(\xi, \alpha) \leq n\}$ *is finite,*
(b) $\rho(\alpha, \gamma) \leq \max\{\rho(\alpha, \beta), \rho(\beta, \gamma)\}$,
(c) $\rho(\alpha, \beta) \leq \max\{\rho(\alpha, \gamma), \rho(\beta, \gamma)\}$.

Proof. Note that $\rho(\alpha, \beta) \geq \rho_1(\alpha, \beta)$, so (a) follows from the corresponding property of ρ_1. The proof of (b) and (c) is by a simultaneous induction on the ordinals α, β and γ.

To prove (b), consider $n = \max\{\rho(\alpha, \beta), \rho(\beta, \gamma)\}$. We have to show that $\rho(\alpha, \gamma) \leq n$. Let

$$\xi_\alpha = \min(C_\gamma \setminus \alpha) \text{ and } \xi_\beta = \min(C_\gamma \setminus \beta).$$

Case 1^b: $\xi_\alpha = \xi_\beta$. Then by the inductive hypothesis,

$$\rho(\alpha, \xi_\alpha) \leq \max\{\rho(\alpha, \beta), \rho(\beta, \xi_\beta)\}.$$

From the definition of $\rho(\beta, \gamma)$ we get that $\rho(\beta, \xi_\beta) \leq \rho(\beta, \gamma) \leq n$, so replacing $\rho(\beta, \xi_\beta)$ by $\rho(\beta, \gamma)$ in the above inequality we get $\rho(\alpha, \xi_\alpha) \leq n$. Consider an ordinal ξ belonging to the set $C_\gamma \cap \alpha = C_\gamma \cap \beta$. By the inductive hypothesis

$$\rho(\xi, \alpha) \leq \max\{\rho(\xi, \beta), \rho(\alpha, \beta)\}.$$

From the definition of $\rho(\beta, \gamma)$ we see that $\rho(\xi, \beta) \leq \rho(\beta, \gamma)$, so replacing $\rho(\xi, \beta)$ with $\rho(\beta, \gamma)$ in the last inequality we get that $\rho(\xi, \alpha) \leq n$. Since

$$|C_\gamma \cap \alpha| = |C_\gamma \cap \beta| \leq \rho(\beta, \gamma) \leq n,$$

referring to the definition of $\rho(\alpha, \gamma)$ we conclude that $\rho(\alpha, \gamma) \leq n$.

Case 2^b: $\xi_\alpha < \xi_\beta$. Then $\xi_\alpha \in C_\gamma \cap \beta$, so

$$\rho(\xi_\alpha, \beta) \leq \rho(\beta, \gamma) \leq n.$$

By the inductive hypothesis

$$\rho(\alpha, \xi_\alpha) \leq \max\{\rho(\alpha, \beta), \rho(\xi_\alpha, \beta)\} \leq n.$$

Similarly, for every $\xi \in C_\gamma \cap \alpha \subseteq C_\gamma \cap \beta$,

$$\rho(\xi, \alpha) \leq \max\{\rho(\xi, \beta), \rho(\alpha, \beta)\} \leq n.$$

Finally $|C_\gamma \cap \alpha| \leq |C_\gamma \cap \beta| \leq \rho(\beta, \gamma) \leq n$. Combining these inequalities we get the desired conclusion $\rho(\alpha, \gamma) \leq n$.

To prove (c), consider now $n = \max\{\rho(\alpha,\gamma), \rho(\beta,\gamma)\}$. We have to show that $\rho(\alpha,\beta) \leq n$. Let ξ_α and ξ_β be as above and let us consider the same two cases as above.

Case 1^c: $\xi_\alpha = \xi_\beta = \bar{\xi}$. Then by the inductive hypothesis

$$\rho(\alpha,\beta) \leq \max\{\rho(\alpha,\bar{\xi}), \rho(\beta,\bar{\xi})\}.$$

This gives the desired bound $\rho(\alpha,\beta) \leq n$, since $\rho(\alpha,\xi_\alpha) \leq \rho(\alpha,\gamma) \leq n$ and $\rho(\beta,\xi_\beta) \leq \rho(\beta,\gamma) \leq n$.

Case 2^c: $\xi_\alpha < \xi_\beta$. Applying the inductive hypothesis again we get

$$\rho(\alpha,\beta) \leq \max\{\rho(\alpha,\xi_\alpha), \rho(\xi_\alpha,\beta)\} \leq n.$$

This completes the proof. □

The following fact shows that the function ρ has a considerably finer coherence property than ρ_1.

Lemma 3.1.3. *If $\alpha < \beta < \gamma$ and if $\rho(\alpha,\beta) > \rho(\beta,\gamma)$, then $\rho(\alpha,\gamma) = \rho(\alpha,\beta)$.*

Proof. Applying Lemma 3.1.2,

$$\rho(\alpha,\gamma) \leq \max\{\rho(\alpha,\beta), \rho(\beta,\gamma)\} = \rho(\alpha,\beta),$$

$$\rho(\alpha,\beta) \leq \max\{\rho(\alpha,\gamma), \rho(\beta,\gamma)\} = \rho(\alpha,\gamma).$$

Note that $\max\{\rho(\alpha,\gamma), \rho(\beta,\gamma)\}$ must be equal to $\rho(\alpha,\gamma)$ rather than $\rho(\beta,\gamma)$ since by the assumption $\rho(\alpha,\beta)$, which is bounded by the maximum, is bigger than $\rho(\beta,\gamma)$. □

Remark 3.1.4. Referring to the standard definitions of *metrics* and *ultra-metrics* the properties in Lemma 3.1.2(b) and (c) could more properly be called *ultra-subadditivities* though we shall keep calling them *subadditivity* properties of the function ρ. Recall the notion of the full lower trace $F(\alpha,\beta)$ given recursively by

$$F(\alpha,\beta) = F(\alpha, \min(C_\beta \setminus \alpha)) \cup \bigcup_{\xi \in C_\beta \cap \alpha} F(\xi,\alpha),$$

with the boundary value $F(\alpha,\alpha) = \{\alpha\}$ for all α (see Definition 2.1.7 above). The function $d : [\omega_1]^2 \longrightarrow \omega$ defined by

$$d(\alpha,\beta) = |F(\alpha,\beta)|$$

is an example of a truly subadditive function, since by Lemma 2.1.8, we have the following inequalities whenever $\alpha < \beta < \gamma$:

(a) $d(\alpha,\gamma) \leq d(\alpha,\beta) + d(\beta,\gamma)$,

(b) $d(\alpha,\beta) \leq d(\alpha,\gamma) + d(\beta,\gamma)$.

Definition 3.1.5. For two distinct countable ordinals α and β, set

$$\alpha <_\rho \beta \text{ iff } \rho(\xi, \alpha) < \rho(\xi, \beta), \tag{3.1.2}$$

where $\xi = \min\{\eta < \min\{\alpha, \beta\} : \rho(\eta, \alpha) \neq \rho(\eta, \beta)\}$; if there is no $\eta < \min\{\alpha, \beta\}$ such that $\rho(\eta, \alpha) \neq \rho(\eta, \beta)\}$, put $\min\{\alpha, \beta\} <_\rho \max\{\alpha, \beta\}$. Clearly, $<_\rho$ is a total ordering on ω_1, so we let $C(\rho)$ denote the uncountable linearly ordered set $(\omega_1, <_\rho)$.

It should be clear that the proof of Lemma 2.2.4 also shows the following corresponding fact about $<_\rho$ in place of $<_{\rho_1}$.

Lemma 3.1.6. *The cartesian square of the linearly ordered set $C(\rho) = (\omega_1, <_\rho)$ is the union of countably many chains.* □

The following fact is also established in the same way as the corresponding results about the orderings $C(\rho_0)$ and $C(\rho_1)$ showing that $C(\rho)$ is yet another realization of the canonical linear ordering satisfying the conclusion of the previous lemma.

Theorem 3.1.7. *Assuming* MA_{ω_1}, *the ordering $C(\rho)$ and its reverse $C(\rho)^*$ serve as a two-element basis for the class of all uncountable linear orderings whose cartesian squares can be covered by countably many chains.* □

There is a common property of the trees associated with the mappings ρ_0, ρ_1, and ρ introduced on the basis of the metric notion of a Lipschitz map that lies behind the common properties of the corresponding linear orderings $C(\rho_0)$, $C(\rho_1)$, and $C(\rho)$. This will be explained in Chapter 4 below.

3.2 Constructing a Souslin tree using ρ

The purpose of this section is to give a particular application of the function ρ to the theory of Cohen forcing.

Definition 3.2.1. Define $\bar{\rho} : [\omega_1]^2 \longrightarrow \omega$ as follows:

$$\bar{\rho}(\alpha, \beta) = 2^{\rho(\alpha,\beta)} \cdot (2 \cdot |\{\xi \leq \alpha : \rho(\xi, \alpha) \leq \rho(\alpha, \beta)\}| + 1).$$

Lemma 3.2.2. *For all $\alpha < \beta < \gamma < \omega_1$,*

(a) $\bar{\rho}(\alpha, \gamma) \neq \bar{\rho}(\beta, \gamma)$,
(b) $\bar{\rho}(\alpha, \gamma) \leq \max\{\bar{\rho}(\alpha, \beta), \bar{\rho}(\beta, \gamma)\}$,
(c) $\bar{\rho}(\alpha, \beta) \leq \max\{\bar{\rho}(\alpha, \gamma), \bar{\rho}(\beta, \gamma)\}$.

Proof. Only (b) and (c) require some argument. Consider first (b). If $\rho(\alpha, \beta) \leq \rho(\beta, \gamma)$, then $\rho(\alpha, \gamma) \leq \max\{\rho(\alpha, \beta), \rho(\beta, \gamma)\} = \rho(\beta, \gamma)$, so

$$\{\xi \leq \alpha : \rho(\xi, \alpha) \leq \rho(\alpha, \gamma)\} \subseteq \{\xi \leq \beta : \rho(\xi, \beta) \leq \rho(\beta, \gamma)\}.$$

Therefore in this case $\bar{\rho}(\alpha,\gamma) \leq \bar{\rho}(\beta,\gamma)$. On the other hand, if $\rho(\beta,\gamma) \leq \rho(\alpha,\beta)$, then $\rho(\alpha,\gamma) \leq \max\{\rho(\alpha,\beta),\rho(\beta,\gamma)\} = \rho(\alpha,\beta)$, so

$$\{\xi \leq \alpha : \rho(\xi,\alpha) \leq \rho(\alpha,\gamma)\} \subseteq \{\xi \leq \alpha : \rho(\xi,\alpha) \leq \rho(\alpha,\beta)\}.$$

So in this case $\bar{\rho}(\alpha,\gamma) \leq \bar{\rho}(\alpha,\beta)$. This checks (b). Checking (c) is similar. □

Lemma 3.2.3. *$\bar{\rho}(\alpha,\beta) \neq \bar{\rho}(\beta,\gamma)$ for all $\alpha < \beta < \gamma < \omega_1$.*

Proof. Suppose the conclusion fails for some triple $\alpha < \beta < \gamma < \omega_1$. Let $n = \bar{\rho}(\alpha,\beta) \neq \bar{\rho}(\beta,\gamma)$ and write it as $n = 2^i(2j+1)$ for some integers i and j. Then $i = \rho(\alpha,\beta) = \rho(\beta,\gamma)$ and

$$|\{\xi \leq \alpha : \rho(\xi,\alpha) \leq i\}| = j = |\{\xi \leq \beta : \rho(\xi,\alpha) \leq i\}|. \tag{3.2.1}$$

Since $i = \rho(\alpha,\beta)$, the ordinal α belongs to the set $\{\xi \leq \beta : \rho(\xi,\alpha) \leq i\}$ and so by the subadditivity of ρ the set $\{\xi \leq \alpha : \rho(\xi,\alpha) \leq i\}$ is an initial segment of the set $\{\xi \leq \beta : \rho(\xi,\alpha) \leq i\}$. Since the two sets have the same cardinality j they must be equal. This is a contradiction, since clearly β belongs to $\{\xi \leq \beta : \rho(\xi,\alpha) \leq i\}$ but not to $\{\xi \leq \alpha : \rho(\xi,\alpha) \leq i\}$. □

We shall also need the following property of $\bar{\rho}$.

Lemma 3.2.4. *Suppose $\eta_\alpha \neq \eta_\beta < \min\{\alpha,\beta\}$ and $\bar{\rho}(\eta_\alpha,\alpha) = \bar{\rho}(\eta_\beta,\beta) = n$. Then $\bar{\rho}(\eta_\alpha,\beta), \bar{\rho}(\eta_\beta,\alpha) > n$.*

Proof. By symmetry, we may assume that $\eta_\alpha < \eta_\beta$. Our assumption $\bar{\rho}(\eta_\alpha,\alpha) = \bar{\rho}(\eta_\beta,\beta)$ yields that for some integers i and j, we have that $\rho(\eta_\alpha,\alpha) = \rho(\eta_\beta,\beta) = i$ and that

$$|\{\xi \leq \eta_\alpha : \rho(\xi,\eta_\alpha) \leq i)\}| = |\{\xi \leq \eta_\beta : \rho(\xi,\eta_\beta) \leq i\}| = j. \tag{3.2.2}$$

Note that $\rho(\eta_\alpha,\eta_\beta) \nleq i$, since otherwise, the set $\{\xi \leq \eta_\alpha : \rho(\xi,\eta_\alpha) \leq i\}$ would be a proper subset of the set $\{\xi \leq \eta_\beta : \rho(\xi,\eta_\beta) \leq i\}$ contradicting (3.2.2). It follows that $k = \rho(\eta_\alpha,\eta_\beta) > i$, and therefore

$$\{\xi \leq \eta_\alpha : \rho(\xi,\eta_\alpha) \leq k\} \supseteq \{\xi \leq \eta_\alpha : \rho(\xi,\eta_\alpha) \leq i\}, \tag{3.2.3}$$

and so in particular, we get that $l = |\{\xi \leq \eta_\alpha : \rho(\xi,\eta_\alpha) \leq k\}| \geq j$. It follows that

$$\bar{\rho}(\eta_\alpha,\eta_\beta) = 2^k(2l+1) > 2^i(2j+1) = \bar{\rho}(\eta_\beta,\beta) = \bar{\rho}(\eta_\alpha,\alpha) = n. \tag{3.2.4}$$

This together with the subadditivity properties of $\bar{\rho}$ yields that

$$n < \bar{\rho}(\eta_\alpha,\eta_\beta) \leq \max\{\bar{\rho}(\eta_\alpha,\alpha),\bar{\rho}(\eta_\beta,\alpha)\} = \bar{\rho}(\eta_\beta,\alpha),$$

$$n < \bar{\rho}(\eta_\alpha,\eta_\beta) \leq \max\{\bar{\rho}(\eta_\alpha,\beta),\bar{\rho}(\eta_\beta,\beta)\} = \bar{\rho}(\eta_\alpha,\beta).$$

This completes the proof. □

Definition 3.2.5. For $p \in \omega^{<\omega}$ define a binary relation $<_p$ on ω_1 by letting $\alpha <_p \beta$ iff $\alpha < \beta$, $\bar\rho(\alpha,\beta) \in |p|$ and

$$p(\bar\rho(\xi,\alpha)) = p(\bar\rho(\xi,\beta)) \text{ for any } \xi < \alpha \text{ such that } \bar\rho(\xi,\alpha) < |p|. \tag{3.2.5}$$

Lemma 3.2.6.

(a) $<_p$ *is a tree-ordering on* ω_1 *of height* $\leq |p|+1$,

(b) $p \subseteq q$ *implies* $<_p \subseteq <_q$.

Proof. This follows immediately from Lemma 3.2.2. □

Definition 3.2.7. For $x \in \omega^\omega$, set

$$<_x = \bigcup\{<_{x\restriction n} : n < \omega\}.$$

Lemma 3.2.8. *For every $p \in \omega^{<\omega}$ there is a partition of ω_1 into finitely many pieces such that if $\alpha < \beta$ belong to the same piece of the partition, then there is $q \supseteq p$ in $\omega^{<\omega}$ such that $\alpha <_q \beta$.*

Proof. For $\alpha < \omega_1$, the set

$$F_{|p|}(\alpha) = \{\xi \leq \alpha : \bar\rho(\xi,\alpha) \leq |p|\}$$

has size $\leq |p|+1$, so if we enrich $F_{|p|}(\alpha)$ to a structure that would code the behavior of $\bar\rho$ on $F_{|p|}(\alpha)$, one can realize only finitely many different isomorphism types. Thus it suffices to show that if $\alpha < \beta < \omega_1$ with $F_{|p|}(\alpha)$, $F_{|p|}(\beta)$ isomorphic, then there is an extension q of p such that $\alpha <_q \beta$.

Consider the graph G of all unordered pairs of integers $< \bar\rho(\alpha,\beta)$ of the form

$$\{\bar\rho(\xi,\alpha), \bar\rho(\xi,\beta)\}$$

for some $\xi < \alpha$. It suffices to show that every connected component of G has at most one point $< |p|$. For if this is true, extend p to a mapping $q : \bar\rho(\alpha,\beta) \longrightarrow \omega$ which is constant on each component of G. Clearly, $\alpha <_q \beta$ for any such q.

Consider an integer $n < |p|$. We claim that the connected component of n is a star with center n which has no other points $< |p|$. This will of course be sufficient for our needs here. Otherwise, pick a simple G-path $n = n_0, n_1, n_2$ which starts with n. Pick ordinals $\xi_0, \xi_1 < \alpha$ and $\gamma_i \in \{\alpha,\beta\}$ for $i = 0,1,2,3$ which witness these connections, or in other words,

$$\bar\rho(\xi_0,\gamma_0) = n_0,\ \bar\rho(\xi_0,\gamma_1) = n_1,\ \bar\rho(\xi_1,\gamma_2) = n_1 \text{ and } \bar\rho(\xi_1,\gamma_3) = n_2. \tag{3.2.6}$$

Since $n_0 \neq n_1$, by Lemma 3.2.2(a), we conclude that $\gamma_0 \neq \gamma_1$. Similarly, we conclude that $\gamma_2 \neq \gamma_3$.

We claim that $\xi_0 \neq \xi_1$. To see this, we suppose $\xi_0 = \xi_1$ and work for a contradiction. Since γ_0 and γ_1 are distinct members of the doubleton $\{\alpha,\beta\}$, either

$\gamma_3 = \gamma_0$, or else $\gamma_3 = \gamma_1$. If $\gamma_3 = \gamma_0$, then $n_2 = \bar{\rho}(\xi_1, \gamma_3) = \bar{\rho}(\xi_0, \gamma_0) = n_0$, a contradiction. On the other hand, if $\gamma_3 = \gamma_1$, then $n_2 = \bar{\rho}(\xi_1, \gamma_3) = \bar{\rho}(\xi_0, \gamma_1) = n_1$ which is also a contradiction.

The non-equality $\xi_0 \neq \xi_1$, the two equalities $\bar{\rho}(\xi_0, \gamma_1) = n_1$ and $\bar{\rho}(\xi_1, \gamma_2) = n_1$, and the property 3.2.2(a) of $\bar{\rho}$, give us the missing non-equality $\gamma_1 \neq \gamma_2$ between two consecutive γ_i's. So, in particular, we have that $\gamma_2 = \gamma_0$.

By our assumption $n_0 < |p|$, the ordinal ξ_0 must belong to $F_{|p|}(\gamma_0)$. However, by our assumption about the isomorphism between the structures $F_{|p|}(\gamma_0)$ and $F_{|p|}(\gamma_1)$, the ordinal $\xi_0 \in F_{|p|}(\gamma_0)$ cannot belong to the root

$$F_{|p|}(\gamma_0) \cap F_{|p|}(\gamma_1).$$

It follows that $\xi_0 \notin F_{|p|}(\gamma_1)$, and therefore,

$$n_1 = \bar{\rho}(\xi_0, \gamma_1) > |p| > n_0.$$

Note that this in particular shows that no immediate G-neighbor of an integer $n = n_0 < |p|$ is also $< |p|$. Applying Lemma 3.2.4 to the two equalities

$$\bar{\rho}(\xi_0, \gamma_1) = n_1 = \bar{\rho}(\xi_1, \gamma_2),$$

we conclude that $\bar{\rho}(\xi_0, \gamma_2) > n_1$. Since $\gamma_2 = \gamma_0$, this gives us

$$n_0 = \bar{\rho}(\xi_0, \gamma_0) = \bar{\rho}(\xi_0, \gamma_2) > n_1,$$

a contradiction. This finishes the proof. □

Theorem 3.2.9. *For every infinite subset $\Gamma \subseteq \omega_1$, the set*

$$G_\Gamma = \{x \in \omega^\omega : \alpha <_x \beta \textit{ for some } \alpha, \beta \in \Gamma\}$$

is a dense open subset of the Baire space.

Proof. This is an immediate consequence of Lemma 3.2.8. □

Definition 3.2.10. For $\alpha < \beta < \omega_1$, let $\alpha <_{\bar{\rho}} \beta$ denote the fact that $\bar{\rho}(\xi, \alpha) = \bar{\rho}(\xi, \beta)$ for all $\xi < \alpha$. Then $<_{\bar{\rho}}$ is a tree-ordering on ω_1 obtained by identifying α with the member $\bar{\rho}(\cdot, \alpha)$ of the tree $T(\bar{\rho})$. Note that $<_{\bar{\rho}} \subseteq <_x$ for all $x \in \omega^\omega$ and that there exists $x \in \omega^\omega$ such that $<_x = <_{\bar{\rho}}$ (e.g., one such x is the identity map id $: \omega \longrightarrow \omega$).

Theorem 3.2.11. *If Γ is an infinite $<_{\bar{\rho}}$-antichain, the set*

$$H_\Gamma = \{x \in \omega^\omega : \alpha \not<_x \beta \textit{ for some } \alpha < \beta \textit{ in } \Gamma\}$$

is a dense-open subset of the Baire space.

Proof. Let $p \in \omega^{<\omega}$ be given. A simple application of Ramsey's theorem gives us a strictly increasing sequence γ_i $(i < \omega)$ of members of Γ such that $\Delta(\gamma_i, \gamma_j) \leq \Delta(\gamma_j, \gamma_k)$[1] for all $i < j < k$. Going to a subsequence, we may assume that either $\Delta(\gamma_i, \gamma_{i+1}) < \Delta(\gamma_{i+1}, \gamma_{i+2})$ for all i or else for some ordinal ξ and all $i < j$, $\Delta(\gamma_i, \gamma_j) = \xi$. In the first case, by Lemma 3.2.2(a), we can find i such that one of the integers $\bar{\rho}(\Delta(\gamma_i, \gamma_{i+1}), \gamma_i)$ or $\bar{\rho}(\Delta(\gamma_i, \gamma_{i+1}), \gamma_{i+1})$ does not belong to the domain of p, so there will be $q \supseteq p$ such that $\bar{\rho}(\gamma_i, \gamma_{i+1}) < |q|$ and $\gamma_i \not<_q \gamma_{i+1}$. This guarantees that $\gamma_i \not<_x \gamma_{i+1}$ for every $x \supseteq q$. In the second case, $\bar{\rho}(\xi, \gamma_i)$ $(i < \omega)$ is a one-to-one sequence of integers so there will be $i < j$ such that neither of the integers $\bar{\rho}(\xi, \gamma_i)$ or $\bar{\rho}(\xi, \gamma_j)$ belongs to the domain of p, so there is $q \supseteq p$ such that $\bar{\rho}(\gamma_i, \gamma_j) < |q|$ and $\gamma_i \not<_q \gamma_j$. Hence, $\gamma_i \not<_x \gamma_j$ for every $x \supseteq q$. □

Definition 3.2.12. For a family $\mathcal{F}$ of infinite $<_{\bar{\rho}}$-antichains, we say that a real $x \in \omega^\omega$ is *$\mathcal{F}$-Cohen* if $x \in G_\Gamma \cap H_\Gamma$ for all $\Gamma \in \mathcal{F}$. We say that x is *$\mathcal{F}$-Souslin* if no member of $\mathcal{F}$ is a $<_x$-chain or a $<_x$-antichain. We say that a real $x \in \omega^\omega$ is *Souslin* if the tree ordering $<_x$ on ω_1 has no uncountable chains nor antichains, i.e., when x is $\mathcal{F}$-Souslin for $\mathcal{F}$ equal to the family of all uncountable subsets of ω_1.

Note that since every uncountable subset of ω_1 contains an uncountable $<_{\bar{\rho}}$-antichain, if a family $\mathcal{F}$ refines the family of all uncountable $<_{\bar{\rho}}$-antichains, then every $\mathcal{F}$-Souslin real is Souslin. The following fact summarizes Theorems 3.2.9 and 3.2.11 and connects the two kinds of reals.

Theorem 3.2.13. *If $\mathcal{F}$ is a family of infinite $<_{\bar{\rho}}$-antichains, then every $\mathcal{F}$-Cohen real is $\mathcal{F}$-Souslin.* □

Corollary 3.2.14. *If the density of the family of all uncountable subsets of ω_1 is smaller than the number of nowhere dense sets needed to cover the real line, then there is a Souslin tree.* □

Remark 3.2.15. Recall that the *density* of a family $\mathcal{F}$ of infinite subsets of some set S is the minimal size of a family $\mathcal{F}_0$ of infinite subsets of S with the property that every member of $\mathcal{F}$ is refined by a member of $\mathcal{F}_0$. A special case of Corollary 3.2.14, when the density of the family of all uncountable subsets of ω_1 is equal to $\aleph_1$, was first observed by T. Miyamoto (unpublished).

Corollary 3.2.16. *Every Cohen real is Souslin.*

Proof. Every uncountable subset of ω_1 in the Cohen extension contains an uncountable subset from the ground model. So it suffices to consider the family $\mathcal{F}$ of all infinite $<_{\bar{\rho}}$-antichains from the ground model. □

If ω_1 is a successor cardinal in the constructible subuniverse, then $\bar{\rho}$ can be chosen to be coanalytic and so the transformation $x \longmapsto <_x$ will transfer combinatorial notions of Souslin, Aronszajn or special Aronszajn trees into the corresponding classes of reals that lie in the third level of the projective hierarchy. This transformation has been explored in several places in the literature (see, e.g., [9], [47]).

[1] For $\alpha < \beta < \omega_1$, $\Delta(\alpha, \beta) = \min\{\xi \leq \alpha : \bar{\rho}(\xi, \alpha) \neq \bar{\rho}(\xi, \beta)\}$.

Remark 3.2.17. We have just seen how the combination of the subadditivity properties (3.2.2(b),(c)) of the coherent sequence $\bar{\rho}_\alpha : \alpha \longrightarrow \omega$ $(\alpha < \omega_1)$ of one-to-one mappings can be used in controlling the finite disagreement between them. It turns out that in many contexts the coherence and the subadditivities are essentially equivalent restrictions on a given sequence $e_\alpha : \alpha \longrightarrow \omega$ $(\alpha < \omega_1)$. For example, the following construction shows that, in the context of finite-to-one mappings $e_\alpha : \alpha \longrightarrow \omega$ $(\alpha < \omega_1)$, coherence does lead us naturally to the two subadditivity properties.

Definition 3.2.18. Given a coherent sequence $e_\alpha : \alpha \longrightarrow \omega$ $(\alpha < \omega_1)$ of finite-to-one mappings, define $\tau_e : [\omega_1]^2 \longrightarrow \omega$ as

$$\tau_e(\alpha, \beta) = \max\{\max\{e(\xi, \alpha), e(\xi, \beta)\} : \xi \leq \alpha, e(\xi, \alpha) \neq e(\xi, \beta)\}.$$ [2]

Lemma 3.2.19. *For every* $\alpha < \beta < \gamma < \omega_1$,

(a) $\tau_e(\alpha, \beta) \geq e(\alpha, \beta)$,

(b) $\tau_e(\alpha, \gamma) \leq \max\{\tau_e(\alpha, \beta), \tau_e(\beta, \gamma)\}$,

(c) $\tau_e(\alpha, \beta) \leq \max\{\tau_e(\alpha, \gamma), \tau_e(\beta, \gamma)\}$.

Proof. To see (a) note that it trivially holds if $e(\alpha, \beta) = 0$ so we can concentrate on the case $e(\alpha, \beta) > 0$. Applying the convention $e(\alpha, \alpha) = 0$ we see that $e(\xi, \alpha) \neq e(\xi, \beta)$ for $\xi = \alpha$. It follows that $\tau_e(\alpha, \beta) \geq \{\max\{e(\alpha, \alpha), e(\alpha, \beta)\} = e(\alpha, \beta)$.

To see (b) let $n = \max\{\tau_e(\alpha, \beta), \tau_e(\beta, \gamma)\}$ and let $\xi \leq \alpha$ be such that $e(\xi, \alpha) \neq e(\xi, \gamma)$. We need to show that $e(\xi, \alpha) \leq n$ and $e(\xi, \gamma) \leq n$. If $e(\xi, \alpha) > n$ then since $\tau_e(\alpha, \beta) \leq n$ we must have $e(\xi, \alpha) = e(\xi, \beta)$ and so $e(\xi, \beta) \neq e(\xi, \gamma)$. It follows that $\tau_e(\beta, \gamma) \geq e(\xi, \beta) > n$, a contradiction. Similarly, $e(\xi, \gamma) > n$ yields $e(\xi, \beta) = e(\xi, \gamma) \neq e(\xi, \alpha)$ and therefore $\tau_e(\alpha, \beta) \geq e(\xi, \beta)) > n$, a contradiction.

The proof of (c) is similar. □

3.3 A Hausdorff gap from ρ

The purpose of this section is to use the ρ-function and give a natural definition of an (ω_1, ω_1^*)-gap in the quotient algebra $\mathcal{P}(\omega)/\mathrm{Fin}$.

Definition 3.3.1. We say that $a : [\omega_1]^2 \longrightarrow \omega$ is *transitive* if for $\alpha < \beta < \gamma < \omega_1$,

$$a(\alpha, \gamma) \leq \max\{a(\alpha, \beta), a(\beta, \gamma)\}.$$

Transitive maps occur quite frequently in set-theoretic constructions. For example, given a sequence A_α $(\alpha < \omega_1)$ of subsets of ω that increases relative

[2] Note the occurence of the boundary value $e(\alpha, \alpha) = 0$ in this formula as well.

to the ordering $\subseteq^*$ of inclusion modulo a finite set, the mapping $a : [\omega_1]^2 \longrightarrow \omega$ defined by

$$a(\alpha, \beta) = \min\{n : A_\alpha \setminus n \subseteq A_\beta\}$$

is a transitive map. The transitivity condition by itself is not nearly as useful as its combination with the other subadditivity property (3.2.2(c)). Fortunately, there is a general procedure that produces a subadditive dominant to every transitive map.

Definition 3.3.2. For a transitive $a : [\omega_1]^2 \longrightarrow \omega$ define $\rho_a : [\omega_1]^2 \longrightarrow \omega$ recursively as follows:

$$\begin{aligned}\rho_a(\alpha, \beta) &= \max\{|C_\beta \cap \alpha|, a(\min(C_\beta \setminus \alpha), \beta), \rho_a(\alpha, \min(C_\beta \setminus \alpha)),\\ &\qquad \rho_a(\xi, \alpha) : \xi \in C_\beta \cap \alpha\}.\end{aligned}$$

Lemma 3.3.3. *For all $\alpha < \beta < \gamma < \omega_1$ and $n < \omega$,*

(a) $\{\xi \leq \alpha : \rho_a(\xi, \alpha) \leq n\}$ *is finite,*
(b) $\rho_a(\alpha, \gamma) \leq \max\{\rho_a(\alpha, \beta), \rho_a(\beta, \gamma)\}$,
(c) $\rho_a(\alpha, \beta) \leq \max\{\rho_a(\alpha, \gamma), \rho_a(\beta, \gamma)\}$,
(d) $\rho_a(\alpha, \beta) \geq a(\alpha, \beta)$.

Proof. The proof of (a), (b) and (c) is quite similar to the corresponding part of the proof of Lemma 3.1.2. This comes of course from the fact that the definition of ρ and ρ_a are closely related. The occurrence of the factor $a(\min(C_\beta \setminus \alpha), \beta)$ complicates a bit the proof that ρ_a is subadditive, and the fact that a is transitive is quite helpful in getting rid of the additional difficulty. The details are left to the interested reader. Given $\alpha < \beta$, for every step $\beta_n \to \beta_{n+1}$ of the minimal walk $\beta = \beta_0 > \beta_1 > \cdots > \beta_k = \alpha$, we have $\rho_a(\alpha, \beta) \geq \rho_a(\beta_n, \beta_{n+1}) \geq a(\beta_n, \beta_{n+1})$ by the very definition of ρ_a. Applying the transitivity of a to this path of inequalities we get the conclusion (d). □

Lemma 3.3.4. $\rho_a(\alpha, \beta) \geq \rho_a(\alpha+1, \beta)$ *whenever* $0 < \alpha < \beta$ *and* α *is a limit ordinal.*

Proof. Recall the assumption (c) about the fixed C-sequence C_ξ ($\xi < \omega_1$) on which all our definitions are based: if ξ is a limit ordinal > 0, then no point of C_ξ is a limit ordinal. It follows that if $0 < \alpha < \beta$ and α is a limit ordinal, then the minimal walk $\beta \to \alpha$ must pass through $\alpha + 1$ and therefore $\rho_a(\alpha, \beta) \geq \rho_a(\alpha + 1, \beta)$. □

Let us now give an application of ρ_a to a classical phenomenon of occurrence of gaps in the quotient algebra $\mathcal{P}(\omega)/\mathrm{fin}$.

Definition 3.3.5. A *Hausdorff gap* in $\mathcal{P}(\omega)/\mathrm{fin}$ is a pair of sequences A_α ($\alpha < \omega_1$) and B_α ($\alpha < \omega_1$) such that

(a) $A_\alpha \subseteq^* A_\beta \subseteq^* B_\beta \subseteq^* B_\alpha$ whenever $\alpha < \beta$, but
(b) there is no C such that $A_\alpha \subseteq^* C \subseteq^* B_\alpha$ for all α.

The following straightforward reformulation shows that a Hausdorff gap is just another instance of a nontrivial coherent sequence

$$f_\alpha : A_\alpha \longrightarrow 2 \ (\alpha < \omega_1)$$

where the domain A_α of f_α is not the ordinal α itself but a subset of ω and that the corresponding sequence of domains A_α $(\alpha < \omega_1)$ is a realization of ω_1 inside the quotient $\mathcal{P}(\omega)/\text{fin}$ in the sense that $A_\alpha \subseteq^* A_\beta$ whenever $\alpha < \beta$.

Lemma 3.3.6. *A pair of ω_1-sequences A_α $(\alpha < \omega_1)$ and B_α $(\alpha < \omega_1)$ form a Hausdorff gap iff the pair*

$$\bar{A}_\alpha = A_\alpha \cup (\omega \setminus B_\alpha) \ (\alpha < \omega_1) \text{ and } \bar{B}_\alpha = \omega \setminus B_\alpha \ (\alpha < \omega_1)$$

has the following properties:

(a) *$\bar{A}_\alpha \subseteq^* \bar{A}_\beta$ whenever $\alpha < \beta$,*
(b) *$\bar{B}_\alpha =^* \bar{B}_\beta \cap \bar{A}_\alpha$ whenever $\alpha < \beta$,*
(c) *there is no B such that $\bar{B}_\alpha =^* B \cap \bar{A}_\alpha$ for all α.* □

From now on we fix a strictly $\subseteq^*$-increasing chain A_α $(\alpha < \omega_1)$ of infinite subsets of ω and let $a : [\omega_1]^2 \longrightarrow \omega$ be defined by

$$a(\alpha, \beta) = \min\{n : A_\alpha \setminus n \subseteq A_\beta\}.$$

Let $\rho_a : [\omega_1]^2 \longrightarrow \omega$ be the corresponding subadditive dominant of a defined above. For $\alpha < \omega_1$, set

$$D_\alpha = A_{\alpha+1} \setminus A_\alpha.$$

Lemma 3.3.7. *The sets $D_\alpha \setminus \rho_a(\alpha, \gamma)$ and $D_\beta \setminus \rho_a(\beta, \gamma)$ are disjoint whenever $0 < \alpha < \beta < \gamma$ and α and β are limit ordinals.*

Proof. This follows immediately from Lemmas 3.3.3 and 3.3.4. □

We are in a position to define a partial mapping $m : [\omega_1]^2 \longrightarrow \omega$ by

$$m(\alpha, \beta) = \min(D_\alpha \setminus \rho_a(\alpha, \beta)),$$

whenever $\alpha < \beta$ and α is a limit ordinal.

Lemma 3.3.8. *The mapping m is coherent, i.e., $m(\alpha, \beta) = m(\alpha, \gamma)$ for all but finitely many limit ordinals $\alpha < \min\{\beta, \gamma\}$.*

Proof. This is by the coherence of ρ_a and the fact that $\rho_a(\alpha, \beta) = \rho_a(\alpha, \gamma)$ already implies $m(\alpha, \beta) = m(\alpha, \gamma)$. □

Lemma 3.3.9. *$m(\alpha, \gamma) \neq m(\beta, \gamma)$ whenever $\alpha \neq \beta < \gamma$ and α, β are limits.*

Proof. This follows from Lemma 3.3.7. □

For $\beta < \omega_1$, set

$$B_\beta = \{m(\alpha, \beta) : \alpha < \beta \text{ and } \alpha \text{ limit}\}.$$

Lemma 3.3.10. *$B_\beta =^* B_\gamma \cap A_\beta$ whenever $\beta < \gamma$.*

Proof. By the coherence of m. □

Note the following immediate consequence of Lemma 3.3.7 and the definition of m.

Lemma 3.3.11. *$m(\alpha, \beta) = \max(B_\beta \cap D_\alpha)$ whenever $\alpha < \beta$ and α is a limit.* □

Lemma 3.3.12. *There is no $B \subseteq \omega$ such that $B \cap A_\beta =^* B_\beta$ for all β.*

Proof. Suppose that such a B exists and for a limit ordinal α let us define $g(\alpha) = \max(B \cap D_\alpha)$. Then by Lemma 3.3.11, $g(\alpha) = m(\alpha, \beta)$ for all $\beta < \omega_1$ and all but finitely many limit ordinals $\alpha < \beta$. By Lemma 3.3.9, it follows that g is a finite-to-one map, a contradiction. □

Theorem 3.3.13. *For every strictly $\subset^*$-increasing chain A_α $(\alpha < \omega_1)$ of subsets of ω, there is a sequence B_α $(\alpha < \omega_1)$ of subsets of ω such that:*

(a) *$B_\alpha =^* B_\beta \cap A_\alpha$ whenever $\alpha < \beta$,*

(b) *there is no B such that $B_\alpha =^* B \cap A_\alpha$ for all α.* □

Remark 3.3.14. For a given countable ordinal α let $h_\alpha : A_\alpha \longrightarrow 2$ be such that $h_\alpha^{-1}(1) = B_\alpha$. Then by Lemma 3.3.10, the corresponding sequence h_α $(\alpha < \omega_1)$ of partial functions is *coherent* in the sense that for every pair $\alpha < \beta < \omega_1$, the set

$$D_h(\alpha, \beta) = \{n \in A_\alpha \cap A_\beta : h_\alpha(n) \neq h_\beta(n)\}$$

is finite. By Lemma 3.3.12, the sequence is *nontrivial* in the sense that there is no $g : \omega \longrightarrow 2$ such that $h_\alpha =^* g \restriction A_\alpha$ for all $\alpha < \omega_1$. Hausdorff's original sequence, when reformulated in this way, has the property that the maps $d_h(\cdot, \beta) : \beta \longrightarrow \omega$ defined by

$$d_h(\alpha, \beta) = |D_h(\alpha, \beta)|$$

are finite-to-one mappings. The P-ideal dichotomy, (see 2.4.11 above) asserts that every coherent and nontrivial sequence must contain a subsequence with Hausdorff's property.

3.4 A general theory of subadditive functions on ω_1

The purpose of this section is a general theory of distance functions on ω_1 which have the following property.

Definition 3.4.1. A *ϱ-function* on an ordinal λ is any function $\varrho : [\lambda]^2 \rightarrow \omega$ with the following three properties:

(1) $\varrho(\alpha,\gamma) \leq \max\{\varrho(\alpha,\beta),\varrho(\beta,\gamma)\}$ for all $\alpha<\beta<\gamma<\lambda$.
(2) $\varrho(\alpha,\beta) \leq \max\{\varrho(\alpha,\gamma),\varrho(\beta,\gamma)\}$ for all $\alpha<\beta<\gamma<\lambda$.
(3) $\{\alpha<\beta : \varrho(\alpha,\beta)\leq n\}$ is finite for all $\beta<\lambda$ and $n<\omega$.

Note that $\lambda=\omega_1$ is the largest ordinal that allows a ϱ-function. We have seen above that the function $\rho : [\omega_1]^2 \longrightarrow \omega$ of Definition 3.1.1 satisfies this definition, but the purpose of this section is to examine ϱ-functions that may not necessarily come from an analysis of the minimal walk. We are of course motivated again by the great potential for application of such a general metric theory of ω_1. For example, having in mind that $\varrho : [\omega_1]^2 \to \omega$ is some sort of metric on ω_1, the following useful definitions are quite natural, where a ϱ-function $\varrho : [\omega_1]^2 \to \omega$ is fixed and where, as usual, we are adopting the convention that

$$\varrho(\alpha,\alpha)=0 \text{ for all } \alpha<\omega_1. \tag{3.4.1}$$

Definition 3.4.2. The *ϱ-number* of a given finite set $F\subseteq\omega_1$ is defined to be

$$p_F = p_\varrho(F) = \max\{\varrho(\alpha,\beta) : \alpha,\beta\in F\}.$$

Definition 3.4.3. For a finite set $F\subseteq\omega_1$ and an integer k, let

$$(F)_k = \{\alpha\leq\max(F) : \varrho(\alpha,\beta)\leq k \text{ for some } \beta\in F\setminus\alpha\}.$$

Note that by (3.4.1), $F\subseteq(F)_k$ and that by 3.4.1(3), $(F)_k$ is also finite. We will say that F is *k-closed* if $(F)_k=F$. We will say that F is *ϱ-closed* if it is k-closed for $k=p_F$.

Note that the operator $F\mapsto(F)_k$ holds all the information about the ϱ-function on which it is based, so one can use it in describing conditions that one can put on such ϱ. One example of this is the following useful property that one can require from a given ϱ-function.

Definition 3.4.4. A ϱ-function $\varrho : [\omega_1]^2\to\omega$ is *smooth* if $\lim_{k\to\omega}\frac{|(\{\alpha\})_k|}{k}=0$ for every countable ordinal α.

Lemma 3.4.5. *There is a smooth ϱ-function on ω_1.*

Proof. We shall define $\varrho : [\omega_1]^2\to\omega$ recursively by the formula

$$\varrho(\alpha,\beta)=\max\{g_\beta(|C_\beta\cap\alpha|),\varrho(\alpha,\min(C_\beta\setminus\alpha)),\varrho(\xi,\alpha) : \xi\in C_\beta\cap\alpha\} \tag{3.4.2}$$

with the boundary condition $\varrho(\alpha,\alpha)=0$, where C_β $(\beta<\omega_1)$ is a C-sequence on ω_1 and where g_β $(\beta<\omega_1)$ is a sequence of nondecreasing maps from ω into ω both to be defined recursively as we go on. However, note that independently of how we choose C_β $(\beta<\omega_1)$ and g_β $(\beta<\omega_1)$, the equation (3.4.2) indeed gives us a function satisfying the conditions in Definition 3.4.1(1), (2), (3) above. Note that smoothness is a condition that needs to be checked only for countable limit ordinals λ, or in other words, we can already make the commitment that $g_{\beta+1}$ is the identity map for all β. Similarly, we can already make the commitment that $C_{\beta+1} = \{\beta\}$ for all countable ordinals β.

Case 1. $\lambda = \delta + \omega$ for some countable limit ordinal δ. Let $g_\lambda(n) = 2^n$ and let $C_\lambda = (\delta, \lambda)$. For $\gamma \leq \lambda$ and $n < \omega$, let

$$F_n^\gamma = (\{\gamma\})_n \cap \gamma = \{\alpha < \gamma : \varrho(\alpha, \gamma) \leq n\}.$$

By our inductive assumption $|F_n^\gamma|/n \to_n 0$ for all (limit) ordinals $\gamma \leq \delta$ and we need to check that this holds also for $\gamma = \lambda$. Note that by the formula (3.4.2) if an ordinal of the form $\delta + k$ belongs to F_n^λ then $k \leq \log_2 n$. It follows that

$$|F_n^\lambda| \leq |F_n^\delta| + 1 + \log_2 n$$

and this, by the inductive hypothesis, implies that $|F_n^\lambda|/n \to_n 0$.

Case 2. $\lambda = \sup_k \lambda_k$ for some strictly increasing sequence (λ_k) of countable limit ordinals. Let $C_\lambda = \{\lambda_k + 1 : k < \omega\}$. By the inductive hypothesis, we can choose a strictly increasing sequence (n_k) of integers such that

$$|F_n^{\lambda_0}|/n + \cdots + |F_n^{\lambda_k}|/n < 2^{-k} \text{ for all } k \text{ and } n \geq n_k. \tag{3.4.3}$$

Fix $\varepsilon > 0$ and choose k such that $2^{-k} < \varepsilon$. We show that $|F_n^\lambda|/n < \varepsilon$ for $n \geq n_k$. Let j be the maximal integer such that $n_k \leq n_j \leq n$. Note that then,

$$F_n^\lambda \subseteq \{\alpha < \lambda : i(\alpha) = |C_\lambda \cap \alpha| \leq j \text{ and } \alpha \in F_n^{\lambda_{i(\alpha)}}\} = F_n^{\lambda_0} \cup \cdots \cup F_n^{\lambda_j}. \tag{3.4.4}$$

It follows that $|F_n^\lambda|/n \leq |F_n^{\lambda_0}|/n + \cdots + |F_n^{\lambda_j}|/n < 2^{-j} \leq 2^{-k} < \varepsilon$, as required. □

Recall that in the theory of metric spaces universal objects such as, for example, the Urysohn space are interesting and useful. The same is true in the context of metric theory of ω_1. Before we introduce the corresponding notion, we need some preliminary definitions.

Definition 3.4.6. A finite *ϱ-model* is a model of the form $(M, <, \varrho_M, p_M)$ where M is a set, $<$ is a total ordering on M, p_M is an integer, and

$$\varrho_M : [M]^2 \to \{0, 1, \ldots, p_M\}$$

is a function with properties 3.4.1(1) and (2) listed above. We also assume that there exists $x < y$ in M such that $\varrho_M(x, y) = p_M$.

Note that for every ϱ-function $\varrho : [\lambda]^2 \to \omega$ and for every ϱ-closed subset M of λ, the structure $(M, <, \varrho \restriction [M]^2, p_M)$ is an example of a ϱ-model. Note also that an initial part M_0 of a ϱ-closed set M is a ϱ-closed set and that its integer p_{M_0} might be smaller than p_M. From this one concludes that an initial part of a ϱ-model is also a ϱ-model with a possibly smaller integer p_M.

Definition 3.4.7. Two ϱ-models $(M_1, <_1, \varrho_1, p_1)$ and $(M_2, <_2, \varrho_2, p_2)$ are *isomorphic* if there is a bijection $\pi : M_1 \to M_2$ such that for all $a, b \in M_1$,

(i) $a <_1 b$ implies $\pi(a) <_2 \pi(b)$,

(ii) $\varrho_1(a, b) = \varrho_2(\pi(a), \pi(b))$,

(iii) $p_1 = p_2$.

We are now ready to introduce the notion of a universal ϱ-function that is quite analogous to the corresponding theory of metric spaces.

Definition 3.4.8. A ϱ-function $\varrho : [\lambda]^2 \to \omega$ defined on some limit ordinal $\lambda \leq \omega_1$ is *universal* if for every finite ϱ-model $(M, <, \varrho_M, p_M)$, every ϱ-closed subset M_0 of λ such that the ϱ-model

$$(M_0, <, \varrho \restriction [M_0]^2, p_{M_0})$$

is isomorphic to an initial segment of $(M, <, \varrho_M, p_M)$, and every ordinal δ such that $\delta + \omega \leq \lambda$, there is a ϱ-closed subset M_1 of $\delta + \omega$ such that:

(a) $(M_1, <, \varrho \restriction [M_1]^2, p_{M_1}) \cong (M, <, \varrho_M, p_M)$,

(b) M_0 is an initial segment of M_1,

(c) $M_1 \setminus M_0 \subseteq [\delta, \delta + \omega)$.

Theorem 3.4.9. *There is a universal ϱ-function on ω_1.*

Proof. We will obtain universal $\varrho : [\omega_1]^2 \to \omega$ by recursively constructing an increasing sequence $\varrho_\lambda : [\lambda]^2 \to \omega$ $(\lambda \in \Lambda)$ of universal ϱ-function. Let $\varrho_0 = \emptyset$, and suppose $\varrho_\lambda : [\lambda]^2 \to \omega$ has been determined for some countable limit ordinal λ. Let C be a subset of λ of order-type ω such that $\lambda = \sup C$. Define

$$\varrho_{\lambda+\omega}(\alpha, \lambda) = \max\{|C \cap \alpha|, \varrho_\lambda(\alpha, \min(C \setminus \alpha)), \varrho_\lambda(\xi, \alpha) : \xi \in C \cap \alpha\}.$$

It can be checked that this defines a function $\varrho_{\lambda+\omega} : [\lambda + 1]^2 \to \omega$ having the properties (1), (2) and (3) of Definition 3.4.1 (which we will refer to throughout this proof). Starting with this extension of ϱ_λ and the assumption that ϱ_λ is universal we build extensions

$$\varrho_{\lambda+\omega} : [\delta]^2 \to \omega \quad (\lambda + 1 \leq \delta < \lambda + \omega)$$

in such a way that at a given stage δ we take care about a particular instance of universality of $\varrho_{\lambda+\omega}$. Thus, modulo some bookkeeping device, it suffices to show how does one deal with the single task of Definition 3.4.8.

Suppose we have already defined an extension $\varrho_{\lambda+\omega} : [\delta]^2 \to \omega$ and that we are given a finite ϱ-model $(M, <, \varrho_M, p_M)$ and a ϱ-closed subset M_0 of δ such that the ϱ-model,

$$(M_0, <, \varrho_{\lambda+\omega} \restriction [M_0]^2, p_{M_0})$$

is isomorphic to a proper initial segment of the ϱ-model $(M, <, \varrho_M, p_M)$. Let

$$l = |M| - |M_0|.$$

We need to extend $\varrho_{\lambda+\omega}$ from $[\delta]^2$ to $[\delta+l]^2$. First of all define $\varrho_{\lambda+\omega}$ on

$$[M_0 \cup [\delta, \delta+l)]^2 \setminus [M_0]^2$$

in such a way that we have the isomorphism

$$(M_0 \cup [\delta, \delta+l), <, \varrho_{\lambda+\omega}, p_M) \cong (M, <, \varrho_M, p_M).$$

Thus, it remains to define $\varrho(\alpha, \gamma)$ for $\alpha \in \delta \setminus M_0$ and $\gamma \in (\delta, \delta+l)$. If $\alpha < \delta$ and $\alpha > \max M_0$, then set

$$\varrho(\alpha, \gamma) = \max\{p_M + 1, \varrho(\alpha, \delta - 1), \varrho(\xi, \alpha) : \xi \in M_0\}. \tag{3.4.5}$$

If $\alpha \leq \max M_0$, then set

$$\varrho(\alpha, \gamma) = \max\{\varrho(\alpha, \min(M_0 \setminus \alpha)), \varrho(\xi, \alpha) : \xi \in M_0 \cap \alpha\}. \tag{3.4.6}$$

It remains to show that $\varrho_{\lambda+\omega} \restriction [\delta+l]^2$ satisfies properties 3.4.1(1) and (2). In order to avoid lengthy formulas, let us simplify the notation as follows,

$$\alpha\beta = \varrho(\alpha, \beta) \text{ and } \alpha\beta \vee \gamma\delta = \max\{\varrho(\alpha, \beta), \varrho(\gamma, \delta)\}. \tag{3.4.7}$$

We also adopt the notation δ^- in place of $\delta - 1$.

So let $\alpha < \beta < \gamma < \delta + l$ be a given triple of countable ordinals.

Case 1. $\alpha < \delta \leq \beta < \gamma < \delta + l$. If $\alpha \in M_0$, then properties 3.4.1(1) and (2) for $\alpha < \beta < \gamma$ follow from the fact that in the definitions of $\alpha\beta$, $\beta\gamma$ and $\alpha\gamma$ we have copied the ϱ-model $(M, <, \varrho_M, p_M)$ which satisfies 3.4.1(1) and (2). If $\alpha \notin M_0$, then in both case $\alpha > \max M_0$ and $\alpha \leq \max M_0$ we conclude that $\alpha\beta = \alpha\gamma$, so 3.4.1(1) and (2) for $\alpha < \beta < \gamma$ follow immediately.

Case 2. $\alpha < \beta < \delta < \gamma < \delta + l$.

Subcase 2.1. $\max M_0 < \delta < \beta$. Consider first the inequality $\alpha\gamma \leq \alpha\beta \vee \beta\gamma$. The quantities $p_M + 1$ and $\xi\alpha$ $(\xi \in M)$ from the definition of $\alpha\gamma$ are all present in the definition of $\beta\gamma$, so it remains only to show that the quantity $\alpha\delta^-$ is bounded by $\alpha\beta \vee \beta\gamma$. Applying the inequality 3.4.1(1) for $\varrho_{\lambda+\omega} \restriction [\delta]^2$, we get

$$\alpha\delta^- \leq \alpha\beta \vee \beta\delta^-,$$

and so we are done as $\beta\delta^-$ shows up in the definition of $\beta\gamma$.

Consider now the inequality $\alpha\beta \leq \alpha\gamma \vee \beta\gamma$. Applying inequality 3.4.1 for $\varrho_{\lambda+\omega} \restriction [\delta]^2$ to the triple $\alpha < \beta < \delta^-$, we get

$$\alpha\beta \leq \alpha\delta^- \vee \beta\delta^-,$$

so we are done also in this case since the quantity on the right-hand side is bounded by $\alpha\gamma \vee \beta\gamma$.

Subcase 2.2. $\alpha \leq \max M_0 < \beta < \delta$. Consider first the subcase when $\alpha \in M_0$. To see that $\alpha\gamma \leq \alpha\beta \vee \beta\gamma$ observe that $\alpha\gamma \leq p_M < p_M + 1 \leq \beta\gamma$. To see that $\alpha\beta \leq \alpha\gamma \vee \beta\gamma$ observe that $\alpha\beta$ appears as a quantity in the definition of $\beta\gamma$. Let us consider the case $\alpha \notin M_0$ and let

$$\alpha' = \min(M_0 \setminus \alpha).$$

The quantity $\alpha\alpha'$ from the definition of $\alpha\gamma$ is bounded by $\alpha\beta \vee \beta\gamma$, since by the inequality 3.4.1(2) for $\varrho_{\lambda+\omega} \restriction [\delta]^2$, we have that

$$\alpha\alpha' \leq \alpha\beta \vee \alpha'\beta,$$

and $\alpha'\beta$ appears in the definition of $\beta\gamma$. Since the sequence of quantities

$$\xi\alpha \quad (\xi \in M_0 \cap \alpha)$$

appear also in the definition of $\beta\gamma$, this shows that $\alpha\gamma \leq \alpha\beta \vee \beta\gamma$.

Let us check now the inequality $\alpha\beta \leq \alpha\gamma \vee \beta\gamma$. Apply the inequality 3.4.1(1) that is valid for $\varrho_{\lambda+\omega} \restriction [\delta]^2$ to the triple $\alpha < \alpha' < \beta$, and get

$$\alpha\beta \leq \alpha\alpha' \vee \alpha'\beta.$$

This finishes the checking, since $\alpha\alpha' \leq \alpha\gamma$ and $\alpha'\beta \leq \beta\gamma$.

Subcase 2.3. $\alpha < \beta \leq \max M_0$. If $\alpha, \beta \in M_0$, then the inequalities 3.4.1(1), (2) for $\alpha < \beta < \gamma$ follow from the fact that in the definitions of $\alpha\beta$, $\beta\gamma$ and $\alpha\gamma$ we copied the ϱ-model $(M, <, \varrho_M, p_M)$.

Subcase 2.3.1. $\alpha \subset M_0$ and $\beta \notin M_0$. Consider the inequality $\alpha\gamma \leq \alpha\beta \vee \beta\gamma$. This follows from the fact that

$$\alpha\gamma \leq p_M < \beta\beta', \text{ where } \beta' = \min(M_0 \setminus \beta)$$

and the fact that in the definition of $\beta\gamma$ the quantity $\beta\beta'$ appears. The inequality $\alpha\beta \leq \alpha\gamma \vee \beta\gamma$ in this subcase follows from the fact that the quantity $\alpha\beta$ appears in the definition of $\beta\gamma$.

Subcase 2.3.2. $\alpha \notin M_0$ and $\beta \in M_0$. Consider the inequality $\alpha\gamma \leq \alpha\beta \vee \beta\gamma$. Let

$$\alpha' = \min(M_0 \setminus \alpha).$$

We need to bound the quantities $\alpha\alpha'$ and $\xi\alpha$ ($\xi \in M_0 \cap \alpha$) by $\alpha\beta \vee \beta\gamma$. Apply the inequality 3.4.1(2) that is valid for $\varrho_{\lambda+\omega} \restriction [\delta]^2$ to $\alpha < \alpha' \leq \beta$, and get

$$\alpha\alpha' \leq \alpha\beta \vee \alpha'\beta.$$

Since $\alpha'\beta \leq p_M$ and $\alpha\alpha' > p_M$ we conclude that $\alpha\alpha' \leq \alpha\beta$ as required. Similarly note that

$$\xi\alpha \leq \xi\beta \vee \alpha\beta = \alpha\beta,$$

since $\xi\beta \leq p_M$ while $\alpha\beta > p_M$. It remains to check the inequality $\alpha\beta \leq \alpha\gamma \vee \beta\gamma$ in this subcase. As before, note that

$$\alpha\beta \leq \alpha\alpha' \vee \alpha'\beta,$$

and that $\alpha\beta > \alpha'\beta$ since $\alpha'\beta \leq p_M$ while $\alpha\beta > p_M$. It follows that $\alpha\beta \leq \alpha\alpha' \leq \alpha\gamma$, as required.

Subcase 2.3.3. $\alpha \notin M_0$ and $\beta \notin M_0$ (and $\alpha, \beta \leq \max M_0$). So in this subcase both quantities $\alpha\gamma$ and $\beta\gamma$ are defined according to the second definition. Let

$$\alpha' = \min(M_0 \setminus \alpha) \text{ and } \beta' = \min(M_0 \setminus \beta).$$

Note that $\alpha' \leq \beta'$. We first check the inequality $\alpha\gamma \leq \alpha\beta \vee \beta\gamma$. Consider first the quantity $\alpha\alpha'$ that appears in the definition of $\alpha\gamma$. If $\alpha' = \beta'$, then

$$\alpha\alpha' \leq \alpha\beta \vee \beta\beta' \leq \alpha\beta \vee \beta\gamma,$$

as $\beta\beta'$ appears in the definition of $\beta\gamma$. Suppose that $\alpha' < \beta'$ i.e., that $\alpha' < \beta$. Then

$$\alpha\alpha' \leq \alpha\beta \vee \alpha' \leq \alpha\beta \vee \beta\gamma$$

as $\alpha'\beta$ appears as a quantity in the definition of $\beta\gamma$. Consider now the quantity $\xi\alpha$ for $\xi \in M_0 \cap \alpha$. Note that

$$\xi\alpha \leq \alpha\beta \vee \xi\beta \leq \alpha\beta \vee \beta\gamma,$$

as $\xi\beta$ appears in the definition of $\beta\gamma$.

It remains to check the inequality $\alpha\beta \leq \alpha\gamma \vee \beta\gamma$ in this subcase. If $\alpha' = \beta'$, then we get that

$$\alpha\beta \leq \alpha\beta' \vee \beta\beta' \leq \alpha\gamma \vee \beta\gamma,$$

as the quantity $\alpha\beta' = \alpha\alpha'$ appears in $\alpha\gamma$ while $\beta\beta'$ appears in $\beta\gamma$. If $\alpha' < \beta'$, i.e., $\alpha' < \beta \leq \beta'$, then we get that

$$\alpha\beta \leq \alpha\alpha' \vee \alpha'\beta \leq \alpha\gamma \vee \beta\gamma,$$

since $\alpha\alpha'$ appears in $\alpha\gamma$ and $\alpha'\beta$ appears in $\beta\gamma$.

This finishes our checking that the extension $\varrho_{\lambda+\omega} \upharpoonright [\delta+l]^2$ remains to satisfy the conditions 3.4.1(1),(2) and (3). Note that

$$\varrho(\alpha,\gamma) > p_M \text{ for all } \alpha \in \delta \setminus M_0 \text{ and } \gamma \in [\delta, \delta+l),$$

we conclude that the set $M_0 \cup [\delta, \delta+l)$ is $\varrho_{\lambda+\omega}$-closed. It follows that the extension $\varrho_{\lambda+\omega} \upharpoonright [\delta, \delta+l]^2$ has the subset

$$M_1 = M_0 \cup [\delta, \delta+1) \subseteq \delta + l,$$

of its domain as a $\varrho_{\lambda+\omega}$-closed set, while the corresponding ϱ-model

$$(M_1, <, \varrho_{\lambda+\omega} \upharpoonright [M_1]^2, p_M)$$

is isomorphic to the given ϱ-model $(M, <, \varrho_M, p_M)$. This finishes the recursive construction of a universal ϱ-function $\varrho : [\omega_1]^2 \to \omega$. □

Many ϱ-functions on ω_1 including the universal one lack the following property frequently useful in applications.

Definition 3.4.10. A ϱ-function $\varrho : [\omega_1]^2 \to \omega$ is *unbounded* if for every $k < \omega$ and every infinite set $A \subseteq \omega_1$ there exist $\alpha < \beta$ in A such that $\varrho(\alpha, \beta) > k$.

It should be clear that there is an unbounded ϱ-function on ω_1. For example, the function $\bar{\rho}$ of Definition 3.2.1 is clearly unbounded. In fact there is a general procedure of changing an arbitrary ϱ-function to an unbounded one via the formula,

$$\widetilde{\varrho}(\alpha,\beta) = \max\{\varrho(\alpha,\beta),\ |\{\xi \leq \alpha : \varrho(\xi,\alpha) \leq \varrho(\alpha,\beta)\}|\}. \tag{3.4.8}$$

Lemma 3.4.11. *Suppose $\varrho : [\omega_1]^2 \to \omega$ is an unbounded ϱ-function. Then for every infinite family A of pairwise-disjoint finite subsets of ω_1, all of some fixed size n and for every k, there exist an infinite subfamily $B \subseteq A$ such that for all $a \neq b$ in B and all $\alpha \in a$ and $\beta \in b$, we have that $\varrho(\alpha, \beta) > k$.*

Proof. Fix an integer k and a nonprincipal ultrafilter $\mathcal{U}$ on A. A simple diagonalization procedure reduces the conclusion of the lemma to showing that

$$(\forall i, j < n)(\mathcal{U}a)(\mathcal{U}b)\ \ \varrho(a(i), b(j)) > k.^3 \tag{3.4.9}$$

Suppose otherwise that there exist $i, j < n$ such that

$$(\mathcal{U}a)(\mathcal{U}b)\ \ \varrho(a(i), b(j)) \leq k. \tag{3.4.10}$$

Note that by the property (3) of ϱ, we must have that

$$\gamma(i) - \lim_{a \to \mathcal{U}} a(i) \leq \lim_{a \to \mathcal{U}} a(j) = \gamma(j), \tag{3.4.11}$$

[3]Here $a(i)$ and $b(j)$ denote the ith element of a and jth element of b, respectively, relative to their increasing enumerations.

or else, (3.4.10) would fail. Let

$$X = \{a(i) : (\mathcal{U}b)\ \varrho(a(i), b(j)) \leq k\}$$

and let $Y = X \cap \gamma(i)$. Then Y is, in particular, infinite. We claim that $\varrho(\alpha, \beta) \leq k$ for all α and β in Y, contradicting the assumption that ϱ is unbounded. To see this consider $\alpha < \beta$ in Y. Then by (3.4.10) and the fact that

$$\lim_{a \to \mathcal{U}} a(j) = \gamma(j) > \beta > \alpha, \tag{3.4.12}$$

we conclude that there must be b in A such that $b(j) > \beta > \alpha$ and $\varrho(\alpha, b(j)) \leq k$ and $\varrho(\beta, b(j)) \leq k$. Using the subadditivity of ϱ, this gives us $\varrho(\alpha, \beta) \leq k$, as required. □

Further analysis of ϱ-functions on ω_1 requires the following natural notion.

Definition 3.4.12. Given a ϱ-function on ω_1 and $n < \omega$, we let G_ϱ^n denote the corresponding graph whose vertex set is ω_1 and where an unordered pair $\{\alpha, \beta\}$ forms an edge if and only if $\varrho(\alpha, \beta) = n$.

Note the following immediate consequence of Lemma 3.2.2(a) that bears on this notion.

Lemma 3.4.13. *If ϱ is equal to $\bar{\rho}$ of Definition* 3.2.1, *then all graphs G_ϱ^n are acyclic.* □

Using deeper properties of $\bar{\rho}$ one has the following information about the graphs G_ϱ^n.

Lemma 3.4.14. *If ϱ is equal to $\bar{\rho}$ of Definition* 3.2.1, *then for a given positive integer n, the connected component of the graph G_ϱ^n is a star whose center is its minimum.*

Proof. Consider a path $\alpha_0, \alpha_1, \ldots, \alpha_k$ of some fixed graph $G_{\bar{\rho}}^n$. Note that by Lemma 3.2.3 for no $i < k-1$ do we have the inequalities

$$\alpha_i < \alpha_{i+1} < \alpha_{i+2} \text{ or } \alpha_i > \alpha_{i+1} > \alpha_{i+2}.$$

On the other hand, by Lemma 3.2.2(a), the path does not contain three consecutive points $\alpha_i, \alpha_{i+1}, \alpha_{i+2}$ with property

$$\alpha_i < \alpha_{i+1} > \alpha_{i+2}.$$

So, if α_j is the minimal point of the path, then on one hand $k - j \leq 1$, and on the other hand $j \leq 1$. It follows that the path has length at most 2, as required. □

Corollary 3.4.15. *The complete graph on ω_1 vertices can be split into countably many graphs, none of which has paths of length* 3. □

Remark 3.4.16. It follows, in particular, that the complete graph on ω_1 vertices can be split into countably many acyclic graphs. In fact, ω_1 is the largest vertex-set that allows such a decomposition. This is an old result of Erdős and Kakutani [30]. Similar argument shows the following result that complements Corollary 3.4.15.

Lemma 3.4.17. *For every partition of the complete graph on ω_2 into countably many subgraphs G_n, one of the graphs G_n contains an infinite path.*

Proof. Let $c : [\omega_2]^2 \longrightarrow \omega$ be a given countable edge-partition of the complete graph $(\omega_2, [\omega_2]^2)$. The conclusion of the lemma will follow if we can find an integer $n < \omega$, an increasing sequence α_i $(i < \omega)$ of countable ordinals, and an increasing sequence β_j $(j < \omega)$ of uncountable ordinals $< \omega_2$ such that

$$c(\alpha_i, \beta_j) = n \text{ for all } i \leq j < \omega. \tag{3.4.13}$$

Since then $\beta_0, \alpha_0, \beta_1, \alpha_1, \dots, \beta_i, \alpha_i, \dots$ would form an infinite path of the graph $(\omega_2, c^{-1}(n))$, as required. Find an $n < \omega$ such that the set B of all $\beta \in [\omega_1, \omega_2)$ for which the corresponding set

$$A_\beta = \{\alpha < \omega_1 : c(\alpha, \beta) = n\}$$

is unbounded in ω_1 is unbounded in ω_2. We shall show that this n works. Find $\beta_0 \in B$ with the property that $\{\beta \in B : c(\alpha, \beta) = n\}$ is unbounded in ω_2 for every $\alpha \in A_{\beta_0}$. Pick an arbitrary $\alpha_0 \in A_{\beta_0}$ and set

$$B_0 = \{\beta \in B : \beta > \beta_0 \text{ and } c(\alpha_0, \beta) = n\}.$$

Then B_0 is an unbounded subset of ω_2 so we can find $\beta_1 \in B_0$ with the property that $\{\beta \in B_0 : c(\alpha, \beta) = n\}$ is unbounded in ω_2 for every $\alpha \in A_{\beta_1}$. Pick an arbitrary $\alpha_1 \in A_{\beta_1}$ such that $\alpha_1 > \alpha_0$ and form the set

$$B_1 = \{\beta \in B_0 : \beta > \beta_1 \text{ and } c(\alpha_1, \beta) = n\}.$$

Then B_1 is unbounded in ω_2, so we can proceed further and find β_1, α_1, B_2, β_2, α_2, B_3, etc. Note that the construction is giving us the desired equalities of (3.4.13). □

Remark 3.4.18. One may wonder if the previous proof can be augmented so that we also get the equalities

$$c(\alpha_i, \beta_j) = n \text{ for all } j < i < \omega. \tag{3.4.14}$$

besides those of (3.4.13). Theorem 9.3.1 below shows that, in fact, this is not possible to achieve.

Clearly no graph G_ϱ^n associated to a universal function ϱ can be acyclic, so one can view $\bar{\rho}$ and universal ϱ as extreme points of a rich spectrum of ϱ-functions relative to the structure theory of the corresponding graphs G_ϱ^n. So let us introduce some interesting classes between these two extremes. First of all note the following general fact about ϱ-functions.

Lemma 3.4.19. *Suppose that ϱ is a ϱ-function and that n is a positive integer. Let F be a ϱ-closed set whose ϱ-number is equal to n. Then there is a decomposition*

$$F = F_0 \cup F_1 \cup F_2, \tag{3.4.15}$$

where F_0 is an initial segment of F consisting of points that are isolated in G^n_ϱ while every point of F_1 is connected to every point of F_2.

Proof. Note that by our assumption $n > 0$, since in case $n = 0$ the restriction of G^n_ϱ on the set F is a complete graph, and so while we can put $F_0 = \emptyset$ and $F_1 = F_2 = F$, the decomposition is not as required since the pieces are not pairwise-disjoint. Let E be the connected component of the point $\delta = \max F$ relative to the graph which G^n_ϱ induces on F. Note that by the subadditivity properties of ϱ the component E has at least one more point besides δ. By our assumption that G^n_ϱ is bipartite so we can decompose E as the union of two nonempty subsets F_1 and F_2 such that all edges of G^n_ϱ inside E are between a point of F_1 and a point of F_2. We assume that δ belongs to F_2. Let $\gamma = \max F_1$. Recall the tree-ordering $<_{n-1}$ on ω_1 associated to ϱ and the integer $n-1$ as follows,

$$\alpha <_{n-1} \beta \text{ iff } \alpha < \beta \text{ and } \varrho(\alpha, \beta) \leq n-1. \tag{3.4.16}$$

Let $P_{n-1}(\gamma) = \{\xi : \xi \leq_{n-1} \gamma\}$, and similarly, let $P_{n-1}(\delta) = \{\xi : \xi \leq_{n-1} \delta\}$. Then

$$F = P_{n-1}(\gamma) \cup P_{n-1}(\delta),$$

and moreover, the set $F_0 = P_{n-1}(\gamma) \cap P_{n-1}(\delta)$ is an initial segment of both $P_{n-1}(\gamma)$ and $P_{n-1}(\delta)$. It follows that F_0 is also an initial segment of F. Note that

$$\varrho(\alpha, \beta) \leq n-1 \text{ for every } \{\alpha, \beta\} \in [F]^2 \text{ with } \{\alpha, \beta\} \cap F_0 \neq \emptyset.$$

From this, we conclude that every point of the set F_0 is an isolated point of the graph G^n_ϱ induces on F. Since every point of the connected component E is connected to at least one more point of E, we conclude that E is disjoint from F_0. Since

$$F_1 \subseteq P_{n-1}(\gamma) \text{ and } F_2 \subseteq P_{n-1}(\delta),$$

we conclude that every point of F_1 is $<_{n-1}$-incomparable to every point of F_2. This precisely means that every point of F_1 is G^n_ϱ-connected to every point of F_2. This completes the proof. □

Corollary 3.4.20. *If ϱ is a ϱ-function and if n is a positive integer, then the graph G^n_ϱ has no odd cycles.* □

Definition 3.4.21. For a ϱ-closed set F with ϱ-number equal to n, we call the decomposition (3.4.15) of Lemma 3.4.19, the *G^n_ϱ-decomposition* of F, or more simply the *ϱ-decomposition* of F. The ρ-decomposition (3.4.15) is *increasing* if $F_0 < F_1 < F_2$.

Definition 3.4.22. We call a ϱ-function $\varrho : [\omega_1]^2 \to \omega$ a *uniform increasing bipartite ϱ-function* if:

(1) Every ϱ-closed set H is finite and is included in a maximal ϱ-closed F which is also finite and which has the same ϱ-number as H, and moreover, every such set F admits an increasing ϱ-decomposition.

(2) Maximal ϱ-closed sets F and the pieces F_0, F_1 and F_2 of their increasing ϱ-decompositions have cardinalities k_n, l_n, and l_n, respectively, that depend only on the ϱ-number n of F.

Theorem 3.4.23. *There is a uniform increasing bipartite ϱ-function on ω_1.* □

Remark 3.4.24. We do not give a proof of Theorem 3.4.23 here, though it bears a considerable similarity with the proof of Theorem 3.4.9 above. It would be of interest, however, to determine the possible sequences $(k_n)_n$ and $(l_n)_n$ that witness uniformity of a given uniform increasing bipartite ϱ-function on ω_1. Some results to that effect are given by Velleman [126] though using a quite different terminology. The relationship between the theory of ϱ-functions and Velleman's theory of 'morasses' was fully explained by Morgan [79].

3.5 Conditional weakly null sequences based on subadditive functions

In this section we mention some applications of the concept of ϱ-function on ω_1 satisfying the inequalities of Lemma 3.1.2. As we shall see, any such a semi-distance function ϱ can be used as a coding procedure in constructions of long weakly null sequences in Banach spaces with no unconditional basic subsequences.

Example 3.5.1. *A weakly null sequence with no unconditional basic subsequence.* We describe a weakly null normalized sequence $(x_\xi)_{\xi<\omega_1}$ of members of some Banach space X which contains no infinite unconditional basic subsequence[4]. Fix an $0 < \varepsilon < 1$ and an infinite set M of positive integers such that $1 \in M$ and

$$\sum \left\{ \sqrt{\frac{m}{n}} : m, n \in M, m < n \right\} \leq \varepsilon. \tag{3.5.1}$$

Fix also a natural bijection $\ulcorner . \urcorner : \mathrm{HF} \to M$, where HF denotes the family of all hereditarily finite sets, so that in particular $\ulcorner \emptyset \urcorner = 1$. Recall that the *$\varrho$-number* of

[4]Recall that a *basic sequence* in a Banach space X is a seminormalized sequence $(x_n)_{n\in\omega}$ with the property that every x in its closed linear span $\overline{\mathrm{span}}\{x_n : n \in \omega\}$ can be uniquely represented as a sum $\sum_{n\in\omega} a_n x_n$. Note that this is equivalent to saying that the norms of projection operators $P_k : \overline{\mathrm{span}}\{x_n : n \in \omega\} \to \overline{\mathrm{span}}\{x_n : n < k\}$ are uniformly bounded. When the norms of projections $P_M : \overline{\mathrm{span}}\{x_n : n \in \omega\} \to \overline{\mathrm{span}}\{x_n : n \in M\}$ over all subsets M of ω are uniformly bounded, the sequence (x_n) is said to be *unconditional*. The natural basis (e_n) of any of the sequence-spaces ℓ_p for $p \in [1, \infty]$ is unconditional. A typical example of a basic sequence that is not unconditional is the *summing basis* $e_0, e_0 + e_1, e_0 + e_1 + e_2, \ldots$ of c_0 or the summing basis $e_0 + e_1 + e_2 + \cdots + e_n + \cdots$, $e_1 + e_2 + \cdots + e_n + \cdots$, $e_2 + \cdots + e_n + \cdots$, $\ldots$ in c.

a finite set $F \subseteq \omega_1$ is the integer

$$\varrho(F) = \max\{\varrho(\alpha, \beta) : \alpha, \beta \in F, \alpha \leq \beta\}.$$

For $F \subseteq \omega_1$ and $k \in \omega$, let

$$(F)_k = \{\alpha \leq \max(F) : \varrho(\alpha, \beta) \leq k \text{ for some } \beta \in F \text{ with } \beta \geq \alpha\}.$$

We say that F is *k-closed* if $(F)_k = F$. The *ϱ-closure* of F is the set $(F)_\varrho = (F)_k$, where $k = \varrho(F)$. This leads us to the function

$$\sigma : [[\omega_1]^{<\omega}]^{<\omega} \longrightarrow M$$

defined by letting $\sigma(\emptyset) = \ulcorner \emptyset \urcorner$ and

$$\sigma(F_0, F_1, \ldots, F_n) = \ulcorner ((\pi_E'' F_0, \pi_E'' F_1, \ldots, \pi_E'' F_n), \varrho(\bigcup_{i=0}^{n} F_i)) \urcorner,$$

where $E = (\bigcup_{i=0}^n F_i)_\varrho$ and where $\pi_E : E \rightarrow |E|$ is the unique order-preserving map between the sets of ordinals E and $|E| = \{0, 1, \ldots, |E| - 1\}$. We say that a sequence $(E_i)_{i<n}$ of finite subsets of ω_1 is *special* if

1. $E_i < E_j$[5] for $i < j < n$,
2. $|E_0| = 1$ and $|E_j| = \sigma(E_0, E_1, \ldots, E_{j-1})$ for $0 < j < n$.

Let $c_{00}(\omega_1)$ be the normed space of all finitely supported maps from ω_1 into the reals. Let $(e_\xi)_{\xi<\omega_1}$ be the basis of $c_{00}(\omega_1)$ and let $(e_\xi)^*_{\xi<\omega_1}$ be the corresponding sequence of biorthogonal functionals. To a finite set $E \subseteq \omega_1$, we associate the following vector and functional on $c_{00}(\omega_1)$,

$$x_E = \frac{1}{|E|^{1/2}} \sum_{\alpha \in E} e_\alpha \text{ and } \phi_E = \frac{1}{|E|^{1/2}} \sum_{\alpha \in E} e^*_\alpha. \tag{3.5.2}$$

Given a special sequence $(E_i)_{i<n}$ of finite subsets of ω_1, the corresponding *special functional* is defined by $\sum_{i<n} \phi_{E_i}$. Let $\mathcal{F}$ be the family of all special functionals of $c_{00}(\omega_1)$. This family induces the following norm on $c_{00}(\omega_1)$,

$$\| x \| = \sup\{\langle f, x \rangle : f \in \mathcal{F}\}. \tag{3.5.3}$$

The Banach space X is the completion of the normed space $(c_{00}(\omega_1), \| \cdot \|)$. Note that $\| e_\alpha \| = 1$ and that $(e_\alpha)_{\alpha<\omega_1}$ is a weakly null sequence in X. To see this note that given a finite set G with $|G| \in M$ and a special sequence $(E_i)_{i<n}$, if

[5] Recall, that for two sets of ordinals E and F, by $E < F$ we denote the fact that every ordinal from E is smaller than every ordinal from F. Sequences of sets of ordinals of this sort are called *block sequences*.

$\phi = \sum_{i<n} \phi_{E_i}$ is the corresponding special functional, then the inner product between ϕ and the average $|G|^{-1} \sum_{\alpha \in G} e_\alpha = |G|^{-1/2} x_G$ is not bigger than

$$|G|^{-1/2}(1 + \sum_{|G| \neq m \in M} \min((m/|G|)^{1/2}, (|G|/m)^{1/2})$$

and therefore not bigger than $|G|^{-1/2}(1+\varepsilon)$ by our assumption (3.5.1). It follows that the averages $\| \; |G|^{-1} \sum_{\alpha \in G} e_\alpha \; \|$ tend to 0 as $|G| \in M$ goes to infinity, so $(e_\alpha)_{\alpha<\omega_1}$ must be weakly null in X. To show that no infinite subsequence of $(e_\alpha)_{\alpha<\omega_1}$ is unconditional it suffices to show that for every set $G \subseteq \omega_1$ of order type ω there is a seminormalized *block-subsequence*[6] of $(e_\alpha)_{\alpha \in G}$ which represents the summing basis in c. More precisely, we shall show that if $(E_i)_{i<\omega}$ is any infinite special sequence of finite subsets of G and if for $i < \omega$ we let $v_i = x_{E_i}$ as defined in (3.5.2), then for every $n < \omega$ and every sequence $(a_i)_{i \leq n} \subseteq [-1, +1]$ of scalars.

$$\max_{0 \leq k \leq n} \mid \sum_{i=0}^{k} a_i \mid \leq \| \sum_{i=0}^{n} a_i v_i \| \leq (3+\varepsilon) \max_{0 \leq k \leq n} \mid \sum_{i=0}^{k} a_i \mid . \tag{3.5.4}$$

To see the first inequality, note that according to (3.5.2) the absolute value of the inner product between the special functional $\phi = \sum_{i=0}^{n} \phi_{E_i}$ and the vector $x = \sum_{i=0}^{n} a_i v_i$ is equal to $\mid \sum_{i=0}^{n} a_i \mid$ giving us the first inequality. To see the second inequality, consider a special functional $\psi = \sum_{i=0}^{m} \phi_{F_i}$ where $(F_i)_{i \leq m}$ is some other special sequence of finite sets. In order to estimate the inner product $\langle \psi, x \rangle$ we need to know how the two special sequences interact and this requires properties of the function ϱ that controls them. To see this, let $l \leq \min(m,n)$ be maximal with the property that $|E_l| = |F_l|$, and let $E = (\bigcup_{i<l} E_i)_\varrho$ and $F = (\bigcup_{i<l} F_i)_\varrho$. By the definition of a special sequence and $|E_l| = |F_l|$, we know that:

3. $\varrho(\bigcup_{i<l} E_i) = \varrho(\bigcup_{i<l} F_i)$,
4. $\pi''_E E_i = \pi''_E F_i$ for $i < l$.

It follows that E and F have the same cardinality and the same ϱ-numbers, and therefore by the properties of ϱ listed in Lemma 3.1.2 their intersection $H = E \cap F$ is their initial segment. Hence, π_E and π_F agree on H. Let k be the maximal integer such that $E_k \subseteq H$. Then $E_i = F_i$ for all $i \leq k$, and using again the properties of ϱ listed in Lemma 3.1.2, $E_i \cap F_j = \emptyset$ for $k+1 < i, j < l$ (see Figure 3.2). Combining all this we see that $|\langle \psi, x \rangle|$ is bounded by

$$\left| \sum_{i=0}^{k} a_i + a_{k+1} \langle \phi_{F_{k+1}}, x_{E_{k+1}} \rangle + a_l \langle \phi_{F_l}, x_{E_l} \rangle \right| + \sum \{ (\frac{m}{n})^{1/2} : m, n \in M, m < n \}.$$

[6] i.e., a seminormalized sequence $(x_i)_{i<\omega}$ of finitely supported vectors of the linear span of $(e_\alpha)_{\alpha \in G}$ such that $\mathrm{supp}(x_i) < \mathrm{supp}(x_j)$for $i < j$.

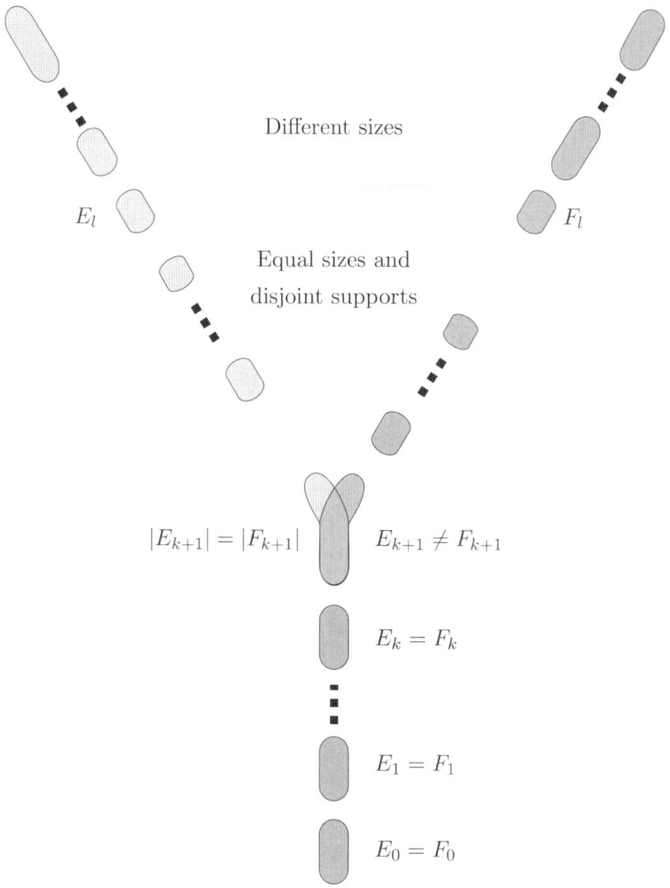

Figure 3.2: The interference of two special functionals.

Referring to (3.5.1) we get that $|\langle\psi, x\rangle|$ is bounded by the right hand side of the second inequality of (3.5.4).

Remark 3.5.2. The first example of a weakly null sequence (of length ω) with no unconditional basic subsequence was constructed by Maurey and Rosenthal [74]. Sixteen years later Gowers and Maurey [39] solved the basic unconditional sequence problem by constructing an infinite-dimensional separable reflexive Banach space with no infinite unconditional basic sequence. The basic new ingredient in the construction of [39] is the idea behind the Schlumprecht space [89] which can in general be described as follows in terms of two increasing sequences (m_k) and (n_k) of positive integers defined by the recursive formulas:

$$m_k = 2^{4^k} \text{ and } n_0 = 4 \text{ and } n_{k+1} = (4n_k)^{\log_2 m_{k+1}^3}. \tag{3.5.5}$$

Definition 3.5.3. For an infinite ordinal γ let $\mathcal{G}_\gamma$ be the minimal subset of $c_{00}(\gamma)$ with the following properties:

(1) $\pm e^*_\alpha \in \mathcal{G}_\gamma$ for all $\alpha < \gamma$,

(2) $(1/m_j)\sum_{i<n_j} \phi_i \in \mathcal{G}_\gamma$ for every non-negative integer j and every block sequence $\phi_0 < \phi_1 < \cdots < \phi_{n_j-1}$ of members of $\mathcal{G}_\gamma$,

(3) $\mathcal{G}_\gamma$ is closed under restrictions to intervals of γ,

(4) $\mathcal{G}_\gamma$ is closed under rational convex combinations.

Let $T_\gamma[m_j^{-1}, n_j]$ be the completion of $c_{00}(\gamma)$ equipped with the norm

$$\| x \| = \sup_{\phi\in\mathcal{G}_\gamma} \langle \phi, x\rangle.$$

Remark 3.5.4. Schlumprecht's original space [89] is really the space $T_\omega[1/m_j, n_j]$ where $m_j = \log_2(j+1)$ and $n_j = j$, though we will find it more convenient to work with $T_\omega[1/m_j, n_j]$ for the particularly chosen sequences (m_j) and (n_j) as in (3.5.5) or their subsequences. Schlumprecht's space was the first known example of an arbitrary distortable reflexive Banach space, and his arguments form an important part of basically all latter constructions of implicit norms including the ones that we choose to present below.

The following result of [4] relates to Example 3.5.1 given at the beginning of this section in the similar way the result of [39] relates to the example of [74]. While the basic idea of adding conditionality to the norm is supplemented by the use of the ϱ-function, substantial new work is still needed to implement the ideas of [89] and [39] to this setting.

Theorem 3.5.5. *There is a reflexive Banach space X_{ω_1} with a transfinite Schauder basis of length ω_1 containing no infinite unconditional basic sequence.* □

The space X_{ω_1} is the completion of $c_{00}(\omega_1)$ under the norm given by the formula (3.5.3), where now $\mathcal{F} = \mathcal{F}_{\omega_1}$ is chosen as the minimal set of finitely supported functionals on $c_{00}(\omega_1)$ such that:

1. $\mathcal{F} \supseteq \{e^*_\alpha : \alpha < \omega_1\}$ and $\mathcal{F}$ is symmetric and closed under restriction on intervals of ω_1.
2. For every $(\phi_i)_{i<n_{2k}} \subseteq \mathcal{F}$ such that $\text{supp}(\phi_i) < \text{supp}(\phi_j)$ for $i < j < n_{2k}$, the functional $m_{2k}^{-1}\sum_{i<n_{2k}} \phi_i$ belongs to $\mathcal{F}$.
3. For every *special sequence* $(\phi_i)_{i<n_{2k+1}} \subseteq \mathcal{F}$ the corresponding *special functional* $m_{2k+1}^{-1}\sum_{i<n_{2k+1}} \phi_i$ belongs to $\mathcal{F}$.
4. $\mathcal{F}$ is closed under rational convex combinations.

The notion of a *special sequence of functionals* is defined analogously to that Example 3.5.1 as follows. First of all we say that $\phi \in \mathcal{F}$ is of type 0 if $\phi = \pm e^*_\alpha$ for some α and of type I if it is obtained from a simpler one as the sum $m_k^{-1}\sum_{i<n_k} \phi_i$ restricted to some interval; further, we call the integer m_k, the *weight* of ϕ, and

denote it $w(\phi)$. Note that a given functional of type I can have more than one representation and therefore more than one weight. We say that ϕ is of type II if it is a rational convex combination of functionals of type 0 and type I. The definition of a special sequence of functionals will be again based on ϱ. For this we need an injection

$$\ulcorner . \urcorner : \mathrm{HF} \to \{2k : k \text{ odd } \},$$

such that if $(\phi_i, w_i, p_i)_{i<j}$ is a finite sequence of triples where $w_i, p_i \in \omega$ and where ϕ_i's are finite partial maps from ω into $\mathbb{Q}$, then $\ulcorner(\phi_i, w_i, p_i)_{i<j}\urcorner$ is bigger than the maximum of support of any of the ϕ_i's, the square of any integer of the form w_i or p_i, and the ratio of the square of any nonzero value of a ϕ_i. A sequence $(\phi_i)_{i<n_{2k+1}}$ of members of $\mathcal{F}$ is a *special sequence of functionals*, if

(a) $\mathrm{supp}(\phi_i) < \mathrm{supp}(\phi_j)$ for $i < j < n_{2k+1}$,

(b) each ϕ_i is of type I and of some fixed weight $w(\phi_i) = m_{2k_i}$ with k_0 even and satisfying $m_{2k_0} > n_{2k+1}^2$,

(c) for $0 < j < n_{2k+1}$ there is an increasing sequence $(p_i)_{i<j}$ of integers such that $p_i \geq \varrho(\bigcup_{i'<i} \mathrm{supp}(\phi_{i'}))$ and $w(\phi_j) = m_{\ulcorner(\pi''_E\phi_i, w(\phi_i), p_i)_{i<j}\urcorner}$,[7] where $E = (\bigcup_{i<j} \mathrm{supp}(\phi_i))_\varrho$.

The condition 1 from the definition of $\mathcal{F}$ provides that $(e_\alpha)_{\alpha<\omega_1}$ is a bimonotone Schauder basis of X_{ω_1}, the condition 3 is responsible for the conditional structure of the norm of X_{ω_1} in a similar manner as Example 3.5.1, but in order to come to the point where the idea can be applied, one needs the unconditional structure given by the closure property 2 of $\mathcal{F}$. This requires a rather involved tree-analysis of the functionals of $\mathcal{F}$ that gives the required norm estimates, an analysis that is beyond the scope of this presentation. We refer the reader to [4] for the details. Here we just list some of the most interesting properties of X_{ω_1}. For example, the space X_{ω_1} is reflexive, has no infinite unconditional basic sequence, and is not isomorphic to any proper subspace or a nontrivial quotient. The basis $(e_\alpha)_{\alpha<\omega_1}$ allows us to talk about natural subspaces of X_{ω_1}. For example, given an infinite interval $I \subseteq \omega_1$, one can consider the corresponding restriction

$$X_I = \overline{\mathrm{span}}(\{e_\alpha : \alpha \in I\}).$$

Then X_I are all reflexive Banach spaces with transfinite Schauder basis $(e_\alpha)_{\alpha\in I}$. Particularly interesting are the separable spaces of the form

$$X_\gamma = \overline{\mathrm{span}}(\{e_\alpha : \alpha < \gamma\})$$

for γ a countable limit ordinal. It turns out that, choosing an appropriate function ϱ, all these separable restrictions of X_{ω_1} not only have transfinite Schauder bases but also Schauder bases of length ω.

[7]The collapse $\psi_i = \pi''_E\phi_i$ is defined by the formula $\psi(\pi_E(\alpha)) = \phi(\alpha)$.

Lemma 3.5.6. *Suppose the construction of X_{ω_1} is based on a ϱ-function $\varrho : [\omega_1]^2 \to \omega$. Fix a countable limit ordinal γ. For $n < \omega$, let*

$$F_n^\gamma = (\{\gamma\})_n \cap \gamma = \{\alpha < \gamma : \varrho(\alpha, \gamma) \leq n\},$$

and let

$$P_{F_n^\gamma} : X_\gamma \to X_{F_n^\gamma}$$

be the projection from X_γ onto $X_{F_n^\gamma} = \overline{\text{span}}(\{e_\alpha : \alpha \in F_n^\gamma\})$. Then the sequence of projections $(P_{F_n^\gamma})_n$ is uniformly bounded by $1 + \sup_k |(\{\gamma\})_k|/k$. □

Clearly, there exist ϱ-functions on ω_1 for which the sequence $(\frac{|(\{\gamma\})_k|}{k})$ is bounded for every γ. For example, the function $\bar{\rho}$ of Definition 3.2.1 is one such ϱ-function. Note also that every smooth ϱ-function constructed above in Lemma 3.4.5 also has the sequence $(\frac{|(\{\gamma\})_k|}{k})$ bounded for every countable ordinal γ and smoothness is exactly what is needed to turn the sequence of projections into a Schauder basis for X_γ of length ω.

Corollary 3.5.7. *If the construction of X_{ω_1} is based on a smooth ϱ-function on ω_1, then every subspace of X_{ω_1} of the form X_I, for $I \subseteq \omega_1$ a countable interval, has a Schauder basis of length ω. In fact, a Schauder basis of X_I of length ω is obtained by re-enumerating e_α $(\alpha \in I)$.* □

If we base X_{ω_1} on a universal ϱ we get a rather interesting property that every finite-dimensional subspace Y of X_{ω_1} can be isomorphically embedded into any subspace X_I of X_{ω_1} spanned by an infinite interval of the transfinite Schauder basis $(e_\alpha)_{\alpha<\omega_1}$ by an operator T such that $\| T \| \| T^{-1} \| \leq 4 + \epsilon$.

Definition 3.5.8. A transfinite sequence $(e_\alpha)_{\alpha<\gamma}$ is *nearly subsymmetric* if there is a positive constant C such that for every $\varepsilon > 0$ and every finite block-sequence F_i $(i \leq k)$ of finite subsets of γ and for every sequence $I_0 \leq I_1 \leq \cdots \leq I_k$ of infinite intervals of γ there is a sequence $G_i \subseteq I_i$, $|G_i| = |F_i|$ $(i \leq k)$ such that the natural isomorphism $T : X_F \to X_G$, where $F = \bigcup_{i \leq k} F_i$ and $G = \bigcup_{i \leq k} G_i$ has the property that $\| T \| \| T^{-1} \| \leq C + \varepsilon$.

Theorem 3.5.9. *If X_{ω_1} is based on a universal ϱ-function, then its Schauder basis $(e_\alpha)_{\alpha<\omega_1}$ is nearly subsymmetric with constant* 4. □

Taking the restrictions $X_I = \overline{\text{span}}(\{e_\alpha : \alpha \in I\})$ for an appropriate family of countable infinite intervals $I \subseteq \omega_1$, one obtains an uncountable family of separable reflexive spaces which are 'asymptotic versions' of each other. This is a new phenomenon in the separable Banach-space theory not revealed by the previous methods.

The basis $(e_\alpha)_{\alpha<\omega_1}$ allows us also to talk about diagonal operators of X_{ω_1}. It turns out that all bounded diagonal operators D of X_{ω_1} are given by a sequence $(\lambda_\alpha)_{\alpha<\omega_1}$ of eigenvalues that have the property that

$$\lambda_\alpha = \lambda_\beta \text{ whenever } \alpha \leq \beta < \alpha + \omega.$$

Moreover, there are only countably many different eigenvalues among $(\lambda_\alpha)_{\alpha<\omega_1}$. Recall, that an operator between two Banach spaces is *strictly singular* if it is not an isomorphism when restricted to any infinite-dimensional subspace of its domain. It turns out that every bounded operator T on X_{ω_1} has the form $D_T + S$, where D_T is a diagonal-step operator with countably many eigenvalues and where X is a strictly singular operator. There is a natural candidate for D_T. The problem is to show that it is a bounded operator. The original way of handling this problem is again based on an universal ϱ-function. To see this, suppose we are given a finitely supported normalized vector x such that $D_T(x)$ is very large in comparison with $T(x)$. The vector x has a natural decomposition $x = x_1 + \cdots + x_n$ such that

$$D_T(x) = \lambda_1 x_1 + \cdots + \lambda_n x_n,$$

where $\lambda_1, \ldots, \lambda_n$ are eigenvalues of T. The universality of ϱ via Theorem 3.5.9 guaranties that x_i's can be simultaneously moved by a $(4+\varepsilon)$-isomorphism somewhere very close to eigenvectors with eigenvalues λ_i's. This will give us a contradiction. It follows that, in particular, every bounded linear operator on X_{ω_1} is a multiple of the identity plus the operator with separable range[8]. It turns out that we can actually identify the space $\mathcal{D}(X_{\omega_1})$ of bounded diagonal operators as the dual of the James space $J_{T_0}(\omega_1)$ over the particular reflexive Banach space $T_0 = T_\omega[1/m_{2j}, n_{2j}]$ and ω_1 as the corresponding ordered set. So let us recall the functor $(X, A) \mapsto J_X(A)$ as introduced in [4].

Definition 3.5.10. Let X be a reflexive Banach space with a 1-subsymmetric basis $(e_k)_{k<\omega}$, and let A be a set of ordinals naturally ordered. Then $J_X(A)$ is the completion of $c_{00}(A)$ under the norm given by

$$\| x \|_{J_X(A)} = \sup\left\{ \left\| \sum_{k<n} \Big(\sum_{\xi\in I_k} x(\xi) \Big) e_k \right\|_X : n < \omega,\ I_0 < \cdots < I_{n-1} \text{ intervals of } A \right\}.$$

The natural Hamel basis $(v_\alpha)_{\alpha\in A}$ of $c_{00}(A)$ is a 1-subsymmetric Schauder basis of $J_X(A)$ which is not unconditional since for example l_1 does not embed into $J_X(A)$. Note also that for an interval I of A the functional $I^*(x) = \sum_{\xi\in I} x(\xi)$ belongs to the dual $J_X^*(A)$ and in fact $\| I^* \| = 1$. Moreover, the dual space $J_X^*(A)$ is generated in norm by the family of functionals of the form I^*, where I is an initial interval of A. It follows that the norm dimension of the dual space $J_X^*(A)$ is equal to the cardinality of the set A. It turns out that the following concept is quite useful in describing the phenomena encountered inside the space X_{ω_1}.

Definition 3.5.11. Let X be a Banach space with a Schauder basis $(x_n)_{n<\omega}$ and let Y be a Banach space with a transfinite Schauder basis $(y_\alpha)_{\alpha<\gamma}$. We say that X is *finitely interval representable* in Y if there is a constant $C > 0$ such that for

[8] A considerably easier example of a space with this property will be given in some detail in Theorem 5.3.10 below.

every finite sequence $I_0 \leq I_1 \leq \cdots \leq I_{n-1}$[9] of infinite intervals of γ there exist $z_k \in \operatorname{span}\{y_\alpha : \alpha \in I_k\}$ $(k < n)$ such that the natural isomorphism

$$H : \operatorname{span}(x_k)_{k<n} \mapsto \operatorname{span}(z_k)_{k<n}$$

has the property that $\| H \| \| H^{-1} \| \leq C$.

Remark 3.5.12. The possible relevance of this concept becomes more clear if one recalls that in Example 3.5.1 above we have really proved that the James-like space $J_{c_0}(\omega)$ is finitely representable in an arbitrary infinite subsequence of the weakly null sequence $(e_\alpha)_{\alpha<\omega_1}$.

The following is the main technical result about the concept of finite interval representability inside our space X_{ω_1}.

Theorem 3.5.13. *The James space J_{T_0}*[10] *is finitely interval representable inside an arbitrary closed linear span of a transfinite normalized block sequence $(y_\alpha)_{\alpha<\gamma}$ of X_{ω_1}.* □

Corollary 3.5.14. *The Banach space X_{ω_1} contains no infinite basic unconditional sequence.* □

Definition 3.5.15. A Banach space X is *indecomposable* if for every topological[11] decomposition $X = Y \oplus Z$ one of the spaces Y of Z must be finite-dimensional. We say that X is *hereditarily indecomposable* if every closed subspace of X is indecomposable.

Corollary 3.5.16. *Every infinite-dimensional closed subspace of X_{ω_1} contains an infinite-dimensional hereditarily indecomposable subspace.* □

Definition 3.5.17. A bounded operator $S : X \mapsto Y$ is *strictly singular* if $S \restriction Z$ is not an isomorphic embedding for every infinite-dimensional subspace Z of X. Let $\mathcal{S}(X,Y)$ denote the collection of all bounded strictly singular operators from X into Y.

The following result characterizes the operator space of an arbitrary closed subspace X of X_{ω_1} spanned by a transfinite normalized block sequence $(x_\alpha)_{\alpha<\gamma}$ of vectors.

Theorem 3.5.18. $\mathcal{L}(X)/\mathcal{S}(X) \cong \mathcal{L}(X, X_{\omega_1})/\mathcal{S}(X, X_{\omega_1}) \cong J^*_{T_0}(\Lambda^0(\gamma))$ *for every closed subspace X of X_{ω_1} generated by a transfinite basic block sequence $(x_\alpha)_{\alpha<\gamma}$, where $\Lambda^0(\gamma)$ denote the set of all limit ordinals $\lambda \leq \gamma$ that have the form $\beta + \omega$ for some ordinal $\beta < \lambda$.* □

Corollary 3.5.19. $\mathcal{L}(X_{\omega_1})/\mathcal{S}(X_{\omega_1}) \cong J^*_{T_0}(\omega_1)$. □

[9] For two intervals of ordinals I and J, we let $I \leq J$ if either $I = J$ or else every ordinal from I is smaller than every ordinal from J

[10] Here, $J_{T_0} = J_{T_0}(\omega)$ for $T_0 = T[1/m_{2j}, n_{2j}]$, where (m_j) and (n_j) are the sequences of positive integers as in (3.5.5).

[11] I.e., $Y \cup Z$ spans X and $d(S_Y, S_Z) > 0$.

Corollary 3.5.20. *No closed subspace X of X_{ω_1} generated by a transfinite basic block sequence is isomorphic to its proper subspace and no such a subspace X of X_{ω_1} can have a nontrivial quotient.* □

Corollary 3.5.21. *X_{ω_1} is a reflexive Banach space that is not isomorphic to any of its proper subspaces and X_{ω_1} has no nontrivial quotients.* □

In fact one has a more informative result that lies behind Theorem 3.5.18. To see this fix a bounded operator $T : X \mapsto X_{\omega_1}$, where X is a closed linear span of some transfinite normalized block sequence $(x_\alpha)_{\alpha<\omega_1}$ of vectors of X_{ω_1}. Recall that $\Lambda(\gamma)$ is the set of limit ordinals $\leq \gamma$ and that $\Lambda^0(\gamma)$ is the set of isolated members of $\Lambda(\gamma)$. Then to the given operator T we associate the function

$$\xi_T : \Lambda^0(\gamma) \to \mathbb{R}$$

by letting $\xi_T(\beta) = \xi$ if $\lim_{n\to\infty} \| T(y_n) - \xi y_n \| = 0$ for every $(3, \varepsilon)$-rapidly increasing sequence block subsequence $(y_n)_{n<\omega}$ of $(x_\alpha)_{\alpha<\omega_1}$ whose supports relative to $(x_\alpha)_{\alpha<\omega_1}$ increasingly converge to β. Here a $(3, \varepsilon)$-*rapidly increasing sequence* is what lies behind Schlumprecht's idea of recovering a biorthogonal block sequence of vectors and functionals inside an arbitrary infinite block sequence of vectors in spaces that incorporated his norming set of functionals, which in our case is captured by the property 2 of $\mathcal{F}_{\omega_1}$. Now, given an *arbitrary* mapping $\xi : \Lambda^0(\gamma) \to \mathbb{R}$, define the corresponding diagonal operator $D_\xi : X \to X$ by letting

$$D_\xi(x_\alpha) = \xi(\alpha + \omega)x_\alpha.$$

For a bounded operator $X \mapsto X_{\omega_1}$ let $D_T : X \to X_{\omega_1}$ be defined as $D_T = D_{\xi_T}$. The following, whose proof is based on Theorem 3.5.13, gives more precise information about operator spaces than Theorem 3.5.18.

Theorem 3.5.22. *Suppose X is a closed linear span of a transfinite normalized block sequence of vectors of X_{ω_1}. Then for every bounded operator $T : X \to X_{\omega_1}$, the corresponding diagonal operator $D_T : X \to X_{\omega_1}$ is bounded as well and the difference $T - D_T$ is strictly singular.* □

While Theorem 3.5.18 gives us a fairly complete description of operator spaces of closed linear spans of transfinite basic block sequences of vectors of X_{ω_1}, one can construct subspaces of X_{ω_1} that reveal some quite different and interesting new phenomena about operator spaces such as the following.

Theorem 3.5.23. *There is a closed subspace Y of X_{ω_1} such that every bounded operator $T : Y \to Y$ is a strictly singular perturbation of a scalar multiple of identity, while operator space $\mathcal{L}(Y, X_{\omega_1})$ is quite rich in the sense that*

$$\mathcal{L}(Y, X_{\omega_1}) = J^*_{T_0} \oplus \mathcal{S}(Y, X_{\omega_1}).$$ □

The method of adding the conditional structure to norms using ϱ-functions is actually quite flexible. To this effect, we give yet another construction of a conditional norm using ϱ. It will utilize the following notion.

Definition 3.5.24. Fix a ϱ-function $\varrho : [\omega_1]^2 \to \omega$. For two subsets s and t of ω_1, set

$$\varrho(s,t) = \min\{\varrho(\alpha,\beta) \,:\, \alpha \in s,\ \beta \in t\}.$$

Given an integer k we say that s and t are *k-separated* if $\varrho(s,t) \geq k$. A sequence $(s_i)_i$ is k-separated if it is pairwise k-separated, i.e., s_i and s_j are k-separated for every $i \neq j$. The sequence $(s_i)_i$ is *separated* if it is $|\bigcup_i s_i|$-separated. These notions are naturally transferred to vectors of $X^0_{\omega_1}$ via their supports.

The following results show that separated sequences appear in profusion.

Lemma 3.5.25. *Let $(A_i)_{i<n}$ be a block sequence of subsets of ω_1, each of them of order-type ω. Then for every block sequence $(F_\alpha)_{\alpha\in\bigcup_{i<n}A_i}$ of finite sets of countable ordinals and every integer k, there are $\alpha_i \in A_i$ $(i<n)$ such that $(F_{\alpha_i})_{i<n}$ is k-separated.*

Proof. This is done by induction on n. Fix all data as in the statement for $n>1$. Then let $\alpha_{n-1} = \min A_{n-1}$. Since the set

$$(F_{\alpha_{n-1}})_k = \{\beta < \omega_1 \,:\, \varrho(\beta,\gamma) \leq k \text{ for some } \gamma \in (F_{\alpha_{n-1}} \setminus \beta)\}$$

is (by property (iii) of ϱ) finite, one easily obtains infinite sets

$$B_i \subseteq A_i \ (i < n-1)$$

such that $F_\alpha \cap (F_{\alpha_{n-1}})_k = \emptyset$ for every $\alpha \in B_i$ and all $i < n-1$. By inductive hypothesis, there are $\alpha_i \in B_i$ $(i<n-1)$ such that $(F_{\alpha_i})_{i<n-1}$ is k-separated. Then $(F_{\alpha_i})_{i<n}$ is the desired k-separated sequence. □

Corollary 3.5.26. *Let n be an integer and let $(F^i_\alpha)_{\alpha<\omega_1}$ be a block sequence of finite sets of countable ordinals for every $i<n$. Then there are $\alpha_0 < \cdots < \alpha_{n-1}$ such that $(F^i_{\alpha_i})_{i<n}$ is a separated block sequence.*

Proof. Let A be an uncountable set such that $|F^i_\alpha| = k_i$ for every $i \in A$ and every $i<n$. Now for each $i<n$, let $A_i \subseteq A$ be of order type ω and such that $A_i < A_j$ and $F^i_\alpha < F^j_\beta$ if $i<j<n$ and $\alpha \in A_i$, $\beta \in A_j$ are such that $\alpha < \beta$. Then apply the previous proposition to $(E_\alpha)_{\alpha\in A_0\cup\cdots\cup A_{n-1}}$ and $\sum_{i<n} k_i$, where $E_\alpha = F^i_\alpha$ for the unique $i<n$ such that $\alpha \in A_i$. □

We are now ready to define the new conditional norm on $c_{00}(\omega_1)$ for which the corresponding space behaves quite differently than the space X_{ω_1} considered above.

Definition 3.5.27. Let K be the minimal subset of $c_{00}(\omega_1)$ satisfying the following conditions:

(i) $e^*_\gamma \in K$ for all $\gamma < \omega_1$, K is *symmetric* (i.e., $\phi \in K$ implies $-\phi \in K$), and K is closed under the restriction on intervals of ω_1.

(ii) For every *separated* block sequence $(\phi_i)_{i=1}^d \subseteq K$, with $d \leq n_{2j}$, one has that the combination $(1/m_{2j})\sum_{i=1}^{n_{2j}} \phi_i \in K$.

(iii) For every separated *special* sequence $(\phi_i)_{i=1}^d \subseteq K$ with $d \leq n_{2j+1}$ one has that $\phi = (1/m_{2j+1}) \sum_{i-1}^{n_{2j+1}} \phi_i$ is in K. The functional ϕ is called a *special functional.*

(iv) K is rationally convex.

Whenever $\phi \in K$ is of the form $\phi = (1/m_j) \sum_{i<d} \phi_i$ given in (ii) or (iii) above we say that ϕ has a *weight* $w(\phi) = m_j$. Finally, the norm on $c_{00}(\omega_1)$ is defined as

$$\| x \| = \sup\{\phi(x) = \Sigma_{\alpha<\omega_1} \phi(\alpha) \cdot x(\alpha) : \phi \in K\},$$

and we let $X_{\omega_1}^0$ be the completion of the normed space $(c_{00}, \| \cdot \|)$.

Let us list some properties of $X_{\omega_1}^0$ that follow directly from this definition.

Remark 3.5.28.

(a) It is clear that the norming set K presented here is a subset of the norming set of the Banach space X_{ω_1} considered above. So, for $x \in c_{00}(\omega_1)$, the norm of x in the space $X_{\omega_1}^0$ is not bigger than the norm of x in X_{ω_1}.

(b) There is a natural notion of *complexity* of elements ϕ of K: Either $\phi = \pm e_\alpha^*$, or ϕ is a rational convex combination $\phi = \sum_{i<k} r_i f_i$ of elements $(f_i)_{i<k}$ of K, or $\phi = (1/m_j) \sum_{i<d} f_i$ for a separated block sequence $(f_i)_{i<d}$ in K with $d \leq n_j$. And in this latter case we say that $w(\phi) = m_j$ is a *weight* of ϕ.

(c) The property (i) makes the natural Hamel basis $(e_\alpha)_{\alpha<\omega_1}$ of $c_{00}(\omega_1)$ a *transfinite bimonotone Schauder basis* of $X_{\omega_1}^0$, or in other words, $X_{\omega_1}^0$ is a closed linear span of $(e_\alpha)_{\alpha<\omega_1}$, and for every interval $I \subseteq \omega_1$ the corresponding projection

$$P_I : X_{\omega_1}^0 \to X_I^0 = \overline{\text{span}}(\{e_\alpha : \alpha \in I\})$$

has norm 1. Let us set $P_\gamma = P_{[0,\gamma]}$ and $X_\gamma^0 = X_{[0,\gamma]}^0$ for $\gamma < \omega_1$.

(d) The basis $(e_\alpha)_{\alpha<\omega_1}$ is *shrinking* in the sense that the subbasis $(e_{\alpha_n})_{n<\omega}$ is shrinking in the usual sense[12] for every increasing sequence $(\alpha_n)_{n<\omega}$ of countable ordinals. It follows that $(e_\alpha)_{\alpha<\omega_1}$ is an uncountable weakly-null sequence, i.e., for every $x^* \in (X_{\omega_1}^0)^*$ the numerical sequence $(x^*(e_\alpha))_{\alpha<\omega_1}$ belongs to $c_0(\omega_1)$. This last property readily implies that if $T : X_{\omega_1}^0 \to X_{\omega_1}^0$ is a bounded operator, then for every uncountable subset A of ω_1 and every countable ordinal γ one has that $P_\gamma(T(e_\alpha)) = 0$ for all but countably many $\alpha \in A$.

The property (ii) of the norming set is responsible for the existence of so-called semi-normalized averages in the span of every uncountable block sequence of $X_{\omega_1}^0$. This in combination with the third (iii) and fourth (iv) properties of the norming set K is behind a rather strong control on operators that is expressible using the following new concept.

[12] I.e., the sequence $\{e_{\alpha_n}^* : n < \omega\}$ of biorthogonal functionals is the Schauder basis of the dual space $(\overline{\text{span}}(\{e_{\alpha_n} : n < \omega\})^*$.

Definition 3.5.29. An operator $T : Y \to X$ is an ω_1-*singular operator* if it has the property that $T \restriction Z$ is not an isomorphic embedding of Z into X for every nonseparable closed subspace Z of Y.

Remark 3.5.30. Observe that strictly singular and separable range operators are ω_1-singular. While the strictly singular operators and operators with separable ranges form closed ideals of the Banach algebra $\mathcal{L}(X)$ of all bounded operators from X into X, it is not clear if for general Banach spaces X this is also the case for the family $\mathcal{S}_{\omega_1}(X)$ of ω_1-singular operators of X. Indeed it is not even clear if $\mathcal{S}_{\omega_1}(X)$ is closed under sums. We shall show however that for subspaces X of our space $X^0_{\omega_1}$ one does have that the $\mathcal{S}_{\omega_1}(X)$ form a closed ideal in the algebra $\mathcal{L}(X)$.

Theorem 3.5.31. *Every operator $T : Y \to X^0_{\omega_1}$ from a closed subspace Y of $X^0_{\omega_1}$ into $X^0_{\omega_1}$ has the form*

$$T = \lambda I_{Y,X^0_{\omega_1}} + S,$$

where λ is a scalar, $I_{Y,X^0_{\omega_1}}$ is the inclusion operator from Y into $X^0_{\omega_1}$ and where S is an ω_1-singular operator. □

Here are some of the other main properties of the space $X^0_{\omega_1}$.

Theorem 3.5.32. *The space $X^0_{\omega_1}$ is* c_0-saturated *in the sense that every infinite-dimensional closed subspace of $X^0_{\omega_1}$ contains an isomorphic copy of c_0.* □

Corollary 3.5.33. *The space $X^0_{\omega_1}$ is not* ω-distortable, *or more precisely, for every $\varepsilon > 0$ and every equivalent norm $\| \cdot \|'$ on $X^0_{\omega_1}$ there is an infinite-dimensional subspace Y of $X^0_{\omega_1}$ such that $\| x \|' \leq (1+\varepsilon) \| y \|'$ for every $x, y \in Y$ with $\| x \| = \| y \| = 1$.* □

Theorem 3.5.34. *The space $X^0_{\omega_1}$ is* arbitrarily ω_1-distortable, *or more precisely, for every positive real number M there is an equivalent norm $\| \cdot \|'$ on $X^0_{\omega_1}$ such that for every* non-separable *closed subspace Y of $X^0_{\omega_1}$ there exist $x, y \in Y$ such that $\| x \| = \| y \| = 1$ but $\| x \|' \geq M \| y \|'$.* □

Note that by Theorem 3.5.32, every infinite-dimensional closed subspace of $X^0_{\omega_1}$ contains an *infinite* basic sequence. However, on the non-separable level, we have a quite different behavior as the following result shows.

Theorem 3.5.35. *The space $X^0_{\omega_1}$ contains no uncountable unconditional basic sequence.* □

This can be made more precise using the following natural variation of the standard notion of decomposability.

Definition 3.5.36. A Banach space X is ω_1-*indecomposable* if for every topological decomposition $X = Y \oplus Z$ one of the spaces Y of Z must be separable. We say that X is *hereditarily ω_1-indecomposable* if every closed subspace of X is ω_1-indecomposable.

This kind of indecomposability corresponds to the notion of ω_1-singularity for operators introduced above. Using this notion one can have the following sufficient condition for being ω_1-hereditarily indecomposable.

Lemma 3.5.37. *Suppose that X has the property that for every subspace Y of X every bounded operator $T : Y \to X$ is of the form $T = \lambda I_{Y,X} + S$ for some scalar λ and a ω_1-singular operator S. Then X is ω_1-hereditarily indecomposable.*

Proof. Otherwise, fix two non-separable subspaces Y and Z of X whose union spans X and which have the property that

$$d(S_Y, S_Z) > 0.$$

It follows that the two natural projections $P_Y : Y \oplus Z \to Y$ and $P_Z : Y \oplus Z \to Z$ are both bounded. Fix a real number λ such that

$$T = I_{Y,X} \circ P_Y = \lambda I_{Y \oplus Z,X} + S$$

with S an ω_1-singular operator. Since $T^2 = T$, we have that

$$(\lambda^2 - \lambda) I_{Y \oplus Z,X} = ((1 - 2\lambda) I_{Y \oplus Z,X} - S) \circ S. \tag{3.5.6}$$

Since it is clear that $U \circ S$ is ω_1-singular if S is ω_1-singular, it follows from (3.5.6) that $\lambda^2 = \lambda$. Without loss of generality, we may assume that $\lambda = 1$ (if $\lambda = 0$ we replace Y by Z in the preceding argument). Since $P_Y \restriction Z = 0$, we obtain that $S = -I_{Z,X}$, a contradiction. □

This leads us to the following more precise version of Theorem 3.5.35.

Theorem 3.5.38. *The space $X^0_{\omega_1}$ is ω_1-hereditarily indecomposable.* □

Remark 3.5.39. Recall that V. Ferenczi [34] has shown that if a complex Banach spaces X is hereditarily indecomposable, then every operator from a subspace Y of X into X is a scalar multiple of the inclusion operator plus a strictly singular operator. We do not know if the analogous result is true for ω_1-singular operators or, in other words, if the converse implication of Lemma 3.5.37 is true in the case of complex Banach spaces.

The reader is referred to [4] and [71] for more information about this new method of using ϱ-functions in order to impose conditional structure on norms of Banach spaces under construction. The following question now naturally suggests itself.

Question 3.5.40. What is the minimal cardinal θ_u with property that every weakly null sequence of length θ_u contains an infinite unconditional subsequence?

We have seen above that $\theta_u \geq \omega_2$. From a result of Ketonen [58] we infer that θ_u is not bigger than the first Ramsey cardinal. Not much more about θ_u seems to be known.

Chapter 4

Coherent Mappings and Trees

4.1 Coherent mappings

In this and the following section we present a general study of the following notion already encountered at several places above.

Definition 4.1.1. A mapping $a : [\omega_1]^2 \longrightarrow \omega$ is *coherent* if for every $\alpha < \beta < \omega_1$ there exist only finitely many $\xi < \alpha$ such that $a(\xi, \alpha) \neq a(\xi, \beta)$, or in other words, $a_\alpha =^* a_\beta \restriction \alpha$[1]. We say that a is *nontrivial* if there is no $h : \omega_1 \longrightarrow \omega$ such that $h \restriction \alpha =^* a_\alpha$ for all $\alpha < \omega_1$.

Examples of coherent nontrivial mappings encountered so far are the mappings ρ_1, ρ_3, and ρ. The general theory of coherent mappings is sometimes conveniently identified with the theory of corresponding trees defined as follows.

Definition 4.1.2. To every coherent mapping $a : [\omega_1]^2 \longrightarrow \omega$ one associates the tree

$$T(a) = \{a_\beta \restriction \alpha : \ \alpha \leq \beta < \omega_1\}$$

and its uniform closure

$$T^*(a) = \{t : \alpha \longrightarrow \omega : \ \alpha < \omega_1 \text{ and } t =^* a_\alpha\}.$$

If the coherent mapping is given as $a : [\omega_1]^2 \longrightarrow k$ where k is some specified positive integer, then the corresponding uniform version of the associated tree is defines as

$$T^*(a) = \{t : \alpha \longrightarrow k : \ \alpha < \omega_1 \text{ and } t =^* a_\alpha\}.$$

An arbitrary tree of height ω_1 is said to be a *coherent tree* if it can be represented as a downward closed subtree of some tree of the form $T(a)$, where $a : [\omega_1]^2 \longrightarrow \omega$ is a coherent mapping.

[1] A mapping $a : [\omega_1]^2 \longrightarrow \omega$ is naturally identified with a sequence a_α $(\alpha < \omega_1)$, where $a_\alpha : \alpha \longrightarrow \omega$ is defined by $a_\alpha(\xi) = a(\xi, \alpha)$.

Clearly, the tree $T(a)$ corresponding to a coherent and nontrivial mapping $a : [\omega_1]^2 \longrightarrow \omega$ is an *Aronszajn tree* (in short, *A-tree*). Note also that the existence of a coherent and nontrivial $a : [\omega_1]^2 \longrightarrow 2$ (such as, for example, the function ρ_3 defined above) is something that corresponds also to the notion of a *Hausdorff gap* in this context. The theory of coherent mappings and coherent trees can therefore be considered as a unified approach towards these two important classical concepts.[2]

Definition 4.1.3. The *support* of a map $a : [\omega_1]^2 \longrightarrow \omega$ is the sequence

$$\text{supp}(a_\alpha) = \{\xi < \alpha : a(\xi, \alpha) \neq 0\} \quad (\alpha < \omega_1)$$

of subsets of ω_1. A set Γ is *orthogonal* to a if $\text{supp}(a_\alpha) \cap \Gamma$ is finite for all $\alpha < \omega_1$. We say that $a : [\omega_1]^2 \longrightarrow \omega$ is *nowhere dense* if there is no uncountable $\Gamma \subseteq \omega_1$ such that $\Gamma \cap \alpha \subseteq^* \text{supp}(a_\alpha)$ for all $\alpha < \omega_1$.

Note that ρ_3 is an example of a nowhere dense coherent map for the simple reason that ω_1 can be covered by countably many sets $\Lambda + n$ $(n < \omega)$ that are orthogonal to ρ_3. Recall, that here Λ denotes the set of all countable limit ordinals including 0 and that $\Lambda + n$ denote its translate $\{\lambda + n : \lambda \in \Lambda\}$. The following immediate fact shows that ρ_3 is indeed a prototype of a nowhere dense and coherent map $a : [\omega_1]^2 \longrightarrow \omega$.

Proposition 4.1.4. *Assuming the P-ideal dichotomy, for every nowhere dense and coherent map $a : [\omega_1]^2 \longrightarrow \omega$ the domain ω_1 can be decomposed into countably many sets orthogonal to a.* □

The nowhere dense mappings like ρ_3 occupy a special place in the theory of coherent mappings and one of their special properties is revealed by considering the following notion.

Definition 4.1.5. The *shift* of $a : [\omega_1]^2 \longrightarrow \omega$ is defined to be the mapping

$$a^{(1)} : [\omega_1]^2 \longrightarrow \omega$$

determined by

$$a^{(1)}(\alpha, \beta) = a(\alpha + 1, \hat{\beta}),$$

where $\hat{\beta} = \min\{\lambda \in \Lambda : \lambda \geq \beta\}$. The n-fold iteration of the shift operation is defined recursively by the formula $a^{(n+1)} = (a^{(n)})^{(1)}$.

Theorem 4.1.6. *If a is nontrivial, coherent and orthogonal to Λ, then it holds that $T(a) > T(a^{(1)})$*[3].

---

[2]Similarities between the notion of a Hausdorff gap and the notion of an A-tree have been further explained recently in the two papers of Talayco ([102], [103]), where it is shown that they naturally correspond to first cohomology groups over a pair of very similar spaces.

[3]For two trees S and T, by $S \leq T$ we denote the fact that there is a strictly increasing map $f : S \longrightarrow T$. Let $S < T$ whenever $S \leq T$ and $T \nleq S$ and let $S \equiv T$ whenever $S \leq T$ and $T \leq S$.

Proof. By the assumption $a \perp \Lambda$, the subtree S of $T(a)$ consisting of all t's that take the value 0 at every limit ordinal in their domains is an uncountable initial part of $T(a)$. Note that every node of $T(a)$ is equal to the restriction of some finite change of a node from S. For a node t at some limit level λ of S, let $t^{(1)} : \lambda \longrightarrow \omega$ be determined by the formula $t^{(1)}(\alpha) = t(\alpha+1)$. Let $S^{(1)} = \{t^{(1)} : t \in S \restriction \lambda\}$. Every node of $T(a^{(1)})$ is equal to the restriction of some finite change of a node of $S^{(1)}$. Note also that $t \longmapsto t^{(1)}$ is a one-to-one map on $S \restriction \Lambda$ and that its inverse naturally extends to a strictly increasing map from $T(a^{(1)})$ into T. This shows that $T(a^{(1)}) \leq T(a)$.

It remains to establish $T(a) \not\leq T(a^{(1)})$. Note that if there is a strictly increasing $g : T(a) \longrightarrow T(a^{(1)})$, then there is one that is moreover level-preserving. For example $f(t) = g(t) \restriction |t|$ is one such map. So let $f : T(a) \longrightarrow T(a^{(1)})$ be a given level-preserving map. For a countable limit ordinal ξ, set

$$F_\xi = (\mathrm{supp}(a_\xi) \cap \Lambda) \cup \{\alpha < \xi : a_\xi^{(1)}(\alpha) \neq f(a_\xi)(\alpha)\}. \tag{4.1.1}$$

Then F_ξ is a finite set of ordinals $< \xi$, so by the Pressing Down Lemma there is a stationary set $\Gamma \subseteq \Lambda$ and $F \subseteq \omega_1$ such that $F_\xi = F$ for all $\xi \in \Gamma$. Let $\alpha_0 = \max(F) + 1$. Shrinking Γ if necessary, assume that for some $s, t : \alpha_0 \longrightarrow 2$,

$$a_\xi \restriction \alpha_0 = s \text{ and } f(a_\xi) \restriction \alpha_0 = t \text{ for all } \xi \in \Gamma. \tag{4.1.2}$$

Choose $\xi < \eta$ in Γ such that $a_\xi \neq a_\eta \restriction \xi$ and let $\delta = \Delta(a_\xi, a_\eta)(= \min\{\alpha < \xi : a_\xi(\alpha) \neq a_\eta(\alpha)\})$. Then by (4.1.1) and (4.1.2), δ is not a limit ordinal and

$$\delta - 1 = \Delta(a_\xi^{(1)}, a_\eta^{(1)}) = \Delta(f(a_\xi), f(a_\eta)). \tag{4.1.3}$$

It follows that a_ξ and a_η split in $T(a)$ after their images $f(a_\xi)$ and $f(a_\eta)$ split in $T(a^{(1)})$, which cannot happen if f is strictly increasing. This finishes the proof. □

Corollary 4.1.7. *If a is nontrivial, coherent and orthogonal to $\Lambda + n$ for all $n < \omega$, then $T(a^{(n)}) > T(a^{(m)})$ whenever $n < m < \omega$.*

Proof. Note that if a is orthogonal to $\Lambda + n$ for all $n < \omega$, then so is every one of its finite shifts $a^{(m)}$. □

Corollary 4.1.8. $T(\rho_3^{(n)}) > T(\rho_3^{(m)})$ *whenever* $n < m < \omega$. □

Remark 4.1.9. It follows that coherent trees are not well-quasi-ordered under any quasi-ordering that is stronger than $\leq$, and so in particular under the ordering $\leq_1$[4] considered by Laver [68] when he proved that the class of σ-scattered trees is w.q.o. under $\leq_1$. This in particular shows that the result of [68] does not extend to any larger class of trees that would include the class of coherent trees.

---

[4]Recall that $S \leq_1 T$ if there is a strictly increasing map $f : S \to T$ with the property that $f(s \wedge t) = f(s) \wedge f(t)$ for all $s, t \in S$.

We now turn our attention to a metric theory of coherent mappings that will give us an interesting Σ_1-definition of an ultrafilter on ω_1. For this we need a piece of notation.

Notation 4.1.10. To every $a : [\omega_1]^2 \longrightarrow \omega$ we associate the corresponding Δ-function $\Delta_a : [\omega_1]^2 \longrightarrow \omega$ as follows:

$$\Delta_a(\alpha, \beta) = \min\{\xi < \alpha : a(\xi, \alpha) \neq a(\xi, \beta)\}$$

with the convention that $\Delta_a(\alpha, \beta) = \alpha$ whenever $a(\xi, \alpha) = a(\xi, \beta)$ for all $\xi < \alpha$. Given this notation, it is natural to let

$$\Delta_a(\Gamma) = \{\Delta_a(\alpha, \beta) : \alpha, \beta \in \Gamma, \alpha < \beta\}$$

for an arbitrary set $\Gamma \subseteq \omega_1$.

Lemma 4.1.11. *Suppose that a is nontrivial and coherent and that every uncountable subset of $T(a)$ contains an uncountable antichain. Then for every pair Σ and Ω of uncountable subsets of ω_1 there exists an uncountable subset Γ of ω_1 such that $\Delta_a(\Gamma) \subseteq \Delta_a(\Sigma) \cap \Delta_a(\Omega)$.*

Proof. For each limit ordinal $\xi < \omega_1$ we fix $\alpha(\xi) \in \Sigma$ and $\beta(\xi) \in \Omega$ above ξ and let

$$F_\xi = \{\eta < \xi : a(\eta, \alpha(\xi)) \neq a(\eta, \xi) \text{ or } a(\eta, \beta(\xi)) \neq a(\eta, \xi)\}.$$

By the Pressing Down Lemma there is a stationary set $\Gamma \subseteq \omega_1$ and a finite set F such that $F_\xi = F$ for all $\xi \in \Gamma$. Let $\eta_0 = \max(F) + 1$. By the assumption about $T(a)$, find an uncountable $\Gamma_0 \subseteq \Gamma$ such that a_ξ $(\xi \in \Gamma_0)$ is an antichain of $T(a)$. Moreover, we may assume that for some s, t and u in $T(a)$:

$$a_{\alpha(\xi)} \restriction \eta_0 = s, a_{\beta(\xi)} \restriction \eta_0 = t \text{ and } a_\xi \restriction \eta_0 = u \text{ for all } \xi \in \Gamma_0. \tag{4.1.4}$$

It follows that for all $\xi < \eta$ in Γ_0:

$$\Delta_a(\alpha(\xi), \alpha(\eta)) = \Delta_a(\xi, \eta) = \Delta_a(\beta(\xi), \beta(\eta)). \tag{4.1.5}$$

So in particular, $\Delta_a(\Gamma_0) \subseteq \Delta_a(\Sigma) \cap \Delta_a(\Omega)$. □

Notation 4.1.12. For $a : [\omega_1]^2 \longrightarrow \omega$, set

$$\mathcal{U}(a) = \{A \subseteq \omega_1 : A \supseteq \Delta_a(\Gamma) \text{ for some uncountable } \Gamma \subseteq \omega_1\}.$$

Note that Lemma 4.1.11 is saying that $\mathcal{U}(a)$ is a uniform filter on ω_1 for every nontrivial coherent $a : [\omega_1]^2 \longrightarrow \omega$ for which $T(a)$ contains no Souslin subtrees. It turns out that under some very mild assumption, $\mathcal{U}(a)$ is in fact a uniform ultrafilter on ω_1.

Theorem 4.1.13. *Under MA_{ω_1}, the filter $\mathcal{U}(a)$ is an ultrafilter for every nontrivial and coherent $a : [\omega_1]^2 \longrightarrow \omega$.*

Proof. Let A be a given subset of ω_1 and let $\mathcal{P}$ be the collection of all finite subsets p of ω_1 such that $\Delta_a(p) \subseteq A$. If $\mathcal{P}$ is a ccc poset, then an application of MA_{ω_1} would give us an uncountable $\Gamma \subseteq \omega_1$ such that $\Delta_a(\Gamma) \subseteq A$, showing thus that A belongs to $\mathcal{U}(a)$. So let us consider the alternative when $\mathcal{P}$ is not a ccc poset. Let $\mathcal{X}$ be an uncountable subset of $\mathcal{P}$ consisting of pairwise incompatible conditions of $\mathcal{P}$. Going to an uncountable Δ-subsystem of $\mathcal{X}$ and noticing that the root does not contribute to the incompatibility, one sees that without loss of generality, we may assume that the members of $\mathcal{X}$ are in fact pairwise-disjoint. Thus for each limit ordinal $\xi < \omega_1$ we can fix p_ξ in $\mathcal{X}$ lying entirely above ξ. For $\xi < \omega_1$, let

$$F_\xi = \{\eta < \xi : a(\eta, \xi) \neq a(\eta, \alpha) \text{ for some } \alpha \in p_\xi\}.$$

By the Pressing Down Lemma find a stationary set $\Gamma \subseteq \omega_1$, a finite set $F \subseteq \omega_1$, an integer n and $t_0, \ldots, t_{n-1}$ in $T(a)$ of height $\eta_0 = \max(F) + 1$ such that for all $\xi \in \Gamma$:

$$F_\xi = F, \tag{4.1.6}$$

$$a_\alpha \restriction \eta_0 = t_i \text{ with } \alpha \text{ the } i\text{th member of } p_\xi \text{ for some } i < n. \tag{4.1.7}$$

By MA_{ω_1} the tree $T(a)$ is special, so going to an uncountable subset of Γ we may assume that a_ξ $(\xi \in \Gamma)$ is an antichain of $T(a)$, and moreover that for every $\xi < \eta$ in Γ:

$$a_\alpha \restriction \xi \nsubseteq a_\beta \restriction \eta \text{ for all } \alpha \in p_\xi \text{ and } \beta \in p_\eta. \tag{4.1.8}$$

It follows that for every $\xi < \eta$ in Γ:

$$\Delta_a(\alpha, \beta) = \Delta_a(\xi, \eta) \text{ for all } \alpha \in p_\xi \text{ and } \beta \in p_\eta. \tag{4.1.9}$$

By our assumption that conditions from $\mathcal{X}$ are incompatible in $\mathcal{P}$, we conclude that for every $\xi < \eta$ in Γ, there exist $\alpha \in p_\xi$ and $\beta \in p_\eta$ such that $\Delta_a(\alpha, \beta) \notin A$. So from (4.1.9) we conclude that $\Delta_a(\Gamma) \cap A = \emptyset$, showing thus that the complement of A belongs to $\mathcal{U}(a)$. □

Remark 4.1.14. One may find Theorem 4.1.13 a bit surprising in view of the fact that it gives us an ultrafilter $\mathcal{U}(a)$ on ω_1 that is Σ_1-definable over the structure $(H_{\omega_2}, \in)$. It is well known that there is no ultrafilter on ω that is Σ_1-definable over the structure $(H_{\omega_1}, \in)$.

4.2 Lipschitz property of coherent trees

Given a tree[5] T, let $\Delta : T^2 \longrightarrow \mathrm{Ord}$ be defined by

$$\Delta(s, t) = \mathrm{otp}\{x \in T : x <_T s \text{ and } x <_T t\}.$$

[5]The notion of a *tree* is to be interpreted in its order-theoretic sense, i.e., a partially ordered set $(T, \leq_T)$ with the property that the predecessors of every point form a well-ordered chain. The order-type of $\{x \in T : x <_T t\}$ is called the height of t in T and denoted by $\mathrm{ht}(t)$. The αth level of T is the set $T_\alpha = \{t \in T : \mathrm{ht}(t) = \alpha\}$. If $\alpha \leq \mathrm{ht}(t)$, then $t \restriction \alpha$ denotes the $s \leq_T t$ such that $\mathrm{ht}(s) = \alpha$. We make the implicit assumption that different nodes of the same level of T have different sets of predecessors. This allows us to define $s \wedge t = \max\{x \in T : x \leq s \text{ and } x \leq t\}$.

One should view Δ as some sort of distance function on T by interpreting inequalities like $\Delta(x, y) > \Delta(x, z)$ as saying that x is closer to y than to z.

Definition 4.2.1. A partial map g from a tree S into a tree T is *Lipschitz*, if g is level-preserving and

$$\Delta(g(x), g(y)) \geq \Delta(x, y)$$

for all $x, y \in \mathrm{dom}(g)$.

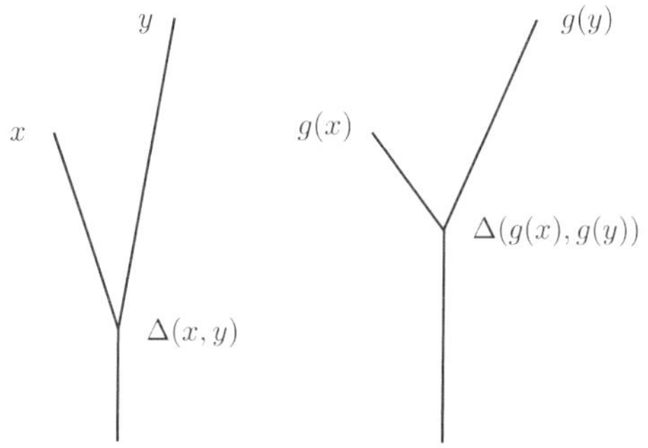

Figure 4.1: Lipschitz map on a tree.

Note that this notion is equivalent to the notion of a strictly increasing level-preserving map, when the domain of g is a downward closed subset of S, but is otherwise more general. Any partial Lipschitz map g from S into T, however, naturally extends to a strictly increasing level-preserving map $\hat{g}$ on the downward closure of $\mathrm{dom}(g)$ in S, defined by letting $\hat{g}(x) = g(s) \upharpoonright \mathrm{ht}(x)$ for some (equivalently, for all) $s \in \mathrm{dom}(g)$ extending x.

Definition 4.2.2. A *Lipschitz tree* is any Aronszajn tree T with the property that every level-preserving map from an uncountable subset of T into T is Lipschitz on an uncountable subset of its domain.

Remark 4.2.3. It should be noted that if we restrict ourselves to countably branching trees of height ω_1, there are essentially no Lipschitz trees outside the class of Aronszajn trees. A tree of height ω_1 in which every node has extensions to all higher levels cannot have an uncountable chain if it is to satisfy the requirement of Definition 4.2.2. Using a bit of PFA one can also show that a Lipschitz tree must have all levels countable.

The following fact shows that coherent trees that we have been encountering so far all satisfy the strong metric-type condition about partial level preserving maps and that therefore Lipschitz trees appear in profusion.

Lemma 4.2.4. *A coherent tree T is Lipschitz if and only if every uncountable subset of T contains an uncountable antichain.*

Proof. The converse implication follows from Lemma 4.2.6 below, so let us concentrate on proving the direct implication. Let T be a given coherent downwards closed subtree of the tree of all maps from countable ordinals into ω with the property that every uncountable subset of T contains an uncountable antichain. Thus, the height of a given node t of T is equal to its domain. Similarly, for two different nodes $s, t \in T$ of the same height α,

$$\Delta(s,t) = \text{otp}\{x \in T : x <_T s \text{ and } x <_T t\} = \min\{\xi < \alpha : s(\xi) \neq t(\xi)\}.$$

Consider a partial level-preserving map f from an uncountable subset X of T into T. By our assumption, we may assume that both domain and the range of f are antichains of T. Moreover, we may assume that X contains no two different nodes of the same height. For $t \in X$, set

$$D_t = \{\xi < \text{ht}(t) : t(\xi) \neq f(t)(\xi)\}.$$

Applying the Δ-system lemma and shrinking X, if necessary, we may assume that D_t $(t \in X)$ forms a Δ-system with root D. Shrinking X even further, we may assume that:

(1) $D_s \setminus D < D_t \setminus D$ whenever $s, t \in X$ are such that $\text{ht}(s) < \text{ht}(t)$,

(2) $s \upharpoonright (\max(D) + 1) = t \upharpoonright (\max(D) + 1)$ for all $s, t \in X$.

Note that if $D_t = D$ for all $t \in X$, then (1) and (2) ensure that

$$\Delta(s,t) = \Delta(f(s), f(t)) \text{ for all } s \neq t \text{ in } X, \tag{4.2.1}$$

and so, in particular, the map f is Lipschitz. Thus we may assume that $D_t \setminus D \neq \emptyset$ for all $t \in X$. For $t \in X$, let $\delta_t = \min(D_t \setminus D)$. By our assumption about T, we can find an uncountable subset Y of X such that both sets

$$\{t \upharpoonright \delta_t : t \in Y\} \text{ and } \{f(t) \upharpoonright \delta_t : t \in Y\}$$

are antichains of T. It follows again that $\Delta(s,t) = \Delta(f(s), f(t))$ for all $s \neq t$ in X, as required. □

The following property of Lipschitz trees will be quite frequently and implicitly used in this and the following section.

Lemma 4.2.5. *Suppose T is a Lipschitz tree, n is a positive integer, and that A is an uncountable family of pairwise-disjoint n-element subsets of T. Then there exists an uncountable $B \subseteq A$ such that $\Delta(a_i, b_i) = \Delta(a_j, b_j)$ for all $a \neq b$ in B and $i, j < n$.*

Proof. Replacing the family A by a family of the form $a \upharpoonright \xi_a$ $(a \in A)$ and applying Lemma 4.2.4, we may assume that a given member of A is in fact a subset of some level of T. Fix $i, j < n$ and apply Definition 4.2.2 to the partial level-preserving map

$$a_i \longmapsto a_j \ (a \in A)$$

obtaining an uncountable $A_0 \subseteq A$ such that

$$\Delta(a_i, b_i) \leq \Delta(a_j, b_j) \text{ for all } a, b \in A_0.$$

Applying Definition 4.2.2 to the inverse map $a_j \longmapsto a_i$ $(a \in A_0)$ will give us an uncountable $A_1 \subseteq A$ such that

$$\Delta(a_i, b_i) = \Delta(a_j, b_j) \text{ for all } a, b \in A_1.$$

Repeating this procedure successively for every pair $i, j < n$, we reach the conclusion of Lemma 4.2.5. □

Lemma 4.2.6. *Every uncountable subset of a Lipschitz tree T contains an uncountable antichain. More generally, every family A of pairwise-disjoint finite subsets of T contains an uncountable subfamily B such that $\cup B$ is an antichain of T.*

Proof. Let X be a given uncountable subset of some Lipschitz tree T. It suffices to find an uncountable antichain in the downwards closure of X in T, so we may in fact assume that X is already downward closed in T. We may also assume that no $x \in X$ is an end-node of T and in fact that every splitting node $x \in X$ has two successors x_0 and x_1 in X that split at x and have the same height in T. Applying the fact that T is a Lipschitz tree to the partial map $f(x_0) = x_1$ (x a splitting node of X) gives us an uncountable set Y of splitting nodes of X such that f is Lipschitz on $Y_0 = \{y_0 : y \in Y\}$. It follows that Y_0 is an uncountable antichain of T. The second part of the lemma follows from the first and the fact that finite powers of T are Lipschitz as well (see Lemma 4.2.5). This finishes the proof. □

The following result shows that under some mild axiomatic assumption the class of Lipschitz trees in fact coincides with the class of coherent trees, the main object of our study here.

Lemma 4.2.7. *Under MA_{ω_1}, every Lipschitz tree T is isomorphic to a coherent downwards closed subtree of the tree of all maps from countable ordinals into ω.*

Proof. Define $\mathcal{P}$ to be the poset of all finite partial functions p from $T \times \omega_1$ into ω such that the following holds for all $x \neq y$ in $\mathrm{dom}_0(p)$[6]:

[6]Here and below, $\mathrm{dom}_0(p) = \{t \in T : (t, \alpha) \in \mathrm{dom}(p) \text{ for some } \alpha\}$ and $\mathrm{dom}_1(p) = \{\xi \in \omega_1 : (t, \xi) \in \mathrm{dom}(p) \text{ for some } t \in T\}$.

(1) $p(x, \xi) = p(y, \xi)$ for $\xi < \Delta(x, y)$,
(2) $p(x, \xi) \neq p(y, \xi)$ for $\xi = \Delta(x, y)$[7].

We let p extend q, if p extends q as a function, and

(3) $p(x, \xi) = p(y, \xi)$ for all $x, y \in \mathrm{dom}_0(q)$ and $\xi < \mathrm{ht}(x), \mathrm{ht}(y)$ with the property that $\xi \notin \mathrm{dom}_1(q)$.

It is clear that a sufficiently generic filter will give us the desired embedding, and so we concentrate on showing that $\mathcal{P}$ satisfies the countable chain condition.

So let p_δ ($\delta < \omega_1$) be a given sequence of elements of $\mathcal{P}$. Let a_δ be the projection of $\mathrm{dom}_0(p_\delta)$ onto the δth level of T and let

$$h(\delta) = \max\{\Delta(s, t) + 1 : s, t \in a_\delta, s \neq t\}.$$

Then there is a stationary set Γ of countable limit ordinals on which the mapping h, as well as the mapping

$$\delta \longmapsto \mathrm{dom}(p_\delta) \cap (T \restriction \delta) \times \delta$$

is constant. Let α and F be these constant values respectively. We may assume that all a_δ's project to the same set a on the αth level and all p_δ's generate isomorphic structures over α, a and F. Thus, in particular we want the isomorphism between the p_δ's ($\delta \in \Gamma$) to respect a fixed enumeration $a_\delta(i)$ ($i < n$) of a_δ, where n is the common cardinality of these sets. As in the previous proof, we find an uncountable subset Σ of Γ such that for all $\gamma \neq \delta$ in Σ:

(4) $a_\gamma(i)$ and $a_\delta(j)$ are incomparable for all $i, j < n$,
(5) $\Delta(a_\gamma(i), a_\delta(i)) = \Delta(a_\gamma(j), a_\delta(j))$ for all $i, j < n$.

We claim that if $\gamma \neq \delta$ are in Σ, then p_γ and p_δ are compatible in $\mathcal{P}$. By (5), we have an ordinal β smaller than both γ and δ such that

$$\Delta(a_\gamma(i), a_\delta(i)) = \beta \text{ for all } i < n.$$

Define $r \in \mathcal{P}$ by letting its domain be

$$\mathrm{dom}(p_\gamma) \cup \mathrm{dom}(p_\delta) \cup \{(t, \beta) : t \in \mathrm{dom}_0(p_\gamma) \cup \mathrm{dom}_0(p_\delta), \mathrm{ht}(t) > \beta\}$$

and letting $r(t, \beta) = 0$ if $t \in \mathrm{dom}_0(p_\gamma)$, $\mathrm{ht}(t) > \beta$ and $r(t, \beta) = 1$ if $t \in \mathrm{dom}_0(p_\delta)$, $\mathrm{ht}(t) > \beta$. Note that r is indeed a member of $\mathcal{P}$, as it clearly satisfies the conditions (1) and (2) above. It is also easily checked that r extends both p and q, i.e., $r(x, \xi) = r(y, \xi)$ for all $x, y \in \mathrm{dom}_0(p)$ or $x, y \in \mathrm{dom}_0(q)$ and ξ is equal to β, the only new member of $\mathrm{dom}_1(r)$. This finishes the proof. □

[7] In (1) and (2) we are making the implicit requirement that for every $\xi \in \mathrm{dom}_1(p)$ and $x \neq y \in \mathrm{dom}_0(p)$, if $\xi \leq \Delta(x, y)$, then $(x, \xi) \in \mathrm{dom}(p)$ iff $(y, \xi) \in \mathrm{dom}(p)$, and moreover, that always $(x, \Delta(x, y)$ and $(y, \Delta(x, y))$ belong to $\mathrm{dom}(p)$.

Definition 4.2.8. For a given tree T of height ω_1,

$$\mathcal{U}(T) = \{A \subseteq \omega_1 : A \supseteq \Delta(X) \text{ for some uncountable } X \subseteq T\}.$$

Then one has the following analogue of Lemma 4.1.11 which in fact can be deduced from it, though for a future motivation we include the more direct proof.

Lemma 4.2.9. *The family $\mathcal{U}(T)$ is a uniform filter on ω_1 for every Lipschitz tree T. In particular, $\mathcal{U}(T)$ is a uniform filter on ω_1 for every coherent tree T with the property that every uncountable subset of T has an uncountable antichain.*

Proof. Given two uncountable subsets X and Y of T, we need to find an uncountable subset Z of T such that

$$\Delta(X) \cap \Delta(Y) \supseteq \Delta(Z).$$

By the assumption about T, it is clear that we may replace X and Y by two level-sequences x_δ $(\delta \in \Gamma)$ and y_δ $(\delta \in \Gamma)$ indexed by the same uncountable set Γ of limit ordinals. Moreover, we may assume that the x_δ's and y_δ's are all pairwise incomparable (see Lemma 4.2.4). Apply Lemma 4.2.5 to the subset (x_δ, y_δ) $(\delta \in \Gamma)$ of $T \otimes T$ and obtain an uncountable set $\Sigma \subseteq \Gamma$ such that

$$\Delta(x_\gamma, x_\delta) = \Delta(y_\gamma, y_\delta) \text{ for all } \gamma, \delta \in \Sigma, \gamma \neq \delta.$$

So we can take Z to be any of the sets $\{x_\delta : \delta \in \Sigma\}$ or $\{y_\delta : \delta \in \Sigma\}$. This finishes the proof. □

One also has the following analogue of Theorem 4.1.13 which can be in fact deduced from it, but again, we give the direct proof.

Theorem 4.2.10. *Under MA_{ω_1}, the filter $\mathcal{U}(T)$ is an ultrafilter for every coherent tree T with no uncountable branches.*

Proof. Let Γ be a given subset of ω_1. We need to find an uncountable subset X of T such that $\Delta(X)$ is included in either Γ or its complement. Let $\mathcal{P}_\Gamma$ be the poset of all finite subsets p of T that take at most one point from a given level of T such that

$$\Delta(p) = \{\Delta(x, y) : x, y \in p, x \neq y\} \subseteq \Gamma.$$

If $\mathcal{P}_\Gamma$ satisfies the countable chain condition, then a straightforward application of (*) gives an uncountable $X \subseteq T$ such that $\Delta(X) \subseteq \Gamma$. So let us consider the alternative that $\mathcal{P}_\Gamma$ fails to satisfy this condition. Let p_δ $(\delta < \omega_1)$ be a sequence of pairwise incomparable members of $\mathcal{P}_\Gamma$. Going to a subsequence, we may assume that every node of a given p_δ has height at least δ, so we can define $a_\delta =$ the projection of p_δ on the δth level of T. For $\delta \in \Lambda$, let

$$h(\delta) = \{\Delta(x, y) + 1 : x, y \in a_\delta, x \neq y\}.$$

Find stationary $\Omega \subseteq \Lambda$ such that h is constant on Ω. Let $\bar{\xi}$ be the constant value of h. Going to an uncountable subset of Ω, we may assume that all a_δ ($\delta \in \Omega$) are of equal size n and that they are given with an enumeration $a_\delta(i)$ ($i < n$). Moreover, we may assume that $a_\gamma(i) \restriction \bar{\xi} = a_\delta(i) \restriction \bar{\xi}$ for all $\gamma, \delta \in \Omega$. Applying Lemmas 4.2.5 and 4.2.6, we obtain an uncountable $\Sigma \subseteq \Omega$ such that for all $\gamma \neq \delta$ in Σ:

(1) $a_\gamma(i)$ and $a_\delta(j)$ are incomparable for all $i, j < n$,

(2) $\Delta(a_\gamma(i), a_\delta(i)) = \Delta(a_\gamma(j), a_\delta(j))$ for all $i, j < n$.

It follows that for all $\gamma \neq \delta$ in Σ:

(3) $\Delta(p_\gamma \cup p_\delta) = \Delta(p_\gamma) \cup \Delta(p_\delta) \cup \{\Delta(a_\gamma(0), a_\delta(0))\}$.

Since $\Delta(p_\gamma)$ and $\Delta(p_\delta)$ are subsets of Γ, $\Delta(a_\gamma(0), a_\delta(0)) \notin \Gamma$ must hold. This gives rise to an uncountable set $X = \{a_\delta(0) : \delta \in \Sigma\}$ with the property that $\Delta(X) \cap \Gamma = \emptyset$. This finishes the proof. □

The definition of $\mathcal{U}(T)$ can be relativized to any countable elementary submodel M of $(H_{\omega_2}, \in)$ as follows.

Definition 4.2.11. For a coherent tree T of height ω_1 and a countable elementary submodel M of H_{ω_2} with $T \in M$, let $\mathcal{U}_M(T)$ be the collection of all subsets A of ω_1 for which one can find uncountable $X \subseteq T$ with $X \in M$ and $t \in T_{M \cap \omega_1}$ belonging to the downward closure of X such that $A \supseteq \Delta(t, X) = \{\Delta(t, x) : x \in X\}$.

It should be clear that the basic idea of the proof of Lemma 4.2.9 also gives the following.

Lemma 4.2.12. *For every coherent tree T and every countable elementary submodel M of H_{ω_2} such that $T \in M$, the family $\mathcal{U}_M(T)$ is a nonprincipal filter concentrating on the set $M \cap \omega_1$.* □

From now on, we fix a special coherent subtree T of the complete binary tree $2^{<\omega_1}$ that is closed under restrictions and finite changes of its elements. We also fix a subset K of T and use K^c to denote its complement in T. Our goal is to find an uncountable subset X of T such that either $\wedge(X) \subseteq K$[8] or $\wedge(X) \subseteq K^c$. Note that in Theorem 4.2.10, assuming MA_{ω_1}, we have achieved this in the case when K is the union of levels of T. The general case requires a further analysis and a stronger axiom. For $J \in \{K, K^c\}$ and $t \in T$, the corresponding fiber is denoted by

$$J(t) = \{\xi < \omega_1 : t \restriction \xi \in J\}.$$

For a subset X of T and $J \in \{K, K^c\}$, let $J(X) = \bigcup_{t \in X} J(t)$. The filters of the form $\mathcal{U}_M(T)$ are used to define the following important notion.

Definition 4.2.13. A countable elementary submodel M of H_{ω_2} *rejects* a finite subset F of T if $K^c(F) \in \mathcal{U}_M(T)$.

---

[8]Recall that for $x, y \in T$ by $x \wedge y$ one denotes the maximal common predecessor of x and y, and for a subset X of T one puts $\wedge(X) = \{x \wedge y : x, y \in X, x \neq y\}$.

Remark 4.2.14. Note that if an elementary submodel M of H_{ω_2} such that $\{T, K\} \subseteq M$ rejects a singleton $\{t\}$, then there is an uncountable antichain X of T such that $\wedge(X) \subseteq K^c$. Note also that if M and N are two countable elementary submodels containing T and K such that $N \subseteq M$ and $M \cap \omega_1 = N \cap \omega_1$, and if N rejects F, then so does M. This follows from the inclusion $\mathcal{U}_N(T) \subseteq \mathcal{U}_M(T)$.

The following lemma suggests the direction of further analysis as it is definitely saying something about the poset of all finite approximations to an uncountable subset X of T such that $\wedge(X) \subseteq K$.

Lemma 4.2.15. *Suppose we have an uncountable family A of pairwise-disjoint finite subsets of T, all of some fixed size n such that the collection $\mathcal{M}$ of all countable elementary submodels of H_{ω_2} which do not reject any member of the family A is a stationary subset of $[H_{\omega_2}]^\omega$. Then there exist $a \neq b$ in A such that $a(i) \wedge b(i) \in K$ for all $i < n$.*

Proof. Using Lemma 4.2.4 and going to an uncountable subfamily of A, we may assume that

$$\Delta(a_i, b_i) = \Delta(a_j, b_j) \text{ for all } a \neq b \text{ in } A \text{ and } i, j < n. \tag{4.2.2}$$

Since $\mathcal{M}$ is stationary, we can find $M \in \mathcal{M}$ such that $\{A, T, K\} \subseteq M$. Pick a $b \in A$ such that the height of every node from b is bigger than $\delta = M \cap \omega_1$. Let $X = \{a(0) : a \in A\}$ and let $t = b(0) \restriction \delta$. Then $\Delta(t, X)$ is a typical generator of $\mathcal{U}_M(T)$ and by our assumption that M does not reject b, we conclude that

$$\Delta(t, X) \cap \delta \nsubseteq K^c(b). \tag{4.2.3}$$

Applying (4.2.2) and (4.2.3), we conclude that there must be $a \in A \cap M$ such that $a(i) \wedge b(i) \in K$ for all $i < n$. □

Corollary 4.2.16. *Assuming* MA_{ω_1}, *if there is an uncountable subset Y of T such that the collection $\mathcal{M}$ of all countable elementary submodels of H_{ω_2} which do not reject any finite subset of Y is stationary in $[H_{\omega_2}]^\omega$, then there is uncountable $X \subseteq Y$ such that $\wedge(X) \subseteq K$.*

Proof. By Lemma 4.2.15, the poset of all finite $F \subseteq Y$ such that $\wedge(F) \subseteq K$ satisfies the countable chain condition. □

Let $\theta_0 = \aleph_2 (= (\aleph_1)^+)$ and let $\mathcal{C}_0$ be the collection of all closed unbounded subsets of $[H_{\theta_0}]^\omega$ consisting of countable elementary submodels M of $(H_{\theta_0}, \in)$ such that $\{T, K\} \subseteq M$. Similarly, let $\theta_1 = (2^{\aleph_1})^+$ and let $\mathcal{C}_1$ be the collection of all closed unbounded subsets of $[H_{\theta_1}]^\omega$ consisting of countable elementary submodels M of $(H_{\theta_1}, \in)$ such that $\{T, K\} \subseteq M$. Finally, let $\theta_2 = (2^{2^{\aleph_1}})^+$.

Lemma 4.2.17. *Assume the Proper Forcing Axiom, and suppose a_ξ $(\xi \in \Gamma)$ is a given sequence of finite subsets of T indexed by some set $\Gamma \subseteq \omega_1$ such that $a_\xi \subseteq T_\xi$ for all $\xi \in \Gamma$. Then there is $C \in \mathcal{C}_1$ such that for every $M \in C$, if $\xi = M \cap \omega_1 \in \Gamma$ then there is $D \in M \cap \mathcal{C}_0$ such that every element of $D \cap M$ rejects a_ξ or no element of $D \cap M$ rejects a_ξ.*

Proof. Let $\mathcal{P}$ be the poset of all continuous $\in$-chains $p = \langle N_\xi^p : \xi \leq \alpha_p \rangle$ of countable elementary submodels N_ξ^p of $(H_{\theta_1}, \in)$ containing all the relevant objects such that, if $N_\xi^p \cap \omega_1 = \delta_\xi$ belongs to Γ and there $D \in \mathcal{C}_0 \cap N_\xi^p$ such that no $N \in D \cap N_\xi^p$ rejects a_{δ_ξ}, then there is $\nu < \xi$ such that N_η^p does not reject a_{δ_ξ} for all $\eta \in (\nu, \xi)$; otherwise, there is $\nu < \xi$ such that N_η^p rejects a_{δ_ξ} for all $\eta \in (\nu, \xi)$. It is easily checked that $\mathcal{P}$ ordered by end-extension is a proper poset, so an application of PFA gives us a continuous $\in$-chain $\langle N_\xi : \xi < \omega_1 \rangle$ of countable elementary submodels N_ξ of $(H_{\theta_1}, \in)$ with the property that if $N_\xi = \delta_\xi$ belongs to Γ, then either every term in a tail of the sequence $\langle N_\eta : \eta < \xi \rangle$ does not reject a_{δ_ξ}, or else, every member of a tail of the sequence $\langle N_\eta : \eta < \xi \rangle$ rejects a_{δ_ξ} depending on whether $D \in \mathcal{C}_0 \cap N_\xi$ with no member of $D \cap N_\xi$ rejecting a_{δ_ξ}, or not.

The conclusion of the lemma would follow if we show that for every countable elementary submodel M of $(H_{\theta_2}, \in)$ containing all the objects accumulated so far and having the property that $\delta = M \cap \omega_1$ belongs to Γ, there is $D \in M \cap \mathcal{C}_0$ such that every element of $D \cap M$ rejects a_δ or no element of $D \cap M$ rejects a_δ. We may assume that the second alternative fails, i.e., that for every $D \in \mathcal{C}_0 \cap M$ there is $N \in D \cap M$ rejecting a_δ. Since $N_\delta \subseteq M$, our assumption about M and a simple elementarity argument shows that for every $D \in \mathcal{C}_0 \cap N_\delta$ there is $N \in D \cap N_\delta$ rejecting a_δ. It follows that there is $\nu_0 < \delta$ such that N_ξ rejects a_δ for all $\xi \in (\nu_0, \delta)$.

Let D_0 be the collection of all countable elementary submodels N of $(H_{\theta_0}, \in)$ such that ν_0 and $\langle N_\xi \cap H_{\theta_0} : \xi < \omega_1 \rangle$ belong to N. Clearly, D_0 is a closed and unbounded subset of $[H_{\theta_0}]^\omega$. We show that every $N \in D_0 \cap M$ rejects a_δ. So, consider an $N \in D_0 \cap M$ and let $\nu = N \cap \omega_1$. Then $\nu \in (\nu_0, \delta)$ and so N_ν rejects a_δ. Since $N_\nu \subseteq N$, we conclude that N also rejects a_δ. □

We are now in a situation to prove the following result of J.T. Moore [77].

Theorem 4.2.18. *Assuming the Proper Forcing Axiom, for every coherent tree T and every subset K of T there is an uncountable subset X of T such that either $\wedge(X) \subseteq K$ or $\wedge(X) \cap K = \emptyset$.* □

Proof. It suffices to concentrate only on our fixed special binary coherent tree T. We shall assume that there is no uncountable $X \subseteq T$ such that $\wedge(X) \subseteq K^c$ and we shall show how to force something similar to the hypothesis of Corollary 4.2.16 so that its proof would give us uncountable $X \subseteq T$ such that $\wedge(X) \subseteq K$. There is a natural partial ordering $\mathcal{P}_K$ for achieving this, the collection of all pairs $p = (F_p, \mathcal{N}_p)$ such that:

(1) F_p is a finite subset of T,

(2) $\mathcal{N}_p$ is a finite $\in$-chain of countable elementary submodels of $(H_{\theta_2}, \in)$ containing all the relevant objects,

(3) for every $N \in \mathcal{N}_p$ there is $D \in \mathcal{C}_0 \cap N$ such that F_p is not rejected by any member of $D \cap N$.

We consider $\mathcal{P}_K$ as a poset ordered by coordinatewise inclusions. We shall prove that $\mathcal{P}_K$ is a proper partial order. This will finish the proof, since we are assuming that no countable elementary submodel N of H_{θ_0} rejects a singleton $\{t\}$, and any condition of the form $(\{t\}, \{M\})$ with $t \in T_{M\cap\omega_1}$ forces that the union of the first coordinates of the generic filter is uncountable.

Suppose that for some large enough regular cardinal $\theta > \theta_2$ there is some countable elementary submodel M of $(H_\theta, \in)$ containing all relevant objects and there is some $p \in \mathcal{P}_K \cap M$ which cannot be extended to an M-generic condition of $\mathcal{P}_K$. In particular, the condition

$$q = (F_p, \mathcal{N}_p \cup \{M \cap H_{\theta_2}\})$$

is not M-generic. So there is a dense-open subset $\mathcal{D}$ of $\mathcal{P}_K$ such that $\mathcal{D} \in M$ and there is extension r of q such that no $\bar{r} \in \mathcal{D} \cap M$ is compatible with r. Extending r, we may assume that it belongs to $\mathcal{D}$, so our assumption that the standard Fubini-type argument (see, e.g., [111],§8) that tries to build a copy $\bar{r}$ of r that belongs to $\mathcal{D} \in M$ and that is compatible with r must fail. This results in two stationary sets $\mathcal{S} \subseteq [H_{\theta_1}]^\omega$ and $\mathcal{T} \subseteq [H_{\theta_2}]^\omega$ and two mappings $N \mapsto a_N$ and $N \mapsto b_N$ with domains $\mathcal{S}$ and $\mathcal{T}$, respectively, and ranges included in $[T]^{<\omega}$ such that:

(4) for every $N \in \mathcal{S}$ there is a countable elementary submodel M of $(H_{\theta_2}, \in)$ such that $N = M \cap H_{\theta_1}$,

(5) for every $N \in \mathcal{S}$ there is $D \in \mathcal{C}_0 \cap N$ such that no model from $D \cap N$ rejects the finite set a_N,

(6) for every $N \in \mathcal{T}$ there is $D \in \mathcal{C}_0 \cap N$ such that no model from $D \cap N$ rejects the finite set b_N,

(7) for every $N_1 \in \mathcal{T}$ and every $N_0 \in \mathcal{S} \cap N_1$ there is no $D \in \mathcal{C}_0 \cap N_0$ with property that no member of $D \cap N_0$ rejects $a_{N_0} \cup b_{N_1}$.

We may assume that $\mathcal{S} \in N$ for all $N \in \mathcal{T}$. Furthermore, we may assume that for every $N \in \mathcal{T}$, the set b_N has some fixed size n and that $b_N \subseteq T_{M\cap\omega_1}$. Moreover, applying the Pressing Down Lemma and shrinking $\mathcal{T}$, we may assume that for some fixed $D_0 \in \mathcal{C}_0$ and some fixed countable ordinal ν_0 and some fixed n-element subset of T_{ν_0}, we have the following conditions on every $N \in \mathcal{T}$:

(8) $D_0 \in N$ and no member of $D_0 \cap N$ rejects b_N,

(9) $s \restriction [\nu_0, N \cap \omega_1) = t \restriction [\nu_0, N \cap \omega_1)$ for all $s, t \in b_N$,

(10) $b_N \restriction \nu_0 = F_0$.

Let U be the collection of all n-element subsets a of T such that for some ordinal $\xi > \nu_0$, the set

$$\mathcal{T}[a] = \{N \in \mathcal{T} : b_N \restriction \xi = a\}$$

is stationary. Note that the natural coordinatewise ordering makes U a tree which is also coherent, being isomorphic to a subtree of $\{t \in T : t \restriction \nu_0 = F_0(0)\}$. We shall now see that there is a natural ccc poset which forces an uncountable antichain A of U with the property that for all $a \neq b$ in A there exist $i < n$ such that $a(i) \wedge b(i) \notin K$. Since no member of D_0 rejects any member of A this would contradict Lemma 4.2.15, and so the proof of Theorem 4.2.18 would be finished.

Let $\mathcal{Q}$ be the collection of all finite antichains q of U that meet a given level of U in at most one point such that

$$\{a(i) \wedge b(i) : i < n\} \nsubseteq K \text{ for all } a \neq b \text{ in } q. \tag{4.2.4}$$

We order $\mathcal{Q}$ by inclusion and we check that it satisfies the countable chain condition. So, let A be a given uncountable subset of $\mathcal{Q}$. By a Δ-system lemma, we may assume that A forms a Δ-system. Since the root can't contribute to incompatibility between two conditions from A, we may actually assume that it is empty and that in fact A is an uncountable family of pairwise-disjoint antichains a of U satisfying the condition 4.2.4 and all having some fixed size m. We may view an element a of A as a sequence $a(i)$ $(i < n)$ which enumerates a according to the height in U. Since each $a(i)$ is in turn an n-element subset of some level of T, we can order it lexicographically as $a(i)(j) = a(i, j)$ $(j < n)$.

Choose $N \in \mathcal{S}$ such that ν_0 and A are elements of N and let M be a countable elementary submodel of $(H_{\theta_2}, \in)$ such that $N = M \cap H_{\theta_1}$. Let $\delta = N \cap \omega_1$ and fix an element b of A, all of whose members are of height $> \delta$. Fix also $C_0 \in \mathcal{C}_0$ such that no member of $C_0 \cap N$ rejects a_N. For each $i < m$, we choose $M_i \in \mathcal{T}$ such that $N \in M_i$ and b_i is a restriction of b_{M_i}. Since every finite subset of T_δ is equal to the δth member of a ω_1-sequence belonging to M, applying Lemma 4.2.17 to the restrictions of b_{M_i}'s to the δth level of T, and intersecting finitely many closed and unbounded subsets of $[H_{\theta_0}]^\omega$, and then relying on the condition (7), we can find $N_0 \in C_0 \cap N$ such that

$$\bigcap_{i<m} K^c(a_N \cup b_i) \in \mathcal{U}_{N_0}(T). \tag{4.2.5}$$

Let $\delta_0 = N_0 \cap \omega_1$. Pick $\nu_1 \in (\nu_0, \delta_0)$ such that elements of $\cup b$ agree on the interval $[\nu_1, \delta_0)$, and moreover, if two different nodes of $\cup b$ have different restrictions on δ, then they also have different restrictions on ν_1. Using the elementarity of N_0, we can choose a sequence $\langle b_\xi : \xi < \omega_1 \rangle \in N_0$ of elements of A such that $b_{\delta_0} \restriction \delta_0 = b \restriction \delta_0$ and such that for all $\xi < \omega_1$, the nodes of $\cup b_\xi$ are of heights $\geq \xi$, they all agree on the interval $[\nu_1, \xi)$, and if two of nodes of $\cup b_\xi$ have different restrictions on ξ, then they also have different restrictions on ν_1.

Using coherence of T, we find uncountable subset Γ of ω_1 in N_0 such that the node $t = b(0,0) \restriction \delta_0$ is in the downwards closure of the set $X = \{b_\xi(0,0) : \xi \in \Gamma\}$, and such that

$$\Delta(t, X) \subseteq \bigcap_{i<m} K^c(a_N \cup b_i). \tag{4.2.6}$$

Since N_0 belongs to $\mathcal{C}_0 \cap N$, it does not reject a_N, and so in particular, $\Delta(t, X)$ is not a subset of $K^c(a_N)$. So there must be $\gamma \in \Gamma \cap N_0$ such that

$$a_N(k) \restriction \Delta(t, b_\gamma(0,0)) \in K \text{ for all } k < |a_N|. \tag{4.2.7}$$

Using this and (4.2.6), we conclude that for all $i < m$ there is $j < n$ such that,

$$b(i,j) \restriction \Delta(t, b_\gamma(0,0)) = b(i,j) \wedge b_\gamma(i,j) \notin K. \tag{4.2.8}$$

We shall show that $b \cup b_\gamma$ satisfies the requirement (4.2.4) and this will finish the proof. Consider $i_0 \neq i_1 < m$. If $b(i_0) \restriction \delta_0 \neq b(i_1) \restriction \delta_0$, then there is $j < n$ such that $b_\gamma(i_0, j) \wedge b_\gamma(i_1, j) \notin K$. Then $\Delta(b(i_1, j), b_\gamma(i_1, j)) \geq \nu_1 > \Delta(b_\gamma(i_0, j), b_\gamma(i_1, j))$, and so we must have that

$$\Delta(b_\gamma(i_0, j), b(i_1, j)) = \Delta(b_\gamma(i_0, j), b_\gamma(i_1, j)), \tag{4.2.9}$$

and therefore, $b_\gamma(i_0, j) \wedge b(i_1, j) = b_\gamma(i_0, j) \wedge b_\gamma(i_1, j) \notin K$. If, on the other hand, $b(i_0) \restriction \delta_0 = b(i_1) \restriction \delta_0$, then for all $j < m$,

$$\Delta(b_\gamma(i_0, j), b(i_1, j)) = \Delta(b_\gamma(i_0, j), b(i_0, j)) = \Delta(b_\gamma(0,0), b(0,0)). \tag{4.2.10}$$

By (4.2.8), there is $j < m$ such that

$$b_\gamma(i_0, j) \wedge b(i_0, j) = b(i_0, j) \restriction \Delta(b_\gamma(0,0), b(0,0)) \notin K. \tag{4.2.11}$$

Hence, $b_\gamma(i_0, j) \wedge b(i_1, j) \notin K$. This finishes the proof. □

Remark 4.2.19. It turns out that the conclusion of Theorem 4.2.18 is a principle of independent interest, so for future reference, we give it a name *Coherent Tree Axiom* and denote it, in short, as CTA[9]. Note that if T contains an uncountable branch, the conclusion of CTA holds trivially, so the axiom is really about coherent trees with no uncountable branches. It is also possible to show that, under MA_{ω_1}, the axiom does not weaken if we restrict our attention to a fixed coherent tree with no uncountable branches such as for example $T(\rho_3)$.

We finish this section with few concepts and facts about the general class of A-trees[10].

Definition 4.2.20. A level-preserving partial map $f : S \to T$ between two trees S and T is *pattern preserving* if for every triple $x, y, z \in \mathrm{dom}(f)$,

$$\Delta(x,y) > \Delta(x,z) \text{ iff } \Delta(f(x), f(y)) > \Delta(f(x), f(z)). \tag{4.2.12}$$

[9]The statement of a stronger axiom with T an arbitrary A-tree appears first in a remark on page 79 of [1]. The relevance of the coherent version of the axiom became evident only after the paper [123]. Needless to say, the deep arguments of [77] are built around the assumption that the tree T is coherent.

[10]Here, *A-tree* stands for an *Aronszajn tree*, an uncountable tree that has all levels countable as well as all branches.

Remark 4.2.21. Note that if $X = \mathrm{dom}(f)$ and $Y = \mathrm{rang}(f)$ are antichains, this is in particular saying that the corresponding subtrees $\wedge(X) \subseteq S$ and $\wedge(Y) \subseteq T$ are isomorphic. So the following fact will be quite useful in transferring statements like CTA from coherent trees to other A-trees.

Lemma 4.2.22. *Assuming the Proper Forcing Axiom, every level-preserving partial map $f : S \to T$ between two A-trees S and T of uncountable domain is pattern preserving on an uncountable subset of its domain.*

Proof. Let $\mathcal{P}$ be the set of all pairs $p = (X_p, C_p)$ of finite subsets of $\mathrm{dom}(f)$ and ω_1, respectively, such that 4.2.12 holds for all $x, y, z \in X_p$ and such that for all $x, y \in X_p$ and $\xi \in C_p$,

$$\Delta(x, y) < \xi \text{ iff } \Delta(f(x), f(y)) < \xi. \tag{4.2.13}$$

We let $\mathcal{P}$ be ordered by coordinatewise inclusions. To see that $\mathcal{P}$ is a proper partial ordering, note that if M is a countable elementary submodel of some large-enough structure of the form $(H_\theta, \in)$, then for every $p \in \mathcal{P} \cap M$, the extension $q = (X_p, C_p \cup \{M \cap \omega_1\})$ is an M-generic condition of $\mathcal{P}$. Note also that a condition of the form $(\{x\}, \{\xi\})$, where $\xi = M \cap \omega_1$ and $x \in \mathrm{dom}(f) \setminus M$ for such a model M, forces the unions of the first projection of the generic filter to uncountable. Hence an application of the Proper Forcing Axiom gives us the conclusion of the lemma. □

Lemma 4.2.23. *Assuming the Proper Forcing Axiom, every level-preserving map f from an uncountable subset of a coherent A-tree T into an A-tree U has an uncountable $X \subseteq \mathrm{dom}(f)$ such that either*

(1) $\Delta(x, y) \geq \Delta(f(x), f(y))$ *for all* $x \neq y \in X$, *or*
(2) $\Delta(x, y) < \Delta(f(x), f(y))$ *for all* $x \neq y \in X$.

Proof. By Lemma 4.2.22, we may assume that f preserves the splitting patterns. Choose an unbounded subset D of ω_1 and an uncountable subset X of $\mathrm{dom}(f)$ such that

$$\Delta(x, y) < \delta \ \text{ iff } \ \Delta(f(x), f(y)) < \delta$$

for all $x \neq y \in X$ and $\delta \in D$. This can be done using the poset of finite approximations to X and D which is easily seen to satisfy the countable chain condition using the fact that f preserves patterns. Taking the closure, we may assume that D is actually closed and unbounded. Moreover, we may assume that both X and its image $f''X$ are antichains in the respective trees.

Refining the set X, we may assume that the downward closure $T(X)$ of X in T, the downward closure $U(f''X)$ of $f''X$ in U, and the downward closure $S(X)$ of the graph of $f \restriction X$ in the product tree $T \otimes U$ [11] are all binary. That this can be assumed follows from the easily checked fact that the poset of finite subsets

[11] Recall that $T \otimes U = \{(x, y) \in T \times U : \mathrm{ht}(x) = \mathrm{ht}(y)\}$ is a tree when ordered coordinatewise.

of an A-tree whose downward closures are binary satisfies the countable chain condition. Refine X still further so that we can assume that the heights of any pair of comparable splitting nodes of one of these three downwards closures are separated by a member of the closed and unbounded set D. This again follows from the fact that the poset of finite subsets p of X, with the property that the comparable splitting nodes generated by p inside these three trees are separated by D, satisfies the countable chain condition. Then to every splitting node $t = x \wedge y$ of $T(X)$ there corresponds a unique splitting node $u_t = f(x) \wedge f(y)$ and this association is independent of the choice of $x \neq y \in X$ used to represent t.

Let K be the collection of all splitting nodes t of $T(X)$ such that $\mathrm{ht}(t) \leq \mathrm{ht}(u_t)$. Applying CTA, we find an uncountable subset Y of X such that

$$\wedge(Y) \subseteq K \text{ or } \wedge(Y) \cap K = \emptyset. \tag{4.2.14}$$

Then in the first case, we have the alternative (1) while in the second case we have the alternative (2) of the lemma. □

Remark 4.2.24. The conclusion of Lemma 4.2.23 in particular says that for every level-preserving partial map f from an uncountable subset of a coherent A-tree T into an A-tree U, there is an uncountable subset X of the domain of f such that either the restriction $f \upharpoonright X$ or its inverse is a Lipschitz map. We take the conclusion of Lemma 4.2.23 as an axiom of independent interest since we are going to use it at several places below. So we introduce a special name for it, the *Lipschitz Map Axiom* and denote it as LMA .

4.3 The global structure of the class of coherent trees

In this section we study the class of coherent trees relative to the following quasi-ordering relation.

Definition 4.3.1. For two trees S and T, by $S \leq T$ we denote the fact that there is a strictly increasing map $f : S \longrightarrow T$. Let $S < T$ whenever $S \leq T$ and $T \nleq S$ and let $S \equiv T$ whenever $S \leq T$ and $T \leq S$. When $S \equiv T$, we will say that the trees S and T are *equivalent trees*.

Remark 4.3.2. Note that if there is a strictly increasing map $f : S \longrightarrow T$ from a tree S into a tree T, then there is such a map $g : S \longrightarrow T$ that is moreover *level preserving*, i.e., a strictly increasing map with the property that $\mathrm{ht}(f(x)) = \mathrm{ht}(x)$ for all $x \in S$. It suffices to take $g(x) = f(x) \upharpoonright \mathrm{ht}(x)$. Note that strictly increasing and level preserving maps from a downwards closed subtree of S into T are nothing else than *Lipschitz maps* for this sort of domains. It is therefore natural to expect some structure theory of the quasi-ordered class $(\mathcal{C}, \leq)$, where $\mathcal{C}$ denotes the class of *coherent trees* of height ω_1, or equivalently, the class of *Lipschitz trees* introduced above in the previous section.

The basic fact which shows that coherent trees might have a special structure under $\leq$ is the following.

Lemma 4.3.3. *Under* MA_{ω_1}, *every coherent tree* T *is* irreducible *in the sense that* $T \leq U$ *for every uncountable downwards closed subtree* U *of* T.

Proof. Let U be an uncountable downwards closed subset of T. We assume that T has no uncountable branches and that every $t \in U$ has extensions in U in all levels above $\mathrm{ht}(t)$ and will use MA_{ω_1} to produce a map witnessing $T \leq U$. Let $\mathcal{P}$ be the poset of all partial finite level-preserving *Lipschitz maps* from T into U, i.e., maps p with the property that

$$\Delta(p(x), p(y)) \geq \Delta(x, y) \tag{4.3.1}$$

for all $x, y \in \mathrm{dom}(p)$. Let us show that $\mathcal{P}$ satisfies the countable chain condition. So let p_ξ $(\xi < \omega_1)$ be a given sequence of elements of $\mathcal{P}$. Using the Δ-system lemma and arguments about coherent trees already encountered in the previous section, we can find $\xi < \eta < \omega_1$, an integer k, two finite subsets D and E of ω_1, and an ordinal $\delta > \max(D \cup E)$ such that $p_\xi \cup p_\eta$ is a function:

(1) $\Delta(\mathrm{dom}(p_\xi) \cup \mathrm{dom}(p_\eta))$[12] $= D \cup \{\delta\}$,
(2) $\Delta(\mathrm{rang}(p_\xi) \cup \mathrm{rang}(p_\eta)) = E \cup \{\delta\}$,
(3) $|\mathrm{dom}(p_\xi) \setminus D| = |\mathrm{dom}(p_\eta) \setminus D|$,
(4) $\Delta(x, y) = \Delta(p_\xi(x), p_\eta(y))$, when $x \in \mathrm{dom}(p_\xi) \setminus D$ and $y \in \mathrm{dom}(p_\eta) \setminus D$ occupy the same position in the increasing enumerations of these sets.

It should be clear now that $p_\xi \cup p_\eta$ satisfies the Lipschitz condition (4.3.1) and therefore is a member of $\mathcal{P}$.

For $x \in T$, let

$$\mathcal{D}_x = \{p \in \mathcal{P} : x \in \mathrm{dom}(p)\}.$$

By our assumption on U, each $\mathcal{D}_x$ is a dense subset of $\mathcal{P}$, so an application of MA_{ω_1} gives us the conclusion of the lemma. □

It follows that a coherent A-tree T is equivalent to all of its uncountable downward closed subtrees. The following concept already appearing in the previous section (though in a different guise) is quite useful in further analysis of the class of Lipschitz tree, or equivalently, the class of coherent A-trees.

Definition 4.3.4. Given a pair of trees S and T of height ω_1 and having fixed *level sequences* $s_\alpha \in S_\alpha$ $(\alpha < \omega_1)$ and $t_\alpha \in T_\alpha$ $(\alpha < \omega_1)$, in order to make it clear where the Δ-function is being applied, we use the notation

$$\Delta_s(\alpha, \beta) = \Delta(s_\alpha, s_\beta) \text{ and } \Delta_t(\alpha, \beta) = \Delta(t_\alpha, t_\beta).$$

---

[12] Recall that coherent trees are subtrees of trees of the form $T(a)$ for a a coherent map, so the notation $\Delta(x, y)$ and $\Delta(X)$ can be taken from the previous section. However, this could be as easily defined for general trees T : For $x, y \in T$, we let $\Delta(x, y)$ be the order type of the set of all common (not necessarily strict) predecessors of x and y, and $\Delta(X) = \{\Delta(x, y) : x, y \in X\}$ for $X \subseteq T$.

As a result of this notation, for $\Gamma \subseteq \omega_1$, we let

$$\Delta_s(\Gamma) = \{\Delta_s(\alpha,\beta) : \alpha,\beta \in \Gamma, \alpha \neq \beta\} \text{ and } \Delta_t(\Gamma) = \{\Delta_t(\alpha,\beta) : \alpha,\beta \in \Gamma, \alpha \neq \beta\}.$$

For example, using this notation (fixing arbitrary level sequences $s_\alpha \in S_\alpha$ $(\alpha < \omega_1)$ and $t_\alpha \in T_\alpha$ $(\alpha < \omega_1)$), and using the arguments from the previous sections, we arrive at the following useful description of $\mathcal{U}(T)$.

Lemma 4.3.5. *For every $\Gamma \in \mathcal{U}(T)$ and uncountable $\Omega \subseteq \omega_1$ there is an uncountable $\Sigma \subseteq \Omega$ such that $\Delta_t(\Sigma) \subseteq \Gamma$.* □

It follows that sets of the form $\Delta_t(\Sigma)$ for uncountable $\Sigma \subseteq \omega_1$ generate the filter $\mathcal{U}(T)$. Using these concepts one also gets a useful reformulation of the quasi-ordering relation $\leq$ in the class of coherent A-trees.

Lemma 4.3.6. *Assuming MA_{ω_1}, the following are equivalent for every pair S and T of coherent A-trees:*

(a) $S \leq T$,

(b) *there is an uncountable $\Gamma \subseteq \omega_1$ such that $\Delta_s(\alpha,\beta) \leq \Delta_t(\alpha,\beta)$ for all $\alpha,\beta \in \Gamma$, $\alpha \neq \beta$,*

(c) *for every uncountable $\Sigma \subseteq \omega_1$ there is an uncountable $\Gamma \subseteq \Sigma$ such that $\Delta_s(\alpha,\beta) \leq \Delta_t(\alpha,\beta)$ for all $\alpha,\beta \in \Gamma$, $\alpha \neq \beta$.*

Proof. To deduce (b) from (a), suppose we are given a strictly increasing level-preserving map $f : S \longrightarrow T$. Apply Lemma 4.2.5 to $(t_\delta, f(s_\delta))$ $(\delta \in \omega_1)$ and obtain an uncountable $\Gamma \subseteq \omega_1$ such that for all $\gamma \neq \delta$:

(1) t_γ and t_δ are incomparable,

(2) $f(s_\gamma)$ and $f(s_\delta)$ are incomparable,

(3) $\Delta(t_\gamma, t_\delta) = \Delta(f(s_\gamma), f(s_\delta))$.

Clearly, this Γ satisfies (b). Similarly one shows that (a) implies (c). Note that clause (b) simply says that the map $s_\delta \longmapsto t_\delta$ $(\delta \in \Gamma)$ is Lipschitz and it therefore extends to a strictly increasing map from the downward closure S_0 of the set $\{s_\delta : \delta \in \Gamma\}$ in S. By Lemma 4.3.3, $S \leq S_0 \leq T$. This shows that (b), and therefore the stronger (c), implies (a) and finishes the proof. □

The next lemma gives us a convenient reformulation of the inequality $T \not\leq S$.

Lemma 4.3.7. *Assuming MA_{ω_1}, the following are equivalent for every pair S and T of coherent A-trees:*

(a) $T \not\leq S$,

(b) *there is an uncountable $\Gamma \subseteq \omega_1$ such that $\Delta_s(\alpha,\beta) < \Delta_t(\alpha,\beta)$ for all $\alpha,\beta \in \Gamma$, $\alpha \neq \beta$,*

(c) *for every uncountable $\Sigma \subseteq \omega_1$ there is an uncountable $\Gamma \subseteq \Sigma$ such that $\Delta_s(\alpha,\beta) < \Delta_t(\alpha,\beta)$ for all $\alpha,\beta \in \Gamma$.*

Proof. To see that (a) implies (b), let $\mathcal{P}$ be the poset of all finite $p \subseteq \omega_1$ such that

(1) $\Delta_t(\alpha, \beta) \leq \Delta_s(\alpha, \beta)$ for all $\alpha, \beta \in p$, $\alpha \neq \beta$.

If $\mathcal{P}$ would satisfy the countable chain condition, an application of MA_{ω_1} would give us an uncountable set $\Gamma \subseteq \omega_1$ such that $\Delta_t(\alpha, \beta) \leq \Delta_s(\alpha, \beta)$ for all $\alpha, \beta \in \Gamma$, $\alpha \neq \beta$ which by Lemma 4.3.6 would give us $T \leq S$, contradicting (a). So let p_δ ($\delta \in \omega_1$) be a given sequence of pairwise incompatible members of $\mathcal{P}$. We may assume that $\min(p_\delta) \geq \delta$ for all $\delta \in \omega_1$. For $\delta \in \omega_1$, let

$$a_\delta = \{s_\xi \upharpoonright \delta : \xi \in p_\delta\} \text{ and } b_\delta = \{t_\xi \upharpoonright \delta : \xi \in p_\delta\}.$$

For $\delta \in \Lambda$, let $h(\delta)$ be the maximum of all ordinals that have the form $\Delta(x, y) + 1$, $x, y \in a_\delta$, $x \neq y$ and $\Delta(x, y) + 1$, $x, y \in b_\delta$, $x \neq y$. Find a stationary $\Omega \subseteq \Lambda$ such that h is constant on Ω and let $\bar{\xi}$ be the constant value. Shrinking Ω, we may assume that all a_δ ($\delta \in \Omega$) are of some fixed size m, and that all b_δ ($\delta \in \Omega$) are of some fixed size n. Let $a_\delta(i)$ ($i < m$) and $b_\delta(j)$ ($j < n$) be fixed enumerations. Applying Lemma 4.2.5, we can find an uncountable $\Sigma \subseteq \Omega$ such that for all $\gamma \neq \delta$ in Σ:

(2) $a_\gamma(i) \upharpoonright \bar{\xi} = a_\delta(i) \upharpoonright \bar{\xi}$ for all $i < m$,
(3) $b_\gamma(i) \upharpoonright \bar{\xi} = b_\delta(i) \upharpoonright \bar{\xi}$ for all $i < n$,
(4) $a_\gamma(i)$ and $a_\delta(j)$ are incomparable for all $i, j < m$,
(5) $b_\gamma(i)$ and $b_\delta(j)$ are incomparable for all $i, j < m$,
(6) $\Delta(a_\gamma(i), a_\delta(i)) = \Delta(a_\gamma(j), a_\delta(j))$ for all $i, j < m$,
(7) $\Delta(b_\gamma(i), b_\delta(i)) = \Delta(b_\gamma(j), b_\delta(j))$ for all $i, j < n$.

Consider $\gamma \neq \delta$ in Σ. Then $p_\gamma \cup p_\delta$ fails to satisfy condition (1), i.e., there exist $\xi \in p_\gamma$ and $\eta \in p_\delta$ such that $\Delta(s_\xi, s_\eta) < \Delta(t_\xi, t_\eta)$. Let $s_\xi \upharpoonright \gamma = a_\gamma(i)$, $s_\eta \upharpoonright \delta = a_\delta(j)$, $t_\xi \upharpoonright \gamma = b_\gamma(k)$ and $t_\eta \upharpoonright \delta = b_\delta(l)$. Using (2),(3),(4) and the fact that p_γ and p_δ satisfy (1), we conclude that $i = j$ and $k = l$ and therefore, by (6) and (7), we have the following:

(8) $\Delta(a_\gamma(0), a_\delta(0)) = \Delta(a_\gamma(i), a_\delta(i)) < \Delta(b_\gamma(k), b_\delta(k))$
$= \Delta(b_\gamma(0), b_\delta(0))$.

Of course, we may assume that the enumerations of a_δ and b_δ are given in a way such that if $\xi(\delta) = \min(p_\delta)$, then $s_{\xi(\delta)} \upharpoonright \delta = a_\delta(0)$ and $t_{\xi(\delta)} \upharpoonright \delta = b_\delta(0)$ for all δ in Σ. Let $\Gamma = \{\xi(\delta) : \delta \in \Gamma\}$. Then Γ satisfies clause (b), finishing thus the proof that (a) implies (b). Similarly, one proves that (a) in fact implies (c). The implication from (b) to (a) follows from Lemma 4.3.6. This finishes the proof. □

Definition 4.3.8. The *uniform closure* of a given downward closed subtree T of $\omega^{<\omega_1}$ is defined to be

$$T^* = \{x : \alpha \longrightarrow \omega : \alpha < \omega_1 \text{ and } x =^* t \text{ for some } t \in T_\alpha\}.$$

If the tree T is given to us as a downward closed subtree of some tree of the form $k^{<\omega_1}$ where $k < \omega$ then it is natural to take its uniform closure T^* inside the set $k^{<\omega_1}$ as well in which case every node of T^* will branch into exactly k immediate successors. In all these cases, we say that T is a *uniform tree* if $T = T^*$.

Uniform trees are natural whenever one considers the isomorphism relation on some class of trees. The following lemma explains why in the context of coherent trees it suffices to study only the quasi-ordering $\leq$ and the associated equivalence relation $\equiv$.

Lemma 4.3.9. *Under* MA_{ω_1}, *two uniform coherent A-trees of the same branching degree* $k \leq \omega$ *are isomorphic if and only if they are equivalent.*

Proof. If $S \equiv T$, then by Lemma 4.3.6, there is an uncountable set $\Gamma \subseteq \omega_1$ such that Δ_s and Δ_t agree on pairs from Γ. Let $\mathcal{P}$ be the poset of all partial finite level-preserving isomorphisms p from S into T which extend to isomorphisms between downward closures of $\mathrm{dom}(p)$ in S and $\mathrm{rang}(p)$ in T. Using the agreement of Δ_s and Δ_t on Γ and the argument from the proof of Lemma 4.3.7, one shows that $\mathcal{P}$ satisfies the countable chain condition. The uniformity of the trees S and T are used to show that all sets of the form

$$\mathcal{D}_x = \{p \in \mathcal{P} : x \in \mathrm{dom}(p)\} \ (x \in S),$$

$$\mathcal{E}_y = \{p \in \mathcal{P} : y \in \mathrm{rang}(p)\} \ (y \in T)$$

are dense-open in $\mathcal{P}$. A filter of $\mathcal{P}$ which intersects all these sets gives us the required isomorphism between S and T. □

Finally, we arrive at the following result that reveals another unexpected structure in the class of coherent trees.

Theorem 4.3.10. *Assuming* MA_{ω_1}, *every two coherent trees are comparable.*

Proof. Suppose we are given a pair S and T of coherent A-trees such that $T \nleq S$. By Lemma 4.3.7, there is an uncountable $\Gamma \subseteq \omega_1$ such that $\Delta_s(\alpha, \beta) < \Delta_t(\alpha, \beta)$ for all $\alpha, \beta \in \Gamma$, $\alpha \neq \beta$. By Lemma 4.3.6, we conclude that $T \leq S$. This completes the proof. □

Remark 4.3.11. While under MA_{ω_1}, the class of coherent trees is totally ordered by $\leq$, Corollary 4.1.8 gives us that this chain of trees is not well ordered. This should be compared with an old result of Ohkuma [81] that the class of all scattered trees is well ordered by $\leq$ and a result of Laver [68] that the class of σ-scattered trees is w.q.o. under $\leq_1$. It turns out that the class of all A-trees is not totally ordered under $\leq$ as the following result from [123] shows.

Theorem 4.3.12. *There exist A-trees S and T such that* $S \nleq T$ *and* $T \nleq S$. *In fact, there is an uncountable family* T_i $(i \in I)$ *of A-trees such that* $T_i \nleq T_j$ *for* $i \neq j$. □

The following concept (already appearing in the previous section though in a different guise) will be quite useful in our further analysis of the class of coherent A-trees (or, equivalently, the class of Lipschitz trees).

Definition 4.3.13. Two trees S and T are Δ-*equivalent* on some set of ordinals Γ, if there are level-sequences $s_\alpha \in S_\alpha$ $(\alpha \in \Gamma)$ and $t_\alpha \in T_\alpha$ $(\alpha \in \Gamma)$ such that for every unordered triple α, β, γ of elements of Γ,

$$\Delta(s_\alpha, s_\beta) > \Delta(s_\alpha, s_\gamma) \text{ iff } \Delta(t_\alpha, t_\beta) > \Delta(t_\alpha, t_\gamma). \tag{4.3.2}$$

Lemma 4.3.14. *Assuming the Proper Forcing Axiom, every pair of coherent A-trees S and T are Δ-equivalent on some uncountable set of levels.*

Proof. Fix level-sequences $s_\alpha \in S_\alpha$ $(\alpha \in \omega_1)$ and $t_\alpha \in T_\alpha$ $(\alpha \in \omega_1)$. Let $\mathcal{P}$ be the set of all pairs $p = (\Gamma_p, C_p)$ of finite subsets of ω_1 such that S and T are equivalent on Γ_p as witnessed by the fixed level-sequences and such that for all $\alpha, \beta \in \Gamma_p$ and $\xi \in C_p$,

$$\Delta(s_\alpha, s_\beta) < \xi \text{ iff } \Delta(t_\alpha, t_\beta) < \xi. \tag{4.3.3}$$

We let $\mathcal{P}$ be ordered by coordinatewise inclusions. To see that $\mathcal{P}$ is a proper partial ordering, note that if M is a countable elementary submodel of some large-enough structure of the form $(H_\theta, \in)$, then for every $p \in \mathcal{P} \cap M$, the extension $q = (\Gamma_p, C_p \cup \{M \cap \omega_1\})$ is an M-generic condition of $\mathcal{P}$. Note also that a condition of the form $(\{\xi\}, \{\xi\})$, where $\xi = M \cap \omega_1$ for such a model M, forces the unions of both projections of the generic filter to be uncountable. Hence an application of the Proper Forcing Axiom gives us the conclusion of Lemma 4.3.14. □

Let us now turn our attention to the transformation $T \longmapsto \mathcal{U}(T)$ and show that it provides a complete invariant of the quasi-ordered class $(\mathcal{C}, \leq)$ of coherent A-trees.

Theorem 4.3.15. *Under* MA_{ω_1}, *two coherent A-trees are equivalent if and only if the corresponding ultrafilters are equal.*

Proof. Using Lemma 4.3.3, we may work with a pair of trees of the form $T(a)$ and $T(b)$ where a and b are a pair of coherent and nontrivial mappings, and prove that the condition $T(a) \equiv T(b)$ is equivalent to the condition $\mathcal{U}(a) = \mathcal{U}(b)$. Choose a pair of strictly increasing mappings

$$f : T(a) \longrightarrow T(b) \text{ and } g : T(b) \longrightarrow T(a).$$

Replacing f by the mapping $t \longmapsto f(t) \restriction \mathrm{dom}(t)$ and replacing g by the mapping $t \longmapsto g(t) \restriction \mathrm{dom}(t)$, we may assume that f and g are also level-preserving. Recall our notation $a_\delta(\xi) = a(\xi, \delta)$ and $b_\delta(\xi) = b(\xi, \delta)$ that gives us particular representatives a_δ and b_δ of the δth level of $T(a)$ and $T(b)$ respectively. For $\delta \in \Lambda$, set

$$F_\delta = \{\xi < \delta : a_\delta(\xi) \neq g(f(a_\delta))(\xi) \text{ or } b_\delta(\xi) \neq f(a_\delta)(\xi)\}.$$

Then F_δ is a finite subset of δ. Find a stationary subset Σ of Λ and $F \subseteq \omega_1$ such that $F_\delta = F$ for all $\delta \in \Sigma$. Find now an uncountable $\Gamma \subseteq \Sigma$ such that a_δ $(\delta \in \Gamma)$

is an antichain of $T(a)$, b_δ $(\delta \in \Gamma)$ is an antichain of $T(b)$, and the mapping

$$\delta \longmapsto (a_\delta \restriction \bar{\xi}, f(a_\delta) \restriction \bar{\xi}, g(f(a_\delta)) \restriction \bar{\xi}, b_\delta \restriction \bar{\xi})$$

is constant on Γ, where $\bar{\xi} = \max(F) + 1$. It follows that for $\gamma \neq \delta$ in Γ,

$$\Delta(a_\gamma, a_\delta) = \Delta(g(f(a_\gamma)), g(f(a_\delta))), \tag{4.3.4}$$

$$\Delta(b_\gamma, b_\delta) = \Delta(f(a_\gamma), f(a_\delta)). \tag{4.3.5}$$

Since f is strictly increasing, we must have $\Delta(a_\gamma, a_\delta) \leq \Delta(f(a_\gamma), f(a_\delta)) = \Delta(b_\gamma, b_\delta)$. Since g is strictly increasing, we have that

$$\Delta(b_\gamma, b_\delta) = \Delta(f(a_\gamma), f(a_\delta)) \leq \Delta(g(f(a_\gamma)), g(f(a_\delta))) = \Delta(a_\gamma, a_\delta).$$

It follows that $\Delta(a_\gamma, a_\delta) = \Delta(b_\gamma, b_\delta)$ for all $\gamma \neq \delta$ in Γ. From the proof of Lemma 4.1.11 we conclude that $\Delta_a(\Omega)$ $(\Omega \subseteq \Gamma, \Omega$ uncountable) generates the filter $\mathcal{U}(a)$ and similarly that $\Delta_b(\Omega)$ $(\Omega \subseteq \Gamma, \Omega$ uncountable) generates the ultrafilter $\mathcal{U}(b)$. It follows that $\mathcal{U}(a) = \mathcal{U}(b)$.

Suppose now that $\mathcal{U}(a) = \mathcal{U}(b)$. By symmetry, it suffices to show that $T(a) \leq T(b)$. Note again that the argument in the proof of Lemma 4.1.11 shows that there is a stationary set $\Gamma \subseteq \Lambda$ such that Δ_a and Δ_b coincide on $[\Gamma]^2$. Let $\mathcal{P}$ be the poset of all finite partial strictly increasing level-preserving mappings p from $T(a)$ into $T(b)$ which (uniquely) extend to strictly increasing level-preserving mappings on the downward-closures of their domains. By MA_{ω_1}, it suffices to show that $\mathcal{P}$ satisfies the countable chain condition. So let p_δ $(\delta \in \Gamma)$ be a given sequence of conditions of $\mathcal{P}$. For $\delta \in \Gamma$, set

$$E_\delta = \{\xi < \delta : a_\delta(\xi) \neq t(\xi) \text{ or } b_\delta(\xi) \neq p(t)(\xi) \\ \text{for some } t \in \mathrm{dom}(p_\delta) \text{ of height} \geq \delta\}.$$

Then E_δ is a finite subset of δ for each $\delta \in \Gamma$. Find a stationary set $\Sigma \subseteq \Gamma$ and $E \subseteq \omega_1$ such that $E_\delta = E$ for all $\delta \in \Sigma$. Shrinking Σ, we may assume that the mapping $\delta \longmapsto p_\delta \restriction \delta$ is constant on Σ. Let $\bar{\eta}$ be the minimal ordinal strictly above $\max(E)$ and the height of every node of $\mathrm{dom}(p_\delta) \restriction \delta$ for some (all) $\delta \in \Sigma$. For $\delta \in \Sigma$, let $\hat{p}_\delta$ be the extension of p_δ on the downward-closure of its domain in $T(a)$. Find a stationary $\Omega \subseteq \Sigma$ such that each of the projections $\delta \longmapsto \hat{p}_\delta \restriction \bar{\eta}$, $\delta \longmapsto a_\delta \restriction \bar{\eta}$ and $\delta \longmapsto b_\delta \restriction \bar{\eta}$ is constant on Ω. Find $\gamma < \delta$ in Ω such that $a_\gamma \neq a_\delta \restriction \gamma$, $b_\gamma \neq b_\delta \restriction \gamma$ and $\hat{p}_\gamma$ and $\hat{p}_\delta$ are isomorphic via the isomorphism that fixes their common restriction to the $\bar{\eta}$th levels of $T(a)$ and $T(b)$. It suffices to check that $q = p_\gamma \cup p_\delta$ satisfies

$$\Delta(q(x), q(y)) \geq \Delta(x, y) \tag{4.3.6}$$

for every $x \in \mathrm{dom}(p_\gamma)$ and $y \in \mathrm{dom}(p_\delta)$. If x and y correspond to each other in the isomorphism between $\hat{p}_\gamma$ and $\hat{p}_\delta$, then either they belong to the common part of p_γ and p_δ in which case (4.3.6) follows by the assumption that p_γ and p_δ are strictly increasing level-preserving, or we have

$$\Delta(x, y) = \Delta(a_\gamma, a_\delta) = \Delta(b_\gamma, b_\delta) = \Delta(p_\gamma(x), p_\delta(y)). \tag{4.3.7}$$

If x is different from the copy $\bar{y} \in \mathrm{dom}(p_\gamma)$ of $y \in \mathrm{dom}(p_\delta)$, then $\Delta(x, y) = \Delta(x, \bar{y})$ and $\Delta(p_\gamma(x), p_\delta(y)) = \Delta(p_\gamma(x), p_\gamma(y))$, so we conclude that (4.3.6) follows from the fact that p_γ is strictly increasing level-preserving. This finishes the proof. □

It turns out that the chain formed by the class of coherent trees in $\leq$ sequence is discrete and in fact there is a natural shift operator giving the immediate successor $T^{(1)}$ of a given coherent tree T.

Definition 4.3.16. For an integer m and a tree T, we let $T^{(m)}$ be its mth *shift*, the downward-closure of $\{t^{(m)} : t \in T \restriction \Lambda\}$, where for a limit node t of T, we let $t^{(m)}$ be the function with the same domain λ as t defined by

$$t^{(m)}(\xi) = t(\xi - m),$$

when $\xi - m$ exists; otherwise (i.e., when m is positive and the largest limit ordinal $\leq \xi$ is less than m steps away), we let $t^{(m)}(\xi) = 0$.

Remark 4.3.17. Note that a positive shift $T^{(m)}$ of any coherent tree T is coherent and that the map $t \longmapsto t^{(m)}$ is a strictly increasing map from T into $T^{(m)}$. It follows that $T \leq T^{(m)}$ for all $m \geq 0$. It is for this reason that we choose to switch the direction of the shift operator on coherent mappings defined in the previous section. Note also that for non-negative integers m and n,

$$T^{(m+n)} = (T^{(m)})^{(n)}$$

holds. Therefore we have that $T^{(m)} \leq T^{(n)}$ for every pair of non-negative integers m and n such that $m \leq n$.

Lemma 4.3.18. *Suppose T is a coherent A-tree in which every uncountable subset contains an uncountable antichain. Then $T^{(m)} < T^{(n)}$ for every pair of non-negative integers m and n such that $m < n$.*

Proof. Suppose $m < n$ are non-negative integers and consider a level-preserving map $f : T^{(n)} \longrightarrow T^{(m)}$. For each countable ordinal δ, pick a representative t_δ from the δth level of T and let $s_\delta \in T_\delta$ be such that

$$f(t_\delta^{(n)}) = s_\delta^{(m)}.$$

Working as before, we find an uncountable set Γ of countable limit ordinals such that $\Delta(t_\gamma, t_\delta) = \Delta(s_\gamma, s_\delta)$ for all $\gamma, \delta \in \Gamma$, $\gamma \neq \delta$. Choose $\gamma \neq \delta$ in Γ such that

$$\alpha = \Delta(t_\gamma, t_\delta) = \Delta(s_\gamma, s_\delta)$$

is smaller than both γ and δ (i.e., t_γ is incomparable to t_δ and s_γ is incomparable to s_δ). Then

$$\Delta(t_\gamma^{(n)}, t_\delta^{(n)}) = \alpha + n > \alpha + m = \Delta(s_\gamma^{(m)}, s_\delta^{(m)}).$$

This shows that f is not a strictly increasing map, finishing the proof. □

The following lemma reveals that in the realm of coherent A-trees, the shift $T^{(1)}$ is indeed the minimal tree above T.

Lemma 4.3.19. *Under* MA_{ω_1}*, for every pair S and T of coherent trees, the inequality $S < T$ implies $S^{(1)} \leq T$.*

Proof. Choose representatives $s_\delta \in S_\delta$ $(\delta < \omega_1)$ and $t_\delta \in T_\delta$ $(\delta < \omega_1)$. Working as before, we show that there is an uncountable set Γ of countable limit ordinals such that $\Delta(s_\gamma, s_\delta) < \Delta(t_\gamma, t_\delta)$ for all $\gamma \neq \delta$ in Γ. We have already seen above during the course of the proof of Lemma 4.3.18 that

$$\Delta(s_\gamma^{(1)}, s_\delta^{(1)}) = \Delta(s_\gamma, s_\delta) + 1 \leq \Delta(t_\gamma, t_\delta)$$

for all $\gamma \neq \delta$ in Γ. This is sufficient for helping us in showing that the poset of all finite partial level-preserving Lipschitz maps from $S^{(1)}$ into T satisfies the countable chain condition. An application MA_{ω_1} will now give us the desired conclusion $S^{(1)} \leq T$. □

It follows that (under MA_{ω_1}) for every coherent tree T, the chain

$$T^{(n)} \ (n \in \mathbb{N})$$

of positive shifts is really an $\mathbb{N}$-chain, i.e., its convex closure inside the class of coherent A-trees is isomorphic to $\mathbb{N}$ as an ordered set. The case of negative shifts is a bit more subtle though we shall see now that they also behave as expected.

Definition 4.3.20. A tree T is *orthogonal* to a set of ordinals Γ, if there is an uncountable subset X of T such that $\Delta(x, y) \notin \Gamma$ for all $x, y \in X$, $x \neq y$.

The proof of the following lemma is similar to the proof of Theorem 4.1.6 from the previous section.

Lemma 4.3.21. *Suppose that $n < m \leq 0$ and that T is a coherent A-tree which is orthogonal to $\Lambda + k$ for all $0 \leq k \leq |n|$. Then $T^{(m)}$ and $T^{(n)}$ are also coherent and $T^{(m)} \not\leq T^{(n)}$.* □

The following result gives us a sufficient condition for the existence of a $\mathbb{Z}$-chain in the class of coherent A-trees under the total quasi-ordering $\leq$.

Theorem 4.3.22. *Under* MA_{ω_1}*, for every coherent A-tree T which is orthogonal to $\Lambda + k$ for all $k \geq 0$, the shifts $T^{(n)}$ $(n \in \mathbb{Z})$ form a family of coherent A-trees with the following properties:*

(1) $T^{(m+n)} \equiv (T^{(m)})^{(n)}$,

(2) $T^{(n)} < T^{(m)}$ *iff* $n < m$,

(3) *no coherent tree S satisfies $T^{(n)} < S < T^{(n+1)}$ for some $n \in \mathbb{Z}$.* □

Definition 4.3.23. Suppose that g is a partial map from ω_1 into ω_1 and that T is a given tree represented as a downward-closed subtree of the tree of functions

$$\{t : \alpha \to \omega : \alpha < \omega_1\}.$$

Then the *g-shift* of T, denoted by $T^{(g)}$ is the downwards-closure of

$$\{t^{(g)} : t \in T \restriction \Omega\},$$

where $\Omega = \{\delta < \omega_1 : g''\delta \subseteq \delta\}$ and $t^{(g)}$ is defined by

$$t^{(g)}(\xi) = t(g(\xi))$$

if $\xi \in \mathrm{dom}(g)$; otherwise $t^{(g)}(\xi) = 0$.

Note that above, we considered the shifts $T^{(g)}$ for maps of the form $g(\xi) = \xi - m$, but it should be clear that the arguments from that special case are sufficient to give us the following two facts.

Lemma 4.3.24. *If g is a partial strictly increasing map on ω_1 and if* $\mathrm{rang}(g) \in \mathcal{U}(T)$ *for some Lipschitz tree T, then the g-shift $T^{(g)}$ is also a Lipschitz tree.* □

Lemma 4.3.25. *Suppose that T is a Lipschitz tree and g is a strictly increasing partial map on ω_1 such that* $\mathrm{rang}(g) \in \mathcal{U}(T)$. *If g is regressive*[13] *then $T^{(g)} \not\leq T$. On the other hand, if g is expanding*[14] *then $T \not\leq T^{(g)}$.* □

Remark 4.3.26. Note that if g is a strictly increasing regressive partial map on ω_1 such that $\mathrm{rang}(g) = \omega_1$, then $T < T^{(g)}$ holds for every Lipschitz tree T. This observation can be used to construct both strictly increasing and strictly decreasing ω_1-sequences of Lipschitz trees.

Theorem 4.3.27. *Assuming the Proper Forcing Axiom, for every pair S and T of Lipschitz trees, there is a strictly increasing partial map g on ω_1 such that $S \equiv T^{(g)}$.*

Proof. By Lemma 4.3.14, we have an uncountable $\Gamma \subseteq \omega_1$ on which S and T are Δ-equivalent for some choice of level-sequences $s_\alpha \in S_\alpha$ $(\alpha \in \Gamma)$ and $t_\alpha \in T_\alpha$ $(\alpha \in \Gamma)$. Define

$$g : \{\Delta(s_\alpha, s_\beta) : \alpha, \beta \in \Gamma, \alpha \neq \beta\} \longrightarrow \{\Delta(t_\alpha, t_\beta) : \alpha, \beta \in \Gamma, \alpha \neq \beta\}$$

by letting $g(\Delta(s_\alpha, s_\beta)) = \Delta(t_\alpha, t_\beta)$. By (4.3.2), this is a well-defined strictly increasing partial map on ω_1 such that $\mathrm{dom}(g) \in \mathcal{U}(S)$ and $\mathrm{rang}(g) \in \mathcal{U}(T)$. Let

$$\Omega = \{\delta < \omega_1 : g''\delta \subseteq \delta\}.$$

Assuming that $\{s_\alpha : \alpha \in \Gamma\}$ and $\{t_\alpha : \alpha \in \Gamma\}$ form antichains, then replacing each s_α and t_α by their extensions on the $\delta(\alpha)$th level of S and T respectively, where

[13] I.e., $g(\xi) < \xi$ for all $\xi \in \mathrm{dom}(g)$.

[14] I.e., $g(\xi) > \xi$ for all $\xi \in \mathrm{dom}(g)$.

$\delta(\alpha) = \min(\Omega \setminus \alpha+1)$, and finally replacing Γ with $\{\delta(\alpha) : \alpha \in \Gamma\}$, we may assume that Γ is actually a subset of Ω. By Lemma 4.3.24, the shift $T^{(g)}$ is a Lipschitz tree and $t_\alpha^{(g)}$ $(\alpha \in \Gamma)$ is its level-sequence such that

$$\Delta(t_\alpha^{(g)}, t_\beta^{(g)}) = \Delta(s_\alpha, s_\beta) \text{ for all } \alpha, \beta \in \Gamma, \alpha \neq \beta. \tag{4.3.8}$$

By Lemma 4.3.6 we conclude that $S \equiv T^{(g)}$. This finishes the proof. □

Corollary 4.3.28. *Assuming the Proper Forcing Axiom, for every pair S and T of coherent A-trees, there is a strictly increasing partial map g on ω_1 which maps $\mathcal{U}(S)$ into $\mathcal{U}(T)$.*

Proof. By Theorems 4.3.15 and 4.3.27 we may assume that in fact $S = T^{(g)}$ for some strictly increasing partial map g on ω_1 with $\operatorname{rang}(g) \in \mathcal{U}(T)$. Now note that $\operatorname{dom}(g) \in \mathcal{U}(T^{(g)})$ and that $g''A \in \mathcal{U}(T)$ for every $A \in \mathcal{U}(T^{(g)})$. □

The following result shows that the class of coherent trees has a very special relationship with the class of all A-trees.

Theorem 4.3.29. *Assuming the Proper Forcing Axiom, every coherent tree is comparable with every Aronszajn tree.*

Proof. By Lemma 4.2.23, we can concentrate on showing that LMA together with MA_{ω_1} yield the conclusion of the theorem. Let T be a given coherent A-tree and let S be any other A-tree. Let $\mathcal{P}$ be the poset of all finite partial level-preserving Lipschitz maps from S into T. If $\mathcal{P}$ satisfies the countable chain condition, then an application of MA_{ω_1} would give us a total strictly increasing map $f : S \longrightarrow T$ witnessing thus the relation $S \leq T$. So we are left with the alternative that $\mathcal{P}$ fails to satisfy the countable chain condition. Let $\mathcal{X}$ be an uncountable family of pairwise incomparable members of $\mathcal{P}$. We may assume that $\mathcal{X}$ forms a Δ-system. Removing the root, which obviously does not contribute to the incomparability between members of $\mathcal{X}$, we may in fact assume that $\operatorname{dom}(p)$ $(p \in \mathcal{X})$ is a family of pairwise-disjoint finite subsets of S, all of some fixed size n. For $p \in \mathcal{X}$, let $s_i(p)$ $(i < n)$ enumerate $\operatorname{dom}(p)$ and let $t_i(p) = p(s_i(p))(i < n)$ enumerate its range. Applying the Δ-System Lemma again, we obtain an uncountable subfamily $\mathcal{Y}$ of $\mathcal{X}$ such that the families of finite subtrees

$$\{\Delta(s_i(p), s_j(p)) : i < j < n\} \ (p \in \mathcal{Y}),$$

and

$$\{\Delta(t_i(p), t_j(p)) : i < j < n\} \ (p \in \mathcal{Y})$$

both form Δ-systems. Shrinking the family $\mathcal{Y}$ even further, we may assume that if α is the least upper bound of the roots of these two Δ-systems, then for all $p \neq q$ in $\mathcal{Y}$:

$$s_i(p) \upharpoonright \alpha = s_i(q) \upharpoonright \alpha \text{ and } t_i(p) \upharpoonright \alpha = t_i(q) \upharpoonright \alpha \text{ for all } i < n.$$

Since any pair $p \neq q$ of conditions from $\mathcal{Y}$ are incompatible in $\mathcal{P}$, it follows that there exists $i = i(p,q) < n$ such that:

$$\Delta(s_i(p), s_i(q)) > \Delta(t_i(p), t_i(q)).$$

Apply the LMA successively n times and obtain an uncountable subset $\mathcal{Z}$ of $\mathcal{Y}$ such that for all $i < n$, the map

$$s_i(p) \longmapsto t_i(p) \ (p \in \mathcal{Z})$$

or its inverse is Lipschitz. If for every $i < n$, the map $s_i(p) \longmapsto t_i(p)$ $(p \in \mathcal{Z})$ is Lipschitz, we would obtain two compatible members of $\mathcal{X}$ contradicting our initial assumption about $\mathcal{X}$. Hence, there must be $i < n$ so that the inverse map

$$g : t_i(p) \longmapsto s_i(p) \ (p \in \mathcal{Z})$$

is Lipschitz. Let T_0 be the downward-closure of $\{t_i(p) : p \in \mathcal{Z}\}$ in T. Then T_0 is uncountable and g extends to a Lipschitz map $\hat{g} : T_0 \longrightarrow S$ which is in particular strictly increasing. It follows that $T_0 \leq S$. Since T is irreducible, we have that $T \leq T_0$, and therefore $T \leq S$. This finishes the proof. □

The following analogue of Lemma 4.3.19 shows that the shift transformation is giving us a true successor operation valid in the class of all trees.

Theorem 4.3.30. *Assuming the Proper Forcing Axiom, no tree lies strictly between a coherent tree T and its shift $T^{(1)}$.*

Proof. Suppose there is a tree S such that $T < S < T^{(1)}$. Clearly, we may assume that S is Aronszajn. During the course of the proof of Theorem 4.3.29, we have seen that, since the poset of all finite Lipschitz maps from S into T fails to satisfy the countable chain condition (since we are assuming MA_{ω_1}), an application of LMA to an uncountable family of pairwise incomparable members of this poset will provide us with an uncountable set $\Gamma \subseteq \Lambda$ and level-sequences $s_\gamma \in S_\gamma$ $(\gamma \in \Gamma)$, $t_\gamma \in T_\gamma$ $(\gamma \in \Gamma)$ such that

$$\Delta(s_\gamma, s_\delta) > \Delta(t_\gamma, t_\delta) \text{ for all } \gamma \neq \delta \text{ in } \Gamma.$$

Let S_0 be the downward closure of $\{s_\gamma : \gamma \in \Gamma\}$ in S. Consider now the poset of all finite partial Lipschitz maps from $T^{(1)}$ into S_0. By MA_{ω_1}, this poset cannot satisfy the countable chain condition either. So applying LMA again to an uncountable family of pairwise incompatible members of this poset would give us uncountable level-sequences $r_\gamma \in S_\gamma$ $(\gamma \in \Sigma)$, $q_\gamma^{(1)} \in T_\gamma^{(1)}$ $(\gamma \in \Sigma)$ such that

$$\Delta(q_\gamma^{(1)}, q_\delta^{(1)}) > \Delta(r_\gamma, r_\delta) \text{ for all } \gamma \neq \delta \text{ in } \Sigma.$$

For each $\gamma \in \Sigma$, choose $\xi(\gamma) \in \Gamma$ such that

$$\xi(\gamma) \geq \gamma \text{ and } r_\gamma = s_{\xi(\gamma)} \upharpoonright \gamma.$$

We may also assume that $\{q_\gamma : \gamma \in \Sigma\}$, $\{q_\gamma^{(1)} : \gamma \in \Sigma\}$ and $\{r_\gamma : \gamma \in \Sigma\}$ are all antichains in their respective trees. For each $\gamma \in \Sigma$, find $p_{\xi(\gamma)} \in T_{\xi(\gamma)}$ such that $p_{\xi(\gamma)} \upharpoonright \gamma = q_\gamma$ and therefore

$$p_{\xi(\gamma)}^{(1)} \upharpoonright \gamma = q_\gamma^{(1)}.$$

Applying the already standard coherent tree argument, we get an uncountable $\Sigma_0 \subseteq \Sigma$ such that

$$\Delta(t_{\xi(\gamma)}, t_{\xi(\delta)}) = \Delta(p_{\xi(\gamma)}, p_{\xi(\delta)}) \text{ for all } \gamma \neq \delta \text{ in } \Sigma_0.$$

Combining all this, we get the following for all $\gamma \neq \delta$ in Σ_0:

$$\Delta(p_{\xi(\gamma)}^{(1)}, p_{\xi(\delta)}^{(1)}) = \Delta(q_\gamma^{(1)}, q_\delta^{(1)}) > \Delta(r_\gamma, r_\delta) = \Delta(s_{\xi(\gamma)}, s_{\xi(\delta)})$$
$$> \Delta(t_{\xi(\gamma)}, t_{\xi(\delta)}) = \Delta(p_{\xi(\gamma)}, p_{\xi(\delta)}).$$

It follows that for $\gamma \neq \delta$ in Σ_0:

$$\Delta(p_{\xi(\gamma)}^{(1)}, q_{\xi(\delta)}^{(1)}) \geq \Delta(p_{\xi(\gamma)}, p_{\xi(\delta)}) + 2,$$

a contradiction. This finishes the proof. □

Remark 4.3.31. The very special properties of the discrete chain of coherent trees described above might give us an impression that the real reason behind the so fine structure theory lies in the fact that the class of trees is rather small. The results that follow show, however, that this is not the case.

Lemma 4.3.32. *Assuming* MA_{ω_1}, *for every A-tree S, there is a coherent A-tree T such that* $S \leq T$.

Proof. Let $\mathcal{P}$ be the set of all finite partial functions p from $S \times \omega_1$ into ω such that:

(1) $\xi < \mathrm{ht}(x)$ for all $(x, \xi) \in \mathrm{dom}(p)$,
(2) $p(x, \xi) = p(y, \xi)$ for all $(x, \xi), (y, \xi) \in \mathrm{dom}(p)$ with $\xi < \Delta(x, y)$.

We let p extend q if p extends q as a function and

(3) $p(x, \xi) = p(y, \xi)$ for all $(x, \xi), (y, \xi) \in \mathrm{dom}_0(q)$ and $\xi < \mathrm{ht}(x), \mathrm{ht}(x)$ such that $\xi \notin \mathrm{dom}_1(q(x)) \cup \mathrm{dom}_1(q(y))$[15],
(4) $p(x, \xi) \neq q(x, \eta)$ for all $(x, \eta) \in \mathrm{dom}(q)$ and $(x, \xi) \in \mathrm{dom}(p) \setminus \mathrm{dom}(q)$.

A simple Δ-system argument (contained in the proofs of Lemmas 4.3.3 and 4.2.7 above) shows that $\mathcal{P}$ satisfies the countable chain condition, so an application of MA_{ω_1} gives us a map g from $S \times \omega_1$ into ω so that its fibers $g_x(\xi) = g(x, \xi)$ are total maps from $\mathrm{ht}(x)$ into ω for all $x \in S$ and such that:

[15]Recall the notations $\mathrm{dom}_0(p) = \{x : (x, \eta) \in \mathrm{dom}(p) \text{ for some } \eta\}$, $\mathrm{dom}_1(p) = \{\eta : (x, \eta) \in \mathrm{dom}(p) \text{ for some } x\}$, and $\mathrm{dom}_1(p(x)) = \{\xi < \mathrm{ht}(x) : (x, \xi) \in \mathrm{dom}(p)\}$.

(5) $g_x : \mathrm{ht}(x) \longrightarrow \omega$ is a finite-to-one map for all $x \in S$,

(6) $\Delta(x, y) \leq \Delta(g_x, g_y)$ for all $x, y \in S$,

(7) $\{\xi : g_x(\xi) \neq g_y(\xi)\}$ is finite for all $x, y \in S$.

It follows that the downwards-closure T of $\{g_x : x \in S\}$ is a coherent A-tree and that $x \longmapsto g_x$ is a Lipschitz map from S into T sufficient for witnessing the relation $S \leq T$. This finishes the proof. □

Lemma 4.3.33. *Assuming* MA_{ω_1}, *for every* A*-tree* S, *there is a coherent tree* T *such that* $T \leq S$.

Proof. Let $\mathcal{P}$ be the set of all finite partial functions p from $S \times \omega_1$ into ω such that:

(1) $\xi < \mathrm{ht}(x)$ for all $(x, \xi) \in \mathrm{dom}(p)$,

(2) for every pair x and y of incomparable nodes from $\mathrm{dom}_0(p)$, there is $\xi \leq \Delta(x, y)$ with $(x, \xi), (y, \xi) \in \mathrm{dom}(p)$ and $p(x, \xi) \neq p(y, \xi)$.

We let p extend q if p extends q as a function and

(3) $p(x, \xi) = p(y, \xi)$ for all $(x, \xi), (y, \xi) \in \mathrm{dom}_0(q)$ and $\xi < \mathrm{ht}(x), \mathrm{ht}(x)$ such that $\xi \notin \mathrm{dom}_1(q(x)) \cup \mathrm{dom}_1(q(y))$,

(4) $p(x, \xi) \neq q(x, \eta)$ for all $(x, \eta) \in \mathrm{dom}(q)$ and $(x, \xi) \in \mathrm{dom}(p) \setminus \mathrm{dom}(q)$.

To prove that $\mathcal{P}$ satisfies the countable chain condition, we start with an uncountable subset $\mathcal{X}$ of $\mathcal{P}$ and perform the Δ-system argument from the proof of Lemma 4.3.3, obtaining two conditions p and q in $\mathcal{X}$ such that for some $\bar{\xi} < \alpha < \beta$, $v_0, \ldots, v_n \in S_{\bar{\xi}}$, $s_0, \ldots, s_n \in S_\alpha$ and $t_0, \ldots, t_n \in S_\beta$ we have:

(5) every node of $\mathrm{dom}_0(p)$ is either of height $< \bar{\xi}$ or it extends some s_i $(i \leq n)$,

(6) every node of $\mathrm{dom}_0(q)$ is either of height $< \bar{\xi}$ or it extends some t_i $(i \leq n)$,

(7) $\mathrm{dom}_1(p) \subseteq \bar{\xi} \cup (\alpha, \beta)$ and $\mathrm{dom}_1(q) \subseteq \bar{\xi} \cup (\beta, \omega_1)$,

(8) $v_i \neq v_j$ for $i \neq j \leq n$,

(9) s_i and t_i extend v_i but are incomparable for all $i \leq n$,

(10) p and q are isomorphic conditions via an isomorphism that is the identity on $\bar{\xi}$, v_i $(i \leq n)$ and maps s_i to t_i for all $i \leq n$.

We claim that such p and q can be amalgamated into a condition r of $\mathcal{P}$ that extends them both. Let

$$\xi = \min\{\Delta(s_i, t_i) : i \leq n\}.$$

Then $\bar{\xi} \leq \xi < \alpha$. Let

$$k = \max(\mathrm{rang}(p)) = \max(\mathrm{rang}(q)).$$

Let dom(r) be equal to the union of dom(p), dom(q) and the following two sets:

$$D = \{(x, \xi) : x \in \mathrm{dom}_0(p), \mathrm{ht}(x) \geq \alpha\},$$
$$E = \{(y, \xi) : y \in \mathrm{dom}_0(q), \mathrm{ht}(y) \geq \beta\}.$$

Define r by giving it constant value $k+1$ on D and constant value $k+2$ on E. Note that r satisfies (1) and (2) as well as conditions (3) and (4) for extending both p and q.

Applying MA_{ω_1} to $\mathcal{P}$ and an appropriate family of dense-open subsets of $\mathcal{P}$ gives us a partial map $g : S \times \omega_1 \longrightarrow \omega$ so that if $g_x(\xi) = g(x, \xi)$ then:

(18) g_x is a finite-to-one map from ht(x) into ω for all $x \in S$,

(19) $\Delta(g_x, g_y) \leq \Delta(x, y)$ for all $x, y \in S$,

(20) $\{\xi : g_x(\xi) \neq g_y(\xi)\}$ is finite for all $x, y \in S$.

It follows that the downwards-closure T of $\{g_x : x \in S\}$ is a coherent tree and that $g_x \longmapsto x$ is a partial Lipschitz map from T into S, sufficient for witnessing the relation $T \leq S$. This finishes the proof. □

Theorem 4.3.34. *Assuming* MA_{ω_1}, *there is no maximal Aronszajn tree.*

Proof. Given an Aronszajn tree S by Lemma 4.3.32 we find a Lipschitz tree T such that $S \leq T$. By Lemma 4.3.18, $T < T^{(1)}$, so in particular $S < T^{(1)}$. □

Theorem 4.3.35. *Assuming the Proper Forcing Axiom, for every coherent tree T there is a coherent tree S such that $S < T$.*

Proof. Fix a level-sequence $t_\alpha \in T_\alpha$ $(\alpha \in \omega_1)$ in a given Lipschitz tree T. Let $\Delta_t : [\omega_1]^2 \longrightarrow \omega_1$ be the corresponding distance function $\Delta_t(\alpha, \beta) = \Delta(t_\alpha, t_\beta)$. For $\Gamma \subseteq \omega_1$, let

$$\Delta_t(\Gamma) = \{\Delta_t(\alpha, \beta) : \alpha, \beta \in \Gamma, \alpha \neq \beta\}.$$

Let $\mathcal{P}$ be the set of all pairs $p = (f_p, \Gamma_p)$ such that:

(21) Γ_p is a finite subset of ω_1,

(22) f_p is a finite partial strictly increasing map from ω_1 into ω_1 which can be extended to a total strictly increasing and continuous map $f : \omega_1 \longrightarrow \omega_1$ so that rang(f) is disjoint from $\Delta_t(\Gamma_p)$ and separates[16] the points of $\Delta_t(\Gamma_p)$.

We order $\mathcal{P}$ by coordinatewise inclusion. To show that $\mathcal{P}$ is proper, consider a countable elementary submodel M of some large-enough structure of the form $(H_\theta, \in)$ such that M contains $\mathcal{P}, T$ and the level-sequence t_α $(\alpha \in \omega_1)$. For a given $p \in \mathcal{P} \cap M$, let

$$q = (f_p \cup \{\langle \delta, \delta \rangle\}, \Gamma_p),$$

where $\delta = M \cap \omega_1$. We claim that q is an M-generic condition of $\mathcal{P}$. To show this, consider a dense-open subset $\mathcal{D}$ of $\mathcal{P}$ such that $\mathcal{D} \in M$ and an extension r of q. We

[16] i.e., between every two members of $\Delta_t(\Gamma_p)$, there is a member of rang(f).

need to show that r is compatible with a member of $\mathcal{D} \cap M$. Extending r, we may assume $r \in \mathcal{D}$. Let v_i $(i \leq n)$ be a one-to-one enumeration of $\{t_\alpha \restriction \delta : \alpha \in \Gamma_p \setminus \delta\}$. Let $\bar{p} = r \restriction M$. Then $\bar{p} \in \mathcal{P} \cap M$ and so we can find an extension $\bar{f} : \omega_1 \longrightarrow \omega_1$ of $f_{\bar{p}}$ satisfying (22) for $\bar{p}$ such that $\bar{f} \in M$. Let $\bar{\xi} \in (\max(\Gamma_{\bar{p}}), \delta)$ be a fixed point of $\bar{f}$. Find a copy $\bar{r}$ of r in $\mathcal{D} \cap M$ such that if we let $\bar{\delta}$ and $\bar{v}_i$ $(i \leq n)$ be its versions of δ and v_i $(i \leq n)$, then

(23) $v_i \restriction \bar{\xi} = \bar{v}_i \restriction \bar{\xi}$ for all $i \leq n$,

(24) v_i and $\bar{v}_i$ are incomparable for all $i \leq n$,

(25) $\Delta(v_0, \bar{v}_0) = \cdots = \Delta(v_n, \bar{v}_n)$.

It is clear that we can combine the function $\bar{f}$ with the normal functions witnessing (22) for $\bar{r}$, r and obtain a strictly increasing continuous function $f : \omega_1 \longrightarrow \omega_1$ which fixes $\bar{\xi}$ and witnesses (22) simultaneously for $\bar{r}$, r and moreover, the ordinal $\Delta(v_0, \bar{v}_0)$ is not in its range. Since

$$\Delta_t(\Gamma_{\bar{r}} \cup \Gamma_r) = \Delta_t(\Gamma_{\bar{r}}) \cup \Delta_t(\Gamma_r) \cup \{\Delta(v_0, \bar{v}_0)\},$$

this shows that $(f_{\bar{r}} \cup f_r, \Gamma_{\bar{r}} \cup \Gamma_r)$ is a member of $\mathcal{P}$ witnessing the compatibility of $\bar{r}$ and r.

Applying the Proper Forcing Axiom to $\mathcal{P}$ gives us an uncountable $\Gamma \subseteq \omega_1$ and a closed unbounded set $C \subseteq \omega_1$ such that $C \cap \Delta_t(\Gamma) = \emptyset$ and C separates the points of $\Delta_t(\Gamma)$. For $\delta < \omega_1$, let δ^+ be the minimal point of C above δ. Define

$$C_0 = \{\delta \in C : (\delta, \delta^+) \cap \Delta_t(\Gamma) \neq \emptyset\}.$$

Note that for $\delta \in C$, there is only one point of $\Delta_t(\Gamma)$ in the interval (δ, δ^+). Call this point $g(\delta)$. This defines a strictly increasing map g from C_0 onto $\Delta_t(\Gamma)$. So in particular, $\mathrm{rang}(g) \in \mathcal{U}(T)$. Let $S = T^{(g)}$ (see Definition 4.3.23). From Lemma 4.3.24 we conclude that S is a Lipschitz tree. Since clearly $g(\delta) > \delta$ for all $\delta \in C_0$, we conclude that $T \not\leq S$ from Lemma 4.3.25. This finishes the proof. □

Corollary 4.3.36. *Assuming the Proper Forcing Axiom, there is no minimal Aronszajn tree.*

Proof. This follows from Lemma 4.3.33 and Theorem 4.3.35. □

Corollary 4.3.37. *Assuming the Proper Forcing axiom, the class $\mathcal{C}$ of all coherent Aronszajn trees is a discrete chain which is coinitial and cofinal in the quasi-ordered class $(\mathcal{A}, \leq)$ of all Aronszajn trees. Moreover, neither the chain $\mathcal{C}$ nor the class $\mathcal{A}$ has a maximum nor a minimum.* □

4.4 Lexicographically ordered coherent trees

The results of the previous section suggest that coherent trees are likely to be relevant to questions surrounding the class of Aronszajn trees and therefore in many questions about uncountable structures. In this section we give one example of such a phenomenon by considering the basis problem for uncountable linear orderings. We start by analysing the lexicographical orderings of coherent trees.

Theorem 4.4.1. *The cartesian square of a lexicographically ordered coherent tree that admits a strictly increasing mapping into the rationals*[17] *can be covered by countably many chains.*

Proof. Let T be a given special coherent tree. Identifying a node $t : \alpha \to \omega$ of T with the characteristic function $\chi_{\Gamma(t)}$ of its graph

$$\Gamma(t) = \{(\xi, t(\xi)) : \xi < \alpha\},$$

the natural bijection between $\alpha \times \omega$ and the ordinal $\omega\alpha$ will move $\chi_{\Gamma(t)}$ to a function $\bar{t} : \omega\alpha \to 2$. Since the transformation $t \mapsto \bar{t}$ preserves the lexicographical ordering and since the downward-closure of $\{\bar{t} : t \in T\}$ remains special, we may assume that our tree T itself consists of maps whose ranges are included in $\{0,1\}$. By our assumption, the tree T can be decomposed into countably many antichains. So let $a : T \longrightarrow \omega$ be a fixed map such that $a(s) \neq a(t)$, whenever $s <_T t$ and such that a is one-to-one on the levels of T. It suffices to decompose the subset

$$\{(s,t) \in T^2 : \mathrm{ht}(s) \leq \mathrm{ht}(t)\}$$

of the cartesian square of $(T, <_{\mathrm{lex}})$ into countably many chains. Note that the existence of a allows us to replace (modulo a countable decomposition) every such ordered pair (s,t) by the pair $(s, t \restriction \mathrm{ht}(s))$. It follows that it suffices to show that if $f : D \longrightarrow T$ is a level-preserving map from a subset X of T which intersects a given level of T in at most one point, then its graph $\Gamma_f = \{(x, f(x)) : x \in X\}$ can be decomposed into countably many cartesian chains.

For $t \in X$, let

$$D_t = \{\xi \in \omega_1 : \xi = \mathrm{ht}(t) \text{ or } \xi < \mathrm{ht}(t) \text{ and } t(\xi) \neq f(t)(\xi)\}.$$

Then by the coherence of the tree T, it follows that D_t is a finite set of countable ordinals for all $t \in X$. Let $p_t : D_t \longrightarrow \omega$ and $q_t : D_t \longrightarrow \omega$ be the functions defined by

$$p_t(\xi) = a(t \restriction \xi) \text{ and } q_t(\xi) = a(f(t) \restriction \xi).$$

It suffices to show that if for some $s <_{\mathrm{lex}} t$ in X, we have that

$$t \restriction D_t \cong s \restriction D_s,\ p_t \restriction D_t \cong p_s \restriction D_s, \text{ and } q_t \restriction D_t \cong q_s \restriction D_s,$$

then $f(s) <_{\mathrm{lex}} f(t)$.

[17]Recall that trees admitting strictly increasing maps into the rationals are in the literature frequently called *special*

Note that from the isomorphisms between the parameters p_s and p_t it follows in particular that $a(s) = a(t)$, and similarly from the isomorphism between parameters q_s and q_t we can conclude that $a(f(s)) = a(f(t))$. It follows that s and t as well as $f(s)$ and $f(t)$ are incomparable in T.

Claim. $\Delta(s,t) = \Delta(f(s), f(t))$.

Proof. Let us first establish the inequality

$$\Delta(f(s), f(t)) \geq \Delta(s,t).$$

To this end, let $\alpha = \Delta(s,t)$. Then $\alpha < \mathrm{ht}(s), \mathrm{ht}(t)$, since s and t are incomparable. By the properties of our parameters s and t, we have that $D_s \cap \alpha = D_t \cap \alpha$ and that $f(s)$ and $f(t)$ agree on this set. From the definition of D_s and D_t, we conclude that $f(s)$ and $f(t)$ must agree below α, and so $\Delta(f(s), f(t)) \geq \alpha$. A completely symmetric argument will give us the other inequality

$$\Delta(s,t) \geq \Delta(f(s), f(t)).$$ □

Let $\alpha = \Delta(s,t) = \Delta(f(s), f(t))$. Note that α cannot belong to the intersection of D_s and D_t since then it would occupy the same position in these sets, and so using the isomorphism $t \restriction D_t \cong s \restriction D_s$, we would have that $s(\alpha) = t(\alpha)$, contradicting the assumption that $\alpha = \Delta(s,t)$. So, we are left with considering the following three cases.

Case 1. Suppose first that $\alpha \in D_s \setminus D_t$. Then

$$s(\alpha) < t(\alpha) = f(t)(\alpha),$$

and since $f(t)(\alpha)$ can take only two values 0 and 1, we conclude that $f(t)(\alpha) = 1$. Since $f(s)(\alpha) \neq f(t)(\alpha)$, using the same reasoning, we conclude that $f(s)(\alpha) = 0$, giving as the desired inequality $f(s) <_{\mathrm{lex}} f(t)$.

Case 2. Suppose now that $\alpha \in D_t \setminus D_s$. Then

$$f(s)(\alpha) = s(\alpha) < t(\alpha),$$

and so as before, we must have that $f(s)(\alpha) = 0$. Since $f(s)(\alpha) \neq f(t)(\alpha)$, we must have that $f(t)(\alpha) = 1$, giving us again the desired inequality $f(s) <_{\mathrm{lex}} f(t)$.

Case 3. Finally, suppose that $\alpha \notin D_t \cup D_s$. Then

$$f(s)(\alpha) = s(\alpha) < t(\alpha) = f(t)(\alpha)$$

giving us the desired inequality $f(s) <_{\mathrm{lex}} f(t)$. □

Corollary 4.4.2. *Assuming* MA_{ω_1}, *the cartesian square of a lexicographically ordered coherent Aronszajn tree is the union of countably many chains.* □

Let us now see how this is related to *the basis problem* for uncountable linear orderings, or equivalently and more naturally, *co-basis problem* for the class of countable linear orderings. Thus, one would like to have a finite list $\mathcal{K}$ of linear orderings with the property that an arbitrary linear ordering L is countable if and only if $K \not\leq L$ for all $K \in \mathcal{K}$. Note that if such a co-basis $\mathcal{K}$ exists and if it is moreover a minimal such co-basis, then every ordering K in $\mathcal{K}$ will have to be rather canonical or at least *minimal* as an uncountable linear ordering[18] Clearly, ω_1 and ω_1^* will be members of any co-basis for the countable linear orderings. Another member of the co-basis is found by considering the following notion.

Definition 4.4.3. A linearly ordered set L is *separable* if it contains a countable dense subset D with the property that for all $x <_L y$ in L there is $d \in D$ such that $x \leq_L d \leq_L y$.

Clearly, for every uncountable separable linear ordering L there is an uncountable set of reals X such that $X \leq L$. The following result of Baumgartner [10] solves this part of the co-basis problem for the class of countable linear orderings.

Theorem 4.4.4. *Assuming the Proper Forcing Axiom, every set of reals of size $\aleph_1$ embeds into any uncountable separable linear ordering.* □

Recall, that a set of reals B of cardinality $\aleph_1$ can indeed be canonically chosen on the basis of the given C-sequence C_α $(\alpha < \omega_1)$. For example we could take $B = B(\rho_0)$ to be the set $T(\rho_0)$ as a subset of the Cantor cube $2^{\mathbb{Q}}$ ordered lexicographically. It follows that under the Proper Forcing Axiom any such set of reals B will appear as a member of a co-basis for the class of countable linear orderings, if such a co-basis exists at all. It remains to concentrate on the orthogonal to $\{\omega_1, \omega_1^*, B\}$.

Definition 4.4.5. A linearly ordered set L is *Aronszajn*[19] if it is uncountable and if it contains no uncountable subordering which is well ordered, conversely well ordered, or separable.

It remains to find a basis for the class of Aronszajn orderings. The following result of J.T. Moore [77] accomplishes this.

Theorem 4.4.6. *Assuming the Proper Forcing Axiom, every Aronszajn ordering contains an uncountable subordering whose cartesian square can be covered by countably many chains.*

Proof. To attain the conclusion, we shall actually need only the Coherent Tree Axiom (see Remark 4.2.19) and a bit of PFA restricted to orderings of size at

[18]Recall that an uncountable linear ordering K is minimal, if it has the property that every linear ordering L such that $L < K$ must be countable.

[19]The name comes from the fact that any partition tree of such an ordering is Aronszajn and that a lexicographical ordering of any Aronszajn tree is an Aronszajn ordering (see, for example, [105]).

most $\aleph_1$. So let L be a given Aronszajn ordering and let T be one of its binary partition trees (see [105]). Thus L can naturally be identified with the subset $\{\{x\} : x \in L\}$ of T. Choose an uncountable subset X of $T(\rho_3)$ which intersects a given level of T in at most one point and a level-preserving map

$$f : X \to L \subseteq T.$$

By Lemma 4.2.22, going to an uncountable subset of X, we may assume that f preserves splitting patterns. Applying LMA and going to an uncountable subset of X, we may assume that f is Lipschitz (the case that f^{-1} is Lipschitz is considered similarly). Let $U(f)$ be the downwards-closure of the graph of f in the tree-product $T(\rho_3) \otimes T$. Shrinking X if necessary we may assume that $U(f)$ is also a binary tree. This is so because the poset of finite subsets of the graph of f whose downward-closures are binary satisfies the countable chain condition. Now, choose an uncountable subset Y of X and an unbounded subset D of ω_1, such that

$$[\Delta(x, y), \Delta(f(x), f(y))] \cap D = \emptyset \text{ for all } x \neq y \in Y. \tag{4.4.1}$$

That this can be done follows from the fact that the natural poset of finite approximations to D and T is easily seen to satisfy the countable chain condition using the fact that f preserves splitting patterns. Taking the closure of D we may assume that D is closed and unbounded. Going to an uncountable subset of Y, we may assume that the heights of every pair of comparable splitting nodes of $U(f \restriction Y)$ are separated by a member of D. That this can be done follows from the fact that the poset of finite subsets of Y with this property satisfies the countable chain condition. It follows that to every splitting node $s = x \wedge y$ of the subtree $S(Y)$ of $T(\rho_3)$ generated by taking the downwards-closure of Y, there corresponds a splitting node $t_s = f(x) \wedge f(y)$ and the association $s \mapsto t$ does not depend on which x and y we pick in Y to represent s. Let K be the collection of all splitting nodes $s = x \wedge y$ of $S(Y)$ such that $x <_{\text{lex}} y$ iff $f(x) <_L f(y)$. Note again that the definition does not depend on the x and y in Y we choose to represent s since the reason for the inequality $f(x) <_L f(y)$ relies on the fact that $f(x)$ must lie in the left and $f(y)$ in the right immediate successor of the splitting node $t_s = f(x) \wedge f(y)$. By CTA, there is an uncountable subset Z of Y such that either

(1) $\{x \wedge y : x, y \in Z, x \neq y\} \subseteq K$, or
(2) $\{x \wedge y : x, y \in Z, x \neq y\} \cap K = \emptyset$.

In the first case, the restriction $f : (Z, <_{\text{lex}}) \to L$ is strictly increasing and in the second case, it is strictly decreasing. In any case, by Corollary 4.4.2, the image $f''Z$ is an uncountable subordering of L whose cartesian square can be covered by countably many chains. □

Combining Theorems 4.4.4 and 4.4.6 and Corollary 2.1.13, we obtain the following.

Corollary 4.4.7. *Assuming the Proper Forcing Axiom, a linearly ordered set L is countable if and only if it contains no isomorphic copy of any ordering from the list ω_1, ω_1^*, B, $C(\rho_0)$, and $C(\rho_0)^*$.* □

As indicated above the role of B in this result can be played by any set of reals of size $\aleph_1$ and the role of $C(\rho_0)$ in this result can be played by $C(\rho_1)$, $C(\rho_3)$, or $C(a)$, where $a : [\omega_1]^2 \to \omega$ is any nontrivial coherent mapping. Here, as before, we let $C(a)$ be the linearly ordered set $(\omega_1, <_a)$, where

$$\alpha <_a \beta \text{ if and only if } a(\xi, \alpha) < a(\xi, \beta)$$

for $\xi = \min\{\eta : a(\eta, \alpha) \neq a(\eta, \beta)\}$. By Corollary 4.4.2, the cartesian square of any such ordering can be covered by countably many chains.

4.5 Stationary C-lines

Recall that a *C-line* is any uncountable linear ordering A whose cartesian square A^2 can be covered by countably many chains. We have seen above many constructions of such objects based on different characteristics of the walk on ω. The purpose of this section is to examine a particular class of C-lines from a topological viewpoint. So, let A be a given C-line and let D be a countable subset of A. Then the complement of D in A is split into a family T_D of convex sets, the equivalence classes of the relation in which x and y are equivalent if no member of D separates them. The members of T_D will also be called the *complementary intervals of* D. Note that T_D must be countable or else A would contain a subset order-isomorphic to a set of reals. For two countable subsets A_0 and A_1, we say that A_1 *properly extends* A_0 if $A_0 \subseteq A_1$ and A_1 has points in everyone of the complementary intervals of A_0.

Definition 4.5.1. A *proper decomposition* of a C-line A is a sequence A_α $(\alpha < \omega_1)$ of countable subsets of A such that A_β properly extends A_α if $\alpha < \beta$, and such that $A_\delta = \bigcup_{\alpha<\delta} A_\alpha$ if δ is a limit ordinal.

Note that *every* ω_1-sequence of properly extending countable subsets of A must have union equal to A. On the other hand, trivially no properly extending sequence can have length $> \omega_1$, so there is only one possibility for the length of any proper decomposition $\{A_\alpha\}$ of A, it must be equal to ω_1. Given such a proper decomposition $\{A_\alpha\}$, let

$$T_A = \bigcup_{\alpha<\omega_1} T_{A_\alpha}.$$

This is the *partition tree* of A associated with a proper decomposition $\{A_\alpha\}$ of A. Needless to say this is an Aronszajn tree. Note also that T uniquely determines the proper decomposition $\{A_\alpha\}$ of A, so their roles will be frequently interchanged below. We shall also simplify the notation and write T_α for T_{A_α}.

Definition 4.5.2. If δ is a countable limit ordinal, then we say that a proper decomposition $\{A_\alpha\}$ of a C-line A is *continuous*[20] *at* δ if every complementary interval of A_δ has a minimal element.

This leads us to the following central notion of this section.

Definition 4.5.3. We shall say that A is a *stationary* [21] C-line if and only if every proper decomposition $\{A_\alpha\}$ of A is (left) continuous at some countable limit ordinal.

Note that if A is a stationary C-line, then the set of continuities of every proper decomposition of A is a stationary subset of ω_1, explaining thus our choice of the name for this concept. It turns out that this stationarity of a C-line A viewed as a topological space with its natural topology is the key to many of its covering properties, but let us first establish that these lines exist.

Theorem 4.5.4. *The tree $T^*(\rho_1)$ ordered lexicographically is a homogeneous stationary C-line.*

Proof. First of all recall that

$$T^*(\rho_1) = \{t : \alpha \longrightarrow \omega : \quad \alpha < \omega_1, \ t =^* \rho_{1\alpha}\}$$

ordered lexicographically is a dense subordering of the homogeneous continuum $\widetilde{A}(\rho_1)$ and that the proof of Lemma 2.2.9 actually shows that $T^*(\rho_1)$ is homogeneous as well. In order to show that $T^*(\rho_1)$ is a C-line, it suffices to decompose the set

$$W = \{(s,t) \in (T^*(\rho_1))^2 : \mathrm{ht}(s) \le \mathrm{ht}(t)\}$$

into countably many chains. To a given pair (s,t) of W with $\alpha = \mathrm{ht}(s) \le \beta = \mathrm{ht}(t)$, we associate a hereditarily finite set that codes the behavior of the restrictions of s and t on the finite set

$$F_{st} = \{\xi \le \alpha : s(\xi) \ne t(\xi)\}.$$

Then working as in the proof of Lemma 2.2.4, we show that this parametrization with hereditarily finite sets will give us a countable chain decomposition of W. Thus, $T^*(\rho_1)$ ordered lexicographically is a homogeneous C-line. To see that it is stationary, note that sets of the form

$$T^*(\rho_1) \restriction \lambda = \{t : \alpha \longrightarrow \omega : \quad \alpha < \lambda, \ t =^* \rho_{1\alpha}\},$$

for limit ordinals $\lambda < \omega_1$ constitute a proper decomposition of $T^*(\rho_1)$. This decomposition is (left) continuous at every such λ, since a typical complementary interval to $T^*(\rho_1) \restriction \lambda$ has the form $I_t = \{x \in T^*(\rho_1) : t \le x\}$ for some $t \in T^*(\rho_1)$ of length λ and since clearly $t = \min I_t$. $\square$

[20] This notion should be more accurately called δ-continuous from the *left* since one may also define its right analogue by requiring that every convex set from T_{A_δ} has a maximal element.
[21] Or, more precisely, *left-stationary*

Recall that E. Borel's notion of *strong measure zero* is a metric notion that has several closely related topological notions due to K. Menger, F. Rothberger and others. For example, *Rothberger's property* C'' is a purely topological notion which in the class of metric spaces is slightly stronger than the *property* C, i.e., being of strong measure zero. It says that for every sequence $\{\mathcal{U}_n\}$ of open covers of a space X one can choose $U_n \in \mathcal{U}_n$ for each n such that $X = \bigcup_{n=0}^{\infty} U_n$. A considerably stronger covering property of a topological space is obtained via the following notion.

Definition 4.5.5. An *ω-cover* $\mathcal{U}$ of a set X is a family of subsets of X with the property that for every finite subset F of X there is $U \in \mathcal{U}$ such that $F \subseteq U$.

Theorem 4.5.6. *For every stationary C-line C and every open ω-cover $\mathcal{U}$ of C there is a sequence $\{U_n\} \subseteq \mathcal{U}$ such that $C = \liminf U_n$.*[22]

Proof. Let $\{C_\alpha\}$ be a fixed proper decomposition of C and let T be the corresponding partition tree of complementary intervals to the sets C_α. Choose a countable elementary submodel M of some large enough structure containing all the relevant objects such that, if δ is its intersection with ω_1, then $\{C_\alpha\}$ is continuous at δ. Let $\{a_{2n}\}_{n=1}^{\infty}$ be an enumeration of all elements of C_δ and let $\{a_{2n-1}\}_{n=1}^{\infty}$ be an enumeration of all minimums of the complementary intervals of C_δ. The theorem will be proved once we show that for every n there is an open set U_n in $\mathcal{U} \cap M$ which includes

$$\{a_2, a_4, \ldots, a_{2n}\} \cup \left(\bigcup_{i=1}^{n} t_{2i-1} \right),$$

where t_{2i-1} is the complementary interval of C_δ whose minimum is equal to a_{2i-1}. Applying the fact that $\mathcal{U}$ is a $2n$-cover of C, choose U in $\mathcal{U}$ containing $\{a_1, a_2, \ldots, a_{2n}\}$ and for each $1 \leq i \leq n$ choose $x_i < a_{2i-1}$ in C_δ $(= C \cap M)$ such that $(x_i, a_{2i-1}] \subseteq U$. Let Y be the set of all $(y_1, \ldots, y_n)$ in C^n such that $x_i < y_i$ $(1 \leq i \leq n)$ and such that

$$\{a_2, a_4, \ldots, a_{2n}\} \cup \left(\bigcup_{i=1}^{n} (x_i, y_i] \right)$$

is covered by a single element of $\mathcal{U}$. Then $Y \in M$ and $(a_1, a_3, \ldots, a_{2n-1}) \in Y$. Since C^n is the union of countably many chains, there is a chain $Y_0 \subseteq Y$ such that $Y_0 \in M$ and $(a_1, a_3, \ldots, a_{2n-1}) \in Y_0$.

Claim. There is $(y_1, \ldots, y_n)$ in $Y_0 \cap M$ such that $a_{2i-1} < y_i$ for all $1 \leq i \leq n$.

Note that to prove the claim, it suffices to find such $(y_1, \ldots, y_n)$ with the property that $a_1 < y_1$ because the other inequalities will follow from this one and the fact that Y_0 is a chain of C^n. So, let c be the supremum (taken in the Dedekind completion of C) of the set of all first coordinates of elements of Y_0. Note that

[22] Recall that $\liminf U_n = \bigcup_m \bigcap_{n>m} U_n$.

since C contains no uncountable well-ordered subsets, this supremum is equal to the supremum of the projection of $Y_0 \cap M$ on the first coordinate. Note also that $c \in M$ and that $a_1 \leq c$. Since a_1 is not a member of M, the equality is impossible, so we have $a_1 < c$. Since $c = \sup \pi_1''(Y_0 \cap M)$, there is $(y_1, \ldots, y_n)$ in $Y_0 \cap M$ such that $a_1 < y_1$ and this proves the claim.

Choose $(y_1, \ldots, y_n)$ as in the claim. The proof of the theorem is finished, once we show that t_{2i-1} is included in $(x_i, y_i]$ for all $1 \leq i \leq n$ or equivalently that $\sup t_{2i-1} \leq y_i$ for all $1 \leq i \leq n$. But this is clearly so since $a_i = \min t_{2i-1} < y_i$ and since the point y_i, being a member of C_δ, cannot split the complementary interval t_{2i-1}. This completes the proof. □

Corollary 4.5.7. *Suppose $\mathcal{F}$ is a family of continuous real-valued functions defined on some stationary C-line C which pointwise accumulates to some continuous real-valued function f on C. Then there is a sequence $(f_n) \subseteq \mathcal{F}$ such that* $\lim f_n = f$.

Proof. Clearly, we may assume that $f(x) = 0$ for all $x \in C$. Fix a sequence (x_n) of pairwise distinct elements of C. Let $\mathcal{U}$ be the set of all subsets of C of the form

$$U(f, n) = \{x \in X : x \neq x_n \text{ and } |f(x)| < 2^{-n}\}$$

where $f \in \mathcal{F}$ and $n \in \mathbb{N}$. By our assumption about $\mathcal{F}$ and f, the family $\mathcal{U}$ is an open ω-cover of C. By Theorem 4.5.6, there exist $\{U(f_i, n_i)\} \subseteq \mathcal{U}$ such that

$$\liminf U(f_i, n_i) = C.$$

It is easily checked that $n_i \to \infty$ which in turn gives $f_i \to f$. This completes the proof. □

Definition 4.5.8. The *left arrow topology* $\leftarrow$ on a given C-line C is the one determined by intervals of the form $(x, y]$ for some $x < y$ in C. Similarly, one defines the *right arrow topology* $\rightarrow$ generated by intervals of the form $[x, y)$.

It should be clear that the proof of Theorem 4.5.6 also establishes the following slightly stronger conclusion.

Theorem 4.5.9. *For every stationary C-line C and every ω-cover $\mathcal{U}$ of C consisting of sets that are open relative to the left arrow topology $\leftarrow$ of C there is a sequence $\{U_n\} \subseteq \mathcal{U}$ such that $C = \liminf U_n$.* □

Corollary 4.5.10. *Suppose $\mathcal{F}$ is a family of continuous real-valued functions defined on the arrow space $(C, \leftarrow)$, where C is a stationary C-line. Suppose further that some continuous real-valued function f on C is a pointwise accumulation point of $\mathcal{F}$. Then there is a sequence $(f_n) \subseteq \mathcal{F}$ such that* $\lim f_n = f$. □

At this point one may ask for the existence of a C-line that has proper decompositions that are stationary both from the left and from the right so that both of its arrow topologies $\leftarrow$ and $\rightarrow$ enjoy the strong covering property that involves the notion of an ω-cover. For this, it suffices to constrict a C-line A with proper decomposition $\{A_\alpha\}$ such that for a stationary set of α's the complementary intervals to A_α have minimums as well as maximums. This could be done, for example, by considering the following extension of $T^*(\rho_1)$,

$$T^*(\rho_1)^+ = T^*(\rho_1) \cup \{t^\frown 1^\omega : t \in T^*(\rho_1)\},$$

where for $t \in T^*(\rho_1)$ of length α, by $t^\frown 1^\omega$, we denote the sequence s of length $\alpha + \omega$ extending t and having the property that $s(\alpha + n) = 1$ for all $n < \omega$. It is clear that the proof of Theorem 4.5.4 shows that $T^*(\rho_1)^+$ is indeed a C-line and that its natural proper decomposition is continuous from both sides, or in other words, that for the complementary intervals

$$I_t^+ = \{x \in T^*(\rho_1)^+ : t \leq x\},$$

we have that $t = \min I_t^+$ and $t^\frown 1^\omega = \max I_t^+$. This proves the following result.

Theorem 4.5.11. *There is a C-line C with the property that, if $\mathcal{U}$ is an ω-cover of C consisting of sets that are all open relative to the left arrow topology of C or sets that are all open relative to the right arrow topology on C, then there is a sequence $(U_n) \subseteq \mathcal{U}$ such that $C = \liminf U_n$.* □

Chapter 5

The Square-bracket Operation on Countable Ordinals

5.1 The upper trace and the square-bracket operation

Recall that a *walk* from a countable ordinal β to a smaller ordinal α along the fixed C-sequence C_ξ $(\xi < \omega_1)$ is a finite decreasing sequence

$$\beta = \beta_0 > \beta_1 > \cdots > \beta_n = \alpha,$$

where $\beta_{i+1} = \min(C_{\beta_i} \setminus \alpha)$ for all $i < n$. Recall also the notion of the *upper trace* of the minimal walk,

$$\mathrm{Tr}(\alpha, \beta) = \{\beta_0, \beta_1, \ldots, \beta_n\},$$

the finite set of places visited in the minimal walk from β to α. The following simple fact about the upper trace lies at the heart of all known definitions of square-bracket operations, not only on ω_1 but also at higher cardinalities.

Lemma 5.1.1. *For every uncountable subset Γ of ω_1, the union of $\mathrm{Tr}(\alpha, \beta)$ for $\alpha < \beta$ in Γ contains a closed and unbounded subset of ω_1.*

Proof. It suffices to show that the union of traces contains every countable limit ordinal δ such that $\sup(\Gamma \cap \delta) = \delta$. Pick an arbitrary $\beta \in \Gamma \setminus \delta$ and let

$$\beta = \beta_0 > \beta_1 > \cdots > \beta_k = \delta$$

be the minimal walk from β to δ. Let $\gamma < \delta$ be an upper bound of all sets of the form $C_{\beta_i} \cap \delta$ for $i < k$. By the choice of δ there is $\alpha \in \Gamma \cap \delta$ above γ. Then the minimal walk from β to α starts as $\beta_0 > \beta_1 > \cdots > \beta_k$, so in particular δ belongs to $\mathrm{Tr}(\alpha, \beta)$. $\square$

We shall now see that it is possible to pick a single place $[\alpha\beta]$ in $\mathrm{Tr}(\alpha, \beta)$ so that Lemma 5.1.1 remains valid with $[\alpha\beta]$ in place of $\mathrm{Tr}(\alpha, \beta)$. Recall that by Lemma 2.1.18,

$$\Delta(\alpha, \beta) = \min\{\xi \leq \alpha : \rho_0(\xi, \alpha) \neq \rho_0(\xi, \beta)\}$$

is a successor ordinal. We shall be interested in its predecessor,

Definition 5.1.2. $\Delta_0(\alpha, \beta) = \Delta(\alpha, \beta) - 1$.

Thus, if $\xi = \Delta_0(\alpha, \beta)$, then $\rho_0(\xi, \alpha) = \rho_0(\xi, \beta)$ and so there is a natural isomorphism between $\mathrm{Tr}(\xi, \alpha)$ and $\mathrm{Tr}(\xi, \beta)$. We shall define $[\alpha\beta]$ by comparing the three sets $\mathrm{Tr}(\alpha, \beta), \mathrm{Tr}(\xi, \alpha)$ and $\mathrm{Tr}(\xi, \beta)$.

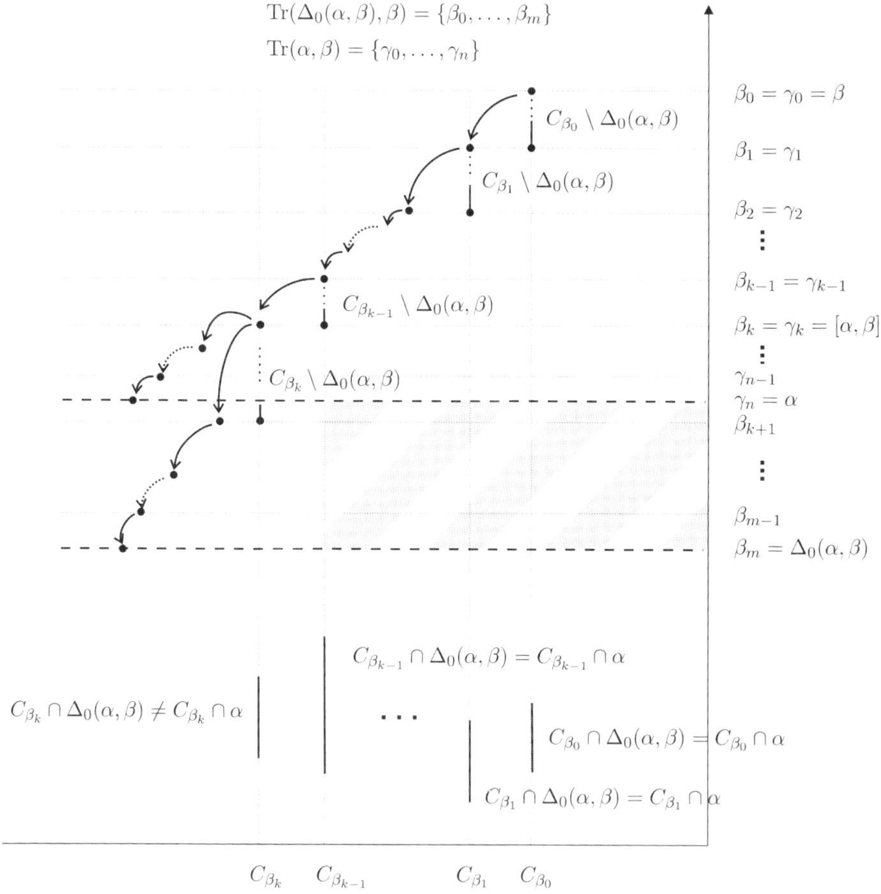

Figure 5.1: The square-bracket operation.

Definition 5.1.3. The *square-bracket operation* on ω_1 is defined as follows (see Figure 5.1):

$$\begin{aligned}[][\alpha\beta] &= \min(\mathrm{Tr}(\alpha,\beta)\cap\mathrm{Tr}(\Delta_0(\alpha,\beta),\ \beta))\\ &= \min(\mathrm{Tr}(\Delta_0(\alpha,\beta),\ \beta)\setminus\alpha).\end{aligned}$$

Next, recall the function $\rho_0 : [\omega_1]^2 \to \omega^{<\omega}$ which codes the walk along the fixed C-sequence C_ξ $(\xi < \omega_1)$ and which is given by the recursive formula

$$\rho_0(\alpha,\beta) = \langle|C_\beta\cap\alpha|\rangle^\frown\rho_0(\alpha,\min(C_\beta\setminus\alpha)),$$

with the boundary value $\rho_0(\alpha,\alpha)=\emptyset$. Recall also the definition of the corresponding tree

$$T(\rho_0) = \{\rho_0(\cdot,\beta)\restriction\alpha : \alpha\le\beta<\omega_1\}.$$

As before, for $\gamma<\omega_1$, we shorten the notation $\rho_0(\cdot,\gamma)$ to $(\rho_0)_\gamma$ of the corresponding fiber-mapping $\gamma\mapsto\omega^{<\omega}$ defined by $(\rho_0)_\gamma(\alpha)=\rho_0(\alpha,\gamma)$.

Lemma 5.1.4. *For every uncountable subset* Γ *of* ω_1, *the set of all ordinals of the form* $[\alpha\beta]$ *for some* $\alpha<\beta$ *in* Γ *contains a closed and unbounded subset of* ω_1.

Proof. For $t\in T(\rho_0)$ let

$$\Gamma_t = \{\gamma\in\Gamma : (\rho_0)_\gamma \text{ end-extends } t\}.$$

Let S be the collection of all $t\in T(\rho_0)$ for which Γ_t is uncountable. Clearly, S is a downward-closed uncountable subtree of $T(\rho_0)$. The lemma is proved once we prove that every countable limit ordinal $\delta>0$ with the following two properties can be represented as $[\alpha\beta]$ for some $\alpha<\beta$ in Γ:

$$\sup(\Gamma_t\cap\delta)=\delta \text{ for every } t\in S \text{ of length } <\delta, \tag{5.1.1}$$

$$\text{every } t\in S \text{ of length } <\delta \text{ has two incomparable successors in } S \text{ both of length } <\delta. \tag{5.1.2}$$

Fix such a δ and choose $\beta\in\Gamma\setminus\delta$ such that $(\rho_0)_\beta\restriction\delta\in S$ and consider the minimal walk from β to δ:

$$\beta=\beta_0>\beta_1>\cdots>\beta_k=\delta.$$

Let $\gamma<\delta$ be an upper bound of all sets of the form $C_{\beta_i}\cap\delta$ for $i<k$. Since the restriction $t=(\rho_0)_\beta\restriction\gamma$ belongs to S, by (5.1.2) we can find one of its end-extensions s in S which is incomparable with $(\rho_0)_\beta$. It follows that for $\alpha\in\Gamma_s$, the ordinal $\Delta_0(\alpha,\beta)$ has the fixed value

$$\xi=\min\{\xi<|s| : s(\xi)\neq\rho_0(\xi,\beta)\}-1.$$

Note that $\xi\ge\gamma$, so the walk $\beta\to\delta$ is a common initial part of walks $\beta\to\xi$ and $\beta\to\alpha$ for every $\alpha\in\Gamma_s\cap\delta$. Hence if we choose $\alpha\in\Gamma_s\cap\delta$ above $\min(C_\delta\setminus\xi)$ (which we can by (5.1.1)), we get that the walks $\beta\to\xi$ and $\beta\to\alpha$ never meet after δ. In other words for any such α, the ordinal δ is the minimum of $\mathrm{Tr}(\alpha,\beta)\cap\mathrm{Tr}(\xi,\beta)$, or equivalently δ is the minimum of $\mathrm{Tr}(\xi,\beta)\setminus\alpha$. □

It should be clear that the above argument can easily be adjusted to give us the following slightly more general fact about the square-bracket operation.

Lemma 5.1.5. *For every uncountable family A of pairwise-disjoint finite subsets of ω_1, all of the same size n, the set of all ordinals ξ of the form*

$$\xi = [a(1)b(1)] = [a(2)b(2)] = \cdots = [a(n)b(n)]$$ [1]

for some $a \neq b$ in A contains a closed and unbounded subset of ω_1. □

It turns out that the square-bracket operation can be used in constructions of various mathematical objects of complex behavior where all known previous constructions needed the Continuum Hypothesis or stronger enumeration principles. The usefulness of $[\cdot\cdot]$ in these constructions is based on the fact that $[\cdot\cdot]$ reduces the quantification over uncountable subsets of ω_1 to the quantification over closed unbounded subsets of ω_1. For example composing $[\cdot\cdot]$ with a unary operation $* : \omega_1 \longrightarrow \omega_1$ which takes each of the values stationary many times, one gets the following fact about the mapping $c(\alpha, \beta) = [\alpha\beta]^*$.

Theorem 5.1.6. *There is a mapping $c : [\omega_1]^2 \longrightarrow \omega_1$ which takes all the values from ω_1 on any square $[\Gamma]^2$ of some uncountable subset Γ of ω_1.* □

This result gives a definitive limitation on any form of a Ramsey Theorem for the uncountable that tries to say something about the behaviors of restrictions of colorings on an uncountable square. Recall that in Section 2.3.1 we have seen that there are mappings of this sort with complex behavior on sets of the form

$$\Gamma_0 \otimes \Gamma_1 = \{\{\alpha_0, \alpha_1\} : \alpha_0 \in \Gamma_0, \alpha_1 \in \Gamma_1, \alpha_0 < \alpha_1\},$$

where Γ_0 and Γ_1 are two uncountable subsets of ω_1. It turns out that the original square-bracket operation has a simple behavior on $\Gamma_0 \times \Gamma_1$ for every pair Γ_0 and Γ_1 of sufficiently thin uncountable subsets of ω_1, a property that turns out to be quite useful in applications. The property is made precise as follows.

Lemma 5.1.7. *For every uncountable family A of pairwise-disjoint subsets of ω_1, all of the same size n, there exist an equivalence relation $\sim$ on $n = \{0, 1, \ldots, n-1\}$, an uncountable $B \subseteq A$, and a mapping $h_b : n \times n \to \omega_1$ for each $b \in B$ such that for all $a < b$* [2] *in B, we have $[a(i)b(j)] = h_b(i, j)$ whenever $i \nsim j$. Moreover, for every uncountable $C \subseteq B$, the set of ordinals δ for which there exist $a < b$ in C such that $[a(p)b(q)] = \delta$ for all $i, j < n$ with $i \sim j$ contains a closed and unbounded subset of ω_1.*

Proof. Identifying an ordinal α with $(\rho_0)_\alpha$, we can think of A as a family of n-tuples of nodes of the tree $T(\rho_0)$. For a limit ordinal $\delta < \omega_1$, choose a_δ in A such

[1] For a finite set x of ordinals of size n we use the notation $x(1)$, $x(2)$, ..., $x(n)$ or $x(0)$, $x(1)$, ..., $x(n-1)$, depending on the context, for the enumeration of x according to the natural ordering on the ordinals.

[2] For a pair a and b of sets of ordinals, $a < b$ denotes the fact that $\alpha < \beta$ for all $\alpha \in a$ and $\beta \in b$.

that every node from a_δ has height $\geq \delta$. Let c_δ be the projection of a_δ on the δth level of $T(\rho_0)$. Let $h(\delta) < \delta$ be such that the projection of c_δ on the $h(\delta)$th level has size equal to the size of c_δ, i.e.,

$$\Delta(s,t) < h(\delta) \text{ for all } s \neq t \text{ in } c_\delta. \tag{5.1.3}$$

Find a stationary set $\Gamma \subseteq \omega_1$ such that h takes a constant value γ_0 on Γ and that the γ_0th projection takes the constant value on c_δ ($\delta \in \Gamma$). Fix $\delta < \varepsilon$ in Γ and consider $\alpha \in a_\delta$ and $\beta \in a_\varepsilon$ that project to different elements of the γ_0th level of $T(\rho_0)$. Then $\Delta(\alpha, \beta)$ is independent of α and β or δ and ε as it is equal to $\Delta((\rho_0)_\alpha \restriction \gamma_0, (\rho_0)_\beta \restriction \gamma_0)$. So the trace $\mathrm{Tr}(\Delta_0(\alpha,\beta),\beta)$ depends only on β and the projection $(\rho_0)_\alpha \restriction \gamma_0$. Thus, going to a subset of Γ, we may assume that its size depends only on the position of β in some a_ε, and that the minimal point of $\mathrm{Tr}(\Delta_0(\alpha,\beta),\beta) \setminus \alpha$ is the same for all α in some a_δ, for $\delta < \varepsilon$, having a fixed projection on the γ_0th level of $T(\rho_0)$. This is the content of the first part of the conclusion of Lemma 5.1.7. To see the second part of the conclusion, we shrink Γ even further to assure that the union of c_δ ($\delta \in \Gamma$) is an antichain of $T(\rho_0)$. Now note that for two typical $\delta < \varepsilon$ in Γ, $[\alpha\beta] = [\alpha'\beta']$ for all $\alpha, \alpha' \in a_\delta$ and $\beta, \beta' \in a_\varepsilon$ with the property that $(\rho_0)_\alpha \restriction \gamma_0 = (\rho_0)_{\alpha'} \restriction \gamma_0 = (\rho_0)_\beta \restriction \gamma_0 = (\rho_0)_{\beta'} \restriction \gamma_0$. So the rest of the proof reduces to the proof of Lemma 5.1.5 above. □

Remark 5.1.8. To see the point of Lemma 5.1.7, consider the projection $[\alpha\beta]^*$ of the square-bracket operation generated by a mapping $*$ from ω_1 onto ω rather than ω_1. Then the sequence $h_b^* : n \times n \longrightarrow \omega$ ($b \in B$) of corresponding compositions will have the same constant value $h : n \times n \longrightarrow \omega$ modulo, of course, shrinking B to some uncountable subset. Noting that the proof of Lemma 5.1.7 applied to a typical A will actually result in the equality $=$ as the equivalence relation $\sim$, we conclude that if α and β are coming from two different positions of some a_δ and a_ε in B, then $[\alpha\beta]^*$ will depend on the positions themselves rather than α and β. It fact, there is a natural variation of the square-bracket operation (see the mapping b of [109, p. 287]), where, in Lemma 5.1.7, we always have the finest possible partition $I_i = \{i\}$ ($i = 1, \ldots, n$) for *every* uncountable family A of pairwise-disjoint n-tuples of countable ordinals. We shall reproduce this variation in Section 5.4 below, where we shall see also one of its applications to the quadratic form theory.

Note that the basic C-sequence C_α ($\alpha < \omega_1$) which we have fixed at the beginning of this chapter can be used to actually define a unary operation $* : \omega_1 \longrightarrow \omega_1$ which takes each of the ordinals from ω_1 stationarily many times. So the projection $[\alpha\beta]^*$ can actually be defined in our basic structure $(\omega_1, \omega, \vec{C})$. We are now at the point to see that our basic structure is actually rigid.

Lemma 5.1.9. *The algebraic structure* $(\omega_1, [\cdot\cdot], *)$ *has no nontrivial automorphisms.*

Proof. Let h be a given automorphism of $(\omega_1, [\cdot\cdot], *)$. If the set Γ of fixed points of h is uncountable, h must be the identity map. To see this, consider a $\xi < \omega_1$.

By the property of the map $c(\alpha, \beta) = [\alpha\beta]^*$ stated in Theorem 5.1.6, there exist $\gamma < \delta$ in Γ such that $[\gamma\delta]^* = \xi$. Applying h to this equation we get

$$h(\xi) = h([\gamma\delta]^*) = (h([\gamma\delta]))^* = [h(\gamma)h(\delta)]^* = [\gamma\delta]^* = \xi.$$

It follows that $\Delta = \{\delta < \omega_1 : h(\delta) \neq \delta\}$ is in particular uncountable. Shrinking Δ and replacing h by h^{-1}, if necessary, we may safely assume that $h(\delta) > \delta$ for all $\delta \in \Delta$. Consider a $\xi < \omega_1$ and let S_ξ be the set of all $\alpha < \omega_1$ such that $\alpha^* = \xi$. By our choice of $*$ the set S_ξ is stationary. By Lemma 5.1.5 applied to the family $A = \{\{\delta, h(\delta)\} : \delta \in \Delta\}$, we can find $\gamma < \delta$ in Δ such that $[\gamma\delta] = [h(\gamma)h(\delta)]$ belongs to S_ξ, or in other words,

$$[\gamma\delta]^* = [h(\gamma)h(\delta)]^* = \xi.$$

Since $[h(\gamma)h(\delta)]^* = h([\gamma\delta]^*)$ we conclude that $h(\xi) = \xi$. Since ξ was an arbitrary countable ordinal, this shows that h is the identity map. □

We give now an application of this rigidity result to a problem in Model Theory about the quantifier Qx = 'there exist uncountably many x' and its higher-dimensional analogues $Q^n x_1 \cdots x_n$ = 'there exist an uncountable n-cube many $x_1, \cdots, x_n$'. By a result of Ebbinghaus and Flum [26] (see also [82]) every model of every sentence of $L(Q)$ has nontrivial automorphisms. However we shall now see that this is no longer true about the quantifier Q^2, for example.

Example 5.1.10. The sentence ϕ will talk about one unary relation N, one binary relation $<$ and two binary functional symbols C and E. It is the conjunction of the following seven sentences:

(ϕ_1) $Qx \quad x = x$.

(ϕ_2) $\neg Qx \quad N(x)$.

(ϕ_3) $<$ is a total ordering.

(ϕ_4) E is a symmetric binary operation.

(ϕ_5) $\forall x < y \quad N(E(x, y))$.

(ϕ_6) $\forall x < y < z \quad E(x, z) \neq E(y, z)$.

(ϕ_7) $\forall x \forall n \{N(n) \rightarrow \neg Q^2 uv[\exists u' < u \exists v' < v(u' \neq v' \wedge E(u', u) = E(v', v) = n) \wedge \forall u' < u \, \forall v' < v(E(u', u) = E(v', v) = n \rightarrow (C(u', v') \neq x \vee C(u, v) \neq x))]\}$.

The model of ϕ that we have in mind is the model $(\omega_1, \omega, <, c, e)$ where $c(\alpha, \beta) = [\alpha\beta]^*$ and $e : [\omega_1]^2 \longrightarrow \omega$ is any mapping such that $e(\alpha, \gamma) \neq e(\beta, \gamma)$ whenever $\alpha < \beta < \gamma$ (e.g., we can take $e = \bar{\rho}_1$ or $e = \bar{\rho}$). The sentence ϕ_7 simply says that for every $\xi < \omega_1$ and every uncountable family A of pairwise-disjoint unordered pairs of countable ordinals, there exist $a \neq b$ in A such that

$$c(\min a, \min b) = c(\max a, \max b) = \xi.$$

This is a consequence of Lemma 5.1.5 and the fact that $S_\xi = \{\alpha : \alpha^* = \xi\}$ is a stationary subset of ω_1. These are the properties of $[\cdot\cdot]$ and $*$ which we have used in the proof of Lemma 5.1.9 in order to prove that $(\omega_1, [\cdot\cdot], *)$ is a rigid structure. So a quite analogous proof will show that any model $(M, N, <, C, E)$ of ϕ must be rigid. □

The crucial property of $[\cdot\cdot]$ stated in Lemma 5.1.5 can also be used to provide a negative answer to the basis problem for uncountable graphs by constructing a large family of pairwise orthogonal uncountable graphs.

Definition 5.1.11. For a subset Γ of ω_1, let $\mathcal{G}_\Gamma$ be the graph whose vertex-set is ω_1 and whose edge-set is equal to

$$\{\{\alpha, \beta\} : [\alpha\beta] \in \Gamma\}.$$

Lemma 5.1.12. *If the symmetric difference between Γ and Δ is a stationary subset of ω_1, then the corresponding graphs $\mathcal{G}_\Gamma$ and $\mathcal{G}_\Delta$ are orthogonal to each other, i.e., they do not contain uncountable isomorphic subgraphs.* □

We have seen above that comparing $[\cdot\cdot]$ with a map $\pi : \omega_1 \longrightarrow I$ where I is some set of mathematical objects/requirements in such a way that each object/requirement is given a stationary preimage, gives us a way to meet each of these objects/requirements in the square of any uncountable subset of ω_1. This observation is the basis of all known applications of the square-bracket operation. A careful choice of I and $\pi : \omega_1 \longrightarrow I$ gives us a projection of the square-bracket operation that can be quite useful. This is exposed in the next section.

5.2 Projecting the square-bracket operation

Definition 5.2.1. Let $\mathcal{G}$ be the collection of all maps $g : 2^m \times 2^m \rightarrow \omega_1$ where m is a positive integer denoted by $m(g)$. Choose a mapping $\pi : \omega_1 \longrightarrow \mathcal{G}$ which takes each value from $\mathcal{G}$ stationarily many times. Choose also a one-to-one sequence r_α $(\alpha < \omega_1)$ of elements of the Cantor set 2^ω. Note that both these objects can actually be defined in our basic structure $(\omega_1, \omega, \vec{C})$. Consider the projection $[\![\cdot\cdot]\!] = [\![\cdot\cdot]\!]_\mathcal{G}$ of the square-bracket operation $[\cdot\cdot]$ defined as follows:

$$[\![\alpha\beta]\!] = \pi([\alpha\beta])(r_\alpha \restriction m(\pi([\alpha\beta])), r_\beta \restriction m(\pi([\alpha\beta]))).$$

Then the property of $[\cdot\cdot]$ stated in Lemma 5.1.5 corresponds to the following property of the projection $[\![\cdot\cdot]\!]$.

Lemma 5.2.2. *For every uncountable family A of pairwise-disjoint finite subsets of ω_1, all of the same size n, and for every n-sequence $\xi_0, \xi_1, \ldots, \xi_{n-1}$ of countable ordinals there exist a and b in A such that $[\![a(i)b(i)]\!] = \xi_i$ for $i = 1, \ldots, n$.*

Proof. Find uncountable $B \subseteq A$ and $m < \omega$ such that for all $b \in B$,

$$r_\alpha \restriction m \neq r_\beta \restriction m \text{ for all } \alpha \neq \beta \text{ in } b.$$

Moreover, we may assume that for some fixed one-to-one sequence $t_0, t_1, \ldots, t_{n-1}$ of elements of 2^m,

$$r_{b(i)} \upharpoonright m = t_i \text{ for all } i < n \text{ and all } b \in B.$$

Find $g : 2^m \times 2^m \to \omega_1$ such that $g(t_i, t_i) = \xi$ for all $i < n$. By our choice of π, the set $\Gamma = \{\gamma < \omega_1 : \pi(\gamma) = g\}$ is stationary, so by Lemma 5.1.5, we can find $a < b$ in B such that for some $\gamma \in \Gamma$,

$$[a(0)b(0)] = [a(1)b(1)] = \cdots = [a(n-1)b(n-1)] = \gamma.$$

It follows that for all $i < n$,

$$[\![a(i)b(i)]\!] = g(r_{a(i)} \upharpoonright m,\ r_{b(i)} \upharpoonright m) = g(t_i, t_i) = \xi,$$

as required. □

The following lemma describes a universality property of the projection $[\![\cdot\cdot]\!] = [\![\cdot\cdot]\!]_{\mathcal{G}}$ of the square-bracket operation $[\cdot\cdot]$.

Lemma 5.2.3. *For every positive integer n, every uncountable subset Γ of ω_1 and every symmetric $n \times n$-matrix M of countable ordinals there is a one-to-one mapping $\phi : n \to \Gamma$ such that $[\![\phi(i)\phi(j)]\!] = M(i,j)$ for $i, j < n$.*

Proof. Let Γ be an uncountable subset of ω_1 and let M be a given $n \times n$ matrix on countable ordinals. Then for each α we can find t_α in the αth level of the tree $T(\rho_0)$ and a finite set $F_\alpha \subseteq \Gamma \setminus \alpha$ of size n such that $t_\alpha \subseteq (\rho_0)_\gamma$ for all $\gamma \in F_\alpha$. Let n_α be the minimal integer such that $r_\gamma \upharpoonright n_\alpha \neq r_\delta \upharpoonright n_\alpha$ whenever $\gamma \neq \delta$ are members of F_α. For a limit ordinal α, let $\xi(\alpha) < \alpha$ be an upper bound of all sets of the form $C_{\gamma_i} \cap \alpha$ where $\gamma \in F_\alpha$, $i < k < \omega$ and $\gamma = \gamma_0 > \cdots > \gamma_k = \alpha$ is the minimal walk from γ to α along the C-sequence fixed at the beginning of this chapter. Then there is a stationary set $\Delta \subseteq \omega_1$, $m < \omega$, $s_0, \ldots, s_{n-1} \in 2^m$ and $\xi < \omega_1$ such that:

$$n_\alpha = m \text{ and } \xi_\alpha = \xi \text{ for all } \alpha \in \Delta, \tag{5.2.1}$$

$$t_\alpha\ (\alpha \in \Delta) \text{ form an antichain of } T(\rho_0), \tag{5.2.2}$$

if γ occupies the ith place in F_α for some $\alpha \in \Delta$, then $r_\gamma \upharpoonright m = s_i$. (5.2.3)

Let $h : 2^m \times 2^m \longrightarrow \omega_1$ be an arbitrary map subject to the requirement that $h : (s_i, s_j) = M(i,j)$ for all $i, j < n$. Let $\Sigma_h = \{\delta < \omega_1 : \pi(\delta) = h\}$. By the choice of π the set Σ_h is stationary in ω_1. By the proof of the basic property Lemma 5.1.4 of the square-bracket operation we can find $\nu \in \Sigma_h$, $\alpha \in \Delta \cap \nu$ and $\beta \in \Delta \setminus \nu$ such that $t_\alpha \upharpoonright \xi = t_\beta \upharpoonright \xi$, $F_\alpha \subseteq \nu$ and $[\alpha\beta] = \nu$. We claim that

$$[\gamma\delta] = \nu \text{ for all } \gamma \in F_\alpha \text{ and } \delta \in F_\beta. \tag{5.2.4}$$

To see this, first note that $\Delta_0(\gamma,\delta) = \Delta_0(\alpha,\beta) = \sigma$ for all $\gamma \in F_\alpha$ and $\delta \in F_\beta$. By the choice of ξ and the fact that $\sigma \geq \xi$, the walk from some $\delta \in F_\beta$ to σ goes through β, so for any $\gamma \in F_\alpha$ and $\delta \in F_\beta$ we have

$$[\gamma\delta] = \min(\mathrm{Tr}(\sigma,\delta) \setminus \gamma) = \min(\mathrm{Tr}(\sigma,\beta) \setminus \alpha) = [\alpha\beta].$$

In fact the proof of Lemma 5.1.4 shows that we can find $\nu_0 \in \Sigma_h$, $\alpha_0 \in \Delta \cap \nu$ and an uncountable set $\Delta_1 \subseteq \Delta \setminus \nu_0$ such that $F_{\alpha_0} \subseteq \nu_0$, $t_{\alpha_0} \restriction \xi = t_\beta \restriction \xi$ and $[\alpha_0\beta] = \gamma_0$ for all $\beta \in \Delta_1$. It follows that

$$[\gamma\delta] = \nu_0 \text{ for all } \gamma \in F_{\alpha_0}, \delta \in F_\beta \text{ and } \beta \in \Delta_1. \tag{5.2.5}$$

So we can repeat the procedure for Δ_1 in place of Δ and get $\nu_1 \in \Sigma_h$, $\alpha_1 \in \Delta_1 \cap \nu_0$ and uncountable $\Delta_2 \subseteq \Delta_1 \setminus \gamma_1$ such that $[\alpha_1\beta] = \nu_1$ for all $\beta \in \Delta_2$, and so on. Repeating this n times we get a sequence $\alpha_0, \ldots, \alpha_{n-1}$ such that for all $i < j < n$,

$$\pi([\gamma\delta]) = h \text{ for all } \gamma \in F_{\alpha_i} \text{ and } \delta \in F_\beta \text{ with } \beta \in \Delta_j. \tag{5.2.6}$$

Define $\phi : n \longrightarrow \omega_1$ by letting $\phi(i)$ be the ith member of F_{α_i}. Going back to the definition of the projection $[\![\cdot\cdot]\!]$ of $[\cdot\cdot]$ one sees that indeed $[\![\phi(i)\phi(j)]\!] = M(i,j)$ holds for all $i < j < n$. □

Projections of the square-bracket operation that have only countably many values are also worth considering. In fact, in later sections of this book, we shall give applications of this kind of projections in cases where the projections with uncountably many values do not seem to fit.

Definition 5.2.4. Let $\mathcal{G}_0$ be the collection of all maps $g : 2^n \times 2^n \to \omega$ where n is a positive integer denoted by $n(g)$. Choose a mapping $\pi : \omega_1 \longrightarrow \mathcal{G}_0$ which takes each value from $\mathcal{G}_0$ stationarily many times. Choose also a one-to-one sequence r_α $(\alpha < \omega_1)$ of elements of the Cantor set 2^ω. Consider the projection

$$[\![\cdot\cdot]\!]_{\mathcal{G}_0} : [\omega_1]^2 \to \omega$$

of the square-bracket operation $[\cdot\cdot]$ defined as follows:

$$[\![\alpha\beta]\!]_{\mathcal{G}_0} = \pi([\alpha\beta])(r_\alpha \restriction m(\pi([\alpha\beta])), r_\beta \restriction m(\pi([\alpha\beta]))).$$

Theorem 5.2.5. *For every uncountable family A of pairwise-disjoint finite subsets of ω_1, all of the same size n, there is uncountable set $B \subseteq A$, an equivalence relation $\sim$ on $n = \{0, 1, \ldots, n-1\}$, and a single mapping $h : n \times n \to \omega$ such that*

$$[\![a(i)b(j)]\!]_{\mathcal{G}_0} = h(i,j) \text{ for all } a < b \text{ in } B \text{ and } i \nsim j.$$

Moreover, for every uncountable $C \subseteq B$ and for every $g : n \times n \to \omega$ there exist $a < b$ in C such that

$$[\![a(i)b(j)]\!]_{\mathcal{G}_0} = g(i,j) \text{ for all } i, j < n \text{ with } i \sim j.$$

Proof. Take uncountable $B \subseteq A$ satisfying the conclusion of Lemma 5.1.7. Since there exist only countably many mappings of the form

$$\pi \circ h_b : n \times n \to \mathcal{G}_0,$$

for $b \in B$, shrinking B, we may assume that for some $h : n \times n \to \mathcal{G}_0$ and all $b \in B$, we have $h_b = h$. Choose an integer m such that for some uncountable $B' \subseteq B$ and some one-to-one sequence t_i $(i < n)$ of elements of 2^m, we have

(a) $m > m(h(i,j))$ for all $i, j < n$,

(b) $r_{b(i)} \upharpoonright m = t_i$ for all $b \in B'$ and $i < n$.

Define $h' : 2^m \times 2^m \to \omega$, by

$$h'(i,j)(s,t) = h(i,j)(t \upharpoonright m(i,j),\ s \upharpoonright m(i,j)).$$

To check that this subset B' of A and the mapping h' satisfy the conclusion of the lemma, consider $a < b$ in B' and $i, j < n$ such that $i \nsim j$. Then by the conclusion of Lemma 5.1.7, we know that $[a(i)b(j)] = h_b(i,j,) = \xi$ for some countable ordinal ξ. Let $g = \pi(\xi)$. Then,

$$[\![a(i)b(j)]\!]_{\mathcal{G}_0} = g(t_i \upharpoonright m(i,j),\ t_j \upharpoonright m(i,j)) = h'(i,j).$$

Consider now an arbitrary $g : n \times n \to \omega$ and an arbitrary uncountable set $C \subseteq B'$. Choose a mapping $f : 2^m \times 2^m \to \omega$ such that $f(t_i, t_j) = g(i,j)$ for $i, j < n$. Let $\Gamma_f = \{\gamma < \omega_1 : \pi(\gamma) = f\}$. Then by our choice of π, the set Γ_f is stationary, so by the conclusion of Lemma 5.1.7, we can find $a < b$ in C and $\gamma \in \Gamma$ such that $[a(i)b(j)] = \gamma$ for all $i, j < n$ such that $i \sim j$. Going back to the definition of $[\![\cdot\cdot]\!]_{\mathcal{G}_0}$, we conclude that

$$[\![a(i)b(j)]\!]_{\mathcal{G}_0} = \pi(\gamma)(r_{a(i)} \upharpoonright m(\pi(\gamma)),\ r_{b(j)} \upharpoonright m(\pi(\gamma))) = f(t_i, t_j) = g(i,j)$$

for all $i, j < n$ such that $i \sim j$. □

For a future use let us state and prove yet another application of the square-bracket operation $[\cdot\cdot]$.

Lemma 5.2.6. *There is a mapping $f : \omega_1^{<\omega} \to \omega_1$ such that, for every sequence τ_γ $(\gamma \in \Gamma)$ of elements of $\omega_1^{<\omega}$ indexed by some uncountable subset Γ of ω_1 such that $\gamma \in \mathrm{rang}(\tau_\gamma)$ for all $\gamma \in \Gamma$, and for every $\xi < \omega_1$, there exist $\gamma < \delta$ in Γ such that $f(\sigma^\frown\upsilon) = \xi$ whenever σ is a subsequence of τ_γ with $\gamma \in \mathrm{rang}(\sigma)$ and υ is a subsequence of τ_δ with $\delta \in \mathrm{rang}(\upsilon)$.*

Proof. Choose a one-to-one sequence r_α $(\alpha < \omega_1)$ of elements of 2^ω. For $\tau \in \omega_1^{<\omega}$, set

$$\Delta(\tau) = \max\{\Delta(r_{\tau(i)}, r_{\tau(j)}) : i < j < |\tau|,\ \tau(i) \neq \tau(j)\},$$

where we stipulate that $\Delta(\tau) = \infty$ whenever τ is a constant sequence. Set $f(\tau) = 0$ for τ a constant sequence; otherwise, set

$$f(\tau) = [\![\tau(i)\tau(j)]\!] \text{ for } (i,j) = \min\{(p,q) : \Delta(r_{\tau(p)}, r_{\tau(q)}) = \Delta(\tau)\}. \tag{5.2.7}$$

Note that the conclusion of the lemma in case the sequences τ_γ for $\gamma \in \Gamma$ are all constant reduces to the case $n = 1$ of Lemma 5.2.2, so we may as well assume that no τ_γ $(\gamma \in \Gamma)$ is a constant sequence. Going to an uncountable subset of Γ, we may assume that for some integers l and m,

$$|\tau_\gamma| = l \text{ and } \Delta(\tau_\gamma) = m \text{ for all } \gamma \in \Gamma,$$

and that $\mathrm{rang}(\tau_\gamma)$ $\gamma \in \Gamma$ forms an increasing Δ-system with root c. Moreover, going to an uncountable subset of Γ, we may also assume that the obviously defined finite structures

$$\mathcal{M}_\gamma = (\mathrm{rang}(\tau_\gamma), <, c, \{\gamma\}, \tau, \{r_{\tau_\gamma(i)} \restriction (m+1) : i < l\}, \dots)\ (\gamma \in \Gamma)$$

are all isomorphic. Applying Lemma 5.2.2, for a given $\xi < \omega_1$, we get $\gamma < \delta$ in Γ such that

$$[\![\tau_\gamma(k)\tau_\delta(k)]\!] = \xi \text{ for all } k < l \text{ with } \tau_\gamma(k), \tau_\delta(k) \notin c. \tag{5.2.8}$$

Let us check that these γ and δ satisfy the conclusion of the lemma. So let σ be a subsequence of τ_γ with $\gamma \in \mathrm{rang}(\sigma)$ and υ be a subsequence of τ_δ with $\delta \in \mathrm{rang}(\upsilon)$. Let $\varsigma = \sigma^\frown\upsilon$. Note that

$$\Delta(\varsigma) \geq \Delta(r_{\sigma(i_0)}, r_{\upsilon(j_0)}) > m, \tag{5.2.9}$$

where i_0 and j_0 are such that $\sigma(i_0) = \gamma$ and $\upsilon(j_0) = \delta$. Let

$$(i,j) = \max\{(p,q) : \Delta(r_{\varsigma(p)}, r_{\varsigma(q)}) = \Delta(\varsigma)\}.$$

Then by (5.2.9), the ordinals $\varsigma(i)$ and $\varsigma(j)$ must not belong to the same set $\mathrm{rang}(\tau_\gamma)$ or $\mathrm{rang}(\tau_\delta)$, and moreover, they must occupy the same positions in the increasing enumerations of the corresponding sets $\mathrm{rang}(\tau_\gamma)$ and $\mathrm{rang}(\tau_\delta)$ to which they belong. Since they are different and since the isomorphism between the structures $\mathcal{M}_\gamma$ and $\mathcal{M}_\delta$ is identity on the root c, none of these two ordinals belongs to c. Applying again the fact that $\mathcal{M}_\gamma$ and $\mathcal{M}_\delta$ are isomorphic, we conclude that there must be an index $k < l$ such that $\tau_\gamma(k) = \varsigma(i)$ and $\tau_\delta(k) = \varsigma(j)$, or vice versa, $\tau_\gamma(k) = \varsigma(j)$ and $\tau_\delta(k) = \varsigma(i)$. Applying (5.2.8), it follows that

$$f(\sigma^\frown\upsilon) = [\![\varsigma(i)\varsigma(j)]\!] = [\![\tau_\gamma(k)\tau_\delta(k)]\!] = \xi,$$

as required. □

5.3 Some geometrical applications of the square-bracket operation

The purpose of this section is to show that the projections $[\![\cdot\cdot]\!]_{\mathcal{G}}$ and $[\![\cdot\cdot]\!]_{\mathcal{G}_0}$ of the square-bracket operation $[\cdot\cdot]$ could be used in constructing interesting examples in the geometry of Banach spaces as well as in constructing some interesting geometrical configurations in finite-dimensional Euclidian spaces.

Theorem 5.3.1. *For every integer $n \geq 2$ there are countably many collections $\mathcal{H}_k$ $(k < \omega)$ of hyperplanes*[3] *of $\mathbb{R}^n$ such that:*

(1) *any n hyperplanes belonging to different $\mathcal{H}_k$'s can have at most one point in common, and*

(2) *for every decomposition $\mathbb{R} = \bigcup_{k<\omega} P_k$ there is $k < \omega$ and hyperplane $H \in \mathcal{H}_k$ which is spanned by n points of the intersection $H \cap P_k$.*

Proof. Choose a set $\Omega \subseteq \mathbb{R}^n$ of cardinality ω_1 such that any n distinct points $x_1, \ldots, x_n$ of Ω are in *general position*, in the sense that they span a unique hyperplane $H(x_1, \ldots, x_n)$ of $\mathbb{R}^n$. Moreover, we assume that the set Ω is chosen in such a way that if $X_1, \ldots, X_n$ is a sequence of n-element subsets of Ω with the property that

$$|X_i \cap X_j| \leq 1 \text{ whenever } i \neq j,$$

then

$$|H(X_1) \cap H(X_2) \cap \cdots \cap H(X_n)| \leq 1.$$

We may assume that the projection $[\![\cdot\cdot]\!]_{\mathcal{G}_0}$ of the square-bracket operation $[\cdot\cdot]$ has domain $[\Omega]^2$ rather than $[\omega_1]^2$. We have seen in the previous section that this projection of the square-bracket operation is quite rich in properties, but in this proof, we shall only use the property of $[\![\cdot\cdot]\!]_{\mathcal{G}_0}$ saying that for every uncountable subset Γ of Ω and every $k < \omega$ there exist x and y in Γ such that $[\![xy]\!]_{\mathcal{G}_0} = k$. So for this particular proof even a simple projection of $[\cdot\cdot]$ would work, such as for example, the composition of $[\cdot\cdot]$ with a partition of ω_1 into countably many pairwise-disjoint stationary sets.

For an n-element subset X of Ω, let $H(X)$ denote the hyperplane of $\mathbb{R}^n$ spanned by X if indeed there is a unique hyperplane of $\mathbb{R}^n$ that contains the set X; otherwise $H(X)$ is left undefined. Let $\mathcal{H}(\mathbb{R}^n)$ denote the collection of all hyperplanes of $\mathbb{R}^n$, and for $k < \omega$, set

$$\mathcal{H}_k = \{H(X) \in \mathcal{H}(\mathbb{R}^n) : X \in [\Omega]^n \text{ and } [\![xy]\!]_{\mathcal{G}_0} = k \text{ for all } x \neq y \text{ in } X\}.$$

Let us check that the sequence $\mathcal{H}_k$ $(k < \omega)$ has the properties (1) and (2). The fact that n hyperplanes coming from distinct $\mathcal{H}_k$'s can have at most one point

[3] A *hyperplane* in $\mathbb{R}^n$ is an arbitrary translate of an $(n-1)$-dimensional subspace of $\mathbb{R}^n$. A set S of points from $\mathbb{R}$ *span* a hyperplane H if H is the only hyperplane that contains all the points of S.

in common is an immediate consequence of the property of the set Ω, since if X and Y are n-element subsets of Ω such that the square-bracket operation takes two different constant values on $[X]^2$ and $[Y]^2$, then $|X \cap Y| \leq 1$. Consider now a partition $\mathbb{R} = \bigcup_{k<\omega} P_k$. Fix a $k < \omega$ such that $\Gamma = \Omega \cap P_k$ is uncountable. By the property of the square-bracket operation and the Dushnik–Miller theorem [25] there must exist an n-element set $X \subseteq \Gamma$ such that $[\![xy]\!]_{\mathcal{G}_0} = k$ for all $x \neq y$ in X. Then $H(X)$ belongs to $\mathcal{H}_k$ and it is spanned by the set $X \subseteq H(X) \cap P_k$. This finishes the proof. □

Remark 5.3.2. Theorem 5.3.1 was discovered by Erdős, Jackson and Mauldin [29] where we refer the reader for further information.

Recall that a *homogeneous polynomial* of degree n on a (typically complex) Banach space X is a mapping $P : X \longrightarrow \mathbb{C}$ of the form

$$P(x) = \Phi(x, x, \ldots, x),$$

where $\Phi : X^n \longrightarrow \mathbb{C}$ is a bounded *symmetric multilinear form* on X. Recall that a multilinear form $\Phi : X^n \longrightarrow \mathbb{C}$ is *bounded* if there is a constant C such that

$$|\Phi(x_0, \ldots, x_{n-1})| \leq C \cdot \| x_0 \| \cdots \| x_{n-1} \| \text{ for all } (x_0, \ldots, x_{n-1}) \in X^n.$$

An arbitrary polynomial on X of degree n is the sum of the form

$$P(x) = P_0(x) + P_1(x)+, \ldots, P_n(x),$$

where for $k \leq n$ the $P_k(x)$ is a homogeneous polynomial of degree k and where $P_n(x)$ is not identically zero. Describing the set $\ker(P)$ of zeros of a polynomial is an old subject (see, e.g., [19]). For example, it is known that an arbitrary polynomial P on a complex infinite-dimensional Banach space X such that $P(0) = 0$ is equal to zero on an infinite-dimensional subspace of X (see, [83]). The following example shows that this result docs not extend to the other dimensions.

Theorem 5.3.3. *There is a homogeneous polynomial $P : \ell_1(\omega_1) \longrightarrow \mathbb{C}$ of degree 2 defined on the complex version of $\ell_1(\omega_1)$ which maps every closed nonseparable subspace of $\ell_1(\omega_1)$ onto $\mathbb{C}$.*

Proof. We shall use the projection $[\![\cdot\cdot]\!]_{\mathcal{G}_0}$ and its properties listed in Theorem 5.2.5 in order to define the symmetric bilinear form Φ on which we base our polynomial P. Recall that $[\![\cdot\cdot]\!]_{\mathcal{G}_0}$ has range ω and by indexing the set D of rational points of the unit disc of $\mathbb{C}$ by ω, we shall actually assume that the range of $[\![\cdot\cdot]\!]_{\mathcal{G}_0}$ is D itself rather than ω. It will also be convenient to extend $[\![\cdot\cdot]\!]_{\mathcal{G}_0}$ on the diagonal by letting $[\![\alpha\alpha]\!]_{\mathcal{G}_0} = 0$ for all $\alpha < \omega_1$. Let $B = \{1_\alpha : \alpha < \omega_1\}$, the Hamel basis of the set $c_{00}(\omega_1)$ of all finitely supported maps from ω_1 into $\mathbb{C}$. The symmetric bilinear form Φ is the unique extension of its restriction on $B \times B$ defined as

$$\Phi(1_\alpha, 1_\beta) = \Phi(1_\beta, 1_\alpha) = [\![\alpha\beta]\!]_{\mathcal{G}_0} \tag{5.3.1}$$

for $\alpha \leq \beta < \omega_1$. Note that $|[\![\alpha\beta]\!]_{\mathcal{G}_0}| \leq 1$ for all α and β. It follows that,

$$|\Phi(x,y)| = \left|\sum_{\alpha,\beta} x(\alpha)y(\beta)[\![\alpha\beta]\!]_{\mathcal{G}_0}\right| \leq \sum_{\alpha} |x(\alpha)| \cdot \sum_{\beta} |y(\beta)| = \| x \|_1 \cdot \| y \|_1 \quad (5.3.2)$$

for all x and y in $\ell_1(\omega_1)$. So Φ is a bounded bilinear form on $\ell_1(\omega_1)$ and so $P(x) = \Phi(x,x)$ is a homogeneous polynomial on $\ell_1(\omega_1)$ of degree 2. We shall show that P maps any non-separable closed subspace of $\ell_1(\omega_1)$ onto $\mathbb{C}$. Consider a closed and non-separable subspace X of $\ell_1(\omega_1)$. A simple consequence of the fundamental theorem of algebra is that every nonzero complex polynomial Q defined on some Banach space over the field $\mathbb{C}$ must map the Banach space onto $\mathbb{C}$. It therefore suffices to find w in X such that $P(w) \neq 0$. Since X is not separable, we can choose a rational $0 < \epsilon \leq 1$, an uncountable subset Y from the unit ball of X, and for each x in Y an ordinal $\alpha_x \in \operatorname{supp}(x)$ such that:

(1) $\alpha_x \neq \alpha_y$ for $x \neq y$ in Y,
(2) $|x(\alpha_x)| \geq \epsilon$ for all x in Y.

For each x in Y choose a finite set $b_x \subseteq \operatorname{supp}(x)$ such that

$$\Sigma_{\alpha \notin b_x} |x(\alpha)| < \epsilon^2/32. \quad (5.3.3)$$

Going to an uncountable subset of Y, we may assume that the family b_x $(x \in Y)$ forms a Δ-system with root b. Let us first consider the case $b = \emptyset$. Thus, we may assume that the family b_x $(x \in Y)$ is a family of pairwise-disjoint n-element subsets of ω_1 for some fixed integer n. Going to an uncountable subfamily of Y, we may assume that for all x and y in Y,

(3) $[\![b_x(i)b_x(j)]\!]_{\mathcal{G}_0} = [\![b_y(i)b_y(j)]\!]_{\mathcal{G}_0}$ for all $i, j < n$,
(4) $|x(b_x(i)) - y(b_y(i))| < \epsilon^2/16n$ for all $i < n$.

Combining this and (5.3.3), we obtain

$$|P(x) - P(y)| < \epsilon^2/4 \text{ for all } x, y \in Y. \quad (5.3.4)$$

Applying Theorem 5.2.5, we find an uncountable set $Z \subseteq Y$, an equivalence relation $\sim$ on $n = \{0, 1, \ldots, n-1\}$, and $h : n \times n \to D$ such that for all x and y in Z with $b_x < b_y$, we have

$$[\![b_x(i)b_y(j)]\!]_{\mathcal{G}_0} = h(i,j) \text{ for all } i, j < n \text{ such that } i \nsim j. \quad (5.3.5)$$

Refining Z, we may assume that for some fixed integer $i_0 < n$ and all x in Z, we have $\alpha_x = b_x(i_0)$. Let $g : n \times n \to D$ be a mapping which agrees with h on pairs (i,j) with property $i \nsim j$ such that $g(i,j) = 0$ for all pairs (i,j) such that $i \sim j$. By (5.3.5) and the second conclusion of Theorem 5.2.5, there exist $x, y \in Z$ such that $b_x < b_y$, and

$$[\![b_x(i)b_y(j)]\!]_{\mathcal{G}_0} = g(i,j) \text{ for all } i, j < n. \quad (5.3.6)$$

Consider now the mapping $g' : n \times n \to D$ which agrees with h on pairs (i, j) with property $i \nsim j$ such that $g'(i, j) = 0$ for all pairs (i, j) with property $i \sim j$ with the exception of the pair (i_0, i_0) when we put $g'(i_0, i_0) = 1$. By (5.3.5) and the second conclusion of Theorem 5.2.5, there exist $x', y' \in Z$ such that $b_{x'} < b_{y'}$, and

$$[\![b_{x'}(i)b_{y'}(j)]\!]_{\mathcal{G}_0} = g'(i, j) \text{ for all } i, j < n. \tag{5.3.7}$$

We claim that $P(x + y) \neq P(x' + y')$. Otherwise, we would have the equation $P(x) + P(y) + 2\Phi(x, y) = P(x') + P(y') + 2\Phi(x', y')$. Hence, by (5.3.4),

$$|\Phi(x, y) - \Phi(x', y')| < 1/2(\epsilon^2/4 + \epsilon^2/4) = \epsilon^2/4. \tag{5.3.8}$$

For $w \in \{x, y, x', y'\}$, let

$$w_0 = w \restriction (\mathrm{supp}(w) \setminus b_w), w_1 = w \restriction (b_w \setminus \{b_w(i_0)\}), \text{ and } w_2 = w \restriction \{b_w(i_0)\}.$$

Thus, $w = w_0 + w_1 + w_2$ for every $w \in \{x, y, x', y'\}$, and so, $\Phi(x, y)$ and $\Phi(x', y')$, have the nine-term expansions

$$\Sigma_{i,j<3}\ \Phi(x_i, y_j) \text{ and } \Sigma_{i,j<3}\ \Phi(x'_i, y'_j),$$

respectively. Note that by the setup above, $|\Phi(x_i, y_j) - \Phi(x'_i, y'_j)| < \epsilon/16$ for all $i, j < 3$ such that $(i, j) \neq (2, 2)$. On the other hand, by (2), (5.3.6) and (5.3.7),

$$|\Phi(x_2, y_2) - \Phi(x'_2, y'_2)| \geq \epsilon^2.$$

Combining all this with (5.3.8), we get that

$$\epsilon^2/4 > |\Phi(x, y) - \Phi(x', y')| \geq \epsilon^2 - 8(\epsilon^2/16),$$

a contradiction. It follows that one of $P(x + y)$ or $P(x' + y')$ must be nonzero and this finishes the proof.

To reduce the general case to the case when the root b of the Δ-system is empty, we start with an uncountable set Y of vectors of X of norm $\leq 1/2$ satisfying (1) and (2) for some $\epsilon > 0$. Find finite sets a_x $(x \in Y)$ such that

$$\Sigma_{\alpha \notin a_x}\ |x(\alpha)| < \epsilon^2/128.$$

Assume that the family forms a Δ-system with root a and such that for all $x, y \in Y$.

$$\Sigma_{\alpha \in a}\ |x(\alpha) - y(\alpha)| < \epsilon^2/128. \tag{5.3.9}$$

For $x \in Z$, set $b_x = a_x \setminus a$. Form an uncountable set Z of elements of the form $z = y - x$ for $x, y \in Y$ such that $b_x < b_y$. We also assume that if $z = y - x$ and $z' = y' - x'$ are members of Z, then $b_x < b_y < b_{x'} < b_{y'}$, or vice versa $b_{x'} < b_{y'} < b_x < b_y$. Note that Z still satisfies (1) and (2) if we put $\alpha_z = \alpha_y$

for $z = y - x \in Z$. Note also that $b_z = b_x \cup b_y$ $(z = y - x \in Z)$ is a family of pairwise-disjoint finite subsets of ω_1 such that

$$\Sigma_{\alpha \notin b_z} |z(\alpha)| < \epsilon^2/32.$$

So we have reached the input of the first part of the proof, i.e., we have reduced the general case to the case when the root of the Δ-system is empty. □

The reader is refereed to [6] for more information on this and other related results. For example, in [6], we show that the above proof can be extended in the following direction.

Theorem 5.3.4. *For every separable Banach space V over the field $\mathbb{C}$, there is a homogeneous polynomial $P : \ell(\omega_1) \longrightarrow V$ of degree 2 defined on the complex version of $\ell(\omega_1)$ which maps every closed non-separable subspace of $\ell(\omega_1)$ onto a somewhere dense subset of V.* □

As one may expect, Banach spaces over the field $\mathbb{C}$ of complex numbers and Banach spaces over the field $\mathbb{R}$ of real numbers behave quite differently when considering zeros of polynomials. For example, on the version of the space ℓ_1 over the field $\mathbb{R}$ there exist homogeneous polynomials of any given even degree that vanish only at 0. It is for this reason that, when studying zero spaces of polynomials in the case of Banach spaces over $\mathbb{R}$, one considers only polynomials of odd degrees. Even under this restriction there is a considerable difference between the real and complex case as the following result from [5] shows.

Theorem 5.3.5. *Let X be an infinite-dimensional Banach space over the field $\mathbb{R}$ of real numbers whose dual X^* is weak*-separable. Then for every odd integer $n > 1$ there is a homogeneous polynomial $P : X \longrightarrow \mathbb{R}$ of degree n which is not zero on any infinite-dimensional subspace of X.* □

On the other hand, we have the following result from [83] for polynomials over the field of complex numbers.

Theorem 5.3.6. *Let X be an infinite-dimensional Banach space over the field $\mathbb{C}$ of complex numbers and let $P : X \longrightarrow \mathbb{C}$ be any polynomial such that $P(0) = 0$. Then there is infinite-dimensional subspace Y of X on which P is identically equal to 0.* □

Note the following immediate corollary of Theorem 5.3.5 which should be compared both to Theorem 5.3.3 and Theorem 5.3.6.

Theorem 5.3.7. *If Γ is an index-set of cardinality at most the continuum (so in particular, if $\Gamma = \omega_1$), then for every odd integer $n > 1$ there is a homogeneous polynomial $P : \ell(\Gamma) \longrightarrow \mathbb{R}$ of degree n which is not zero on any infinite-dimensional subspace of $\ell(\Gamma)$.* □

On the other hand, in the case of the field of complex numbers, we have the following problem.

Question 5.3.8. Let Γ be an index set of cardinality continuum and let us consider the version of the Banach space $\ell(\Gamma)$ over the field $\mathbb{C}$ of complex numbers. Is there a homogeneous polynomial $P : \ell(\Gamma) \longrightarrow \mathbb{C}$ which is not zero on any non-separable subspace of $\ell(\Gamma)$?

Remark 5.3.9. Recall that by the result of [83] there will be always an infinite-dimensional subspace of $\ell(\Gamma)$ on which P is identically equal to 0. Part of the interest in this question comes from the fact that for sets Γ of cardinality continuum there is essentially only one known example of a mapping $f : [\Gamma]^2 \longrightarrow 2$ which is not constant on any square $[A]^2$ of an uncountable subset A of Γ. This is the well-known partition of Sierpinski, obtained by comparing the usual ordering of the reals with a well-ordering. However, it can be shown, see [6], that if $f : [\Gamma]^2 \longrightarrow 2$ is the Sierpinski partition, then the corresponding homogeneous polynomial $P_f : \ell(\Gamma) \longrightarrow \mathbb{C}$ defined by[4]

$$P_f(x) = \sum_{\alpha,\beta\in\Gamma} f(\alpha,\beta)x_\alpha x_\beta$$

does admit a non-separable null subspace. The reader is refereed to papers [5] and [6] where this problem is treated for bigger index-sets Γ where the difference between the fields $\mathbb{R}$ and $\mathbb{C}$ is even more apparent.

We shall now see that the square-bracket operation can also be effectively used in constructing Banach spaces X with some extreme properties of bounded operators $T : X \longrightarrow X$.

Theorem 5.3.10. *There is a non-separable reflexive Banach space E with the property that every bounded linear operator $T : E \longrightarrow E$ can be expressed as $T = \lambda I + S$ where λ is a scalar, I the identity operator of E, and S a bounded operator with separable range.*

Proof. Let $I = 3 \times [\omega_1]^{<\omega}$ and let us identify the index-set I with ω_1, i.e., pretend that the projection $[\![\cdot\cdot]\!] = [\![\cdot\cdot]\!]_{\mathcal{G}}$ given above in Definition 5.2.1 takes its values in I rather than in ω_1. Let

$$[\![\cdot\cdot]\!]_0 : [\omega_1]^2 \longrightarrow 3 \text{ and } [\![\cdot\cdot]\!]_1 : [\omega_1]^2 \longrightarrow [\omega_1]^{<\omega}$$

be the first and second coordinates of $[\![\cdot\cdot]\!] : [\omega_1]^2 \longrightarrow 3 \times [\omega_1]^{<\omega}$, respectively. Let

$$\mathcal{G} = \{G \in [\omega_1]^{<\omega} : [\![\alpha\beta]\!]_0 = 0 \text{ for all } \{\alpha,\beta\} \in [G]^2\},$$

$$\mathcal{H} = \{H \in [\omega_1]^{<\omega} : [\![\alpha\beta]\!]_0 = 1 \text{ for all } \{\alpha,\beta\} \in [H]^2\}.$$

[4] As customary, in this formula, we are identifying f with a symmetric function $f : \Gamma^2 \longrightarrow 2$ which is identically equal to 0 on the diagonal.

Let $\mathcal{K}$ be the collection of all finite sets $\{\{\alpha_i, \beta_i\} : i < k\}$ of pairs of countable ordinals such that for all $i < j < k$:

$$\max\{\alpha_i, \beta_i\} < \min\{\alpha_j, \beta_j\}, \tag{5.3.10}$$

$$[\![\alpha_i\alpha_j]\!]_0 = [\![\beta_i\beta_j]\!]_0 = 2, \tag{5.3.11}$$

$$[\![\alpha_i\alpha_j]\!]_1 = [\![\beta_i\beta_j]\!]_1 = \{\alpha_l : l < i\} \cup \{\beta_l : l < i\}. \tag{5.3.12}$$

The following properties of $\mathcal{G}, \mathcal{H}$ and $\mathcal{K}$ should be clear:

(5.3.13) $\mathcal{G}$ and $\mathcal{H}$ contain all the singletons, are closed under subsets and they are 1-orthogonal to each other in the sense that $\mathcal{G} \cap \mathcal{H}$ contains no doubleton.

(5.3.14) $\mathcal{G}$ and $\mathcal{H}$ are both 2-orthogonal to the family of the unions of members of $\mathcal{K}$.

(5.3.15) If K and L are two distinct members of $\mathcal{K}$, then there are no more than five ordinals α such that $\{\alpha, \beta\} \in K$ and $\{\alpha, \gamma\} \in L$ for some $\beta \neq \gamma$.

(5.3.16) For every sequence $\{\alpha_\xi, \beta_\xi\}$ $(\xi < \omega_1)$ of pairwise-disjoint pairs of countable ordinals there exist arbitrarily large finite sets $\Gamma, \Delta \subseteq \omega_1$ such that $\{\alpha_\xi : \xi \in \Gamma\} \in \mathcal{G}$, $\{\beta_\xi : \xi \in \Gamma\} \in \mathcal{H}$ and $\{\{\alpha_\xi, \beta_\xi\} : \xi \in \Delta\} \in \mathcal{K}$.

For a function x from ω_1 into $\mathbb{R}$, set

$$\|x\|_{\mathcal{H},2} = \sup\{(\Sigma_{\alpha \in H}\ x(\alpha)^2)^{\frac{1}{2}} : H \in \mathcal{H}\},$$

$$\|x\|_{\mathcal{K},2} = \sup\{(\Sigma_{\{\alpha,\beta\} \in K}\ (x(\alpha) - x(\beta))^2)^{\frac{1}{2}} : K \in \mathcal{K}\}.$$

Let $\|\cdot\| = \max\{\|\cdot\|_\infty, \|\cdot\|_{\mathcal{H},2}, \|\cdot\|_{\mathcal{K},2}\}$ and define $\bar{E}_2 = \{x : \|x\| < \infty\}$. Let 1_α be the characteristic function of $\{\alpha\}$. Finally, let E_2 be the closure of the linear span of $\{1_\alpha : \alpha \in \omega_1\}$ inside $(\bar{E}_2, \|\cdot\|)$. The following facts about the norm $\|\cdot\|$ are easy to establish using the properties of the families $\mathcal{G}, \mathcal{H}$ and $\mathcal{K}$ listed above:

$$\text{If } x \text{ is supported by some } G \in \mathcal{G}, \text{ then } \|x\| \leq 2 \cdot \|x\|_\infty. \tag{5.3.17}$$

$$\text{If } x \text{ is supported by } \bigcup K \text{ for some } K \text{ in } \mathcal{K}, \text{ then } \|x\| \leq 10 \cdot \|x\|_\infty. \tag{5.3.18}$$

The role of the seminorm $\|\cdot\|_{\mathcal{H},2}$ is to ensure that every bounded operator $T : E_2 \longrightarrow E_2$ can be expressed as $D + S$, where D is a bounded diagonal operator relative to the basis[5] 1_α $(\alpha < \omega_1)$ and where S has separable range. Namely, if this were not so, we could find a sequence $\{\alpha_\xi, \beta_\xi\}$ $(\xi < \omega_1)$ of pairwise-disjoint pairs from ω_1 such that $T(1_{\alpha_\xi})(\beta_\xi) \neq 0$ for all $\xi < \omega_1$. Thinning the sequence, we may

[5] Indeed it can be shown that 1_α $(\alpha < \omega_1)$ is a 'transfinite basis' of E_2 in the sense of [100]. So every vector x of E_2 has a unique representation as $\Sigma_{\alpha<\omega_1} x(\alpha) 1_\alpha$ and the projection operators $P_\beta : E_2 \to E_2 \restriction \beta$ $(\beta < \omega_1)$ are uniformly bounded.

assume that for some fixed $r > 0$ and all $\xi < \omega_1$, $|T(1_{\alpha_\xi})(\beta_\xi)| \geq r$. By (5.3.16) for every $n < \omega$ we can find a subset Γ of ω_1 of size n such that $G = \{\alpha_\xi : \xi \in \Gamma\} \in \mathcal{G}$ and $H = \{\beta_\xi : \xi \in \Gamma\} \in \mathcal{H}$. Then, by (5.3.17),

$$2 \cdot \|T\| \geq \|T(\chi_G)\| \geq \left(\sum_{\beta \in H} T(\chi_G)(\beta)^2\right)^{\frac{1}{2}} \geq r \cdot \sqrt{n}.$$

Since n is arbitrary, this contradicts the fact that T is bounded. So from now on we may concentrate on showing that every bounded diagonal operator has the form $\lambda I + S$, where S is a bounded operator with separable range.

The role of the seminorm $\|\cdot\|_{\mathcal{K},2}$ is to represent every bounded diagonal operator

$$D(x)(\xi) = \lambda_\xi x(\xi) \ (\xi < \omega_1)$$

as $\lambda I + S$ for some scalar λ and operator S with separable range. This reduces to showing that the sequence λ_ξ $(\xi < \omega_1)$ is eventually constant. If this were false, we would be able to extract a sequence $\{\alpha_\xi, \beta_\xi\}$ $(\xi < \omega_1)$ of pairwise-disjoint pairs of countable ordinals and an $r > 0$ such that $|\lambda_{\alpha_\xi} - \lambda_{\beta_\xi}| \geq r$ for all $\xi < \omega_1$. By (5.3.16), for every $n < \omega$, we can find a subset Δ of ω_1 of size n such that $K = \{\{\alpha_\xi, \beta_\xi\} : \xi \in \Delta\}$ belongs to $\mathcal{K}$. Then, by (5.3.18),

$$\begin{aligned} 10 \cdot \|D\| &\geq \|D(\chi_{\bigcup K})\| \\ &\geq (\Sigma_{\xi \in \Delta} \, (D(\chi_{\bigcup K})(\alpha_\xi) - D(\chi_{\bigcup K})(\beta_\xi))^2)^{\frac{1}{2}} \\ &= (\Sigma_{\xi \in \Delta} \, (\lambda_{\alpha_\xi} - \lambda_{\beta_\xi})^2)^{\frac{1}{2}} \\ &\geq r \cdot \sqrt{n}. \end{aligned}$$

Since n was arbitrary this contradicts our initial assumption that D is a bounded operator.

Note that $\|x\| \leq 2\|x\|_2$ for all $x \in \ell_2(\omega_1)$. It follows that $\ell_2(\omega_1) \subseteq E_2$ and the inclusion is a bounded linear operator. Note also that $\ell_2(\omega_1)$ is a dense subset of E_2. Therefore E_2 is a weak compactly-generated space. For example, $W = \{x \in \ell_2(\omega_1) : \|x\|_2 \leq 1\}$ is a weakly-compact subset of E_2 and its linear span is dense in E_2. To get a reflexive example out of E_2 one uses an interpolation method of Davis, Figiel, Johnson and Pelczynski [17] as follows. Let p_n be the Minkowski functional of the set $2^n W + 2^{-n}\mathrm{Ball}(E_2)$.[6] Let

$$E = \{x \in E_2 : \|x\|_E = (\Sigma_{n=0}^{\infty} \, p_n(x)^2)^{\frac{1}{2}} < \infty\}.$$

By [17, Lemma 1], E is a reflexive Banach space and $\ell_2(\omega_1) \subseteq E \subseteq E_2$ are continuous inclusions. Note that $p_n(x) < r$ iff $x = y + z$ for some $y \in E_2$ and $z \in \ell_2(\omega_1)$ such that $\|y\| < 2^{-n}r$ and $\|z\|_2 < 2^n r$. The following two facts about

[6] In other words, $p_n(x) = \inf\{\lambda > 0 : x \in \lambda B\}$, where $B = 2^n W + 2^{-n}\mathrm{Ball}(E_2)$.

E correspond to the facts (5.3.17) and (5.3.18) and they are used in a similar way in the proof that every bounded operator $T : E \longrightarrow E$ has the form $\lambda I + S$.

If x is the vector supported by some $G \in \mathcal{G}$ which can be split into n blocks such that the kth one has size 2^{4k} and x takes the value 2^{-2k} on each term of the kth block, then $\|x\|_E \leq 2$. (5.3.19)

If $\{\{\alpha_\xi, \beta_\xi\} : \xi \in \Delta\}$ is a member of K so that Δ can be split into n blocks such that the k-th one Δ_k has size 2^{4k} and if x is the vector supported by $\bigcup K$ for $\xi \in \Delta_k, x(\alpha_\xi) = 2^{-2k}$ and $x(\beta_\xi) = 2^{-2k}$, then $\|x\|_E \leq 12$. (5.3.20)

We leave the checking of (5.3.19) and (5.3.20) as well as the corresponding proof that E has 'few' operators to the interested reader. □

Remark 5.3.11. The first example of a reflexive Banach space X with a Schauder basis of length ω_1 on which every bounded linear operator is the sum of a scalar multiple of the identity and an operator with a separable range, is given by Wark [130] who was motivated by a previous and similar construction due to Shelah and Steprans [97], a construction in which the square-bracket operation also plays the key role.

5.4 A square-bracket operation from a special Aronszajn tree

In this section we try to show that the basic idea of the square-bracket operation on ω_1 can perhaps be more easily grasped by working on an arbitrary special Aronszajn tree rather than $T(\rho_0)$. So let $T = \langle T, <_T \rangle$ be a fixed special Aronszajn tree and let

$$a : T \longrightarrow \omega$$

be a fixed map witnessing this, i.e., a mapping with the property that for two different nodes s and t of T, the equality $a(s) = a(t)$ implies that s and t are incomparable in T. We shall also assume that for every $s, t \in T$ the greatest lower bound $s \wedge t$ exists in T. For $t \in T$ and $n < \omega$, set

$$F_n(t) = \{s \leq_T t : s = t \text{ or } a(s) \leq n\}.$$

Finally, for $s, t \in T$ with $ht(s) \leq ht(t)$, let

$$[st]_T = \min\{v \in F_{a(s \wedge t)}(t) : ht(v) \geq ht(s)\}.^7 \tag{5.4.1}$$

The following fact corresponds to Lemma 5.1.4 in the case when $T = T(\rho_0)$.

[7] If $ht(s) \geq ht(t)$ we let $[st]_T = [ts]_T$.

Lemma 5.4.1. *If X is an uncountable subset of T, the set of nodes of T of the form $[st]_T$ for some $s, t \in X$ intersects a closed and unbounded set of levels of T.* □

We do not give a proof of this fact as it is almost identical to the proof of Lemma 5.1.4 which deals with the special case $T = T(\rho_0)$. But one can go further and show that $[\cdot\cdot]_T$ shares all the other properties of the square-bracket operation $[\cdot\cdot]$ described in the previous section. Some of these properties, however, are easier to visualize and prove in the general context. For example, consider the following fact, which in the case $T = T(\rho_0)$ is the essence of Lemma 5.2.3.

Lemma 5.4.2. *Suppose $A \subseteq T$ is an uncountable antichain and that for each $t \in A$ there is given a finite set F_t of its successors. Then for every stationary set $\Gamma \subseteq \omega_1$ there exists an arbitrarily large finite set $B \subseteq A$ such that the height of $[xy]_T$ belongs to Γ whenever $x \in F_s$ and $y \in F_t$ for some $s \neq t$ in B.* □

Let us now examine in more detail the collection of graphs $\mathcal{G}_\Gamma(\Gamma \subseteq \omega_1)$ of Definition 5.1.11 but in the present more general context.

Definition 5.4.3. For $\Gamma \subseteq \omega_1$, let

$$K_\Gamma = \{\{s,t\} \in [T]^2 : ht([st]_T) \in \Gamma\}.$$

Working as in Lemma 5.1.12 one shows that (T, K_Γ) and (T, K_Δ) have no isomorphic uncountable subgraphs whenever the symmetric difference between Γ and Δ is a stationary subset of ω_1, i.e., whenever they represent different members of the quotient algebra $\mathcal{P}(\omega_1)/NS$. In particular, K_Γ contains no square $[X]^2$ of an uncountable set $X \subseteq T$ whenever Γ contains no closed and unbounded subset of ω_1. The following fact is a sort of converse to this.

Lemma 5.4.4. *If Γ contains a closed and unbounded subset of ω_1, then there is a proper forcing notion introducing an uncountable set $X \subseteq T$ such that $[X]^2 \subseteq K_\Gamma$.*

Proof. The forcing notion $\mathbb{P}$ is defined to be the set of all pairs $p = \langle X_p, \mathcal{N}_p \rangle$ where:

$\mathcal{N}_p$ is a finite $\in$-chain of countable elementary submodels of H_{ω_2} containing all the relevant objects. (5.4.2)

X_p is a finite subset of T such that $ht([st]_T) \in \Gamma$ for every pair $\{s,t\}$ of elements of X_p. (5.4.3)

For all $s \neq t$ in X_p there exist $N \in \mathcal{N}_p$ such that $s \in N$ iff $t \notin N$. (5.4.4)

For every $s \neq t$ in X_p, $N \in \mathcal{N}_p$, if $t \notin N$ then the height of $\min(F_{a(s \wedge t)}(t) \setminus N)$ belongs to Γ. (5.4.5)

The ordering of $\mathbb{P}$ is the coordinatewise inclusion. To show that $\mathbb{P}$ is a proper forcing notion, let M be a given countable elementary submodel of some large enough H_θ containing all the relevant objects and let $p \in \mathbb{P} \cap M$. We shall show that

$$q = \langle X_p, \mathcal{N}_p \cup \{M \cap H_{\omega_2}\}\rangle$$

is an M-generic condition extending p. So let $\mathcal{D} \in M$ be a dense-open subset of $\mathbb{P}$ and let r be a given extension of q. Extending r we may assume that $r \in \mathcal{D}$. Let $p_0 = r \cap M$. Note that $p_0 \in \mathbb{P} \cap M$. Let m_0 be the maximum of the union of the following two finite sets of integers

$$\{a(s \wedge t) : s, t \in X_r\},$$

$$\{a(t \restriction \delta) : t \in X_r, \delta = N \cap \omega_1 \text{ for some } N \in \mathcal{N}_p \text{ such that } t \notin N\}.$$

Let k be the size of the projection of the set $X_r \setminus M$ on the level $T_{M\cap\omega_1}$ of the tree T. Let $\mathcal{F}$ be the family of all k-element subsets F of T for which one can find $\bar{r} \in \mathcal{D}$ realizing the same type as r over p_0 such that, if N is the minimal model of $\mathcal{N}_{\bar{r}} \setminus \mathcal{N}_{p_0}$, then

$$F = \{x \restriction (N \cap \omega_1) : x \in X_{\bar{r}} \setminus X_{p_0}\}.$$

Since the projection of $X_{\bar{r}} \setminus M$ onto $T_{M\cap\omega_1}$ belongs to $\mathcal{F}$, the family is a rather large subset of $[T]^k$. It follows that the family $\mathcal{E}$ of all k-element subsets E of T for which we can find $\delta < \omega_1$ such that

$$\mathcal{F}_E = \{F \in \mathcal{F} : \{x \restriction \delta : x \in F\} = E\}$$

is uncountable, is also quite large. In particular, we can find an ordinal $\alpha_0 \in M \cap \omega_1$ above the heights of all nodes from the set

$$\bigcup \{M \cap F_{m_0}(t) : t \in X_r\}$$

and $E_1 \in \mathcal{E} \cap M$ such that

$$\{x \restriction \alpha_0 : x \in E_1\} = \{x \restriction \alpha_0 : x \in X_r \setminus M\}$$

and such that every node of $X_r \setminus M$ is incomparable with every node of E_1. Let

$$m^+ = \max\{a(x \wedge y) : x \in X_r \setminus M, y \in E_1\},$$

$$m^- = \min\{a(x \wedge y) : x \in X_r \setminus M, y \in E_1 \text{ and } a(x \wedge y) > m_0\}.$$

Let β_0 be an ordinal of $M \cap \omega_1$ above α_0 which bounds the heights of all nodes from the set

$$\bigcup \{M \cap F_{m^+}(t) : t \in X_r \setminus M\}.$$

By the definitions of $\mathcal{E}$ and $\mathcal{F}$ there is $\bar{r} \in \mathcal{D} \cap M$ realizing the same type over p_0 as r such that, if N is the minimal model of $\mathcal{N}_{\bar{r}} \setminus \mathcal{N}_{p_0}$, then $N \cap \omega_1 \geq \beta_0$ and the projection

$$F = \{x \restriction (N \cap \omega_1) : x \in X_{\bar{r}} \setminus X_{p_0}\}$$

dominates the set E_1. We claim that

$$\bar{q} = \langle X_{\bar{r}} \cup X_r, \mathcal{N}_{\bar{r}} \cup \mathcal{N}_r \rangle$$

is a condition of $\mathbb{P}$. This will finish the proof of properness of $\mathbb{P}$ since clearly $\bar{q}$ extends both $\bar{r}$ and r. Clearly, only (5.4.3) and (5.4.5) for $\bar{q}$ require some argumentation. We check first (5.4.5), so let s and t be a pair of distinct members of $X_{\bar{r}} \cup X_r$ and let $N \in \mathcal{N}_{\bar{r}} \cup \mathcal{N}_r$ be a model such that $t \notin N$. Let $m = a(s \wedge t)$. If $t \in X_{\bar{r}}$ and if there is $s' \in X_{\bar{r}}$ such that $s' \neq t$ and $s' \wedge t = s \wedge t$, the conclusion that the height of $\min(F_m(t) \setminus N)$ belongs to Γ follows from the fact that $\bar{r}$ satisfies (5.4.5). Similarly one considers the case when $t \in X_r \setminus M$ and when $s' \wedge t = s \wedge t$ for some $s' \in X_r$, $s' \neq t$. For if N belongs to $\mathcal{N}_r$, the conclusion follows from the fact that r satisfies (5.4.5) and if $N \in \mathcal{N}_{\bar{r}} \setminus \mathcal{N}_r$, then by the choice of $\bar{r}$,

$$F_m(t) \setminus N = F_m(t) \setminus M,$$

and so the conclusion that the height of $\min(F_m(t) \setminus N)$ belongs to Γ follows from the fact that r satisfies (5.4.5) for s' and t from X_r and $M \cap H_{\omega_2}$ from $\mathcal{N}_r$. Let us now consider the case $t \in X_r \setminus M$, $s \in X_{\bar{r}} \setminus X_r$ and $s' \wedge t \neq s \wedge t$ for all $s' \in X_r \setminus \{t\}$. Note that in this case we have the following inequalities:

$$m_0 < m^- \leq m = a(s \wedge t) \leq m^+.$$

So by the choice of m_0 the set $F_m(t)$ contains $t \upharpoonright (N' \cap \omega_1)$ for every $N' \in \mathcal{N}_r$. Since every ordinal of the form $N' \cap \omega_1$ ($N' \in \mathcal{N}_r$) belongs to Γ we are done in case N belongs to $\mathcal{N}_r$. So suppose $N \in \mathcal{N}_{\bar{r}} \setminus \mathcal{N}_r$. Then by the choice of the bound β_0 and the choice of $\bar{r}$,

$$F_m(t) \setminus N = F_m(t) \setminus M$$

so the conclusion of (5.4.5) follows again from the fact that $M \cap H_{\omega_2}$ belongs to $\mathcal{N}_r$. Consider now the remaining case $t \in X_{\bar{r}} \setminus X_r$ and $s \in X_r \setminus X_{\bar{r}}$. Note that in this case the model N must belong to $\mathcal{N}_r$, so if $s' \wedge t = s \wedge t$ for some $s' \in X_{\bar{r}}$, $s \neq t$, the conclusion follows from the fact that $\bar{r}$ satisfies the condition (5.4.5). So let us consider the subcase when $s' \wedge t \neq s \wedge t$ for all $s' \in X_{\bar{r}}$, $s' \neq t$. Thus, $s \wedge t$ is the new splitting, so by the choice of α_0 and E_0 we have the inequalities:

$$m_0 < m^- \leq m = a(s \wedge t).$$

Recall that $\bar{r}$ was chosen to realize the same type over p_0 as r so we in particular have that $F_{m_0}(t)$ and therefore its superset $F_m(t)$ contains the projection $t \upharpoonright (N \cap \omega_1)$. It follows that $\min(F_m(t) \setminus N) = t \upharpoonright (N \cap \omega_1)$, so its height is indeed a member of Γ (as $\Gamma \in N$ and Γ contains a closed and unbounded subset of ω_1).

It remains to check (5.4.3) for $\bar{q}$. So let s and t be a given pair of distinct members of $X_{\bar{r}} \cup X_r$. We need to show that the height of $[st]_T$ belongs to Γ. Clearly the only nontrivial case is when, say, $s \in X_{\bar{r}} \setminus X_r$ and $t \in X_r \setminus X_{\bar{r}}$. Suppose first that $s \wedge t = s' \wedge t$ for some $s' \in X_r$, $s' \neq t$. Since $M \cap H_{\omega_2}$ belongs to $\mathcal{N}_r$ and since

r satisfies (5.4.5), the height of $\min(F_{a(s'\wedge t)}(t)\setminus M)$ belongs to Γ. Since the height of s is above the bound β_0 for $F_{m^+}(t)\cap M$,

$$[st]_T = \min(F_{a(s\wedge t)}(t)\setminus M) = \min(F_{a(s'\wedge t)}(t)\setminus M),$$

so we are done in this subcase. Suppose now that $s\wedge t$ is a new splitting, i.e., that

$$m_0 < m = a(s\wedge t) \leq m^+.$$

So again we know that in this subcase, $F_m(t)$ contains the projection $t\restriction(M\cap\omega_1)$. By our choice of the bound β_0 and the condition $\bar{r}$ we know that this projection is the minimal point of $F_m(t)$ whose height is bigger than or equal to the height of s. By the definition of the square-bracket operation, we have

$$[st]_T = t\restriction(M\cap\omega_1).$$

Since $M\cap\omega_1$ belongs to Γ this finishes our checking that $\bar{q}$ satisfies (5.4.3). Moreover, this also finishes the proof that the forcing notion $\mathbb{P}$ is proper.

It is clear that $\mathbb{P}$ forces the square of the union $\dot{X}$ of the X_p's for p belonging to the generic filter to be included in K_Γ. What is not so clear is that $\mathbb{P}$ forces the set $\dot{X}$ to be uncountable as promised. To ensure this we replace $\mathbb{P}$ with its restriction $\mathbb{P}(\leq r)$ to the condition

$$r = \langle\{t\},\{M\cap H_{\omega_2}\}\rangle,$$

where M is an arbitrary countable elementary submodel of some large enough H_θ containing all the relevant objects and where t is any node of T of height at least $M\cap\omega_1$. Note that the above proof of properness of $\mathbb{P}$ starting from $p=\langle\emptyset,\emptyset\rangle$ shows that

$$q = \langle\emptyset,\{M\cap H_{\omega_2}\}\rangle$$

is an M-generic condition of $\mathbb{P}$, so in particular its extension r is also M-generic. It follows that if we let $M[\dot{G}]$ be the name for the set formed by interpreting all $\mathbb{P}$-names belonging to M itself. Then r forces $\dot{X}$ to be an element of the countable elementary submodel $M[\dot{G}]$ of $H_\theta[\dot{G}]$ which contains a point t of height above $M[\dot{G}]\cap\omega_1$. It follows that r forces $\dot{X}$ to be uncountable. This finishes the proof of Lemma 5.4.4. □

Corollary 5.4.5. *The graph K_Γ contains the square of some uncountable subset of T in some ω_1-preserving forcing extension, if and only if Γ is a stationary subset of ω_1.*

Proof. If Γ is disjoint from a closed and unbounded subset, then in any ω_1-preserving forcing extension its complement $\Delta=\omega_1\setminus\Gamma$ will be a stationary subset of ω_1. So by the basic property in Lemma 5.4.1 of the square-bracket operation, no such a forcing extension will contain an uncountable set $X\subseteq T$ such that $[X]^2\subseteq K_\Gamma$. On the other hand, if Γ is a stationary subset of ω_1, going first to some standard ω_1-preserving forcing extension in which Γ contains a closed and unbounded subset of ω_1 and then applying Lemma 5.4.4, we get an ω_1-preserving forcing extension having an uncountable set $X \subseteq T$ such that $[X]^2 \subseteq K_\Gamma$. $\square$

Remark 5.4.6. Corollary 5.4.5 gives us a further indication of the extreme complexity of the class of graphs on the vertex-set ω_1. It also bears some relevance to the recent work of Woodin [135] who, working in his $\mathbb{P}_{\max}$-forcing extension, was able to associate a stationary subset of ω_1 to any partition of $[\omega_1]^2$ into two pieces. So one may view Corollary 5.4.5 as some sort of converse to this, since in the $\mathbb{P}_{\max}$-extension one is able to get a sufficiently generic filter to the forcing notion $\mathbb{P} = \mathbb{P}_\Gamma$ of Lemma 5.4.4 that would give us an uncountable $X \subseteq T$ such that $[X]^2 \subseteq K_\Gamma$. In other words, under a bit of PFA or Woodin's axiom (*), a set $\Gamma \subseteq \omega_1$ contains a closed and unbounded subset of ω_1 if and only if K_Γ contains $[X]^2$ for some uncountable $X \subseteq T$.

5.5 A square-bracket operation from the complete binary tree

The variation of the square operation can also be based on the complete binary tree of height ω rather than a special Aronszajn tree. To define this variation we start with a one-to-one sequence

$$\{r_\xi : \xi < \omega_1\} \subseteq 2^\omega$$

of branches through the complete binary tree $R = 2^{<\omega}$, and a mapping

$$e : [\omega_1]^2 \longrightarrow \omega$$

such that $e(\alpha, \gamma) \neq e(\beta, \gamma)$ whenever $\alpha < \beta < \gamma < \omega_1$.

For $n < \omega$ and $\alpha < \omega_1$, set

$$F_n(\alpha) = \{\xi \leq \alpha : \xi = \alpha \text{ or } e(\xi, \alpha) \leq n\}.$$

For $\alpha < \beta < \omega_1$, set

$$\Delta(\alpha, \beta) = \min\{n < \omega : r_\alpha(n) \neq r_\beta(n)\}.$$

Finally,we are in a position to define the variation of the square-bracket operation on ω_1 by putting for $\alpha < \beta < \omega_1$,

$$[\alpha\beta]_R = \min(F_{\Delta(\alpha,\beta)}(\beta) \setminus \alpha). \tag{5.5.1}$$

Then we have the following fact whose proof is very similar to the proof of Lemma 5.1.4, the corresponding fact for the original square-bracket operation.

Lemma 5.5.1. *If Γ is an uncountable subset of ω_1, the set of all ordinals of the form $[\alpha\beta]_R$ for some $\alpha < \beta$ in Γ contains a closed and unbounded subset of ω_1.* $\square$

A simple analysis of the proof of Lemma 5.1.7 for the original square-bracket operation gives the following result for the variation $[\cdot\cdot]_R$.

Lemma 5.5.2. *For every uncountable family A of pairwise-disjoint subsets of ω_1, all of the same size n, there exist an uncountable $B \subseteq A$, and a mapping $h_b : n \times n \to \omega_1$ for each $b \in B$ such that for all $a < b$ in B, we have that $[a(i)b(j)]_R = h_b(i,j)$ whenever $(i,j) \in n \times n \setminus \Delta$[8]. Moreover, for every uncountable $C \subseteq B$, the set of ordinals δ for which there exist $a < b$ in C such that $[a(i)b(i)]_R = \delta$ for all $i < n$ contains a closed and unbounded subset of ω_1.* $\square$

Let h_ξ $(\xi < \omega_1)$ be a list of all mappings of the form $h : 2^n \times 2^n \to \omega$, where $n = n(h) < \omega$, such that every such mapping appears stationarily often on the list. This leads us to a projection

$$[\cdot\cdot]^0_R : [\omega_1]^2 \to \omega$$

of the square-bracket operation $[\cdot\cdot]_R$, defined as

$$[\alpha\beta]^0_R = h_{[\alpha\beta]_R}(r_\alpha \restriction n(h_{[\alpha\beta]_R}),\ r_\beta \restriction n(h_{[\alpha\beta]_R})).$$

Then working as in the proof of Theorem 5.2.5, we conclude that Lemma 5.5.2 has the following immediate consequence.

Theorem 5.5.3. *For every uncountable family A of pairwise-disjoint subsets of ω_1, all of some fixed size n, there exist an uncountable $B \subseteq A$, and a single mapping $h : n \times n \to \omega$ such that for all $a < b$ in B, we have that*

$$[a(i)b(j)]^0_R = h(i,j) \text{ whenever } (i,j) \in n \times n \setminus \Delta.$$

Moreover, for every uncountable $C \subseteq B$ and for every mapping $g : n \to \omega_1$ there exist $a < b$ in C such that

$$[a(i)b(i)]^0_R = g(i) \text{ for all } i < n. \qquad \square$$

We finish this section with an application of $[\cdot\cdot]^0_R$. Suppose V and W are vector spaces over the same field K. Recall that a mapping $\vartheta : V \times V \to W$ is a *bilinear map* if it is bilinear in each of the variables separately, or in other words, if it satisfies the equalities of the form

$$\vartheta(\Sigma_1^m \lambda_i x_j, \Sigma_1^n \mu_j y_j) = \Sigma_1^m \Sigma_1^n \lambda_i \mu_j \vartheta(x_i, y_j).$$

If $\vartheta(x,y) = \vartheta(y,x)$ then we call ϑ a *symmetric* bilinear map. We say that (V, W, ϑ) is a *non-degenerate* bilinear space, if for every $x \neq 0$ in V there is y in V such that $\vartheta(x,y) \neq 0$.

[8]Here, $\Delta = \{(i,i) : i < n\}$ is the diagonal of $n \times n$.

Theorem 5.5.4. *Suppose V and W are two vectors spaces over the same field K such that W is of countable (including finite) dimension while V has dimension $\aleph_1$. Then there is a symmetric non-degenerate bilinear map $\vartheta : V \times V \to W$ such that $\{\vartheta(x, y) : x, y \in X,\ x \neq y\} = W$ for every uncountable subset X of V. So in particular, there is no uncountable subset X of V that is* biorthogonal *relative to ϑ, i.e., that has the property that $\vartheta(x, y) = 0$ for $x \neq y$ in X.*

Proof. Fix a linear basis $B = \{v_\alpha : \alpha < \omega_1\}$ of V. In order to simplify the notation, we assume that $[\cdot\cdot]^0_R$ takes its values in W rather than ω. Since an arbitrary symmetric mapping from $B \times B$ into W extends to a uniquely determined symmetric bilinear mapping from $V \times V$ into W, it suffices to define $\vartheta \restriction B \times B$. For $\alpha < \omega_1$, let $\vartheta(v_\alpha, v_\alpha)$ be an arbitrary vector of W. For $\alpha \neq \beta$, set

$$\vartheta(v_\alpha, v_\beta) = \vartheta(v_\beta, v_\alpha) = [\alpha\beta]^0_R.$$

Consider an uncountable subset X of V and fix a vector t in W. We need to find $x \neq y$ in X such that $\vartheta(x, y) = t$. Since the field K is countable, going to an uncountable subset of X, we may assume that for some positive integer n and some fixed sequence λ_i $(i < n)$ of nonzero scalars, every vector $x \in X$, has a representation of the form

$$x = \Sigma_{i<n} \lambda_i v_{a_x(i)},$$

for some subset a_x of ω_1 of size n. We may assume that a_x $(x \in X)$ form a Δ-system with some root a.

Consider first the case $a = \emptyset$, or in other words, the case that a_x $(x \in X)$ is a family of pairwise-disjoint sets. By the first part of Theorem 5.5.3, we can find uncountable $Y \subseteq X$ and a mapping $h : n \times n \to W$ such that for every $x, y \in Y$ with the property that $a_x < a_y$, we have

$$[a_x(i)a_y(j)]^0_R = h(i, j) \text{ for all } (i, j) \in n \times n \setminus \Delta. \tag{5.5.2}$$

Changing, h, we may assume that $h(i, i) = 0$ of all $i < n$. Let

$$u = \Sigma_{(i,j)\neq(0,0)} \lambda_i \lambda_j h(i, j).$$

Let $v = t - u$ and let $w = (1/\lambda_0^2)v$. Define $g : n \to W$ by letting $g(i) = 0$ for $i \neq 0$ and $g(0) = w$. By the second part of Theorem 5.5.3, we can find $x, y \in Y$ such that $a_x < a_y$ and

$$[a_x(i)a_y(i)]^0_R = g(i) \quad \text{for all } i < n. \tag{5.5.3}$$

Using (5.5.2) and (5.5.3) and referring to how u, v, and w were defined, we conclude that

$$\vartheta(x, y) = \lambda_0^2[a_x(0)a_y(0)]^0_R + u = \lambda_0^2(1/\lambda_0^2)v + u = t,$$

and so we are done in this case.

Let us now consider the case $a \neq \emptyset$. Fix an element x_0 of X. Find uncountable $Y \subseteq X$ and $w \in W$ such that $\vartheta(x_0, y) = w$ for all $y \in Y$. The family

$$\{y - x_0 : y \in Y\}$$

is an uncountable family of disjointly supported vectors, so by the first part of the proof, we can find $y \neq z$ in Y such that

$$\vartheta(y - x_0, z - x_0) = t - 2w + \vartheta(x_0, x_0).$$

Since $\vartheta(y - x_0, z - x_0) = \vartheta(y, z) - 2w - \vartheta(x_0, x_0)$, it follows that $\vartheta(y, z) = t$, as required.

Let $V^{\perp} = \{x \in V : \vartheta(x, y) = 0 \text{ for all } y \in V\}$. Then by the basic property of ϑ that we have just established, it follows that $V^{\perp}$ is a countable vector subspace of V, so its linear complement V_0 in V is of dimension $\aleph_1$. Note that the restriction $\vartheta \upharpoonright V_0 \times V_0$ is non-degenerate while it still keeps the basic property of ϑ. So, taking a linear isomorphism between V and V_0 and transferring $\vartheta \upharpoonright V_0 \times V_0$ to a bilinear map on $V \times V$, we get the full conclusion of the theorem. □

Remark 5.5.5. As indicated above, the variation $[\cdot\cdot]_R$ of the square-bracket operation appears first as the mapping b of [109, p. 287]. Its application given above in Theorem 5.5.4 is motivated by a similar application of $[\cdot\cdot]_R$ appearing in [12] (see also [96]).

Chapter 6

General Walks and Their Characteristics

6.1 The full code and its application in characterizing Mahlo cardinals

One of the most basic questions frequently asked about set-theoretical trees is the question whether they contain any *cofinal branch*, a branch that intersects each level of the tree. The fundamental importance of this question has already been realized in the work of Kurepa [65] and then later in the works of Erdős and Tarski [32] in their respective attempts to develop the theory of partition calculus and large cardinals. A tree T of height equal to some regular cardinal θ may not have a cofinal branch for a very special reason as the following definition indicates.

Definition 6.1.1. For a tree $T = \langle T, <_T \rangle$, a function $f : T \to T$ is *regressive* if $f(t) <_T t$ for every $t \in T$ that is not a minimal node of T. A tree T of height θ is *special* if there is a regressive map $f : T \longrightarrow T$ with the property that the f-preimage of every point of T can be written as the union of $< \theta$ antichains of T.

This definition in case $\theta = \omega_1$ reduces indeed to the old definition of special tree, a tree that can be decomposed into countably many antichains. More generally we have the following:

Lemma 6.1.2. *If θ is a successor cardinal, then a tree T of height θ is special if and only if T is the union of $< \theta$ antichains.* □

The new definition, however, seems to be the right notion of speciality as it makes sense even if θ is a limit cardinal.

Definition 6.1.3. A tree T of height θ is *Aronszajn* if T has no cofinal branches and if every level of T has size $< \theta$.

Recall the well-known characterization of weakly compact cardinals due to Tarski and his collaborators: a strongly inaccessible cardinal θ is weakly compact if and only if there are no Aronszajn trees of height θ. We supplement this with the following:

Theorem 6.1.4. *The following are equivalent for a strongly inaccessible cardinal θ:*

(1) *θ is Mahlo.*

(2) *There are no special Aronszajn trees of height θ.*

Proof. Suppose θ is a Mahlo cardinal and let T be a given tree of height θ all of whose levels have size $< \theta$. To show that T is not special let $f : T \longrightarrow T$ be a given regressive mapping. By our assumption of θ there is an elementary submodel M of some large enough structure H_κ such that $T, f \in M$ and $\lambda = M \cap \theta$ is a regular cardinal $< \theta$. Note that $T \restriction \lambda$ is a subset of M and since this tree of height λ is clearly not special, there is $t \in T \restriction \lambda$ such that the preimage $f^{-1}(t)$ is not the union of $< \lambda$ antichains. Using the elementarity of M we conclude that $f^{-1}(t)$ is actually not the union of $< \theta$ antichains.

The proof that (2) implies (1) uses the method of minimal walks in a rather crucial way. So suppose to the contrary that our cardinal contains a closed and unbounded subset C consisting of singular strong limit cardinals. Using C, we choose a C-sequence C_α $(\alpha < \theta)$ such that: $C_{\alpha+1} = \{\alpha\}$, $C_\alpha = (\bar{\alpha}, \alpha)$ for α limit such that $\bar{\alpha} = \sup(C \cap \alpha) < \alpha$, but if $\alpha = \sup(C \cap \alpha)$ then take C_α such that:

$$\mathrm{tp}(C_\alpha) = \mathrm{cf}(\alpha) < \min(C_\alpha), \tag{6.1.1}$$

$$\xi = \sup(C_\alpha \cap \xi) \text{ implies } \xi \in C, \tag{6.1.2}$$

$$\xi \in C_\alpha \text{ and } \xi > \sup(C_\alpha \cap \xi) \text{ imply } \xi = \eta + 1 \text{ for some } \eta \in C. \tag{6.1.3}$$

Given the C-sequence C_α $(\alpha < \theta)$ we have the notion of minimal walk along the sequence and various distance functions defined above. In this proof we are particularly interested in the function ρ_0 from $[\theta]^2$ into the set $\mathbb{Q}_\theta$ of all finite sequences of ordinals from θ:

$$\rho_0(\alpha, \beta) = \langle \mathrm{tp}(C_\beta \cap \alpha) \rangle^\frown \rho_0(\alpha, \min(C_\beta \setminus \alpha))$$

with the boundary value $\rho_0(\gamma, \gamma) = 0$ for all γ. We would like to show that the tree[1]

$$T(\rho_0) = \{(\rho_0)_\beta \restriction \alpha : \alpha \leq \beta < \theta\}$$

is a special Aronszajn tree of height θ. Note that the size of the αth level $(T(\rho_0))_\alpha$ of $T(\rho_0)$ is controlled in the following way:

$$|(T(\rho_0))_\alpha| \leq |\{C_\beta \cap \alpha : \alpha \leq \beta < \theta\}| + |\alpha| + \aleph_0. \tag{6.1.4}$$

[1]Recall, that we consider $T(\rho_0)$ ordered by the natural initial segment relation $\sqsubseteq$. Recall also that $\sqsubset$ denotes the strict version of $\sqsubseteq$

So under the present assumption that θ is a strongly inaccessible cardinal, all levels of $T(\rho_0)$ do indeed have size $< \theta$. It remains to define the regressive map

$$f : T(\rho_0) \longrightarrow T(\rho_0) \tag{6.1.5}$$

that will witness speciality of $T(\rho_0)$. Note that it really suffices defining f on all levels whose index belong to our club C of singular cardinals. So let $t = (\rho_0)_\beta \restriction \alpha$ be a given node of T such that $\alpha \in C$ and $\alpha \leq \beta < \theta$. Note that by our choice of the C-sequence every term of the finite sequence of ordinals $\rho_0(\alpha, \beta)$ is strictly smaller than α. So, if we let

$$f(t) = t \restriction \ulcorner \rho_0(\alpha, \beta) \urcorner, \tag{6.1.6}$$

where $\ulcorner \cdot \urcorner$ is a standard coding of finite sequences of ordinals by ordinals, we get a regressive map. To show that f is one-to-one on chains of $T(\rho_0)$, which would be more than sufficient, suppose $t_i = (\rho_0)_{\beta_i} \restriction \alpha_i$ $(i < 2)$ are two nodes such that $t_0 \sqsubset t_1$. Our choice of the C-sequence allows us to deduce the following general fact about the corresponding ρ_0-function whose proof is similar to the proof of Lemma 2.1.18 which treats the case $\theta = \omega_1$.

$$\begin{array}{l}\text{If } \alpha \leq \beta \leq \gamma,\ \alpha \text{ is a limit ordinal, and if } \rho_0(\xi, \beta) = \rho_0(\xi, \gamma) \\ \text{for all } \xi < \alpha, \text{ then } \rho_0(\alpha, \beta) = \rho_0(\alpha, \gamma).\end{array} \tag{6.1.7}$$

Applying this to the triple of ordinals α_0, β_0 and β_1 we conclude that

$$\rho_0(\alpha_0, \beta_0) = \rho_0(\alpha_0, \beta_1). \tag{6.1.8}$$

Now observe another fact about the ρ_0-function whose proof is almost identical to that of case $\theta = \omega_1$, seen above in Lemma 2.1.16.

$$\text{If } \alpha < \beta \leq \gamma, \text{ then } \rho_0(\alpha, \gamma) <_r \rho_0(\beta, \gamma), \tag{6.1.9}$$

where $<_r$ denotes the right lexicographical ordering between finite sequences of ordinals $< \theta$. Applying this to the triple $\alpha_0 < \alpha_1 \leq \beta_1$ we in particular get that,

$$\rho_0(\alpha_0, \beta_1) \neq \rho_0(\alpha_1, \beta_1). \tag{6.1.10}$$

Combining (6.1.8) and (6.1.10), we get that $\rho_0(\alpha_0, \beta_0) \neq \rho_0(\alpha_1, \beta_1)$ and therefore that $f(t_0) \neq f(t_1)$. □

It turns out that the regressive mapping defined during the course of the previous proof leads us to the following purely Ramsey-theoretic characterization of Mahlo cardinals.

Theorem 6.1.5. *A cardinal θ is a Mahlo cardinal if and only if every regressive map f defined on a cube $[C]^3$ of a closed and unbounded subset of θ has an infinite* min-*homogeneous set $X \subseteq C$.*[2]

[2]Recall, that X is min-*homogeneous* for f if $f(\alpha, \beta, \gamma) = f(\alpha', \beta', \gamma')$ for every pair $\alpha < \beta < \gamma$ and $\alpha' < \beta' < \gamma'$ of triples of elements of X such that $\alpha = \alpha'$.

Proof. To see the direct implication, let $f : [\theta]^3 \to \theta$ be a given regressive mapping. Define the tree-ordering $\leq_f$ on θ by letting $\alpha \leq_f \beta$ if $\alpha \leq \beta$ and

$$f(u \cup \{\alpha\}) = f(v \cup \{\beta\}) \text{ for all } u, v \in [\{\xi : \xi <_f \alpha\}]^2 \text{ with } \min u = \min v.$$

Since θ is a strongly inaccessible cardinal, one easily concludes that the levels of the tree $(\theta, <_f)$ have cardinalities $< \theta$. So, using the assumption that θ is a Mahlo cardinal, we can find an inaccessible cardinal $\kappa < \theta$ such that the set of predecessors $\{\xi : \xi <_f \kappa\}$ has cardinality κ. Thus we have reduced the conclusion to the fact that regressive mappings $g : [\kappa]^2 \to \kappa$ for inaccessible cardinals κ have infinite min-homogeneous sets. This fact in turn is proved by considering the corresponding tree-orderings $\leq_g$ and noticing that, by the inaccessibility of κ, the tree $(\kappa, \leq_g)$ must have an infinite chain (and in fact a chain of arbitrary order type $< \kappa$). Now note that any chain of the tree $(\kappa, \leq_g)$ is min-homogeneous relative to g.

For the converse implication, note that the statement about regressive maps implies that θ must be a strongly inaccessible cardinal. So assume θ is a strongly inaccessible non-Mahlo cardinal. Let C be a closed and unbounded subset of θ consisting of strong limit singular cardinals. We shall find a regressive map on $[C]^3$ without an infinite min-homogeneous set. This will give us an opportunity to further expose the tree $T(\rho_0)$ and the regressive map

$$f : T(\rho_0) \restriction C \longrightarrow T(\rho_0)$$

defined above in (6.1.6) in the course of proving Theorem 6.1.4. We have already observed that the length of every branching node t must be of the form $\alpha + 1$ for some α and in fact the following more precise description is true:

$$\text{If } t \text{ is a branching node of } T(\rho_0)\text{, then its length is equal to } \alpha + 1 \text{ for some } \alpha \in C. \tag{6.1.11}$$

To see this, let t be a given branching node of $T(\rho_0)$. By (6.1.7) we know that the length of t is equal to $\alpha + 1$ for some ordinal α. Let

$$\alpha^- = \max(C \cap (\alpha + 1)) \text{ and } \alpha^+ = \min(C \setminus (\alpha + 1)).$$

Let $\beta, \gamma > \alpha + 1$ be two ordinals such that $(\rho_0)_\beta$ and $(\rho_0)_\gamma$ split at t. Let $\beta = \beta_0 > \cdots > \beta_k = \alpha$ and $\gamma = \gamma_0 > \cdots > \gamma_l = \alpha$ be the walks from β to α and γ to α respectively. From the fact that $(\rho_0)_\beta$ and $(\rho_0)_\gamma$ agree below α we conclude that $k = l$ and $C_{\beta_i} \cap \alpha = C_{\gamma_i} \cap \alpha$ for all $i \leq k$. By the choice of the C-sequence and the inequality $\alpha > \alpha^-$, we can deduce a sequence of stronger agreements $C_{\beta_i} \cap \alpha^+ = C_{\gamma_i} \cap \alpha^+$ for all $i \leq k$. However, that would mean that $\rho_0(\alpha + 1, \beta) = \rho_0(\alpha + 1, \gamma)$, contrary to our assumption that $(\rho_0)_\beta$ and $(\rho_0)_\gamma$ split at t. It follows that $\alpha = \alpha^-$, and therefore $\alpha \in C$, which was to be shown.

Going back to the proof of Theorem 6.1.5, recall the definition of the regressive map $f : T(\rho_0) \restriction C \longrightarrow T(\rho_0)$ given above in Theorem 6.1.4:

$$f((\rho_0)_\beta \restriction \alpha) = (\rho_0)_\beta \restriction \ulcorner \rho_0(\alpha, \beta) \urcorner.$$

We extend f to the whole tree $T(\rho_0)$, by letting $f(t) = t \restriction \alpha$, where α is the maximal element of C that is less than the length of t for any node $t \in T(\rho_0)$ whose length does not belong to C. Then we know that f is one-to-one on any chain whose nodes are separated by C. Using the mapping f we shall derive three other regressive maps:

$$p : [C]^3 \longrightarrow \theta, \quad q : [C]^3 \longrightarrow \omega \text{ and } r : [C]^2 \longrightarrow \theta.$$

First note that for $\alpha \neq \beta$ in C the functions $(\rho_0)_\alpha$ and $(\rho_0)_\beta$ are incomparable, so it is natural to denote by $\alpha \wedge \beta$ the node of $T(\rho_0)$ where they split. For $\alpha < \beta < \gamma < \theta$, let

$$p(\alpha, \beta, \gamma) = \text{the length of } f^n(\beta \wedge \gamma), \text{ where } n \text{ is the minimal integer such that } f^n(\beta \wedge \gamma) \sqsubseteq (\rho_0)_\alpha. \tag{6.1.12}$$

$$q(\alpha, \beta, \gamma) = \min\{n : f^n(\beta \wedge \gamma) = \emptyset\}. \tag{6.1.13}$$

$$r(\alpha, \beta) = \ulcorner \langle \rho_0(\xi, \alpha), \rho_0(\xi, \beta) \rangle \urcorner, \text{ where } \xi = \text{length}(\alpha \wedge \beta). \tag{6.1.14}$$

Now we claim that no infinite subset of C can be simultaneously min-homogeneous with respect to all three regressive maps. For, given an infinite increasing sequence $\{\alpha_i\}$ of elements of C, using Ramsey's theorem we may assume that either

$$\alpha_i \wedge \alpha_j = \alpha_k \wedge \alpha_l \text{ for all } i \neq j \text{ and } k \neq l \text{ in } \omega, \text{ or} \tag{6.1.15}$$

$$\alpha_i \wedge \alpha_{i+1} \sqsubset \alpha_{i+1} \wedge \alpha_{i+2} \text{ for all } i < \omega. \tag{6.1.16}$$

Note that (6.1.15) would violate min-homogeneity with respect to the mapping r. If (6.1.16) holds and if the sequence is min-homogeneous with respect to p and q, then another application of Ramsey's theorem would give us an integer m and an ordinal ξ such that for all $i < j < k < \omega$:

$$m = \min\{n : f^n(\alpha_j \wedge \alpha_k) \subseteq (\rho_0)_{\alpha_i}\}, \tag{6.1.17}$$

$$f^m(\alpha_j \wedge \alpha_k) = (\rho_0)_{\alpha_0} \restriction \xi (= t). \tag{6.1.18}$$

By (6.1.11) the chain $\{\alpha_i \wedge \alpha_{i+1} : i \geq 1\}$ is separated by C, so using the basic property of f we conclude that $m > 1$. Then the pre-image $f^{-1}(t)$ includes the set

$$\{f^{m-1}(\alpha_i \wedge \alpha_{i+1}) : i \geq 1\}.$$

By (6.1.17) we have,

$$\alpha_i \wedge \alpha_{i+1} \sqsubset f^{m-1}(\alpha_{i+1} \wedge \alpha_{i+2}) \sqsubseteq \alpha_{i+1} \wedge \alpha_{i+2}, \tag{6.1.19}$$

so $\{f^{m-1}(\alpha_i \wedge \alpha_{i+1}) : i \geq 1\}$ is an infinite chain of $T(\rho_0)$ separated by C, a contradiction. This finishes the proof of Theorem 6.1.5. □

Starting from the case $n = 1$, one can go further and obtain the following characterization of n-Mahlo cardinals.

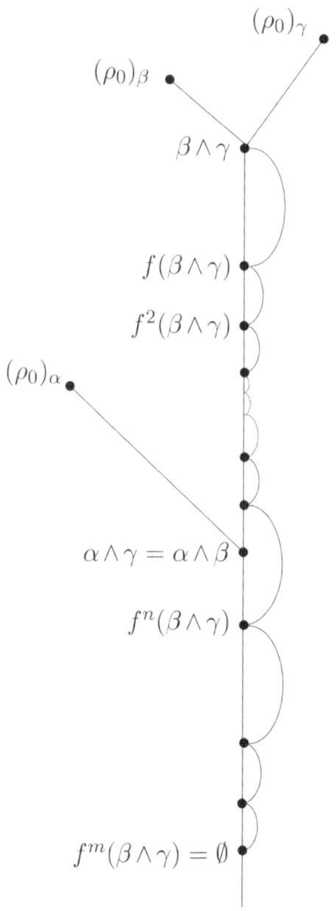

Figure 6.1: The regressive mapping f.

Theorem 6.1.6. *The following are equivalent for an uncountable cardinal θ and a positive integer n:*

(1) *θ is n-Mahlo.*

(2) *For every $\xi < \theta$ and every regressive map defined on $[\Gamma]^{n+2}$, some unbounded subset Γ of θ has a min-homogeneous set of order type ξ.*

(3) *Every regressive map defined on $[C]^{n+2}$ for some closed and unbounded subset C of θ has an infinite min-homogeneous subset.*

Proof. The proof of the direct implication of Theorem 6.1.5 gives us the necessary inductive step for proving the implication from (1) to (2), so it remains only to supply the argument for the implication from (3) to (1).

For a set X of ordinals and $n \in \omega$, let $W(X, n)$ denote the statement that there exist $f : [X]^{n+2} \longrightarrow \omega$ and regressive $g : [X]^{n+3} \longrightarrow \sup(X)$ such that, if $H \subseteq X$ is min-homogeneous for g and $f''[H]^{n+2} = \{k\}$ for some $k \in \omega$, then

$|H| \leq n + k + 5$. Note that in the course of proof of Theorem 6.1.5, we have established $W(C, 0)$ for every closed unbounded subset C of some limit ordinal θ such that C contains no inaccessible cardinals, since the mapping q is really a map from $[C]^2$ into ω. The proof of the theorem is finished once we show that, in general, $W(C, n)$ holds for every set of ordinals C of limit order type that is closed in its supremum and which contains no n-Mahlo cardinals. So let C be a given closed and unbounded subset of some limit ordinal θ and suppose that C contains no $(n+1)$-Mahlo cardinals. We need to show that $W(C, n+1)$ holds. We first show, by induction on $\delta \in C$, that $W(C \cap \delta, n)$ holds for all $\delta \in C$. Clearly, $W(C \cap \delta, n)$ holds for $\delta = \min C$. Suppose that $W(C \cap \delta, n)$ holds and show that $W(C \cap \delta^+, n)$ for $\delta^+ = \min(C \setminus (\delta + 1))$. Let f and g be a pair of mappings witnessing $W(C \cap \delta, n)$ and assume, as we may, that the range of f does not include the value 0. Define

$$f^+ : [C \cap \delta^+]^{n+2} \to \omega$$

by letting $f^+(\alpha_0, \ldots, \alpha_{n+1}) = f(\alpha_0, \ldots, \alpha_{n+1})$ if $\alpha_{n+1} < \delta$ and $f^+(\alpha_0, \ldots, \alpha_{n+1}) = 0$ if $\alpha_{n+1} \geq \delta$. Similarly, define regressive

$$g^+ : [C \cap \delta^+]^{n+3} \to \theta$$

by letting $g^+(\alpha_0, \ldots, \alpha_{n+2}) = g(\alpha_0, \ldots, \alpha_{n+2})$ if $\alpha_{n+2} < \delta$ and $g^+(\alpha_0, \ldots, \alpha_{n+2}) = 0$ if $\alpha_{n+2} \geq \delta$. It is easily seen that the pair f^+ and g^+ of mappings witness $W(C \cap \delta^+, n)$. Assume now that δ is a limit point of C and that $W(C \cap \gamma, n)$ holds for all $\gamma \in C \cap \delta$. For each $\gamma \in C \cap \delta$ fix a pair f_γ and g_γ witnessing $W(C \cap \gamma, n)$. Since δ is not an $(n+1)$-Mahlo cardinal, there is a closed and unbounded set $C' \subseteq C \cap \delta$ containing no n-Mahlo cardinals. By the inductive hypothesis $W(C', n)$ holds, so we can fix a pair of mappings f' and g' witnessing this. For $\alpha \in C \cap \delta$, let

$$\alpha^- = \sup(C' \cap (\alpha + 1)) \text{ and } \alpha^+ = \min(C' \setminus (\alpha + 1)).$$

A *C'-pattern* of an element $\{\alpha_0, \ldots, \alpha_{n+1}\}$ of $[C]^{n+2}$ written in its increasing order is the equivalence relation $E = E(\alpha_0, \ldots, \alpha_{n+1})$ on the index-set $\{0, \ldots, n+1\}$ where iEj if and only if $\alpha_i^- = \alpha_j^-$. We can of course code (enumerate) all C'-patterns by natural numbers and assume that the ranges of the mappings f_γ and f' do not take any of these codes as values. Define

$$f : [C \cap \delta]^{n+2} \to \omega$$

by letting $f(\alpha_0, \ldots, \alpha_{n+1}) = f'(\alpha_0^-, \ldots, \alpha_{n+1}^-)$ if $\alpha_0^- < \cdots < \alpha_{n+1}^-$, and by letting $f(\alpha_0, \ldots, \alpha_{n+1}) = f_{\alpha_0^+}(\alpha_0^-, \ldots, \alpha_{n+1}^-)$ if $E(\alpha_0, \ldots, \alpha_{n+1})$ is the trivial equivalence relation where every pair of integers of its domain are equivalent; in all other cases, $f(\alpha_0, \ldots, \alpha_{n+1}) = k$, where k is the integer which codes the C'-pattern of $\{\alpha_0, \ldots, \alpha_{n+1}\}$. Similarly, define

$$g : [C \cap \delta]^{n+3} \to \theta$$

by letting $g(\alpha_0, \ldots, \alpha_{n+2}) = g'(\alpha_0^-, \ldots, \alpha_{n+2}^-)$ if $\alpha_0^- < \cdots < \alpha_{n+2}^-$, and by letting $g(\alpha_0, \ldots, \alpha_{n+2}) = g_{\alpha_0^+}(\alpha_0^-, \ldots, \alpha_{n+2}^-)$ if $E(\alpha_0, \ldots, \alpha_{n+2})$ is the trivial equivalence relation where every pair of integers of its domain are equivalent; in all other cases, $g(\alpha_0, \ldots, \alpha_{n+2}) = 0$. It is easily checked that the mappings f and g witness $W(C \cap \delta, n)$. This completes our proof that $W(C \cap \delta, n)$ holds for all $\delta \in C$.

In order to prove that $W(C, n+1)$ holds we fix for each $\delta \in C$ a pair of mappings f_δ and g_δ witnessing $W(C \cap \delta, n)$. Define $f : [C]^{n+3} \to \omega$ and regressive $g : [C]^{n+4} \to \theta$ as

$$f(\delta_0, \ldots, \delta_{n+2}) = f_{\delta_{n+2}}(\delta_0, \ldots, \delta_{n+1}) + 1,$$

$$g(\delta_0, \ldots, \delta_{n+2}) = g_{\delta_{n+2}}(\delta_0, \ldots, \delta_{n+1}).$$

Then f and g witness $W(C, n+1)$ finishing thus the proof. □

Remark 6.1.7. Theorems 6.1.5 and 6.1.6 are due to Hajnal, Kanamori and Shelah [44]. An earlier characterization of this sort is due to Schmerl [91] who was the first to consider statements like $W(C, n)$. In particular Schmerl [91] proves the following characterization theorem using a bit different terminology (see also [44]), a theorem which has some interesting metamathematical applications (see [38], [56], and [54]).

Theorem 6.1.8. *The following are equivalent for an uncountable cardinal θ and a non-negative integer n:*

(1) *θ is n-Mahlo.*
(2) *For every positive integer k and every regressive mapping defined on $[\Gamma]^{n+3}$, some unbounded subset Γ of θ has a min-homogeneous set of size k.*
(3) *Every regressive map defined on $[\Gamma]^{n+3}$ for some unbounded subset Γ of θ has a min-homogeneous set of size $n+5$.*

The proof of Theorem 6.1.8 will be given in a sequence of lemmas. We start with the following lemma which has essentially appeared above in the course of the proof of Theorem 6.1.6.

Lemma 6.1.9. *Suppose that $n \geq 3$ is an integer, that $\xi > n$ is an ordinal, and that θ is an infinite limit ordinal. Suppose also that C and Γ are subsets of $\theta \setminus \omega$ with C closed and unbounded and* $\min C \leq \min \Gamma$. *Suppose further that there is a regressive map $f : [C]^n \to \theta$ with no min-homogeneous set of order type ξ. If for each $\gamma < \theta$ there is a regressive map $g_\gamma : [\Gamma \cap \gamma]^n \to \theta$ with no min-homogeneous set of order type ξ, then there is such a regressive map on $[\Gamma]^n$ as well.*

Proof. By our assumption $\min C \leq \min \Gamma$ and the fact that C is closed, for each $\gamma \in \Gamma$ the ordinal $\gamma^- = \sup(C \cap (\gamma + 1))$ belongs to C. For $\gamma < \theta$, let

$$\gamma^+ = \sup(C \setminus (\gamma + 1)).$$

For a given n-tuple $\alpha_0 < \alpha_1 < \cdots < \alpha_{n-1}$ of elements of Γ, one associates the equivalence relation $E_{(\alpha_0,\alpha_1,\ldots,\alpha_{n-1})}$ on $\{0,1,\ldots,n-1\}$ by letting

$$iE_{(\alpha_0,\alpha_1,\ldots,\alpha_{n-1})}j \text{ if and only if } \alpha_i^- = \alpha_j^-.$$

We shall code such equivalence relations by integers $\ulcorner E_{(\alpha_0,\alpha_1,\ldots,\alpha_{n-1})}\urcorner$ in some natural way. Choose a regressive map $g : [C]^n \to \mathrm{Ord}$ which has no min-homogeneous set of order type ν. Similarly, choose regressive mappings

$$f_\gamma : [\Gamma \cap \gamma]^n \to \mathrm{Ord},\ (\gamma < \theta)$$

that have no min-homogeneous set of order type ν. We may assume that the ranges of these regressive mapping avoid the integers that code the equivalence relations of the form $E_{(\alpha_0,\alpha_1,\ldots,\alpha_{n-1})}$. Now, define the regressive mapping $f : [\Gamma]^n \to \mathrm{Ord}$ according to the following three cases.

Case 1. If $\alpha_0^- < \alpha_1^- < \cdots < \alpha_{n-1}^-$, let

$$f(\alpha_0,\alpha_1,\ldots,\alpha_{n-1}) = g(\alpha_0^-,\alpha_1^-,\ldots,\alpha_{n-1}^-).$$

Case 2. If $\alpha_1^- = \cdots = \alpha_{n-1}^-$, let

$$f(\alpha_0,\alpha_1,\ldots,\alpha_{n-1}) = f_{\alpha_1^+}(\alpha_0,\alpha_1,\ldots,\alpha_{n-1}).$$

Case 3. In all other cases, put,

$$f(\alpha_0,\alpha_1,\ldots,\alpha_{n-1}) = \ulcorner E_{(\alpha_0,\alpha_1,\ldots,\alpha_{n-1})}\urcorner.$$

Suppose there is a min-homogeneous subset X of Γ of order-type ξ. Let $\alpha_0 < \alpha_1$ be the least two elements of X. If $\alpha_0^- = \alpha_1^-$, then there will be no $\alpha \in X$ with $\alpha^- > \alpha_1^-$, or else we would get two n-tuples with minimum α_0 that realize two different equivalence relations. It follows that in this case $\alpha \mapsto \alpha^-$ is a constant map on X. This means that Case 2 applies for all the n-tuples from $[X]^n$. So, in this case X would be a min-homogeneous set for $f_{\alpha_1^+}$ of order-type ξ, a contradiction.

Assume now that $\alpha_0^- < \alpha_1^-$. Suppose first that there is some $\alpha \in X$ such that $\alpha^- > \alpha_1^-$. Then the mapping $\alpha \mapsto \alpha^-$ must be strictly increasing on X, or else we would find two n-tuples of elements of X with minimum α_0 realizing two different equivalence relations, a contradiction. In follows that Case 1 applies all the time for n-tuples from $[X]^n$. So in this case $\{\alpha^- : \alpha \in X\}$ would be a subset of C of order-type ξ that is min-homogeneous for the regressive mapping g, a contradiction. It remains to consider the case $\alpha_0^- < \alpha_1^-$ and $\alpha^- = \alpha_1^-$ for all $\alpha \in X$ with $\alpha > \alpha_1$. This means that Case 2 applies again for the n-tuples from $[X]^n$. So, in this case X would be a min-homogeneous set for $f_{\alpha_1^+}$ of order-type ξ, a contradiction. This finishes the proof. □

Lemma 6.1.10. *Suppose that some limit ordinal $\theta > \omega$ has the property that, for every $\gamma < \theta$ and every regressive map $f_\gamma : [\theta \setminus \gamma]^3 \to \theta$, there exists a min-homogeneous set of cardinality 4. Then θ is a strong limit cardinal.*

Proof. Suppose that for some $\gamma < \theta$, the cardinality of 2^γ is bigger than or equal to the cardinality of θ. Fix a one-to-one sequence x_ξ $(\xi < \theta)$ of elements of 2^γ, and define a regressive map $f : [\theta \setminus \gamma]^3 \to \theta$, by

$$f(\xi_0, \xi_1, \xi_2) = \min\{\alpha < \gamma : x_{\xi_0}(\alpha) \neq x_{\xi_1}(\alpha)\}. \tag{6.1.20}$$

It is clear that f has no min-homogeneous set of size 4. □

Lemma 6.1.11. *Suppose that $n > 0$ is an integer, that θ is an inaccessible cardinal, and that C is a closed and unbounded subset of θ consisting of strong limit cardinals that are not $(n-1)$-Mahlo. Then there is a regressive mapping $f : [C]^{n+3} \to \theta$ which has no min-homogeneous set of size $n + 5$.*

Proof. Let $S_1(m, n, \nu)$ denote the statement that, for every inaccessible cardinal θ and every closed and unbounded subset C of θ consisting of strong limit cardinals that are not m-Mahlo, there is a regressive mapping $f : [C]^n \to \theta$ which has no min-homogeneous set of order-type ν. We need to prove that $S_1(n-1, n+3, n+5)$ holds for every positive integer n. For this we will need another such a statement $S_2(m, n, \nu)$ saying that, for every inaccessible cardinal θ, every closed and unbounded subset C of θ consisting of strong limit cardinals that are not m-Mahlo, and every $\gamma < \theta$, there is a regressive mapping $f_\gamma : [C \cap \gamma]^n \to \gamma$ with no min-homogeneous set of order-type ν. First of all note that

$$S_2(m, n, \nu) \text{ implies } S_1(m, n+1, \nu+1). \tag{6.1.21}$$

To see this fix an inaccessible cardinal θ and a closed unbounded set $C \subseteq \theta$ consisting of strong limit cardinals. By $S_2(m, n, \nu)$, for each $\gamma < \theta$, we can fix also a regressive mapping $f_\gamma : [C \cap \gamma]^n \to \theta$ with no min-homogeneous set of order-type ν. Define $f : [C]^{n+1} \to \theta$, by

$$f(\gamma_0, \gamma_1, \dots, \gamma_n) = f_{\gamma_n}(\gamma_0, \gamma_1, \dots, \gamma_{n-1}).$$

It is clear that f is then a regressive mapping with no min-homogeneous set of order-type $\nu + 1$.

Claim. Suppose that $\nu \geq n + 2$, $n \geq 3$. Suppose further that $S_1(k, n, \nu)$ holds for all $k < m$. Then $S_2(m, n, \nu)$ holds.

Proof. Let C be a given closed and unbounded subset of some inaccessible cardinal θ consisting only of strong limit cardinals that are not m-Mahlo. We need to construct for each $\gamma < \theta$ a regressive mapping $f_\gamma : [C \cap \gamma]^n \to \gamma$ with no min-homogeneous set of order-type ν. This will be done by recursion on $\gamma \in C$. Suppose first that

$$\beta = \max(C \cap \gamma) < \gamma.$$

We may assume that 0 is not in the range of f_β. Define $f_\gamma : [C \cap \gamma]^n \to \gamma$ by

$$f_\gamma(\alpha_0, \ldots, \alpha_{n-1}) = f_\beta(\alpha_0, \ldots, \alpha_{n-1})$$

if $\alpha_{n-1} < \beta$; otherwise put $f_\gamma(\alpha_0, \ldots, \alpha_{n-1}) = 0$. It is clear that f_γ is a regressive map with no min-homogeneous set of order-type ν. Suppose now that

$$\sup(C \cap \gamma) = \gamma.$$

Then γ is either a singular strong limit cardinal or an inaccessible cardinal which is not m-Mahlo. In any case, we can select a subset $D \subseteq C \cap \gamma$ that is closed and unbounded in γ and by the hypothesis of the claim (or singularity of γ) a regressive map $g : [D]^n \to \gamma$ which has no min-homogeneous set of order-type γ. For $\alpha < \gamma$, set

$$\alpha^- = \sup(D \cap (\alpha + 1) \text{ and } \alpha^+ = \min(D \setminus (\alpha + 1).$$

As before, a *D-pattern* of an n-tuple $\{\alpha_0, \alpha_1, \ldots, \alpha_{n-1}\}$ of ordinals $< \gamma$ written in increasing order is the equivalence relation $E = E_{(\alpha_0, \alpha_1, \ldots, \alpha_{n-1})}$ on the index set $\{0, 1, \ldots, n-1\}$, where iEj if and only if $\alpha_i^- = \alpha_j^-$. As before we code patterns by positive integers as $\ulcorner E_{(\alpha_0,\alpha_1,\ldots,\alpha_{n-1})} \urcorner$ and assume that neither of our regressive mappings fixed so far takes these integers as values. We define a regressive mapping

$$f : [C \cap \gamma]^n \to \gamma$$

according to the following three cases.

Case 1. If $\alpha_0^- < \alpha_1^- < \cdots < \alpha_{n-1}^-$, let

$$f(\alpha_0, \alpha_1, \ldots, \alpha_{n-1}) = g(\alpha_0^-, \alpha_1^-, \ldots, \alpha_{n-1}^-).$$

Case 2. If $\alpha_0^- = \alpha_1^- = \cdots = \alpha_{n-1}^-$, let

$$f(\alpha_0, \alpha_1, \ldots, \alpha_{n-1}) = f_{\alpha_0^+}(\alpha_0, \alpha_1, \ldots, \alpha_{n-1}).$$

Case 3. If $E_{(\alpha_0,\alpha_1,\ldots,\alpha_{n-1})}$ has at least two but less than n classes, we let

$$f(\alpha_0, \alpha_1, \ldots, \alpha_{n-1}) = \ulcorner E_{(\alpha_0,\alpha_1,\ldots,\alpha_{n-1})} \urcorner.$$

Suppose there is a min-homogeneous subset Γ of $C \cap \gamma$ of order-type ν and let $\alpha = \min \Gamma$. If the value of f at some n-tuple of elements of Γ with minimum α gets assigned according to Case 1, then

$$\Gamma^- = \{\beta^- : \beta \in \Gamma\}$$

has order-type ν and is min-homogeneous for g, a contradiction. If the value of f at some n-tuple of elements of Γ with minimum α gets assigned according to Case 2, then $\Gamma \subseteq C \cap \alpha^+$ and Γ is min-homogeneous for g_{α^+}, a contradiction. If the value of f at some n-tuple of elements of Γ with minimum α gets assigned according to Case 3, then this is true about all such n-tuples. However, our assumptions of $\nu \geq n + 2$ and $n \geq 3$ will give us two such n-tuples realizing different patterns, a contradiction.This completes the proof of the claim. □

Note that the Claim in particular shows that $S_2(0,3,5)$ holds. Applying (6.1.21) and the claim repeatedly, we obtain that $S_1(n-1,n+3,n+5)$ holds for every positive integer n. This finishes the proof of Lemma 6.1.11. □

We are now ready to finish the proof of Theorem 6.1.8. The direct implication from (1) to (2) (and, therefore (3)) uses the partition tree in a similar manner as in the proof of Theorem 6.1.6, so we don't repeat it. So let us show that (3) implies (1). First of all note that if (1) holds, the cardinal θ must be inaccessible by Lemma 6.1.10. So assuming that θ is inaccessible, the implication from (3) towards (1) follows directly from Lemma 6.1.11.

The following refinement of Theorem 6.1.8 from [54] is also worth pointing out.

Theorem 6.1.12. *The following are equivalent for every set Γ of infinite ordinals and a non-negative integer n:*

(1) *$\Gamma \cap \theta$ is unbounded in θ for some n-Mahlo cardinal θ.*

(2) *Every regressive mapping $f : [\Gamma]^{n+3} \to \mathrm{Ord}$ has a min-homogeneous set of size $n+5$.*

Proof. The direction from (1) to (2) follows from Theorem 6.1.8. To see the converse implication, let θ be the minimal ordinal γ with the property that every regressive mapping $f : [\Gamma \cap \gamma]^{n+3} \to \mathrm{Ord}$ has a min-homogeneous set of size $n+5$. Note that θ must be a limit ordinal. We show that θ is in fact an n-Mahlo cardinal. By Lemma 6.1.9, we know that for every closed and unbounded set $C \subseteq \theta$ with properties

$$C \cap \omega = \emptyset \text{ and } \min(C) \leq \min(\Gamma),$$

every regressive map $f : [C]^{n+3} \to \mathrm{Ord}$ has a min-homogeneous set of size $n+5$. Applying this to closed and unbounded sets of the form

$$C_\gamma = \{\min(\Gamma)\} \cup [\gamma, \theta) \text{ for } \gamma < \theta,$$

by Lemma 6.1.10, we conclude that θ must be a strong limit cardinal. If θ were a singular cardinal, then taking a closed unbounded subset D of θ of order-type $\mathrm{cf}(\theta)$ with $\min(D) \geq \mathrm{cf}(\theta)$ and taking a regressive one-to-one mapping on $[D]^{n+3}$ would contradict this property for the closed and unbounded set

$$C = \{\min(\Gamma)\} \cup D.$$

It follows that θ is an inaccessible cardinal, so if θ would not be an n-Mahlo cardinal, we would have that $n > 0$ and that we have contradicted Lemma 6.1.11. This finishes the proof. □

We finish this section by showing how the idea of proof of Theorem 6.1.4 leads us naturally to the following well-known fact, first established by Silver (see [75]) when θ is a successor of a regular cardinal.

Theorem 6.1.13. *If θ is a regular uncountable cardinal which is not Mahlo in the constructible universe, then there is a constructible special Aronszajn tree of height θ.*

Proof. Working in L we choose a closed and unbounded subset C of θ consisting of singular ordinals and a C-sequence C_α $(\alpha < \theta)$ such that

$$C_{\alpha+1} = \{\alpha\}, \tag{6.1.22}$$

$$C_\alpha = (\bar{\alpha}, \alpha) \text{ when } \bar{\alpha} = \sup(C \cap \alpha) < \alpha, \tag{6.1.23}$$

while if α is a limit point of C we take C_α to have the following two properties:

$$\xi = \sup(C_\alpha \cap \xi) \text{ implies } \xi \in C, \tag{6.1.24}$$

$$\xi > \sup(C_\alpha \cap \xi) \text{ implies } \xi = \eta + 1 \text{ for some } \eta \in C. \tag{6.1.25}$$

It is clear that we can choose the C-sequence C_α $(\alpha < \theta)$ to have also the following important property,

$$|\{C_\alpha \cap \xi : \xi \leq \alpha < \theta\}| \leq |\xi| + \aleph_0 \text{ for all } \xi < \theta. \tag{6.1.26}$$

Then from what we have learned above these properties guarantee that the tree $T(\rho_0)$, where ρ_0 is the full code of the walk along the C-sequence C_α $(\alpha < \theta)$, is a constructible special Aronszajn tree of height θ. □

We are also in a position to deduce the following well-known fact.

Theorem 6.1.14. *The following are equivalent for a successor cardinal θ:*

(a) *There is a special Aronszajn tree of height θ.*
(b) *There is a C-sequence C_α $(\alpha < \theta)$ such that $\mathrm{tp}(C_\alpha) \leq \theta^-$ for all α and such that $\{C_\alpha \cap \xi : \alpha < \theta\}$ has size $\leq \theta^-$ for all $\xi < \theta$.*

Proof. If C_α $(\alpha < \theta)$ is a C-sequence satisfying (b) and if ρ_0 is the associated ρ_0-function, then $T(\rho_0)$ is a special Aronszajn tree of height θ. Suppose $<_T$ is a special Aronszajn tree-ordering on θ such that $[\theta^- \cdot \alpha, \theta^- \cdot (\alpha + 1))$ is its αth level. Let C be the closed and unbounded set of ordinals $< \theta$ that are divisible by θ^-. Let $f : \theta \longrightarrow \theta^-$ be such that the f-preimage of every ordinal $< \theta^-$ is an antichain of the tree $(\theta, <_T)$. We choose a C-sequence C_α $(\alpha < \theta)$ such that $C_{\alpha+1} = \{\alpha\}$, $C_\alpha = (\bar{\alpha}, \alpha)$ for α limit with the property that $\bar{\alpha} = \sup(C \cap \alpha) < \alpha$, but if α is a limit point of C we take C_α more carefully as follows: $C_\alpha = \{\alpha_\xi : \xi < \eta\}$ where

$$\alpha_\lambda = \sup\{\alpha_\xi : \xi < \lambda\} \text{ for } \lambda \text{ limit} < \eta, \tag{6.1.27}$$

$$\alpha_0 = \text{the } <_T\text{-predecessor of } \alpha \text{ with minimal } f\text{-image}, \tag{6.1.28}$$

$$\alpha_{\xi+1} = \text{the } <_T\text{-predecessor of } \alpha \text{ with minimal } f\text{-image subject to the requirement that } f(\alpha_{\xi+1}) > f(\alpha_{\zeta+1}) \text{ for all } \zeta < \xi, \tag{6.1.29}$$

$$\eta \text{ is the limit ordinal } \leq \theta^- \text{ where the process stops, i.e., } \sup\{f(\alpha_{\xi+1}) : \xi < \eta\} = \theta^-. \tag{6.1.30}$$

Note that if α and β are two limit points of C and if $\gamma <_T \alpha, \beta$, then $C_\alpha \cap \gamma = C_\beta \cap \gamma$. From this one concludes that the C-sequence is locally small, i.e., that $\{C_\alpha \cap \gamma : \gamma \leq \alpha < \theta\}$ has size $\leq \theta^-$ for all $\gamma < \theta$. □

Corollary 6.1.15. *If $\theta^{<\theta} = \theta$, then there exists a special Aronszajn tree of height θ^+.* □

Corollary 6.1.16. *In the constructible universe, special Aronszajn trees of any regular uncountable non-Mahlo height exist.* □

Remark 6.1.17. In a large portion of the literature on this subject the notion of a special Aronszajn tree of height equal to some successor cardinal θ^+ is somewhat weaker, equivalent to the fact that the tree can be embedded inside the tree

$$S(\theta) = \{f : \alpha \longrightarrow \theta : \alpha < \theta \;\&\; f \text{ is } 1-1\}.$$

One would get our notion of speciality by restricting the tree on successor ordinals, losing thus the frequently useful property of a tree that different nodes of the same limit height have different sets of predecessors. The result Corollary 6.1.15 in this weaker form is due to Specker [101], while the result Corollary 6.1.16 in this weaker form is essentially due to Jensen [52].

6.2 The weight function and its local versions

In this section we assume that $\theta = \kappa^+$ and we fix a C-sequence C_α $(\alpha < \kappa^+)$ such that

$$\operatorname{tp}(C_\alpha) \leq \kappa \text{ for all } \alpha < \kappa^+. \tag{6.2.1}$$

Let $\rho_1 : [\kappa^+]^2 \longrightarrow \kappa$ be defined recursively by

$$\rho_1(\alpha, \beta) = \max\{\operatorname{tp}(C_\beta \cap \alpha), \rho_1(\alpha, \min(C_\beta \setminus \alpha))\}$$

where we stipulate that $\rho_1(\gamma, \gamma) = 0$ for all γ.

Lemma 6.2.1. $|\{\xi \leq \alpha : \rho_1(\xi, \alpha) \leq \nu\}| \leq |\nu| + \aleph_0$ *for all $\alpha < \kappa^+$ and $\nu < \kappa$.*

Proof. Let ν^+ be the first infinite cardinal above the ordinal ν. The proof of the conclusion is by induction on α. So let $\Gamma \subseteq \alpha$ be a given set of order-type ν^+. We need to find $\xi \in \Gamma$ such that $\rho_1(\xi, \alpha) > \nu$. This will clearly be true if there is $\xi \in \Gamma$ such that $\operatorname{tp}(C_\alpha \cap \xi) > \nu$. So, we may assume that

$$\operatorname{tp}(C_\alpha \cap \xi) \leq \nu \text{ for all } \xi \in \Gamma. \tag{6.2.2}$$

Then there must be an ordinal $\alpha_1 \in C_\alpha$ such that the set

$$\Gamma_1 = \{\xi \in \Gamma : \alpha_1 = \min(C_\alpha \setminus \xi)\},$$

has size ν^+. By the inductive hypothesis there is $\xi \in \Gamma_1$ such that (see (6.2.2))

$$\rho_1(\xi, \alpha_1) > \nu \geq \operatorname{tp}(C_\alpha \cap \xi). \tag{6.2.3}$$

It follows that

$$\rho_1(\xi, \alpha) = \max\{\operatorname{tp}(C_\alpha \cap \xi), \rho_1(\xi, \alpha_1)\} = \rho_1(\xi, \alpha_1) > \nu.$$

This finishes the proof. □

Lemma 6.2.2. *If κ is regular, then $\{\xi \leq \alpha : \rho_1(\xi, \alpha) \neq \rho_1(\xi, \beta)\}$ has size $< \kappa$ for all $\alpha < \beta < \kappa^+$.*

Proof. The proof is by induction on α and β. Let $\Gamma \subseteq \alpha$ be a given set of order-type κ. We need to find $\xi \in \Gamma$ such that $\rho_1(\xi, \alpha) = \rho_1(\xi, \beta)$. Let $\gamma = \sup(\Gamma)$ and

$$\gamma_0 = \max(C_\beta \cap \gamma), \beta_0 = \min(C_\beta \setminus \gamma). \tag{6.2.4}$$

Note that by our assumption on κ and the C-sequence, these two ordinals are well defined and

$$\gamma_0 < \gamma \leq \beta_0 < \beta. \tag{6.2.5}$$

By Lemma 6.2.1 and the inductive hypothesis, there is ξ in $\Gamma \cap (\gamma_0, \gamma)$ such that

$$\rho_1(\xi, \alpha) = \rho_1(\xi, \beta_0) > \operatorname{tp}(C_\beta \cap \gamma). \tag{6.2.6}$$

It follows that $C_\beta \cap \gamma = C_\beta \cap \xi$ and $\beta_0 = \min(C_\beta \setminus \xi)$, and so

$$\rho_1(\xi, \beta) = \max\{\operatorname{tp}(C_\beta \cap \xi), \rho_1(\xi, \beta_0)\} = \rho_1(\xi, \beta_0) = \rho_1(\xi, \alpha).$$

This completes the proof. □

Remark 6.2.3. The assumption about the regularity of κ in Lemma 6.2.2 is essential. For example, it can be seen (see [13, p. 72]) that the conclusion of this lemma fails if κ is a singular limit of supercompact cardinals.

Definition 6.2.4. Define $\bar{\rho}_1 : [\kappa^+]^2 \longrightarrow \kappa$ by

$$\bar{\rho}_1(\alpha, \beta) = 2^{\rho_1(\alpha,\beta)} \cdot (2 \cdot \operatorname{tp}\{\xi \leq \alpha : \rho_1(\xi, \beta) = \rho_1(\alpha, \beta)\} + 1).$$

The following fact shows that this stretching of ρ_1 keeps the basic coherence property stated in Lemma 6.2.2 but it gives us also the injectivity property that could be quite useful.

Lemma 6.2.5. *If κ is a regular cardinal, then*

(a) $\bar{\rho}_1(\alpha, \gamma) \neq \bar{\rho}_1(\beta, \gamma)$ *whenever* $\alpha < \beta < \gamma < \kappa^+$,

(b) $|\{\xi \leq \alpha : \bar{\rho}_1(\xi, \alpha) \neq \bar{\rho}_1(\xi, \beta)\}| < \kappa$ *whenever* $\alpha < \beta < \kappa^+$.

Proof. Suppose that $\bar{\rho_1}(\alpha,\gamma) = \bar{\rho_1}(\beta,\gamma)$ for some $\alpha \leq \beta < \gamma < \kappa^+$. Then $\rho_1(\alpha,\gamma) = \rho_1(\beta,\gamma) = \nu$, and

$$\mathrm{tp}\{\xi \leq \alpha : \rho_1(\xi,\gamma) = \nu\} = \mathrm{tp}\{\xi \leq \beta : \rho_1(\xi,\gamma) = \nu\}.$$

This can only happen if $\alpha = \beta$. This proves (a).

To prove (b), for a given $\alpha < \beta < \kappa^+$, set

$$D_{\alpha\beta} = \{\xi \leq \alpha : \rho_1(\xi,\alpha) \neq \rho_1(\xi,\beta)\}.$$

By Lemma 6.2.2, the set $D_{\alpha\beta}$ has cardinality $< \kappa$, so by the regularity of κ,

$$\mu = \sup\{\rho_1(\xi,\eta) : \xi \in D_{\alpha\beta}, \eta \in \{\alpha,\beta\} + 1 < \kappa.$$

It suffices to show that, if some $\xi < \alpha$ is such that $\rho_1(\xi,\alpha) > \mu$ and $\rho_1(\xi,\beta) > \mu$, then $\bar{\rho_1}(\xi,\alpha) = \bar{\rho_1}(\xi,\beta)$. To see this, note that by the definition of μ, we must have that $\rho_1(\xi,\alpha) = \rho_1(\xi,\beta)$ and that, if we let ν denote this equal value, then

$$\{\eta \leq \xi : \rho_1(\eta,\alpha) = \nu\} = \{\eta \leq \xi : \rho_1(\eta,\beta) = \nu\}.$$

So, if we let ζ denote the order type of this set, we have that

$$\bar{\rho_1}(\xi,\alpha) = 2^\nu \cdot (2 \cdot \zeta + 1) = \bar{\rho_1}(\xi,\beta),$$

as required. □

Remark 6.2.6. Note that Lemma 6.2.5 gives an alternative proof of Corollary 6.1.15 since under the assumption $\kappa^{<\kappa} = \kappa$ the tree

$$T(\bar{\rho}_1) = \{\bar{\rho_1}(\cdot,\beta) \restriction \alpha : \alpha \leq \beta < \kappa^+\}$$

will have levels of size at most κ. It should be noted that the coherent sequence $(\bar{\rho})_\alpha$ $(\alpha < \kappa^+)$ of one-to-one mappings is an object of independent interest which can be particularly useful in stepping-up combinatorial properties of κ to κ^+. It is also an object that has interpretations in such areas as the theory of Čech–Stone compactifications of discrete spaces (see, e.g., [131], [16], [86], [64], [23]). We have already noted that if κ is singular, then we may no longer have the coherence property of Lemma 6.2.2. To get this property, one needs to make some additional assumption on the C-sequence C_α $(\alpha < \kappa^+)$, an assumption about the coherence of the C-sequence. This will be a subject of some of the following chapters where we will concentrate on a finer function ρ instead of ρ_1.

We finish this section by exposing local versions of the maximal weight function. We start with a C-sequence C_α $(\alpha < \theta)$[3] on a regular uncountable cardinal θ and an infinite cardinal $\kappa < \theta$ that is not necessarily its cardinal predecessor as above. Let $\Lambda_\kappa : [\theta]^2 \longrightarrow \theta$ be defined by

$$\Lambda_\kappa(\alpha,\beta) = \max\{\xi \in C_\beta \cap (\alpha+1) : \kappa \text{ divides } \mathrm{tp}(C_\beta \cap \xi)\},^4 \tag{6.2.7}$$

[3]As always, we are implicitly assuming that $C_{\alpha+1} = \{\alpha\}$ for all $\alpha < \theta$.

[4]We say that κ *divides* an ordinal γ of $\gamma = \xi \cdot \kappa$ for some ordinal ξ.

where we stipulate that $\max\emptyset = 0$ and that κ divides the ordinal 0. Using this mapping we define characteristic

$$\rho_1^\kappa : [\theta]^2 \to \kappa \tag{6.2.8}$$

by the recursive formula

$$\rho_1^\kappa(\alpha,\beta) = \max\{\rho_1^\kappa(\alpha, \min(C_\beta \setminus \alpha)), |C_\beta \cap [\Lambda_\kappa(\alpha,\beta),\alpha)|\} \tag{6.2.9}$$

with the boundary value $\rho_1^\kappa(\alpha,\alpha) = 0$. Observe that when θ is a successor of some cardinal κ, and if the C-sequence C_α $(\alpha < \theta = \kappa^+)$ is chosen to have the property that $\mathrm{tp}(C_\alpha) \leq \kappa$ for all $\alpha < \theta = \kappa^+$, then ρ_1^κ is nothing else than the old characteristic ρ_1. The following unboundedness property is one of the main reasons for introducing the local version of the maximal weight characteristic.

Theorem 6.2.7. *Suppose that the C-sequence C_α $(\alpha < \theta)$ avoids*[5] *a stationary set $\Gamma \subseteq \theta$ and that $\mathrm{tp}(C_\gamma) \leq \kappa$ for all $\gamma \in \Gamma$. Then for every family A of size θ of pairwise-disjoint subsets of θ such that $\mathrm{cf}(\gamma) > |a|$ for all $\gamma \in \Gamma$ and $a \in A$, and for every $\nu < \kappa$, there is a subfamily $B \subseteq A$ of size θ such that for every pair $a \neq b$ in B, we have $\rho_1^\kappa(\alpha,\beta) > \nu$ whenever $\alpha \in a$ and $\beta \in b$.*

Proof. Choose a continuous $\in$-chain $\mathcal{M}$ of order-type θ consisting of elementary submodels M of $(H_{\theta^+}, \in)$ such that M contains all the relevant objects and such that $M \cap \theta = \delta_M \in \theta$. Let

$$D = \{\delta_M : M \in \mathcal{M}\}.$$

Then D is a closed and unbounded subset of θ, and so we can find $\gamma \in \Gamma \cap D$. Choose $b \in A$ such that $\min(b) > \gamma$. Fix $\beta \in b$. Since our C-sequence avoids Γ the walk $\beta = \beta_0 > \beta_1 > \cdots > \beta_{n-1} > \beta_n = \gamma$ from β to γ passes through $\gamma + 1$, i.e., $\beta_{n-1} = \gamma + 1$. In particular the set $\bigcup_{i=1}^{n-1} C_{\beta_i} \cap \gamma$ is bounded in γ. Since $\mathrm{cf}(\gamma) > |b|$, there is $\eta < \gamma$ such that for all $\beta \in b$,

$$\eta > \max(C_\delta \cap \gamma) \text{ for all } \delta \in \mathrm{Tr}(\gamma,\beta) \setminus \{\gamma\}. \tag{6.2.10}$$

Moreover, we may assume that $\mathrm{tp}(C_\gamma \cap \eta) > \nu$. It follows that $\gamma \in \mathrm{Tr}(\xi,\beta)$ for all $\xi \in [\eta,\gamma)$ and $\beta \in b$, and so in particular, since $\mathrm{tp}(C_\gamma) \leq \kappa$,[6]

$$\rho_1^\kappa(\xi,\beta) \geq \mathrm{tp}(C_\gamma \cap \eta) > \nu \text{ for all } \xi \in [\eta,\gamma) \text{ and } \beta \in b. \tag{6.2.11}$$

Let $M \in \mathcal{M}$ be such that $\delta_M = \gamma$. Then M satisfies the following sentence: For every $\delta < \theta$ there is $b \in A$ with $\min(b) > \delta$ such that $\rho_1^\kappa(\xi,\beta) > \nu$ for all $\xi \in [\eta,\delta)$ and $\beta \in b$. By elementarity of M this sentence is true, and so for each $\delta < \theta$, we can fix $b_\delta \in A$ satisfying it. Let E be the collection of all ordinals $\varepsilon < \theta$ such that $\varepsilon > \eta$ and such that $\sup(b_\delta) < \varepsilon$ for all $\delta < \varepsilon$. Then E is a closed and unbounded subset of θ and the family $B = \{b_\delta : \delta \in E\}$ satisfies the conclusion of the theorem. □

[5]We say that C_α $(\alpha < \theta)$ *avoids* $\Gamma \subseteq \theta$ if $C_\alpha \cap \Gamma = \emptyset$ for all limit ordinals $\alpha < \theta$.
[6]and therefore, $\Lambda_\kappa(\xi,\gamma) = \min(C_\gamma)$ for all $\xi < \gamma$.

6.3 Unboundedness of the number of steps

The purpose of this section is to isolate a condition on given C-sequences

$$C_\alpha \ (\alpha < \theta)$$

defined on a regular uncountable cardinal θ that would correspond to the requirement that the corresponding function $\rho_2 : [\theta]^2 \longrightarrow \omega$, defined recursively by the formulas

$$\rho_2(\alpha, \beta) = \rho_2(\alpha, \min(C_\beta \setminus \alpha)) + 1 \text{ and}$$
$$\rho_2(\alpha, \alpha) = 0,$$

is in some sense nontrivial, and in particular, far from being constant. Without doubt the C-sequence $C_\alpha = \alpha$ $(\alpha < \theta)$ is the most trivial choice and the corresponding ρ_2-function, being constantly equal to 1, gives no information whatsoever about the cardinal θ. The following notion of the triviality of a C-sequence on θ appears to be only marginally different from this one, but we will soon see that it actually captures a natural unboundedness property of the characteristic ρ_2 that counts the number of steps for the corresponding notion of walk.

Definition 6.3.1. A C-sequence C_α $(\alpha < \theta)$ on a regular uncountable cardinal θ is *trivial* if there is a closed and unbounded set $C \subseteq \theta$ such that for every $\alpha < \theta$ there is $\beta \geq \alpha$ with $C \cap \alpha \subseteq C_\beta$.

Theorem 6.3.2. *The following are equivalent for any C-sequence C_α $(\alpha < \theta)$ on a regular uncountable cardinal θ and the corresponding function ρ_2:*

(i) *C_α $(\alpha < \theta)$ is nontrivial.*

(ii) *For every family A of θ pairwise-disjoint finite subsets of θ and every integer n, there is a subfamily B of A of size θ such that $\rho_2(\alpha, \beta) > n$ for all $\alpha \in a$, $\beta \in b$ and $a \neq b$ in B.*

Proof. Note that if a closed and unbounded set $C \subseteq \theta$ witnesses the triviality of the C-sequence, and if for $\alpha < \theta$ we let $\beta(\alpha) \geq \alpha$ be the minimal β such that $C \cap \alpha \subseteq C_{\beta(\alpha)}$, then any disjoint subfamily of $A = \{\{\alpha, \beta(\alpha)\} : \alpha \in C\}$ of size θ violates (ii) for $n = 1$.

Conversely, assume we are given a nontrivial C-sequence C_α $(\alpha < \theta)$. Consider the following statement:

(iii) *For every pair Γ and Δ of unbounded subsets of θ and every integer n, there exist a tail subset Γ_0 of Γ, an unbounded subset Δ_0 of Δ and a regressive strictly increasing map $f : \Delta_0 \longrightarrow \theta$ such that $\rho_2(\alpha, \beta) > n$ for all $\alpha \in \Gamma_0$ and $\beta \in \Delta_0$ with $\alpha < f(\beta)$.*

To get (ii) from (iii) one assumes that the family A consists of finite sets of some fixed size m and performs m^2 successive applications of (iii): this will give us an unbounded $\bar{A} \subseteq A$ and regressive strictly increasing maps $f_{ij} : \pi_j[\bar{A}] \longrightarrow \theta \, (i,j < m)$, where π_j is the projection mapping that takes b to the jth member of b. We have then that for all $a, b \in \bar{A}$ and $i, j < m$:

$$\text{if } \max(a) < f_{ij}(\pi_j(b)), \text{ then } \rho_2(\pi_i(a), \pi_j(b)) > n.$$

Going from there, we get a B as in (ii) by thinning out $\bar{A}$ once again.

Let us prove (iii) from (i) by induction on n. So suppose Γ and Δ are two unbounded subsets of θ and that (iii) is true for $n-1$. Let M_ξ ($\xi < \theta$) be a continuous $\in$-chain of submodels of H_{θ^+} containing all relevant objects such that $\delta_\xi = M_\xi \cap \theta \in \theta$. Let

$$C = \{\xi < \theta : \delta_\xi = \xi\}.$$

Then C is a closed and unbounded subset of θ, so by our assumption (i) on C_α ($\alpha < \theta$) there is $\delta \in C$ such that $C \cap \delta \not\subseteq C_\beta$ for all $\beta \geq \delta$. Pick an arbitrary $\beta \in \Delta$ above δ. Then there is $\xi \in C \cap \delta$ such that $\xi \notin C_\beta$. Then $\bar{\alpha} = \sup(C_\beta \cap \xi) < \xi$. Using the elementarity of the submodel M_ξ we conclude that for every $\alpha \in \Gamma \setminus \bar{\alpha}$ there is $\beta(\alpha) \in \Delta \setminus (\alpha+1)$ such that $\sup(C_{\beta(\alpha)} \cap \alpha) = \bar{\alpha}$. Applying the inductive hypothesis to the two sets

$$\Gamma \setminus \bar{\alpha} \text{ and } \{\min(C_{\beta(\alpha)} \setminus \alpha) : \alpha \in \Gamma \setminus \bar{\alpha}\}$$

we get a tail subset Γ_0 of $\Gamma \setminus \bar{\alpha}$, an unbounded subset Γ_1 of $\Gamma \setminus \bar{\alpha}$ and a strictly increasing map $g : \Gamma_1 \longrightarrow \theta$ such that $g(\gamma) < \min(C_{\beta(\gamma)} \setminus \gamma)$ and

$$\rho_2(\alpha, \min(C_{\beta(\gamma)} \setminus \gamma)) > n-1 \text{ for all } \alpha \in \Gamma_0, \gamma \in \Gamma_1 \text{ with } \alpha < g(\gamma). \tag{6.3.1}$$

Let $\Delta_0 = \{\beta(\gamma) : \gamma \in \Gamma_1\}$ and $f : \Delta_0 \longrightarrow \theta$ be defined by

$$f(\beta(\gamma)) = \min\{\gamma, g(\gamma)\}.$$

Then for $\alpha \in \Gamma_0$ and $\gamma \in \Gamma_1$ with $\alpha < f(\beta(\gamma)) \leq \gamma < \beta(\gamma)$ we have

$$\begin{aligned} \rho_2(\alpha, \beta(\gamma)) &= \rho_2(\alpha, \min(C_{\beta(\gamma)} \setminus \alpha)) + 1 \\ &= \rho_2(\alpha, \min(C_{\beta(\gamma)} \setminus \gamma)) + 1 \\ &> (n-1) + 1 = n. \end{aligned}$$

This finishes the proof. □

The following consequence of Theorem 6.3.2 shows that no tree of the form $T(\rho_0)$ can be Souslin.

Corollary 6.3.3. *Suppose that C_α ($\alpha < \theta$) is a nontrivial C-sequence, let*

$$\rho_0 : [\theta]^2 \longrightarrow \theta^{<\omega}$$

be the full code of the walk along this C-sequence, and let

$$T(\rho_0) = \{\rho_0(\cdot, \beta) \upharpoonright \alpha : \alpha \leq \beta < \kappa^+\}$$

be the corresponding tree. Then every subset of $T(\rho_0)$ of size θ contains an antichain of size θ.

Proof. Consider a subset X of $T(\rho_0)$ of size θ. Clearly, we may assume that X takes at most one point from a given level of $T(\rho_0)$. Replacing X by a set lying inside its downward closure, we may assume that the set consists of successor nodes of $T(\rho_0)$. Let $K \subseteq [\theta]^2$ be such that

$$X = \{\rho_0(\cdot, \beta) \restriction (\alpha + 1) : \{\alpha, \beta\} \in K\}.$$

Shrinking X, we may assume that K consists of pairwise-disjoint pairs. Shrinking X further, we may also assume that ρ_2 is constant on K. Let n be the constant value of $\rho_2 \restriction K$. Applying Theorem 6.3.2(ii) to K and n, we get $K_0 \subseteq K$ of size θ such that $\rho_2(\alpha, \delta) > n$ for all $\{\alpha, \beta\}$ and $\{\gamma, \delta\}$ from K_0 with properties $\alpha < \beta$, $\gamma < \delta$ and $\alpha < \gamma$. Then

$$X_0 = \{\rho_0(\cdot, \beta) \restriction (\alpha + 1) : \{\alpha, \beta\} \in K_0\}$$

is an antichain in $T(\rho_0)$. □

Remark 6.3.4. It should be clear that nontrivial C-sequences exist on any successor cardinal. Indeed, with very little extra work one can show that nontrivial C-sequences exist for some inaccessible cardinals quite high in the Mahlo-hierarchy. To show how close this is to the notion of weak compactness, we will give the following characterization of it which is of independent interest.[7]

Theorem 6.3.5. *The following are equivalent for an inaccessible cardinal θ:*

(i) *θ is weakly compact.*

(ii) *For every C-sequence C_α $(\alpha < \theta)$ there is a closed and unbounded set $C \subseteq \theta$ such that for all $\alpha < \theta$ there is $\beta \geq \alpha$ such that $C_\beta \cap \alpha = C \cap \alpha$.*

Proof. To see the implication from (i) to (ii), let us assume that C_α $(\alpha < \theta)$ is a C-sequence and let $\rho_0 : [\theta]^2 \longrightarrow \mathbb{Q}_\theta$ be the corresponding ρ_0-function. Set

$$T(\rho_0) = \{(\rho_0)_\beta \restriction \alpha : \alpha \leq \beta < \theta\}.$$

Since θ is inaccessible, the levels of $T(\rho_0)$ have size $< \theta$, so by the tree-property of θ (see [55]) the tree $T(\rho_0)$ must contain a cofinal branch b. For each $\alpha < \theta$ fix $\beta(\alpha) \geq \alpha$ such that the restriction of $(\rho_0)_{\beta(\alpha)}$ to α belongs to b. For limit $\alpha < \theta$ let $\gamma(\alpha)$ be the largest member γ appearing in the walk

$$\beta(\alpha) = \beta_0(\alpha) > \beta_1(\alpha) > \cdots > \beta_{k(\alpha)}(\alpha) = \alpha$$

along the C-sequence with the property that $C_\gamma \cap \alpha$ is unbounded in α. (Thus $\gamma = \alpha$ or $\gamma = \beta_{k(\alpha)-1}(\alpha)$.) By the Pressing Down Lemma there is a stationary set

[7] It turns out that every C-sequence on θ being trivial is not quite as strong as the weak compactness of θ. As pointed out to us by Donder and König, one can show this using a model of Kunen [63, §3].

$\Gamma \subseteq \theta$ and an integer k, a sequence t in $\mathbb{Q}_\theta$ and an ordinal $\bar{\alpha} < \theta$ such that for all $\alpha \in \Gamma$:

$$k(\alpha) = k \text{ and } \sup(C_{\beta_i(\alpha)} \cap \alpha) \leq \bar{\alpha} \text{ for all } i < k \text{ for which this set is bounded in } \alpha. \tag{6.3.2}$$

$$\operatorname{tp}(C_{\beta_i(\alpha)} \cap \alpha) = t(i) \text{ for all } i < k \text{ for which this set is bounded in } \alpha. \tag{6.3.3}$$

Shrinking Γ we may assume that either for all $\alpha \in \Gamma$, $\gamma(\alpha) = \alpha$, in which case t is chosen to be a k-sequence, or for all $\alpha \in \Gamma$, $\gamma(\alpha) = \beta_{k-1}(\alpha)$, in which case t is chosen to be a $(k-1)$-sequence. It follows that

$$\rho_0(\xi, \beta(\alpha)) = t^\frown \rho_0(\xi, \gamma(\alpha)) \text{ for all } \alpha \in \Gamma \text{ and } \xi \in [\bar{\alpha}, \alpha). \tag{6.3.4}$$

It follows that for $\alpha < \alpha'$ in Γ the mappings $(\rho_0)_{\gamma(\alpha)}$ and $(\rho_0)_{\gamma(\alpha')}$ have the same restrictions on the interval $[\bar{\alpha}, \alpha)$. So in particular we have that

$$C_{\gamma(\alpha)} \cap [\bar{\alpha}, \alpha) = C_{\gamma(\alpha')} \cap [\bar{\alpha}, \alpha) \text{ whenever } \alpha < \alpha' \text{ in } \Gamma. \tag{6.3.5}$$

Going to a stationary subset of Γ we may assume that all $C_{\gamma(\alpha)}$ $(\alpha \in \Gamma)$ have the same intersection with $\bar{\alpha}$. It is now clear that the union of $C_{\gamma(\alpha)} \cap \alpha$ $(\alpha \in \Gamma)$ is a closed and unbounded subset of θ satisfying the conclusion of (ii).

To prove the implication from (ii) to (i), let $T = (\theta, <_T)$ be a tree on θ whose levels are intervals of θ, or more precisely, there is a closed and unbounded set Δ of limit ordinals in θ such that if δ_ξ $(\xi < \theta)$ is its increasing enumeration, then $[\delta_\xi, \delta_{\xi+1})$ is equal to the ξth level of T. Choose a C-sequence C_α $(\alpha < \theta)$ as follows. First of all let $C_{\alpha+1} = \{\alpha\}$ and $C_\alpha = (\bar{\alpha}, \alpha)$ if α is the limit and $\bar{\alpha} = \sup(\Delta \cap \alpha) < \alpha$. If α is a limit point of Δ, let

$$C_\alpha = \overline{\{\xi < \alpha : \xi <_T \alpha\}}.$$

By (ii) there is a closed and unbounded set $C \subseteq \theta$ such that for all $\alpha < \theta$ there is $\beta(\alpha) \geq \alpha$ with $C_{\beta(\alpha)} \cap \alpha = C \cap \alpha$. Let Γ be the set of all successor ordinals $\xi < \theta$ such that $C \cap [\delta_\xi, \delta_{\xi+1}) \neq \emptyset$. For $\xi \in \Gamma$ let $t_\xi = \min(C \cap [\delta_\xi, \delta_{\xi+1}))$. Then it is seen that $\{t_\xi : \xi \in \Gamma\}$ is a chain of T of size θ. This finishes the proof. $\square$

We have already remarked that every successor cardinal $\theta = \kappa^+$ admits a nontrivial C-sequence C_α $(\alpha < \theta)$. It suffices to take the C_α's to be all of order-type $\leq \kappa$. It turns out that for such a C-sequence the corresponding ρ_2-function has a property that is considerably stronger than 6.3.2(ii).

Theorem 6.3.6. *For every infinite cardinal κ there is a C-sequence on κ^+ such that the corresponding ρ_2-function has the following unboundedness property: for every family A of κ^+ pairwise-disjoint subsets of κ^+, all of size $< \kappa$, and for every $n < \omega$, there exists $B \subseteq A$ of size κ^+ such that $\rho_2(\alpha, \beta) > n$ whenever $\alpha \in a$ and $\beta \in b$ for some $a \neq b$ in B.*

Proof. We shall show that this is true for every C-sequence C_α $(\alpha < \kappa^+)$ with the property that $\mathrm{tp}(C_\alpha) \leq \kappa$ for all $\alpha < \kappa^+$. Let us first consider the case when κ is a regular cardinal. The proof is by induction on n. Suppose the conclusion is true for some integer n and let A be a given family of κ^+ pairwise-disjoint subsets of κ^+, all of size $< \kappa^+$. Let M_ξ $(\xi < \kappa^+)$ be a continuous $\in$-chain of elementary submodels of $H_{\kappa^{++}}$ such that for every ξ the M_ξ contains all the relevant objects and has the property that $\delta_\xi = M_\xi \cap \kappa^+ \in \kappa^+$. Let $C = \{\delta_\xi : \xi < \kappa^+\}$. Then C is a closed and unbounded subset of κ^+. Choose $\xi = \delta_\xi \in C$ of cofinality κ and $b \in A$ such that $\beta > \xi$ for all $\beta \in b$. Then

$$\gamma = \sup\{\max(C_\beta \cap \xi) : \beta \in b\} < \xi.$$

By the elementarity of M_ξ we conclude that for every $\eta \in (\gamma, \kappa^+)$ there is b_η in A above η such that

$$\gamma = \sup\{\max(C_\beta \cap \eta) : \beta \in b_\eta\}.$$

Consider the following family of subsets of κ^+,

$$\hat{a}_\eta = b_\eta \cup \{\min(C_\beta \setminus \eta) : \beta \in b_\eta\} \ (\eta \in (\gamma, \kappa^+)).$$

By the inductive hypothesis there is an unbounded $\Gamma \subseteq \kappa^+$ such that for all $\eta < \zeta$ in Γ, $b_\eta \subseteq \zeta$ and $\rho_2(\alpha, \beta) > n$ for all $\alpha \in \hat{a}_\eta$ and $\beta \in \hat{a}_\zeta$. Let $B = \{b_\eta : \eta \in \Gamma\}$ and let b_η and b_ζ for $\eta < \zeta$ be two given members of B. Then

$$\rho_2(\alpha, \beta) = \rho_2(\alpha, \min(C_\beta \setminus \alpha)) + 1 \geq n + 1 \text{ for all } \alpha \in b_\eta \text{ and } \beta \in b_\zeta.$$

This completes the inductive step and therefore the proof of the theorem when κ is a regular cardinal.

If κ is a singular cardinal, going to a subfamily of A, we may assume that there is a regular cardinal $\lambda < \kappa$ and an unbounded set $\Gamma \subseteq \kappa^+$ with the property that A can be enumerated as a_η $(\eta \in \Gamma)$ such that for all $\eta \in \Gamma$:

$$|a_\eta| < \lambda, \tag{6.3.6}$$

$$\beta > \eta \text{ for all } \beta \in a_\eta, \tag{6.3.7}$$

$$|C_\beta \cap \eta| < \lambda \text{ for all } \beta \in a_\eta. \tag{6.3.8}$$

Choose $\xi = \delta_\xi \in C$ of cofinality λ and $\hat{\eta} \in \Gamma$ above ξ and proceed as in the previous case. This completes the inductive step and the proof of Theorem 6.3.6. □

Note that the combination of the ideas from the proofs of Theorem 6.3.2 and 6.3.6 gives us the following variation.

Theorem 6.3.7. *Suppose that a regular uncountable cardinal θ supports a nontrivial C-sequence and let ρ_2 be the associated function. Then for every integer n and every pair of θ-sized families A_0 and A_1, where the members of A_0 are pairwise-disjoint bounded subsets of θ and the members of A_1 are pairwise-disjoint finite subsets of θ, there exist $B_0 \subseteq A_0$ and $B_1 \subseteq A_1$ of size θ such that for all $a \in B_0$ and $b \in B_1$ such that* $\sup(a) < \min(b)$, *we have that $\rho_2(\alpha, \beta) > n$ for every choice of $\alpha \in a$ and $\beta \in b$.* □

Let us give an interesting application of this result found recently by Gruenhage [41]. Recall that for every perfect[8] surjection $f : X \longrightarrow Y$ from a topological space X onto a topological space Y there is a closed subset H of X on which f is *irreducible*[9]. The set H is obtained as the intersection of a maximal chain $\mathcal{C}$ of closed subsets of X such that $f''K = Y$ for all $K \in \mathcal{C}$. When the map f is closed but not necessarily perfect, such a set may not exist, in which case we associate to it the following natural cardinal characteristic,

$$\sharp(f) = \min\{\mathrm{otp}(\mathcal{C}) : \mathcal{C} \in \mathrm{IC}_f(X) \text{ and } f''(\bigcap \mathcal{C}) \neq Y\}, \tag{6.3.9}$$

where $\mathrm{IC}_f(X)$ denotes the collection of all closed subspaces H of X such that $f''H = Y$ and $|f^{-1}(y) \cap H| = 1$ for every isolated point y of Y. If the set on the right-hand side of (6.3.9) is empty, which happens for example in the case when the map f is perfect, set $\sharp(f) = \infty$. Clearly $\sharp(f)$ is a regular cardinal unless it is equal to ∞. The following result says that mild restrictions on spaces X and Y force $\sharp(f)$ to be rather large, or in other words that when X and Y are relatively small, then every closed surjection $f : X \longrightarrow Y$ will have an irreducible restriction $f : H \longrightarrow Y$ on some closed subspace H of X.

Theorem 6.3.8. *Let $f : X \longrightarrow Y$ be a given closed surjection from a topological space X onto a topological space Y which has no irreducible restriction on a closed subspace H of X. Assume that the space X has the property that every open cover of X has a point-countable open refinement and that the space Y has the property that every open subspace of Y either contains an isolated point or a sequence of open sets whose intersection is not open. Then the corresponding cardinal $\sharp(f)$ is uncountable and it does not support a nontrivial C-sequence.*

Proof. Clearly, $\sharp(f) < \infty$. Let us first show that $\sharp(f) \neq \omega$. Otherwise, there is a decreasing sequence H_n $(n < \omega)$ of closed subspaces of X such that $f''H_n = Y$ for all n, but $f''H_\omega \neq Y$, where $H_\omega = \bigcap_{n<\omega} H_n$. Moreover, we are assuming that $|f^{-1}(y) \cap H_0| = 1$ for every isolated point y of Y, and therefore the set $U = Y \setminus f''H_\omega$ is open, nonempty, and has no isolated points. By our assumption on Y, there is a decreasing sequence U_n $(n < \omega)$ of open subsets of U and a point $y \in \bigcap_{n<\omega} U_n$ such that $y \in \overline{\bigcup_{n<\omega} K_n}$, where $K_n = Y \setminus U_n$ for $n < \omega$. Let

$$G = \bigcup_{n<\omega} (H_n \cap f^{-1}(K_n)).$$

Note that $f''G = \bigcup_{n<\omega} K_n$, and therefore, $y \in \overline{f''G} = f''\overline{G}$. Pick $x \in \overline{G} \cap f^{-1}(y)$. Then for every $n < \omega$,

$$x \in \overline{\bigcup_{m \geq n} (H_m \cap f^{-1}K_m)},$$

[8]Recall that a continuous surjection $f : X \longrightarrow Y$ is *perfect* if it is *closed* (i.e., it maps closed subsets of X onto closed subsets of Y) and it has the property that $f^{-1}(y)$ is compact for all y in Y.

[9]Recall that a surjection $f : H \longrightarrow Y$ is irreducible if $f''K \neq Y$ for every proper closed subset K of H.

and therefore, $x \in H_n$. It follows that $x \in H_\omega \cap f^{-1}(y)$, and so, $y = f(x) \in f''H_\omega$, a contradiction.

It follows that $\theta = \sharp(f)$ is a regular uncountable cardinal. We need to show that θ does not support a nontrivial C-sequence. Suppose it does; fix such a sequence C_α $(\alpha < \theta)$, and consider the corresponding characteristic

$$\rho_2 : [\theta]^2 \longrightarrow \omega$$

that counts the number of steps of the walk along C_α $(\alpha < \theta)$, i.e., a characteristic defined by the recursive formula

$$\rho_2(\alpha, \beta) = \rho_2(\alpha, \min(C_\beta \setminus \alpha)) + 1$$

with the boundary value $\rho_2(\alpha, \alpha) = 0$ for all α. Choose a decreasing sequence H_α $(\alpha < \theta)$ of closed subsets of X such that $f''H_\alpha = Y$ for all $\alpha < \theta$, but $f''H_\theta \neq Y$, where $H_\theta = \bigcap_{\alpha<\theta} H_\alpha$. Moreover we assume that $|f^{-1}(y) \cap H_0| = 1$ for every isolated point y of Y, and therefore the set $Y \setminus f''H_\omega$ is open, nonempty, and has no isolated points. So, as before, on the basis of our assumption on Y, we can find an increasing sequence K_n $(n < \omega)$ of closed subsets of Y and a point $y \in \overline{\bigcup_{n<\omega} K_n} \setminus \bigcup_{n<\omega} K_n$. By our assumption on Y there is a point-countable open cover V_α $(\alpha < \theta)$ of the closed set $f^{-1}(y)$ such that $V_\alpha \subseteq X \setminus H_\alpha$ for all $\alpha < \theta$. For $\alpha < \theta$ and $n < \omega$, set

$$V_{\alpha n} = V_\alpha \setminus f^{-1}(K_n). \tag{6.3.10}$$

Then for every $\beta < \theta$, the open set $\bigcup_{\alpha<\theta} V_{\beta\rho_2\{\alpha,\beta\}}$[10] covers the set $f^{-1}(y)$, and since the mapping f is closed, for each such ordinal $\beta < \theta$, we can select a neighborhood W_β of y such that

$$f^{-1}(W_\beta) \subseteq \bigcup_{\alpha<\theta} V_{\beta\rho_2\{\alpha,\beta\}}. \tag{6.3.11}$$

Using again the fact that f is a closed mapping, we can find an open neighborhood U of y such that

$$f^{-1}(U) \subseteq \bigcup_{\alpha<\theta} (V_\alpha \cap f^{-1}(W_\alpha)). \tag{6.3.12}$$

Fix an $n_0 < \omega$ such that $K_{n_0} \cap U \neq \emptyset$ and pick a point z in this set. For $x \in f^{-1}(z)$, set

$$a_x = \{\alpha < \theta : x \in V_\alpha\}.$$

Then a_x $(x \in f^{-1}(z))$ is a family of countable subsets of θ. Note that $a_x > \beta$ for all $\beta < \theta$ and $x \in H_\beta$. So one can find a subset $S \subseteq f^{-1}(z)$ of cardinality θ such that a_x $(x \in S)$ is a family of pairwise-disjoint countable subsets of θ. Let

$$B = \{\beta < \theta : z \in W_\beta\}.$$

[10]Here, $\rho_2\{\alpha,\beta\} = \rho_2(\alpha,\beta)$ if $\alpha < \beta$, $\rho_2\{\alpha,\beta\} = \rho_2(\beta,\alpha)$ if $\alpha > \beta$, and $\rho_2\{\alpha,\beta\} = 0$ if $\alpha = \beta$.

We claim that B is an unbounded subset of θ. Pick $\alpha < \theta$. Then $f^{-1}(z) \cap H_\alpha \neq \emptyset$, so we can fix a point x_0 in this set. Then

$$x_0 \in f^{-1}(U) \subseteq \bigcup_{\beta<\theta} (V_\beta \cap f^{-1}(W_\beta)).$$

So we can find $\beta < \theta$ such that $x_0 \in V_\beta \cap f^{-1}(W_\beta)$. Then $z = f(x_0) \in W_\beta$. Since $x_0 \in V_\beta$ and since $V_\beta \cap H_\gamma = \emptyset$ for $\gamma > \beta$, we conclude that β is a member of $B \setminus \alpha$, as required.

Applying Theorem 6.3.7 to the family a_x $(x \in S)$, the set B, and the integer n_0, we can find $x \in S$ and $\beta \in B$ such that $a_x < \beta$ and

$$\rho_2(\alpha, \beta) > n_0 \text{ for all } \alpha \in a_x.$$

On the other hand, $f(x) = z \in K_{n_0}$, and so $x \notin V_{\beta n}$ for all $n \geq n_0$. It follows that

$$x \notin \bigcup_{\alpha<\theta} V_{\beta\rho_2\{\alpha,\beta\}},$$

and so on the basis of (6.3.11), we conclude that $z = f(x) \notin W_\beta$, a contradiction. □

Corollary 6.3.9. *Let X and Y be two topological spaces that are not bigger than the first regular uncountable cardinal θ not supporting a nontrivial C-sequence. Suppose that X every open cover of X has a point-countable open refinement and that every open subspace of Y either contains an isolated point or a sequence of open sets whose intersection is not open. Then every closed surjection $f : X \longrightarrow Y$ is irreducible when restricted on some closed subspace H of X.* □

Thus, in particular, if X and Y are smaller than, say, the first ω-Mahlo cardinal, then every closed surjection $f : X \longrightarrow Y$ has an irreducible restriction on a closed subspace of X.

Remark 6.3.10. As shown in [41] there is a closed surjection $f : X \longrightarrow Y$ for a Lindelöf space X which admits no irreducible restriction. It turns out that for that closed surjection f, we have $\sharp(f) = \omega$, so the restriction on the space Y in Theorem 6.3.8, needed to eliminate this case, is in some sense necessary.

Chapter 7

Square Sequences

7.1 Square sequences and their full lower traces

The purpose of this section is to study walks along C-sequences that have the following pleasant coherence property.

Definition 7.1.1. A C-sequence C_α $(\alpha < \theta)$ is a *square sequence* if and only if it is *coherent*, i.e., if it has the property that $C_\alpha = C_\beta \cap \alpha$ whenever α is a limit point of C_β.

Note that the nontriviality conditions appearing in Definition 6.3.1 and Theorem 6.3.5 coincide in the realm of square sequences:

Lemma 7.1.2. *A square sequence C_α $(\alpha < \theta)$ is trivial if and only if there is a closed and unbounded subset C of θ such that $C_\alpha = C \cap \alpha$ whenever α is a limit point of C.* □

To a given square sequence C_α $(\alpha < \theta)$ one naturally associates a tree-ordering $<^2$ on θ as follows:

$$\alpha <^2 \beta \text{ if and only if } \alpha \text{ is a limit point of } C_\beta. \tag{7.1.1}$$

The triviality of C_α $(\alpha < \theta)$ is then equivalent to the statement that the tree $(\theta, <^2)$ has a chain of size θ. In fact, one can characterize the tree-orderings $<_T$ on θ for which there exists a square sequence C_α $(\alpha < \theta)$ such that for all $\alpha < \beta < \theta$:

$$\alpha <_T \beta \text{ if and only if } \alpha \text{ is a limit point of } C_\beta. \tag{7.1.2}$$

Lemma 7.1.3. *A tree-ordering $<_T$ on θ admits a square sequence C_α $(\alpha < \theta)$ satisfying* (7.1.2) *if and only if*

(i) *$\alpha <_T \beta$ can hold only for limit ordinals α and β such that $\alpha < \beta$,*

(ii) $P_\beta = \{\alpha : \alpha <_T \beta\}$ *is a closed subset of* β, *which is unbounded in* β *whenever* $\mathrm{cf}(\beta) > \omega$ *and*

(iii) *minimal as well as successor nodes of the tree* $<_T$ *on* θ *are ordinals of cofinality* ω.

Proof. For each ordinal $\alpha < \theta$ of countable cofinality we fix a subset $S_\alpha \subseteq \alpha$ of order-type ω cofinal with α. Given a tree-ordering $<_T$ on θ with properties (i)–(iii), for a limit ordinal $\beta < \theta$ let P_β^+ be the set of all successor nodes from $P_\beta \cup \{\beta\}$ including the minimal one. For $\alpha \in P_\beta^+$, let α^- be its immediate predecessor in P_β. Finally, set

$$C_\beta = P_\beta \cup \bigcup \{S_\alpha \cap [\alpha^-, \alpha) : \alpha \in P_\beta^+\}.$$

It is easily checked that this defines a square sequence C_β $(\beta < \theta)$ with the property that $\alpha <_T \beta$ holds if and only if α is a limit point of C_β. □

Remark 7.1.4. It should be clear that the proof of Lemma 7.1.3 shows that the exact analogue of this result is true for any cofinality $\kappa < \theta$ rather than ω.

An important result about square sequences is the following fact that can be deduced on the basis of the well-known construction of square sequences in the constructible universe, a construction that is essentially contained in the classical paper of Jensen [52] (see also 1.10 of [109]).

Theorem 7.1.5. *If a regular uncountable cardinal* θ *is not weakly compact in the constructible subuniverse, then there is a nontrivial square sequence on* θ *which is moreover constructible.* □

Corollary 7.1.6. *If a regular uncountable cardinal* θ *is not weakly compact in the constructible subuniverse, then there is a constructible Aronszajn tree on* θ.

Proof. Let C_α $(\alpha < \theta)$ be a fixed nontrivial square sequence which is constructible. Changing the C_α's a bit, we may assume that if β is a limit ordinal with $\alpha = \min C_\beta$, or if $\alpha \in C_\beta$ but $\sup(C_\beta \cap \alpha) < \alpha$, then α must be a successor ordinal in θ. Consider the corresponding characteristic

$$\rho_0 : [\theta]^2 \longrightarrow \theta^{<\omega}$$

of the walk along C_α $(\alpha < \theta)$ defined by the recursive formula

$$\rho_0(\alpha, \beta) = \mathrm{tp}(C_\beta \cap \alpha)^\frown \rho_0(\alpha, \min(C_\beta \setminus \alpha)),$$

where $\rho_0(\gamma, \gamma) = \emptyset$ for all $\gamma < \theta$. Consider also the corresponding tree

$$T(\rho_0) = \{(\rho_0)_\beta \restriction \alpha : \alpha \leq \beta < \theta\}.$$

Clearly $T(\rho_0)$ is constructible. By (6.1.4) the αth level of $T(\rho_0)$ is bounded by the size of the set $\{C_\beta \cap \alpha : \beta \geq \alpha\}$. Since the intersection of the form $C_\beta \cap \alpha$ is determined by its maximal limit point modulo a finite subset of α, we conclude that the αth level of $T(\rho_0)$ has size $\leq |\alpha| + \aleph_0$. Since the sequence C_α $(\alpha < \theta)$ is nontrivial, the proof of Theorem 6.3.5 shows that $T(\rho_0)$ has no cofinal branches. □

Lemma 7.1.7. *Suppose C_α $(\alpha < \theta)$ is a square sequence on θ, $<^2$ the associated tree-ordering on θ. Let $\rho_0 : [\theta]^2 \longrightarrow \theta^{<\omega}$ be the full code of the walk along C_α $(\alpha < \theta)$, and let $T(\rho_0) = \{(\rho_0)_\beta \restriction \alpha : \alpha \leq \beta < \theta\}$ be the associated tree. Then $\alpha \longmapsto (\rho_0)_\alpha$ is a strictly increasing map from the tree $(\theta, <^2)$ into the tree $T(\rho_0)$.*

Proof. If α is a limit point of C_β, then $C_\alpha = C_\beta \cap \alpha$, so the walks $\alpha \to \xi$ and $\beta \to \xi$ for $\xi < \alpha$ get the same code $\rho_0(\xi, \alpha) = \rho_0(\xi, \beta)$. □

Having explained the nontriviality condition on a given square sequence, for the rest of this section, we fix a regular uncountable cardinal θ and a nontrivial square sequence C_α $(\alpha < \theta)$ on θ and study walks along C_α $(\alpha < \theta)$. In particular, we will be interested in the full lower trace associated to these walks. For this it is convenient to introduce the function

$$\Lambda = \Lambda_\omega : [\theta]^2 \longrightarrow \theta$$

given by the formula

$$\Lambda(\alpha, \beta) = \text{maximal limit point of } C_\beta \cap (\alpha + 1),$$

where we stipulate that $\Lambda(\alpha, \beta) = 0$ if $C_\beta \cap (\alpha + 1)$ has no limit points.

Having this function, we can write a recursive trace formula describing *the full lower trace* characteristic

$$\mathrm{F} : [\theta]^2 \longrightarrow [\theta]^{<\omega} \tag{7.1.3}$$

of the walk along C_α $(\alpha < \theta)$ as follows:

$$\mathrm{F}(\alpha, \beta) = \mathrm{F}(\alpha, \min(C_\beta \setminus \alpha)) \cup \bigcup \{\mathrm{F}(\xi, \alpha) : \xi \in C_\beta \cap [\Lambda(\alpha, \beta), \alpha)\}, \tag{7.1.4}$$

with the boundary value $\mathrm{F}(\alpha, \alpha) = \{\alpha\}$ for all α.

Lemma 7.1.8. *For all $\alpha \leq \beta \leq \gamma < \theta$,*

(a) $\mathrm{F}(\alpha, \gamma) \subseteq \mathrm{F}(\alpha, \beta) \cup \mathrm{F}(\beta, \gamma)$,
(b) $\mathrm{F}(\alpha, \beta) \subseteq \mathrm{F}(\alpha, \gamma) \cup \mathrm{F}(\beta, \gamma)$.

Proof. The proof of the lemma is by simultaneous induction. We first check (a). Let

$$\gamma_\alpha = \min(C_\gamma \setminus \alpha),\ \gamma_\beta = \min(C_\gamma \setminus \beta) \text{ and } \lambda = \Lambda(\beta, \gamma). \tag{7.1.5}$$

If $\alpha < \lambda$, then $C_\lambda = C_\gamma \cap \lambda$ and therefore $\mathrm{F}(\alpha, \lambda) = \mathrm{F}(\alpha, \gamma)$. Applying the inductive hypothesis (b) for $\alpha < \lambda \leq \beta$, we get the required conclusion

$$\mathrm{F}(\alpha, \gamma) = \mathrm{F}(\alpha, \lambda) \subseteq \mathrm{F}(\alpha, \beta) \cup \mathrm{F}(\lambda, \beta) \subseteq \mathrm{F}(\alpha, \beta) \cup \mathrm{F}(\beta, \gamma). \tag{7.1.6}$$

So we may assume that $\lambda \leq \alpha$. Then $\Lambda(\alpha,\gamma) = \lambda$. If $\gamma_\alpha < \beta$, applying the inductive hypothesis (b) for $\alpha \leq \gamma_\alpha < \beta$, we have

$$\mathrm{F}(\alpha,\gamma_\alpha) \subseteq \mathrm{F}(\alpha,\beta) \cup \mathrm{F}(\gamma_\alpha,\beta) \subseteq \mathrm{F}(\alpha,\beta) \cup \mathrm{F}(\beta,\gamma), \tag{7.1.7}$$

since $\gamma_\alpha \in C_\gamma \cap [\Lambda(\beta,\gamma),\beta)$. If $\gamma_\alpha \geq \beta$, then $\gamma_\alpha = \gamma_\beta$, and applying the inductive hypothesis (a) for $\alpha < \beta \leq \gamma_\alpha = \gamma_\beta$, we have

$$\mathrm{F}(\alpha,\gamma_\alpha) \subseteq \mathrm{F}(\alpha,\beta) \cup \mathrm{F}(\beta,\gamma_\beta) \subseteq \mathrm{F}(\alpha,\beta) \cup \mathrm{F}(\beta,\gamma). \tag{7.1.8}$$

Consider now $\xi \in C_\gamma \cap [\lambda,\alpha)$. By the inductive hypothesis (b) for $\xi < \alpha \leq \beta$ we have

$$\mathrm{F}(\xi,\alpha) \subseteq \mathrm{F}(\alpha,\beta) \cup \mathrm{F}(\xi,\beta) \subseteq \mathrm{F}(\alpha,\beta) \cup \mathrm{F}(\beta,\gamma), \tag{7.1.9}$$

since $\xi \in C_\gamma \cap [\Lambda(\beta,\gamma),\beta)$. It follows that all the factors of $\mathrm{F}(\alpha,\gamma)$ are included in the union $\mathrm{F}(\alpha,\beta) \cup \mathrm{F}(\beta,\gamma)$, so this completes the checking of (a).

To prove (b) define γ_α, γ_β and λ as above (see 7.1.5) and consider first the case $\lambda \geq \alpha$. Then $C_\lambda = C_\gamma \cap \lambda$ and therefore $\mathrm{F}(\alpha,\lambda) = \mathrm{F}(\alpha,\gamma)$. Applying the inductive hypothesis (a) for $\alpha < \lambda \leq \beta$ we get

$$\mathrm{F}(\alpha,\beta) \subseteq \mathrm{F}(\alpha,\lambda) \cup \mathrm{F}(\lambda,\beta) \subseteq \mathrm{F}(\alpha,\gamma) \cup \mathrm{F}(\beta,\gamma). \tag{7.1.10}$$

So we may assume that $\lambda < \alpha$ in which case we also know that $\Lambda(\alpha,\gamma) = \lambda$. Consider first the case $\gamma_\alpha < \beta$. Applying the inductive hypothesis (a) for $\alpha \leq \gamma_\alpha < \beta$ we have

$$\mathrm{F}(\alpha,\beta) \subseteq \mathrm{F}(\alpha,\gamma_\alpha) \cup \mathrm{F}(\gamma_\alpha,\beta) \subseteq \mathrm{F}(\alpha,\gamma) \cup \mathrm{F}(\beta,\gamma), \tag{7.1.11}$$

since $\gamma_\alpha \in C_\gamma \cap [\Lambda(\beta,\gamma),\beta)$. Suppose now that $\gamma_\alpha \geq \beta$. Then $\gamma_\alpha = \gamma_\beta$ and applying the inductive hypothesis (b) for the triple $\alpha \leq \beta \leq \gamma_\alpha = \gamma_\beta$ we have

$$\mathrm{F}(\alpha,\beta) \subseteq \mathrm{F}(\alpha,\gamma_\alpha) \cup \mathrm{F}(\beta,\gamma_\beta) \subseteq \mathrm{F}(\alpha,\gamma) \cup \mathrm{F}(\beta,\gamma). \tag{7.1.12}$$

This completes the proof. □

Lemma 7.1.9. *For all $\alpha \leq \beta \leq \gamma < \theta$,*

(a) $\rho_0(\alpha,\beta) = \rho_0(\min(\mathrm{F}(\beta,\gamma) \setminus \alpha),\beta)^\frown \rho_0(\alpha,\min(\mathrm{F}(\beta,\gamma) \setminus \alpha))$,
(b) $\rho_0(\alpha,\gamma) = \rho_0(\min(\mathrm{F}(\beta,\gamma) \setminus \alpha),\gamma)^\frown \rho_0(\alpha,\min(\mathrm{F}(\beta,\gamma) \setminus \alpha))$.

Proof. The proof is by induction on α,β and γ. Let

$$\lambda = \Lambda(\beta,\gamma), \gamma_1 = \min(C_\gamma \setminus \alpha) \text{ and } \alpha_1 = \min(\mathrm{F}(\beta,\gamma) \setminus \alpha). \tag{7.1.13}$$

Pick $\xi \in \{\min(C_\gamma \setminus \beta)\} \cup (C_\gamma \cap [\lambda,\beta))$ such that $\alpha_1 \in \mathrm{F}(\xi,\beta)$. Then

$$\alpha_1 = \min(\mathrm{F}(\xi,\beta) \setminus \alpha), \tag{7.1.14}$$

so applying the inductive hypothesis (b) for $\alpha \leq \xi < \beta$ we get

$$\rho_0(\alpha, \beta) = \rho_0(\alpha_1, \beta)^\frown \rho_0(\alpha, \alpha_1) \tag{7.1.15}$$

which checks (a) for $\alpha \leq \beta \leq \gamma$. Suppose first that $\lambda > \alpha$. Note that $\mathrm{F}(\lambda, \beta)$ is a subset of $\mathrm{F}(\beta, \gamma)$, and therefore,

$$\alpha_2 = \min(\mathrm{F}(\lambda, \beta) \setminus \alpha) \geq \alpha_1. \tag{7.1.16}$$

Applying the inductive hypothesis for $\alpha \leq \lambda \leq \beta$, we get

$$\rho_0(\alpha, \beta) = \rho_0(\alpha_2, \beta)^\frown \rho_0(\alpha, \alpha_2), \tag{7.1.17}$$

$$\rho_0(\alpha, \lambda) = \rho_0(\alpha_2, \lambda)^\frown \rho_0(\alpha, \alpha_2). \tag{7.1.18}$$

Combining (7.1.15) and (7.1.17), we infer that $\rho_0(\alpha, \alpha_1)$ is a tail of $\rho_0(\alpha, \alpha_2)$, so (7.1.18) can be written as

$$\rho_0(\alpha, \lambda) = \rho_0(\alpha_2, \lambda)^\frown \rho_0(\alpha_1, \alpha_2)^\frown \rho_0(\alpha, \alpha_1) = \rho_0(\alpha_1, \lambda)^\frown \rho_0(\alpha, \alpha_1). \tag{7.1.19}$$

Since λ is a limit point of C_γ, we have $\rho_0(\alpha, \gamma) = \rho_0(\alpha, \lambda)$, so (7.1.19) gives us the conclusion (b) of Lemma 7.1.9.

Consider now the case $\lambda \leq \alpha$. Then γ_1 is either equal to $\min(C_\gamma \setminus \beta)$ or it belongs to $[\lambda, \beta) \cap C_\gamma$. In any case we have that $\mathrm{F}\{\gamma_1, \beta\} \subseteq \mathrm{F}(\beta, \gamma)$, so

$$\gamma_2 = \min(\mathrm{F}\{\gamma_1, \beta\} \setminus \alpha) \geq \alpha_1. \tag{7.1.20}$$

Applying the inductive hypothesis for α, γ_1 and β we get

$$\rho_0(\alpha, \beta) = \rho_0(\gamma_2, \beta)^\frown \rho_0(\alpha, \gamma_2), \tag{7.1.21}$$

$$\rho_0(\alpha, \gamma_1) = \rho_0(\gamma_2, \gamma_1)^\frown \rho_0(\alpha, \gamma_2). \tag{7.1.22}$$

Combining (7.1.15) and (7.1.21) we see that $\rho_0(\alpha, \alpha_1)$ is a tail of $\rho_0(\alpha, \gamma_2)$, so (7.1.22) can be rewritten as

$$\rho_0(\alpha, \gamma_1) = \rho_0(\gamma_2, \gamma_1)^\frown \rho_0(\alpha_1, \gamma_2)^\frown \rho_0(\alpha, \alpha_1) = \rho_0(\alpha_1, \gamma_1)^\frown \rho_0(\alpha, \alpha_1). \tag{7.1.23}$$

Since $\rho_0(\alpha, \gamma) = \mathrm{tp}(C_\gamma \cap \alpha)^\frown \rho_0(\alpha, \gamma_1)$, (7.1.23) gives us the conclusion (b) of Lemma 7.1.9. This finishes the proof. □

Recall the function $\rho_2 : [\theta]^2 \longrightarrow \omega$ which counts the number of steps in the walk along the fixed C-sequence C_α $(\alpha < \theta)$ which in this section is assumed to be moreover a square sequence:

$$\rho_2(\alpha, \beta) = \rho_2(\alpha, \min(C_\beta \setminus \alpha)) + 1,$$

where we let $\rho_2(\gamma, \gamma) = 0$ for all γ. Thus $\rho_2(\alpha, \beta) + 1$ is simply equal to the cardinality of the trace $\mathrm{Tr}(\alpha, \beta)$ of the minimal walk from β to α.

Lemma 7.1.10. $\sup_{\xi<\alpha}|\rho_2(\xi,\alpha)-\rho_2(\xi,\beta)|<\infty$ *for all* $\alpha<\beta<\theta$.

Proof. By Lemma 7.1.9,

$$\sup_{\xi<\alpha}|\rho_2(\xi,\alpha)-\rho_2(\xi,\beta)|\leq \sup_{\xi\in F(\alpha,\beta)}|\rho_2(\xi,\alpha)-\rho_2(\xi,\beta)|.$$ □

Definition 7.1.11. Set $\mathcal{I}$ to be the set of all countable $\Gamma\subseteq\theta$ such that $\sup_{\xi\in\Delta}\rho_2(\xi,\alpha)=\infty$ for all $\alpha<\theta$ and infinite $\Delta\subseteq\Gamma\cap\alpha$.

Lemma 7.1.12. *$\mathcal{I}$ is a P-ideal of countable subsets of θ.*

Proof. Let Γ_n $(n<\omega)$ be a given sequence of members of $\mathcal{I}$ and fix $\beta<\theta$ such that $\Gamma_n\subseteq\beta$ for all n. For $n<\omega$, set $\Gamma_n^*=\{\xi\in\Gamma_n:\rho_2(\xi,\beta)\geq n\}$. Since Γ_n belongs to $\mathcal{I}$, Γ_n^* is a cofinite subset of Γ_n. Let $\Gamma_\infty=\bigcup_{n<\omega}\Gamma_n^*$. Then Γ_∞ is a member of $\mathcal{I}$ such that $\Gamma_n\setminus\Gamma_\infty$ is finite for all n. □

Theorem 7.1.13. *The P-ideal dichotomy implies that a nontrivial square sequence can exist only on $\theta=\omega_1$.*

Proof. Applying the P-ideal dichotomy to the ideal $\mathcal{I}$ of Definition 7.1.11, we get the following two alternatives (see Definition 2.4.11):

$$\text{there is an uncountable } \Delta\subseteq\theta \text{ such that } [\Delta]^\omega\subseteq\mathcal{I}, \text{ or} \tag{7.1.24}$$

$$\text{there is a partition } \theta=\bigcup_{n<\omega}\Sigma_n \text{ such that } \Sigma_n\perp\mathcal{I} \text{ for all } n. \tag{7.1.25}$$

By Lemma 7.1.10, if (7.1.24) holds, then $\Delta\cap\alpha$ must be countable for all $\alpha<\theta$ and so the cofinality of θ must be equal to ω_1. Since we are working only with regular uncountable cardinals, we see that (7.1.24) gives us that $\theta=\omega_1$ must hold. Suppose now (7.1.25) holds and pick $k<\omega$ such that Σ_k is unbounded in θ. Since $\Sigma_k\perp\mathcal{I}$ we have that $(\rho_2)_\alpha$ is bounded on $\Sigma_k\cap\alpha$ for all $\alpha<\theta$. So there is an unbounded set $\Gamma\subseteq\theta$ and an integer n such that for each $\alpha\in\Gamma$, the restriction of $(\rho_2)_\alpha$ on $\Sigma_k\cap\alpha$ is bounded by n. By Theorem 6.3.2 we conclude that the square sequence C_α $(\alpha<\theta)$ we started with must be trivial. □

Definition 7.1.14. By S_θ we denote the *sequential fan* with θ edges, i.e., the space on $(\theta\times\omega)\cup\{*\}$ with $*$ as the only non-isolated point, while a typical neighborhood of $*$ has the form $\mathcal{U}_f=\{(\alpha,n):n\geq f(\alpha)\}\cup\{*\}$ where $f:\theta\longrightarrow\omega$.

Theorem 7.1.15. *If there is a nontrivial square sequence on θ, then the square of the sequential fan S_θ has tightness* [1] *equal to θ.*

The proof will be given after a sequence of definitions and lemmas.

[1]The *tightness* of a point x in a space X is equal to θ if θ is the minimal cardinal such that, if a set $W\subseteq X\setminus\{x\}$ accumulates to x, then there is a subset of W of size $\leq\theta$ that accumulates to x.

Definition 7.1.16. Given a square sequence C_α ($\alpha < \theta$) and the corresponding function $\rho_2 : [\theta]^2 \longrightarrow \omega$ that counts the number of steps of the minimal walks, we define $d : [\theta]^2 \longrightarrow \omega$ by letting

$$d(\alpha, \beta) = \sup_{\xi \leq \alpha} |\rho_2(\xi, \alpha) - \rho_2(\xi, \beta)|.$$

Lemma 7.1.17. *For all $\alpha \leq \beta \leq \gamma < \theta$,*

(a) $d(\alpha, \beta) \geq \rho_2(\alpha, \beta)$,
(b) $d(\alpha, \gamma) \leq d(\alpha, \beta) + d(\beta, \gamma)$,
(c) $d(\alpha, \beta) \leq d(\alpha, \gamma) + d(\beta, \gamma)$.

Proof. The conclusion (a) follows from the fact that we allow $\xi = \alpha$ in the definition of $d(\alpha, \beta)$. The conclusions (b) and (c) are consequences of the triangle inequalities of the ℓ_∞-norm and the fact that in both inequalities we have that the domain of functions on the left-hand side is included in the domain of functions on the right-hand side. □

Definition 7.1.18. For $\gamma \leq \theta$, let

$$W_\gamma = \{((\alpha, d(\alpha, \beta)), (\beta, d(\alpha, \beta))) : \alpha < \beta < \gamma\}.$$

The following lemma establishes that the tightness of the point $(*, *)$ of S_θ^2 is equal to θ, giving us the proof of Theorem 7.1.15.

Lemma 7.1.19. $(*, *) \in \bar{W}_\theta$ *but* $(*, *) \notin \bar{W}_\gamma$ *for all* $\gamma < \theta$.

Proof. To see that W_θ accumulates to $(*, *)$, let $\mathcal{U}_f^2$ be a given neighborhood of $(*, *)$. Fix an unbounded set $\Gamma \subseteq \theta$ on which f is constant. By Theorem 6.3.2 and Lemma 7.1.17(a) there exist $\alpha < \beta$ in Γ such that $d(\alpha, \beta) \geq f(\alpha) = f(\beta)$. Then $((\alpha, d(\alpha, \beta)), (\beta, d(\alpha, \beta)))$ belongs to the intersection $W_\theta \cap \mathcal{U}_f^2$. To see that for a given $\gamma < \theta$ the set W_γ does not accumulate to $(*, *)$, choose $g : \theta \longrightarrow \omega$ such that

$$g(\alpha) = 2d(\alpha, \gamma) + 1 \text{ for } \alpha < \gamma.$$

Suppose $W_\gamma \cap \mathcal{U}_g^2$ is nonempty and choose $((\alpha, d(\alpha, \beta)), (\beta, d(\alpha, \beta)))$ from this set. Then

$$d(\alpha, \beta) \geq 2d(\alpha, \gamma) + 1 \text{ and } d(\alpha, \beta) \geq 2d(\beta, \gamma) + 1. \tag{7.1.26}$$

It follows that $d(\alpha, \beta) \geq d(\alpha, \gamma) + d(\beta, \gamma) + 1$ contradicting Lemma 7.1.17(c). □

Since $\theta = \omega_1$ admits a nontrivial square sequence, Theorem 7.1.15 leads to the following result of Gruenhage and Tanaka [42].

Corollary 7.1.20. *The square of the sequential fan with ω_1 edges is not countably tight.* □

Question 7.1.21. What is the tightness of the square of the sequential fan with ω_2 edges?

Corollary 7.1.22. *If a regular uncountable cardinal θ is not weakly compact in the constructible subuniverse, then the square of the sequential fan with θ edges has tightness equal to θ.* □

The way we have proved that S^2_θ has tightness equal to θ is of independent interest. We have found a point $(*, *)$ of S^2_θ and a set W_θ of isolated points which accumulates to $(*, *)$ but has no other accumulation points and moreover, no subset of W_θ of size $< \theta$ has accumulation points at all. This is interesting in view of the following result of Dow and Watson [24], where $\mathfrak{C}$ denotes the category of spaces that contain the converging sequence $\omega + 1$ and is closed under finite products, quotients and arbitrary topological sums.

Theorem 7.1.23. *If for every regular cardinal θ there is a space X in $\mathfrak{C}$ which has a set of isolated points of size θ witnessing tightness θ of its unique accumulation point in X, then every space belongs to $\mathfrak{C}$.* □

Theorem 7.1.24. *If every regular uncountable cardinal supports a nontrivial square sequence, then every topological space X can be obtained from the converging sequence after finitely many steps of taking (finite) products, quotients, and arbitrary topological sums.*

Proof. This follows from combining (the proof of) Theorem 7.1.15 and Theorem 7.1.23. □

If κ is a strongly compact cardinal, then the tightness of the product of every pair of spaces of tightness $< \kappa$ is $< \kappa$ and so $\mathfrak{C}$ contains no space of tightness κ or larger. It has been shown in [24] that if the Lebesgue measure extends to a countably additive measure on all sets of reals, then there even is a countable space which does not belong to $\mathfrak{C}$. On the other hand, by Theorem 7.1.5 and Theorem 7.1.24, if no regular cardinal above ω_1 is weakly compact in the constructible subuniverse, then every space can be generated by the converging sequence by taking finite products, quotients and arbitrary topological sums.

Remark 7.1.25. We have seen that the case $\theta = \omega_1$ is quite special when one considers the problem of existence of various nontrivial square sequences on θ. It should be noted that a similar result about the problem of the tightness of S^2_θ is not available. In particular, it is not known whether the P-ideal dichotomy or a similar consistent hypothesis of set theory implies that the tightness of the square of S_{ω_2} is smaller than ω_2. It is interesting that considerably more is known about the dual question, the question of initial compactness of the Tychonoff cube $\mathbb{N}^\theta$. For example, if one defines $B_{\alpha\beta} = \{f \in \mathbb{N}^\theta : f(\alpha), f(\beta) \leq d(\alpha, \beta)\}$ $(\alpha < \beta < \theta)$ one gets an open cover of $\mathbb{N}^\theta$ without a subcover of size $< \theta$. However, for small θ such as $\theta = \omega_2$ one is able to find such a cover of $\mathbb{N}^\theta$ without any additional set-theoretic assumptions and in particular without the assumption that θ carries a nontrivial square sequence.

7.2 Square sequences and local versions of ρ

In this section, we give the analysis of a family of ρ-functions associated with a (nontrivial) square sequence C_α $(\alpha < \theta)$ that lives on some regular uncountable cardinal θ. Recall that an ordinal α *divides* an ordinal γ if there is β such that $\gamma = \alpha\cdot\beta$, i.e., γ can be written as the union of an increasing β-sequence of intervals of type α. Let $\kappa < \theta$ be a fixed infinite cardinal. Let $\Lambda_\kappa : [\theta]^2 \longrightarrow \theta$ be defined by

$$\Lambda_\kappa(\alpha,\beta) = \max\{\xi \in C_\beta \cap (\alpha+1) : \kappa \text{ divides } \mathrm{tp}(C_\beta \cap \xi)\}, \tag{7.2.1}$$

where we stipulate that $\max\emptyset = 0$ and that κ divides the ordinal 0. This translates into the requirements that

$$\Lambda_\kappa(\alpha,\beta) = 0 \text{ if } C_\beta \cap \alpha = \emptyset \text{ and } \Lambda_\kappa(\alpha,\beta) = \min(C_\beta) \text{ if } \mathrm{tp}(C_\beta) \leq \kappa. \tag{7.2.2}$$

Finally we are ready to define the main object of study in this section:

$$\rho^\kappa : [\theta]^2 \longrightarrow \kappa \tag{7.2.3}$$

defined recursively by

$$\begin{aligned}\rho^\kappa(\alpha,\beta) \quad = \quad & \sup\{\mathrm{tp}(C_\beta \cap [\Lambda_\kappa(\alpha,\beta),\alpha)), \rho^\kappa(\alpha, \min(C_\beta \setminus \alpha)), \rho^\kappa(\xi,\alpha) : \\ & \qquad \xi \in C_\beta \cap [\Lambda_\kappa(\alpha,\beta),\alpha)\},\end{aligned}$$

with the boundary condition $\rho^\kappa(\alpha,\alpha) = 0$ for all α.

The following consequence of the coherence property of C_α $(\alpha < \theta)$ will be quite useful.

Lemma 7.2.1. *If α is a limit point of C_β, then $\rho^\kappa(\xi,\alpha)=\rho^\kappa(\xi,\beta)$ for all $\xi < \alpha$.* □

Note that ρ^κ is something that corresponds to the function $\rho : [\omega_1]^2 \longrightarrow \omega$ considered in Definition 3.1.1 (see also Section 7.3) and that the ρ^κ's are simply *local versions* of the key definition. We shall show that they all have the crucial subadditivity properties.

Lemma 7.2.2. *If $\alpha < \beta < \gamma < \theta$, then*

(a) $\rho^\kappa(\alpha,\gamma) \leq \max\{\rho^\kappa(\alpha,\beta), \rho^\kappa(\beta,\gamma)\}$,

(b) $\rho^\kappa(\alpha,\beta) \leq \max\{\rho^\kappa(\alpha,\gamma), \rho^\kappa(\beta,\gamma)\}$.

Proof. The proof is by induction on α, β and γ. We first prove the inductive step for (a). Let $\nu = \max\{\rho^\kappa(\alpha,\beta), \rho^\kappa(\beta,\gamma)\}$. We need to show that $\rho^\kappa(\alpha,\gamma) \leq \nu$. To this end, let

$$\gamma_\alpha = \min(C_\gamma \setminus \alpha) \text{ and } \gamma_\beta = \min(C_\gamma \setminus \beta). \tag{7.2.4}$$

Case 1^a: $\alpha < \Lambda_\kappa(\beta,\gamma)$. Then by Lemma 7.2.1, $\rho^\kappa(\alpha,\gamma) = \rho^\kappa(\alpha,\Lambda_\kappa(\beta,\gamma))$. By the definition of $\rho^\kappa(\beta,\gamma)$ we have

$$\rho^\kappa(\Lambda_\kappa(\beta,\gamma),\beta) \leq \rho^\kappa(\beta,\gamma) \leq \nu. \tag{7.2.5}$$

Applying the inductive hypothesis (b) for the triple $\alpha < \Lambda_\kappa(\beta,\gamma) \leq \beta$ we conclude that $\rho^\kappa(\alpha, \Lambda_\kappa(\beta,\gamma)) \leq \nu$.

Case 2^a**:** $\alpha \geq \Lambda_\kappa(\beta,\gamma)$. Then $\Lambda_\kappa(\beta,\gamma) = \Lambda_\kappa(\alpha,\gamma) = \lambda$.

Subcase $2^a.1$**:** $\gamma_\alpha = \gamma_\beta = \bar{\gamma}$. By the inductive hypothesis (a) for the triple $\alpha < \beta \leq \bar{\gamma}$ we get

$$\rho^\kappa(\alpha,\bar{\gamma}) \leq \max\{\rho^\kappa(\alpha,\beta), \rho^\kappa(\beta,\bar{\gamma})\} \leq \nu, \tag{7.2.6}$$

since $\rho^\kappa(\beta,\bar{\gamma}) \leq \rho^\kappa(\beta,\gamma)$. Note that $[\lambda,\alpha) \cap C_\gamma$ is an initial part of $[\lambda,\beta) \cap C_\gamma$ so its order-type is bounded by $\rho^\kappa(\beta,\gamma)$ and therefore by ν. Consider a $\xi \in [\lambda,\alpha) \cap C_\gamma$. Applying the inductive hypothesis (b) for the triple $\xi < \alpha < \beta$ we conclude that

$$\rho^\kappa(\xi,\alpha) \leq \max\{\rho^\kappa(\xi,\beta), \rho^\kappa(\alpha,\beta)\} \leq \nu, \tag{7.2.7}$$

since $\rho^\kappa(\xi,\beta) \leq \rho^\kappa(\beta,\gamma)$ by the definition of $\rho^\kappa(\beta,\gamma)$. This shows that all ordinals appearing in the formula for $\rho^\kappa(\alpha,\gamma)$ are $\leq \nu$ and so $\rho^\kappa(\alpha,\gamma) \leq \nu$.

Subcase $2^a.2$**:** $\gamma_\alpha < \gamma_\beta$. Then γ_α belongs to $C_\gamma \cap [\lambda,\beta)$ and therefore $\rho^\kappa(\gamma_\alpha,\beta) \leq \rho^\kappa(\beta,\gamma) \leq \nu$. Applying the inductive hypothesis (b) for $\alpha \leq \gamma_\alpha < \beta$ gives us

$$\rho^\kappa(\alpha,\gamma_\alpha) \leq \max\{\rho^\kappa(\alpha,\beta), \rho^\kappa(\gamma_\alpha,\beta)\} \leq \nu. \tag{7.2.8}$$

Given $\xi \in C_\gamma \cap [\lambda,\alpha)$ using the inductive hypothesis (b) for $\xi < \alpha < \beta$ we get that

$$\rho^\kappa(\xi,\alpha) \leq \max\{\rho^\kappa(\xi,\beta), \rho^\kappa(\alpha,\beta)\} \leq \nu, \tag{7.2.9}$$

since $\rho^\kappa(\xi,\beta) \leq \rho^\kappa(\beta,\gamma)$. Since $\mathrm{tp}(C_\gamma \cap [\lambda,\alpha)) \leq \mathrm{tp}(C_\gamma \cap [\lambda,\beta)) \leq \rho^\kappa(\beta,\gamma) \leq \nu$ we see again that all the ordinals appearing in the formula for $\rho^\kappa(\alpha,\gamma)$ are $\leq \nu$.

Let us now concentrate on the inductive step for (b). This time let $\nu = \max\{\rho^\kappa(\alpha,\gamma), \rho^\kappa(\beta,\gamma)\}$ and let γ_α and γ_β be as in (7.2.4). We need to show that $\rho^\kappa(\alpha,\beta) \leq \nu$.

Case 1^b**:** $\alpha < \Lambda_\kappa(\beta,\gamma)$. By Lemma 7.2.1,

$$\rho^\kappa(\alpha, \Lambda_\kappa(\beta,\gamma)) = \rho^\kappa(\alpha,\gamma) \leq \nu. \tag{7.2.10}$$

From the formula for $\rho^\kappa(\beta,\gamma)$ we see that $\rho^\kappa(\Lambda_\kappa(\beta,\gamma),\beta) \leq \rho^\kappa(\beta,\gamma) \leq \nu$. Applying the inductive hypothesis (a) for the triple $\alpha < \Lambda_\kappa(\beta,\gamma) \leq \beta$ we get

$$\rho^\kappa(\alpha,\beta) \leq \max\{\rho^\kappa(\alpha,\Lambda_\kappa(\beta,\gamma)), \rho^\kappa(\Lambda_\kappa(\beta,\gamma),\beta)\} \leq \nu. \tag{7.2.11}$$

Case 2^b**:** $\alpha \geq \Lambda_\kappa(\beta,\gamma)$.

Subcase $2^b.1$**:** $\gamma_\alpha = \gamma_\beta = \bar{\gamma}$. Then $\rho^\kappa(\alpha,\bar{\gamma}) \leq \rho^\kappa(\alpha,\gamma) \leq \nu$ and $\rho^\kappa(\beta,\bar{\gamma}) \leq \rho^\kappa(\beta,\gamma) \leq \nu$. Applying the inductive hypothesis (b) for $\alpha < \beta < \bar{\gamma}$ we get

$$\rho^\kappa(\alpha,\beta) \leq \max\{\rho^\kappa(\alpha,\bar{\gamma}), \rho^\kappa(\beta,\bar{\gamma})\} \leq \nu. \tag{7.2.12}$$

Subcase 2^b.2: $\gamma_\alpha < \gamma_\beta$. Then $\gamma_\alpha \in C_\gamma \cap [\lambda, \beta)$ and so $\rho^\kappa(\gamma_\alpha, \beta)$ appears in the formula for $\rho^\kappa(\beta, \gamma)$ and so we conclude that $\rho^\kappa(\gamma_\alpha, \beta) \leq \rho^\kappa(\beta, \gamma) \leq \nu$. Similarly, $\rho^\kappa(\alpha, \gamma_\alpha) \leq \rho^\kappa(\alpha, \gamma) \leq \nu$. Applying the inductive hypothesis (a) for the triple $\alpha \leq \gamma_\alpha < \beta$ we get

$$\rho^\kappa(\alpha, \beta) \leq \max\{\rho^\kappa(\alpha, \gamma_\alpha), \rho^\kappa(\gamma_\alpha, \beta)\} \leq \nu. \tag{7.2.13}$$

This finishes the proof. □

The following is an immediate consequence of the fact that the definition of ρ^κ is closely tied to the notion of a walk along the fixed square sequence.

Lemma 7.2.3. *$\rho^\kappa(\alpha, \gamma) \geq \rho^\kappa(\alpha, \beta)$ whenever $\alpha \leq \beta \leq \gamma$ and β belongs to the trace of the walk from γ to α.* □

Lemma 7.2.4. *Suppose $\beta \leq \gamma < \theta$ and that β is a limit ordinal > 0. Then $\rho^\kappa(\alpha, \gamma) \geq \rho^\kappa(\alpha, \beta)$ for coboundedly many $\alpha < \beta$.*

Proof. Let $\gamma = \gamma_0 > \gamma_1 > \cdots > \gamma_{n-1} > \gamma_n = \beta$ be the trace of the walk from γ to β. Let $\bar{\gamma} = \gamma_{n-1}$ if β is a limit point of $C_{\gamma_{n-1}}$, otherwise let $\bar{\gamma} = \beta$. Note that by Lemma 7.2.1, in any case we have that

$$\rho^\kappa(\alpha, \beta) = \rho^\kappa(\alpha, \bar{\gamma}) \text{ for all } \alpha < \beta. \tag{7.2.14}$$

Let $\bar{\beta} < \beta$ be an upper bound of all $C_{\gamma_i} \cap \beta$ $(i < n)$ which are bounded in β. Then $\bar{\gamma}$ is a member of the trace of any walk from γ to some ordinal α in the interval $[\bar{\beta}, \beta)$. Applying Lemma 7.2.3 to this fact gives us

$$\rho^\kappa(\alpha, \gamma) \geq \rho^\kappa(\alpha, \bar{\gamma}) \text{ for all } \alpha \in [\bar{\beta}, \beta). \tag{7.2.15}$$

Since $\rho^\kappa(\alpha, \bar{\gamma}) = \rho^\kappa(\alpha, \beta)$ for all $\alpha < \beta$ (see (7.2.14)), this gives us the conclusion of the lemma. □

Lemma 7.2.5. *The set $P_\nu^\kappa(\beta) = \{\xi < \beta : \rho^\kappa(\xi, \beta) \leq \nu\}$ is a closed subset of β for every $\beta < \theta$ and $\nu < \kappa$.*

Proof. The proof is by induction on β. So let $\alpha < \beta$ be a limit point of $P_\nu^\kappa(\beta)$. Let $\beta_1 = \min(C_\beta \setminus \alpha)$ and let $\lambda = \Lambda_\kappa(\alpha, \beta)$. If $\lambda = \alpha$, then $\rho^\kappa(\alpha, \beta) = 0$ and therefore $\alpha \in P_\nu^\kappa(\beta)$. So we may assume that $\lambda < \alpha$. Then

$$\Lambda_\kappa(\xi, \beta) = \lambda \text{ for every } \xi \in [\lambda, \alpha). \tag{7.2.16}$$

Case 1: $\bar{\alpha} = \sup(C_\beta \cap \alpha) < \alpha$. Then $\beta_1 = \min(C_\beta \setminus \xi)$ for every $\xi \in P_\nu^\kappa(\beta) \cap (\bar{\alpha}, \alpha)$. It follows that

$$P_\nu^\kappa(\beta) \cap (\bar{\alpha}, \alpha) \subseteq P_\nu^\kappa(\beta_1), \tag{7.2.17}$$

so α is also a limit point of $P_\nu^\kappa(\beta_1)$. By the inductive hypothesis $\alpha \in P_\nu^\kappa(\beta_1)$, i.e., $\rho^\kappa(\alpha, \beta_1) \leq \nu$. Since $\Lambda_\kappa(\xi, \beta) = \lambda$ for all $\xi \in [\lambda, \alpha)$, we get

$$\mathrm{tp}(C_\beta \cap [\lambda, \alpha)) = \sup\{\mathrm{tp}(C_\beta \cap [\lambda, \xi)) : \xi \in P_\nu^\kappa(\beta) \cap [\lambda, \alpha)\} \leq \nu. \tag{7.2.18}$$

Consider $\eta \in C_\beta \cap [\lambda, \alpha)$. Then $\eta \in C_\beta \cap [\lambda, \xi)$ for some $\xi \in P^\kappa_\nu(\beta) \cap [\bar{\alpha}, \alpha)$ so by the definition of $\rho^\kappa(\xi, \beta)$ we conclude that $\rho^\kappa(\eta, \xi) \leq \rho^\kappa(\xi, \beta) \leq \nu$. By (7.2.17) we also have $\rho^\kappa(\xi, \beta_1) \leq \nu$ so by Lemma 7.2.2(a), $\rho^\kappa(\eta, \beta_1) \leq \nu$. Applying Lemma 7.2.2(b) to $\eta < \alpha < \beta_1$, we conclude that

$$\rho^\kappa(\eta, \alpha) \leq \max\{\rho^\kappa(\eta, \beta_1), \rho^\kappa(\alpha, \beta_1)\} \leq \nu. \tag{7.2.19}$$

This shows that all the ordinals appearing in the formula for $\rho^\kappa(\alpha, \beta)$ are $\leq \nu$ and therefore that $\rho^\kappa(\alpha, \beta) \leq \nu$.

Case 2: α is a limit point of C_β. As in the previous case, to show that $\rho^\kappa(\alpha, \beta) \leq \nu$ we need to have $\mathrm{tp}(C_\beta \cap [\lambda, \alpha))$, $\rho^\kappa(\alpha, \beta_1)$ and $\rho^\kappa(\eta, \alpha)$ with $\eta \in C_\beta \cap [\lambda, \alpha)$ all $\leq \nu$. Note that $\beta_1 = \alpha$ so $\rho^\kappa(\alpha, \beta_1) = 0$. Similarly as in the first case we conclude $\mathrm{tp}(C_\beta \cap [\lambda, \alpha)) \leq \nu$. By Lemma 7.2.1,

$$\rho^\kappa(\xi, \alpha) = \rho^\kappa(\xi, \beta) \leq \nu \text{ for all } \xi \in P^\kappa_\nu(\beta) \cap \alpha. \tag{7.2.20}$$

Consider an $\eta \in C_\beta \cap [\lambda, \alpha)$. Since α is a limit point of $P^\kappa_\nu(\beta)$, there is ξ from this set above η. By (7.2.16) we have that $\eta \in C_\beta \cap [\Lambda_\kappa(\xi, \beta), \xi)$ so $\rho^\kappa(\eta, \xi) \leq \rho^\kappa(\xi, \beta) \leq \nu$. Applying 7.2.2(a) to $\eta < \xi < \alpha$ we get

$$\rho^\kappa(\eta, \alpha) \leq \max\{\rho^\kappa(\eta, \xi), \rho^\kappa(\xi, \alpha)\} \leq \nu. \tag{7.2.21}$$

This finishes the proof. □

For $\alpha < \beta < \theta$ and $\nu < \kappa$, set

$$\alpha <^\kappa_\nu \beta \text{ if and only if } \rho^\kappa(\alpha, \beta) \leq \nu. \tag{7.2.22}$$

Lemma 7.2.6.

(1) $<^\kappa_\nu$ *is a tree ordering on* θ,

(2) $<^\kappa_\nu \subseteq <^\kappa_\mu$ *whenever* $\nu < \mu < \kappa$,

(3) $\in \restriction \theta = \bigcup_{\nu < \kappa} <^\kappa_\nu$.

Proof. This follows immediately from Lemma 7.2.2. □

Recall the notion of a special tree of height θ from Section 6.1, a tree T for which one can find a T-regressive map $f : T \longrightarrow T$ with the property that the preimage of any point is the union of $< \theta$ antichains. By a *tree on* θ we mean a tree of the form $(\theta, <_T)$ with the property that $\alpha <_T \beta$ implies $\alpha < \beta$.

Lemma 7.2.7. *If a tree* T *on* θ *is special, then there is an ordinal-regressive map* $f : \theta \longrightarrow \theta$ *and a closed and unbounded set* $C \subseteq \theta$ *such that* f *is one-to-one on all chains separated by* C.

Proof. Let $g : \theta \longrightarrow \theta$ be a T-regressive map such that for each $\xi < \theta$ the preimage $g^{-1}(\xi)$ can be written as a union of a sequence $A_\delta(\xi)$ $(\delta < \lambda_\xi)$ of antichains, where $\lambda_\xi < \theta$. Let C be the collection of all limits $\alpha < \theta$ with the property that $\lambda_\xi < \alpha$ for all $\xi < \alpha$. Let $\ulcorner \cdot, \cdot \urcorner$ denote the Gödel pairing function which in particular has the property that $\ulcorner \alpha, \beta \urcorner < \delta$ for $\delta \in C$ and α, β, δ. Define ordinal-regressive map $f : \theta \longrightarrow \theta$ by letting $f(\alpha) = \max(C \cup \alpha)$ if $\alpha \notin C$ and for $\alpha \in C$, letting $f(\alpha) = \ulcorner g(\alpha), \gamma \urcorner$ where $\gamma < \lambda_{g(\alpha)} (< \alpha)$ is minimal ordinal with the property that $\alpha \in A_\gamma(g(\alpha))$. Then f is as required. □

By Lemma 7.2.6 we have a sequence $<^\kappa_\nu$ $(\nu < \kappa)$ of tree-orderings on θ. The following lemma tells us that they are frequently quite large orderings.

Lemma 7.2.8. *If $\theta > \kappa$ is not a successor of a cardinal of cofinality κ, then there must be $\nu < \kappa$ such that $(\theta, <^\kappa_\nu)$ is a nonspecial tree on θ.*

Proof. Suppose to the contrary that all trees are special. By Lemma 7.2.7 we may choose ordinal-regressive maps $f_\nu : \theta \longrightarrow \theta$ for all $\nu < \kappa$ and a single closed and unbounded set $C \subseteq \theta$ such that each of the maps f_ν is one-to-one on $<^\kappa_\nu$-chains separated by C. Using the Pressing Down Lemma, we find a stationary set Γ of cofinality κ^+ ordinals $< \theta$ and $\lambda < \theta$ such that $f_\nu(\gamma) < \lambda$ for all $\gamma \in \Gamma$ and $\nu < \kappa$. If $|\lambda|^+ < \theta$, let $\Delta = \lambda$, $\Gamma = \Gamma_0$ and if $|\lambda|^+ = \theta$, represent λ as the increasing union of a sequence Δ_ξ $(\xi < \mathrm{cf}(|\lambda|))$ of sets of size $< |\lambda|$. Since $\kappa \neq \mathrm{cf}(|\lambda|)$ there is a $\bar{\xi} < \mathrm{cf}(|\lambda|)$ and a stationary $\Gamma_0 \subseteq \Gamma$ such that for all $\gamma \in \Gamma_0$, $f_\nu(\gamma) \in \Delta_{\bar{\xi}}$ for κ many $\nu < \kappa$. Let $\Delta = \Delta_{\bar{\xi}}$. This gives us subsets Δ and Γ_0 of θ such that

$$|\Delta|^+ < \theta \text{ and } \Gamma_0 \text{ is stationary in } \theta, \tag{7.2.23}$$

$$\Sigma_\gamma = \{\nu < \kappa : f_\nu(\gamma) \in \Delta\} \text{ is unbounded in } \kappa \text{ for all } \gamma \in \Gamma_0. \tag{7.2.24}$$

Let $\bar{\theta} = \kappa^+ \cdot |\Delta|^+$. Then $\bar{\theta} < \theta$ and so we can find $\beta \in \Gamma_0$ such that $\Gamma_0 \cap C \cap \beta$ has size $\bar{\theta}$. Then there will be $\nu_0 < \kappa$ and $\Gamma_1 \subseteq \Gamma_0 \cap C \cap \beta$ of size $\bar{\theta}$ such that $\rho^\kappa(\alpha, \beta) \leq \nu_0$ for all $\alpha \in \Gamma_1$. By (7.2.24) we can find $\Gamma_2 \subseteq \Gamma_1$ of size $\bar{\theta}$ and $\nu_1 \geq \nu_0$ such that $f_{\nu_1}(\alpha) \in \Delta$ for all $\alpha \in \Gamma_2$. Note that Γ_2 is a $<^\kappa_{\nu_1}$-chain separated by C, so f_{ν_1} is one-to-one on Γ_2. However, this gives us the desired contradiction since the set Δ, in which f_{ν_1} embeds Γ_2 has size smaller than the size of Γ_2. This finishes the proof. □

It is now natural to ask the following question: under which assumption on the square sequence C_α $(\alpha < \theta)$ can we conclude that neither of the trees $(\theta, <^\kappa_\nu)$ will have a branch of size θ?

Lemma 7.2.9. *If the set $\Gamma_\kappa = \{\alpha < \theta : \mathrm{tp}(C_\alpha) = \kappa\}$ is stationary in θ, then none of the trees $(\theta, <^\kappa_\nu)$ has a branch of size θ.*

Proof. Assume that B is a $<^\kappa_\nu$-branch of size θ. By Lemma 7.2.5, B is a closed and unbounded subset of θ. Pick a limit point β of B which belongs to Γ_κ. Pick $\alpha \in B \cap \beta$ such that $\mathrm{tp}(C_\beta \cap \alpha) > \nu$. By definition of $\rho^\kappa(\alpha, \beta)$ we have that

$\rho^\kappa(\alpha,\beta) \geq \text{tp}(C_\beta \cap \alpha) > \nu$ since clearly $\Lambda_\kappa(\alpha,\beta) = 0$. This contradicts the fact that $\alpha <_\nu^\kappa \beta$ and finishes the proof. □

Definition 7.2.10. A square sequence on θ is *special* if the corresponding tree $(\theta, <^2)$ is special, i.e., there is a $<^2$-regressive map $f : \theta \longrightarrow \theta$ with the property that the f-preimage of every $\xi < \theta$ is the union of $< \theta$ antichains of $(\theta, <^2)$.

Theorem 7.2.11. *Suppose $\kappa < \theta$ are regular cardinals such that θ is not a successor of a cardinal of cofinality κ. Then to every square sequence C_α $(\alpha < \theta)$ for which there exist stationarily many α such that* $\text{tp}\, C_\alpha = \kappa$, *one can associate a sequence $C_{\alpha\nu}$ $(\alpha < \theta, \nu < \kappa)$ such that:*

(i) *$C_{\alpha\nu} \subseteq C_{\alpha\mu}$ for all α and $\nu \leq \mu$,*

(ii) *$\alpha = \bigcup_{\nu<\kappa} C_{\alpha\nu}$ for all limit α,*

(iii) *$C_{\alpha\nu}$ $(\alpha < \theta)$ is a nonspecial (and nontrivial) square sequence on θ for all $\nu < \kappa$.*

Proof. Fix $\nu < \kappa$ and define $C_{\alpha\nu}$ by induction on $\alpha < \theta$. So suppose β is a limit ordinal $< \theta$ and that $C_{\alpha\nu}$ is defined for all $\alpha < \beta$. If $P_\nu^\kappa(\beta)$ is bounded in β, let $\bar\beta$ be the maximal limit point of $P_\nu^\kappa(\beta)$ ($\bar\beta = 0$ if the set has no limit points) and let

$$C_{\beta\nu} = C_{\bar\beta\nu} \cup P_\nu^\kappa(\beta) \cup (C_\beta \cap [\max(P_\nu^\kappa(\beta)), \beta)). \tag{7.2.25}$$

If $P_\nu^\kappa(\beta)$ is unbounded in β, let

$$C_{\beta\nu} = P_\nu^\kappa(\beta) \cup \bigcup\{C_{\alpha\nu} : \alpha \in P_\nu^\kappa(\beta)\ \&\ \alpha = \sup(P_\nu^\kappa(\beta) \cap \alpha)\}. \tag{7.2.26}$$

By Lemmas 7.2.2 and 7.2.5, $C_{\beta\nu}$ $(\beta < \theta)$ is well defined and it forms a square sequence on θ. The properties (i) and (ii) are also immediate. To see that for each $\nu < \kappa$ the sequence $C_{\beta\nu}$ $(\beta < \theta)$ is nontrivial, one uses Lemma 7.2.9 and the fact that if α is a limit point of $C_{\beta\nu}$ occupying a place in $C_{\beta\nu}$ that is divisible by κ, then $\alpha <_\nu^\kappa \beta$. By Lemma 7.2.8, or rather its proof, we conclude that there is $\bar\nu < \kappa$ such that $C_{\beta\nu}$ $(\beta < \theta)$ is nonspecial for all $\nu \geq \bar\nu$. This finishes the proof. □

Lemma 7.2.12. *For every pair of regular cardinals $\kappa < \theta$, every special square sequence C_α $(\alpha < \theta)$ can be refined to a special square sequence $\bar C_\alpha$ $(\alpha < \theta)$ with the property that $\{\alpha < \theta : \text{tp}\, \bar C_\alpha = \kappa\}$ is stationary in θ and the sequence $\bar C_\alpha$ $(\alpha < \theta)$ avoids some stationary subset of this set.*

Proof. Let $<^2$ be the tree-ordering associated with C_α $(\alpha < \theta)$ and let $f : \theta \longrightarrow \theta$ be a $<^2$-regressive map such that $f^{-1}(\xi)$ is the union of $< \theta$ antichains of $(\theta, <^2)$ for all $\xi < \theta$.

Case 1: $\kappa = \omega$. By the Pressing Down Lemma there is a stationary set Γ of cofinality ω ordinals $< \theta$ on which f takes some constant value. Thus Γ can be decomposed into $< \theta$ antichains of $(\theta, <^2)$ and so Γ contains a stationary subset

Γ_0 of pairwise $<^2$-incomparable ordinals. For $\alpha \in \Gamma_0$ let $\bar{C}_\alpha$ be any ω-sequence of successor ordinals converging to α. If α is a limit ordinal not in Γ_0 but there is (a unique!) $\bar{\alpha} <^2 \alpha$ in Γ_0, let $\bar{C}_\alpha = C_\alpha \setminus \bar{\alpha}$. In all other cases, let $\bar{C}_\alpha = C_\alpha$. It is easily seen that $\bar{C}_\alpha$ $(\alpha < \theta)$ is as required.

Case 2: $\kappa > \omega$. By Lemma 7.2.7 we may find a closed and unbounded subset C of θ and an ordinal-regressive map $g : C \longrightarrow \theta$ which is one-to-one on $<^2$-chains included in C (and which has the property that for $\alpha \in C$ $g(\alpha)$ is the Gödel pairing of $f(\alpha)$ and the index of the antichain which contains α in the fixed antichain-decomposition of the preimage $f^{-1}(f(\alpha))$. By the argument from Case 1 there is a stationary set Γ of limit points of C consisting of cofinality κ ordinals $< \theta$ such that Γ is an antichain of $(\theta, <^2)$. Given $\alpha \in \Gamma$, let $h(\alpha)$ be the minimal ordinal $\lambda < \theta$ such that $g(\xi) < \lambda$ for unboundedly many limit points ξ of C_α that also belong to C. Since g is regressive and since κ, the cofinality of α, is uncountable, the Pressing Down Lemma gives us that $h(\alpha) < \alpha$. Choose a stationary set $\Gamma_0 \subseteq \Gamma$ such that h takes some constant value λ on Γ_0. Note that since g is one-to one on $<^2$-chains included in C, the constant value λ cannot be a successor ordinal. It follows that λ must be a limit ordinal of cofinality equal to κ. Let λ_ν $(\nu < \kappa)$ be a strictly increasing sequence which converges to λ.

We first define $\bar{C}_\alpha$ for $\alpha \in \Gamma_0$ as a strictly increasing continuous sequence $c_\nu(\alpha)$ $(\nu < \kappa)$ of limit points of C_α that also belong to C and satisfy the following requirements at each $\nu < \kappa$:

$$c_\mu(\alpha) = \sup_{\nu<\mu} c_\nu(\alpha) \text{ with limit } \mu, \tag{7.2.27}$$

$$\begin{aligned} &c_{\nu+1}(\alpha) > c_\nu(\alpha) \text{ is minimal subject to being a limit point} \\ &\text{of } C_\alpha, \text{ belonging to } C, \text{ and being} > \gamma \text{ for every limit point} \\ &\gamma \text{ of } C_\alpha \text{ such that } \gamma \in C \text{ and } g(\gamma) < \lambda_{\nu+1}. \end{aligned} \tag{7.2.28}$$

If α is a limit point of $\bar{C}_\beta$ for some $\beta \in \Gamma_0$, set $\bar{C}_\alpha = \bar{C}_\beta \cap \alpha$. Note that by the canonicity of the definition of $c_\nu(\beta)$ $(\nu < \kappa)$ for $\beta \in \Gamma_0$, we would get the same result, as $\bar{C}_\beta \cap \alpha = \bar{C}_{\bar{\beta}} \cap \alpha$ for any other $\bar{\beta} \in \Gamma_0$ for which α is a limit point of $\bar{C}_{\bar{\beta}}$.

If α is a limit ordinal which is not a limit point of any $\bar{C}_\beta$ for $\beta \in \Gamma_0$ and there is (a unique!) $\bar{\alpha} <^2 \alpha$ in Γ_0, let $\bar{C}_\alpha = C_\alpha \setminus \bar{\alpha}$.

If α is a limit ordinal such that $\alpha <^2 \bar{\alpha}$ for some $\bar{\alpha} \in \Gamma_0$ but α is not a limit point on any $\bar{C}_\beta$ for $\beta \in \Gamma_0$, let $\bar{C}_\alpha = C_\alpha \setminus \max(\bar{C}_{\bar{\alpha}} \cap \alpha)$. Note again that there is no ambiguity in this definition, as

$$\bar{C}_\beta \cap \alpha = \bar{C}_\gamma \cap \alpha$$

holds for every β and γ in Γ_0 such that $\alpha <^2 \beta$ and $\alpha <^2 \gamma$. In all other cases set $\bar{C}_\alpha = C_\alpha$. It should be clear that $\bar{C}_\alpha$ $(\alpha < \theta)$ is a nontrivial square sequence satisfying the conclusion of Lemma 7.2.12. □

Before going further let us mention an unboundedness property of the function ρ^κ whenever it is based on a square sequence satisfying the conclusion of Lemma 7.2.12

Theorem 7.2.13. *Suppose that θ is a regular cardinal such that, for some infinite cardinal $\kappa < \theta$, we base the mapping $\rho^\kappa : [\theta]^2 \to \kappa$ on a square sequence C_α $(\alpha < \theta)$ with the property that the set $S = \{\alpha < \theta : \mathrm{tp}(C_\alpha) = \kappa\}$ is stationary in θ and that the sequence C_α $(\alpha < \theta)$ avoids some stationary subset of S. Then for every $\nu < \kappa$ and every family A of θ pairwise-disjoint subsets of θ that have sizes $< \mathrm{cf}(\kappa)$, there exists $B \subseteq A$ of size θ such that for all $a \neq b$ in B and all $\alpha \in a$, $\beta \in b$ we have $\rho^\kappa(\alpha, \beta) > \nu$.*

Proof. The conclusion follows from Theorem 6.2.7 which claims the same thing for the local version ρ_1^κ of ρ_1, having in mind the obvious inequalities $\rho^\kappa(\alpha, \beta) \geq \rho_1^\kappa(\alpha, \beta)$ for $\alpha < \beta < \theta$. □

We are finally ready to state the main result of this section which follows from Theorem 7.2.11 and Lemma 7.2.12.

Theorem 7.2.14. *If regular uncountable cardinal $\theta \neq \omega_1$ carries a nontrivial square sequence, then it also carries a nontrivial square sequence which is moreover nonspecial.* □

Corollary 7.2.15. *If a regular uncountable cardinal $\theta \neq \omega_1$ is not weakly compact in the constructible subuniverse, then there is a nonspecial Aronszajn tree of height θ.*

Proof. By Theorem 7.1.5, θ carries a nontrivial square sequence C_α $(\alpha < \theta)$. By Theorem 7.2.14, we may assume that the sequence is moreover nonspecial. Let ρ_0 be the associated ρ_0-function and consider the tree $T(\rho_0)$. As in Corollary 7.1.6 we conclude that $T(\rho_0)$ is an Aronszajn tree of height θ. By Lemma 7.1.7 there is a strictly increasing map from $(\theta, <^2)$ into $T(\rho_0)$, so $T(\rho_0)$ must be nonspecial. □

Remark 7.2.16. The assumption $\theta \neq \omega_1$ in Theorem 7.2.14 is essential as there is always a nontrivial square sequence on ω_1, but it is possible to have a situation where all Aronszajn trees on ω_1 are special. For example MA_{ω_1} implies this. In [69], Laver and Shelah have shown that any model with a weakly compact cardinal admits a forcing extension satisfying CH and the statement that all Aronszajn trees on ω_2 are special. A well-known open problem in this area asks whether one can have GCH rather than CH in a model where all Aronszajn trees on ω_2 are special.

7.3 Special square sequence and the corresponding function ρ

The following well-known result of Jensen [52] supplements the corresponding result for weakly compact cardinals listed above as Theorem 7.1.5.

Theorem 7.3.1. *If a regular uncountable cardinal θ is not Mahlo in the constructible subuniverse, then there is a special square sequence on θ which is moreover constructible.* □

Today we know many more inner models with sufficient amount of fine structure necessary for building special square sequences. So the existence of special square sequences, especially at successors of strong-limit singular cardinals, is tied to the existence of some other large cardinal axioms. The reader is referred to standard textbooks in this area such as [55] for the specific information. In this section we give the combinatorial analysis of walks along special square sequences and the corresponding distance functions. Let us start by restating some results of Section 7.2.

Theorem 7.3.2. *Suppose $\kappa < \theta$ are regular cardinals and that θ carries a special square sequence. Then there exist $C_{\alpha\nu}$ $(\alpha < \theta, \nu < \kappa)$ such that:*

(1) *$C_{\alpha\nu} \subseteq C_{\alpha\mu}$ for all α and $\nu < \mu$,*
(2) *$\alpha = \bigcup_{\nu<\kappa} C_\alpha$ for all limit α,*
(3) *$C_{\alpha\nu}$ $(\alpha < \theta)$ is a nontrivial square sequence on θ for all $\nu < \kappa$.*

Moreover, if θ is not a successor of a cardinal of cofinality κ, then each of the square sequences can be chosen to be nonspecial. $\square$

Theorem 7.3.3. *Suppose $\kappa < \theta$ are regular cardinals and that θ carries a special square sequence. Then there exist $<_\nu^\kappa$ $(\nu < \kappa)$ such that:*

(i) *$<_\nu^\kappa$ is a closed tree ordering of θ for each $\nu < \kappa$,*
(ii) *$\in\restriction \theta = \bigcup_{\nu<\kappa} <_\nu^\kappa$,*
(iii) *no tree $(\theta, <_\nu^\kappa)$ has a chain of size θ.* $\square$

Let us now consider the case when θ is a successor cardinal, since in this case some finer special square sequences exist.

Lemma 7.3.4. *The following are equivalent when θ is a successor of some infinite cardinal κ:*

(1) *there is a special square sequence on θ,*
(2) *there is a square sequence C_α $(\alpha < \theta)$ such that $\mathrm{tp}(C_\alpha) \leq \kappa$ for all $\alpha < \theta$.*

Proof. Let D_α $(\alpha < \kappa^+)$ be a given special square sequence. By Lemma 6.1.2 the corresponding tree $(\kappa^+, <^2)$ can be decomposed into κ antichains. So, let $f : \kappa^+ \longrightarrow \kappa$ be a fixed map such that $f^{-1}(\xi)$ is a $<^2$-antichain for all $\xi < \kappa$. Let $\alpha < \kappa^+$ be a given limit ordinal. If D_α has a maximal limit point $\bar{\alpha} < \alpha$, let $C_\alpha = D_\alpha \setminus \bar{\alpha}$. Suppose now that $\{\xi : \xi <^2 \alpha\}$ is unbounded in α and define a strictly increasing continuous sequence $c_\alpha(\xi)$ $(\xi < \nu(\alpha))$ of its elements as follows. Let $c_\alpha(0) = \min\{\xi : \xi <^2 \alpha\}$, $c_\alpha(\eta) = \sup_{\xi<\eta} c_\alpha(\xi)$ for η limit, and $c_\alpha(\xi + 1)$ is the minimal $<^2$-predecessor γ of α such that $\gamma > c_\alpha(\xi)$ and has the minimal f-image among all $<^2$-predecessors that are $> c_\alpha(\xi)$. The ordinal $\nu(\alpha)$ is defined as the place where the process stops, i.e., when $\alpha = \sup_{\xi<\nu(\alpha)} c_\alpha(\xi)$. Let

$$C_\alpha = \{c_\alpha(\xi) : \xi < \nu(\alpha)\}.$$

It is easily checked that this gives a square sequence C_α ($\alpha < \kappa^+$) with the property that $\mathrm{tp}(C_\alpha) \leq \kappa$ for all $\alpha < \kappa^+$. $\square$

Definition 7.3.5. Square sequences C_α ($\alpha < \kappa^+$) that have the property $\mathrm{tp}(C_\alpha) \leq \kappa$ for all $\alpha < \kappa^+$ are usually called $\square_\kappa$-sequences.

So, from now on, we fix a $\square_\kappa$-sequence C_α ($\alpha < \kappa^+$) and concentrate on defining and analysing the corresponding function ρ, the main object of study in this section. Let

$$\Lambda(\alpha, \beta) = \text{maximal limit point of } C_\beta \cap (\alpha + 1) \tag{7.3.1}$$

when such a limit point exists; otherwise $\Lambda(\alpha, \beta) = 0$. The main purpose of this section is to introduce a distance function

$$\rho : [\kappa^+]^2 \longrightarrow \kappa \tag{7.3.2}$$

associated with the $\square_\kappa$-sequence C_α ($\alpha < \kappa^+$) and defined recursively by the formula

$$\begin{aligned}\rho(\alpha, \beta) \quad = \quad & \max\{\mathrm{tp}(C_\beta \cap \alpha), \rho(\alpha, \min(C_\beta \setminus \alpha)), \rho(\xi, \alpha) : \\ & \qquad\qquad\qquad \xi \in C_\beta \cap [\Lambda(\alpha, \beta), \alpha)\},\end{aligned}$$

with the boundary value $\rho(\alpha, \alpha) = 0$ for all α. The following immediate consequence of the coherence property of the $\square_\kappa$-sequence is quite useful.

Lemma 7.3.6. *If α is a limit point of C_β, then $\rho(\xi, \alpha) = \rho(\xi, \beta)$ for all $\xi < \alpha$.* $\square$

For example, using Lemma 7.3.6, one can easily conclude that the mapping ρ is in fact equal to the function ρ^κ defined in Section 7.2. So, we have the following crucial subadditivity properties of ρ.

Lemma 7.3.7. *For all $\alpha \leq \beta \leq \gamma < \kappa^+$,*

(a) $\rho(\alpha, \gamma) \leq \max\{\rho(\alpha, \beta), \rho(\beta, \gamma)\}$,
(b) $\rho(\alpha, \beta) \leq \max\{\rho(\alpha, \gamma), \rho(\beta, \gamma)\}$. $\square$

Clearly $\rho(\alpha, \beta) \geq \rho_1(\alpha, \beta)$, so by Lemma 6.2.1, we have

Lemma 7.3.8. $|\{\xi \leq \alpha : \rho(\xi, \alpha) \leq \nu\}| \leq |\nu| + \aleph_0$ *for all $\alpha < \kappa^+$ and $\nu < \kappa$.* $\square$

The following property is another immediate consequence of the fact that the definition of ρ is closely tied to the notion of a minimal walk along the square sequence (compare with Lemma 7.2.3 above).

Lemma 7.3.9. $\rho(\alpha, \gamma) \geq \max\{\rho(\alpha, \beta), \rho(\beta, \gamma)\}$ *whenever $\alpha \leq \beta \leq \gamma$ and β belongs to the trace of the walk from γ to α.* $\square$

Using Lemmas 7.3.6 and 7.3.9 one proves the following fact exactly as in the case of ρ^κ of Section 7.2 (see the proof of Lemma 7.2.4).

Lemma 7.3.10. *If $0 < \beta \leq \gamma$ and if β is a limit ordinal, then there is $\bar{\beta} < \beta$ such that $\rho(\alpha, \gamma) \geq \rho(\alpha, \beta)$ for all α in the interval $[\bar{\beta}, \beta)$.* □

The proof of the following fact is also completely analogous to the proof of the corresponding fact for the local version ρ^{κ} considered above in Section 7.2 (see the proof of Lemma 7.2.5).

Lemma 7.3.11. *$P_{\nu}(\gamma) = \{\beta < \gamma : \rho(\beta, \gamma) \leq \nu\}$ is a closed subset of γ for all $\gamma < \kappa^{+}$ and $\nu < \kappa$.* □

Since ρ in this case is equal to the local version ρ^{κ}, it is worth restating the following unboundedness property that follows from Theorem 7.2.13, a property that we will be continuously refining as we go on.

Theorem 7.3.12. *It is possible to modify the $\square_{\kappa}$-sequence C_{α} $(\alpha < \kappa^{+})$ in such a way that the corresponding function $\rho : [\kappa^{+}]^2 \rightarrow \kappa$ has the property that for every $\nu < \kappa$ and every family A of pairwise-disjoint subsets of κ^{+} that have sizes $< \mathrm{cf}(\kappa)$ there is a subfamily $B \subseteq A$ of cardinality κ^{+} such that for all $a \neq b$ in B and all $\alpha \in a$, $\beta \in b$ we have $\rho(\alpha, \beta) > \nu$.* □

The discussion of $\rho : [\kappa^{+}]^2 \longrightarrow \kappa$ now splits naturally into two cases depending on whether κ is a regular or a singular cardinal (with the case $\mathrm{cf}(\kappa) = \omega$ of special importance). This is done in the following two sections.

7.4 The function ρ on successors of regular cardinals

In this section κ is a fixed regular cardinal, C_{α} $(\alpha < \kappa^{+})$ a fixed $\square_{\kappa}$-sequence and

$$\rho : [\kappa^{+}]^2 \longrightarrow \kappa$$

is the corresponding ρ-function defined by the recursive formula

$$\begin{aligned} \rho(\alpha, \beta) \quad = \quad & \max\{\mathrm{tp}(C_{\beta} \cap \alpha), \rho(\alpha, \min(C_{\beta} \setminus \alpha)), \rho(\xi, \alpha) : \\ & \qquad \xi \in C_{\beta} \cap [\Lambda(\alpha, \beta), \alpha)\}, \end{aligned}$$

with the boundary condition $\rho(\alpha, \alpha) = 0$ for all α.

For $\nu < \kappa$ and $\alpha < \beta < \kappa^{+}$, set

$$\alpha <_{\nu} \beta \text{ if and only if } \rho(\alpha, \beta) \leq \nu. \tag{7.4.1}$$

The following lemma summarizes some of the properties of these relations and they are all immediate consequences of the corresponding facts about various ρ-function given in previous sections.

Lemma 7.4.1.

(a) *$<_{\nu}$ is a closed tree-ordering of κ^{+} of height $\leq \kappa$ for all $\nu < \kappa$,*

(b) *$<_{\nu} \subseteq <_{\mu}$ whenever $\nu \leq \mu < \kappa$,*

(c) *$\in \restriction \kappa^{+} = \bigcup_{\nu < \kappa} <_{\nu}$.* □

Remark 7.4.2. Recall that by Theorem 7.3.3 for every regular cardinal $\lambda \leq \kappa$ the $\in$-relation of κ^+ can also be written as an increasing union of a λ-sequence $<_\nu$ $(\nu < \lambda)$ of closed tree-orderings. When $\lambda < \kappa$, however, we can no longer insist that the trees $(\kappa^+, <_\nu)$ have heights $\leq \kappa$. Using an additional set-theoretic assumption (compatible with the existence of a $\square_\kappa$-sequence) one can strengthen (a) of Lemma 7.4.1 and have that the height of $(\kappa^+, <_\nu)$ is $< \kappa$ for all $\nu < \kappa$. Without additional set-theoretic assumptions, these trees, however, do have some properties of smallness not covered by Lemma 7.4.1.

Lemma 7.4.3. *If $\kappa > \omega$, then no tree $(\kappa^+, <_\nu)$ has a branch of size κ.*

Proof. Suppose towards a contradiction that some tree $(\kappa^+, <_\nu)$ does have a branch of size κ and let B be one such fixed branch (maximal chain). By Lemmas 7.3.8 and 7.3.11, if $\gamma = \sup(B)$ then B is a closed and unbounded subset of γ of order-type κ. Since κ is regular and uncountable, $C_\gamma \cap B$ is unbounded in C_γ, so in particular we can find $\alpha \in C_\gamma \cap B$ such that $\mathrm{tp}(C_\gamma \cap \alpha) > \nu$. Reading off the definition of $\rho(\alpha, \gamma)$ we conclude that $\rho(\alpha, \beta) = \mathrm{tp}(C_\gamma \cap \alpha) > \nu$. Similarly we can find a $\beta > \alpha$ belonging to the intersection of $\lim(C_\gamma)$ and B. Then $C_\beta = C_\gamma \cap \beta$ so $\alpha \in C_\beta$ and therefore $\rho(\alpha, \beta) = \mathrm{tp}(C_\beta \cap \alpha) > \nu$ contradicting the fact that $\alpha <_\nu \beta$. □

Lemma 7.4.4. *If $\kappa > \omega$, then no tree $(\kappa^+, <_\nu)$ has a Souslin subtree of height κ.*

Proof. Forcing with a Souslin subtree of $(\kappa^+, <_\nu)$ of height κ would produce an ordinal γ of cofinality κ and a closed and unbounded subset B of C_γ forming a chain of the tree $(\kappa^+, <_\nu)$. It is well known that in this case B would contain a ground model subset of size κ contradicting Lemma 7.4.3. □

The unboundedness property of ρ stated in Theorem 7.3.12 should be compared to the following one that fits better the main theme of this section.

Theorem 7.4.5. *If $\kappa > \omega$, then for every $\nu < \kappa$ and every family A of κ pairwise-disjoint finite subsets of κ^+, there exists $A_0 \subseteq A$ of size κ such that for all $a \neq b$ in A_0 and all $\alpha \in a$, $\beta \in b$ we have $\rho(\alpha, \beta) > \nu$.*

Proof. We may assume that for some n and all $a \in A$, we have $|a| = n$. Let $a(0), \ldots, a(n-1)$ enumerate a given element a of A increasingly. Using Lemma 7.4.4 and shrinking A, we may assume that $\{a(i) : a \in A\}$ is an antichain of $(\kappa^+, <_\nu)$ for all $i < n$. For two elements a and b of A, we write $a < b$ if every ordinal from a is smaller than every ordinal from b. Going to a subfamily of A of equal size we may assume that $a < b$ or $b < a$ for every pair a and b of distinct elements of A. We now define a coloring

$$f : [A]^2 \longrightarrow \{0\} \cup \{(i,j) : i, j < n\}$$

as follows. Let $f(a, b) = 0$ if $a \cup b$ is an $<_\nu$-antichain. Otherwise, assuming $a < b$, let $f(a, b) = (i, j)$ where (i, j) is the minimal pair such that $a(i) <_\nu b(j)$. By the

Dushnik–Miller partition theorem ([25]), either there exists $A_0 \subseteq A$ of size κ such that f is constantly equal to 0 on $[A_0]^2$ or there exists a pair (i, j) of integers $< n$ and an infinite $A_1 \subseteq A$ such that f is constantly equal to (i, j) on $[A_1]^2$. The first alternative is what we want, so let us see that the second one is impossible. Otherwise, fix a triple $a < b < c$ of elements of A_1. Then $a(i)$ and $b(i)$ are both $<_\nu$-dominated by $c(j)$. Since $<_\nu$ is a tree-ordering, we must have $a(i) <_\nu b(i)$, contradicting our initial assumption that the set $\{a(i) : a \in A\}$ forms an antichain of the tree $(\kappa^+, <_\nu)$. This completes the proof. □

We shall see later that the unboundedness property of Theorem 7.4.5 can be quite useful in designing forcing notions satisfying good chain conditions. Having such applications in mind, we shall now work on refining further this kind of unboundedness property of the ρ-function.

Lemma 7.4.6. *Suppose $\kappa > 0$, let $\gamma < \kappa^+$ and let $\{\alpha_\xi, \beta_\xi\}$ $(\xi < \kappa)$ be a sequence of pairwise-disjoint elements of $[\kappa^+]^{\leq 2}$. Then there is an unbounded set $\Gamma \subseteq \kappa$ such that $\rho\{\alpha_\xi, \beta_\eta\} \geq \min\{\rho\{\alpha_\xi, \gamma\}, \rho\{\beta_\eta, \gamma\}\}$ for all $\xi \neq \eta$ in Γ.*[2]

Proof. Clearly we may assume that the sequences α_ξ $(\xi < \kappa)$ and β_ξ $(\xi < \kappa)$ are strictly increasing. For definiteness we assume that $\alpha_\xi \leq \beta_\xi$ for all ξ. The case $\alpha_\xi > \beta_\xi$ for all ξ is considered similarly. Let

$$\alpha = \sup \alpha_\xi \text{ and } \beta = \sup \beta_\xi. \tag{7.4.2}$$

Then α and β are ordinals of cofinality κ, so the order-types of C_α and C_β are both equal to κ. It will be convenient to assume that the two sequences $\{\alpha_\xi\}$ and $\{\beta_\xi\}$ are actually indexed by C_α rather than κ. It is then clear we may assume that

$$\beta_\xi \geq \alpha_\xi \geq \xi \text{ for all } \xi \in C_\alpha. \tag{7.4.3}$$

Case 1: $\alpha = \beta$. By Lemma 7.3.10, for each limit point ν of C_α there is $f(\nu) < \nu$ in C_α such that

$$\rho(\xi, \alpha_\nu), \rho(\xi, \beta_\nu) \geq \rho(\xi, \nu) = \rho(\xi, \alpha) \text{ for all } \xi \in [f(\nu), \nu). \tag{7.4.4}$$

Fix a stationary $\Gamma \subseteq \lim(C_\alpha)$ such that f is constant on Γ. Then

$$\rho(\alpha_\xi, \beta_\eta) \geq \rho(\alpha_\xi, \alpha) \text{ for all } \xi < \eta \text{ in } \Gamma, \text{ and} \tag{7.4.5}$$

$$\rho(\beta_\eta, \alpha_\xi) \geq \rho(\beta_\eta, \alpha) \text{ for all } \eta < \xi \text{ in } \Gamma. \tag{7.4.6}$$

Subcase 1^a: $\gamma \geq \alpha$. Using the two subadditive properties of ρ in Lemma 7.3.7 one easily concludes that $\rho(\xi, \alpha) = \rho(\xi, \gamma)$ for any $\xi < \alpha$ such that $\rho(\xi, \alpha) > \rho(\alpha, \gamma)$. So, going to a tail of Γ we may assume that

$$\rho(\alpha_\xi, \alpha) = \rho(\alpha_\xi, \gamma) \text{ and } \rho(\beta_\xi, \alpha) = \rho(\beta_\xi, \gamma) \text{ for all } \xi \in \Gamma. \tag{7.4.7}$$

[2]Here, and everywhere else later in this chapter, the convention is that, $\rho\{\alpha, \beta\}$ is meant to be equal to $\rho(\alpha, \beta)$ if $\alpha < \beta$, equal to $\rho(\beta, \alpha)$ if $\beta < \alpha$, and equal to 0 if $\alpha = \beta$.

Consider $\xi < \eta$ in Γ. Combining (7.4.5) and the first equality of (7.4.7) gives us

$$\rho(\alpha_\xi, \beta_\eta) \geq \rho(\alpha_\xi, \alpha) = \rho(\alpha_\xi, \gamma),$$

which is sufficient for the conclusion of Lemma 7.4.6. Suppose $\xi > \eta$ are chosen from Γ in this order. Combining (7.4.6) and the second equality of (7.4.7) gives us

$$\rho(\beta_\eta, \alpha_\xi) \geq \rho(\beta_\eta, \alpha) = \rho(\beta_\eta, \gamma),$$

which is also sufficient for the conclusion of Lemma 7.4.6.

Subcase 1^b: $\gamma < \alpha$. Using Lemma 7.3.8 and going to a tail of the set Γ we may assume that Γ lies above γ and that

$$\rho(\alpha_\xi, \alpha), \rho(\beta_\xi, \alpha) > \rho(\gamma, \alpha) \text{ for all } \xi \in \Gamma. \tag{7.4.8}$$

Applying the subadditive property Lemma 7.3.7(b) of ρ we get for all $\xi \in \Gamma$:

$$\rho(\gamma, \alpha_\xi) \leq \max\{\rho(\gamma, \alpha), \rho(\alpha_\xi, \alpha)\} = \rho(\alpha_\xi, \alpha), \text{ and} \tag{7.4.9}$$

$$\rho(\gamma, \beta_\xi) \leq \max\{\rho(\gamma, \alpha), \rho(\beta_\xi, \alpha)\} = \rho(\beta_\xi, \alpha). \tag{7.4.10}$$

If $\xi < \eta$ are chosen from Γ in this order, then (7.4.5) and (7.4.9) give us

$$\rho(\alpha_\xi, \beta_\eta) \geq \rho(\alpha_\xi, \alpha) \geq \rho(\gamma, \alpha_\xi),$$

which is sufficient for the conclusion of Lemma 7.4.6. If $\xi > \eta$ are chosen from Γ in this order, then combining (7.4.6) and (7.4.10) gives us

$$\rho(\beta_\eta, \alpha_\xi) \geq \rho(\beta_\eta, \alpha) \geq \rho(\gamma, \beta_\eta),$$

which is again sufficient for the conclusion of Lemma 7.4.6.

Case 2: $\alpha < \beta$. We may assume all $\beta_\xi > \alpha$ for all ξ and working as in Case 1 we find a stationary set Γ of limit points of C_α such that

$$\rho(\alpha_\xi, \beta_\eta) \geq \rho(\alpha_\xi, \alpha) \text{ for all } \xi < \eta \text{ in } \Gamma. \tag{7.4.11}$$

By Lemma 7.3.8 for each $\eta \in \Gamma$ there is $g(\eta) \in C_\alpha$ such that $\rho(\xi, \alpha) > \rho(\alpha, \beta_\eta)$ for all ξ in the interval $[g(\eta), \alpha)$. Using the two subadditive properties of ρ from Lemma 7.3.7, one concludes that $\rho(\xi, \alpha) = \rho(\alpha, \beta_\eta)$ for all $\xi \in [g(\eta), \alpha)$. Intersecting Γ with the closed and unbounded subset of C_α of all ordinals that are closed under the mapping g we may assume

$$\rho(\alpha_\xi, \beta_\eta) = \rho(\alpha_\xi, \alpha) \text{ for all } \xi > \eta \text{ in } \Gamma. \tag{7.4.12}$$

Subcase 2^a: $\gamma < \alpha$. Applying Lemma 7.3.8 again and going to a tail of Γ we may assume that Γ lies above γ and

$$\rho(\alpha_\xi, \alpha) > \rho(\gamma, \alpha) \text{ for all } \xi \in \Gamma. \tag{7.4.13}$$

Applying the subadditive properties of ρ we get

$$\rho(\gamma, \alpha_\xi) \leq \max\{\rho(\gamma, \alpha), \rho(\alpha_\xi, \alpha)\} = \rho(\alpha_\xi, \alpha) \text{ for all } \xi \in \Gamma. \tag{7.4.14}$$

If $\xi < \eta$ are chosen from Γ in this order, then combining the inequalities (7.4.11) and (7.4.14) we get

$$\rho(\alpha_\xi, \beta_\eta) \geq \rho(\alpha_\xi, \alpha) \geq \rho(\gamma, \alpha_\xi),$$

which is sufficient to give us the conclusion of Lemma 7.4.6. If $\xi > \eta$ are chosen from Γ in this order, then combining (7.4.12) and (7.4.14) we get

$$\rho(\alpha_\xi, \beta_\eta) = \rho(\alpha_\xi, \alpha) \geq \rho(\gamma, \alpha_\xi),$$

which is again sufficient for the conclusion of Lemma 7.4.6.

Subcase 2^b: $\gamma \geq \alpha$. Going to a tail of Γ we may assume that $\rho(\alpha_\xi, \alpha) > \rho(\alpha, \gamma)$, so as before the subadditive properties give us that

$$\rho(\alpha_\xi, \alpha) = \rho(\alpha_\xi, \gamma) \text{ for all } \xi \in \Gamma. \tag{7.4.15}$$

If $\xi < \eta$ are chosen from Γ in this order, then combining the inequality (7.4.11) and equality (7.4.15) we have

$$\rho(\alpha_\xi, \beta_\eta) \geq \rho(\alpha_\xi, \alpha) = \rho(\alpha_\xi, \gamma),$$

which is sufficient to give us the conclusion of Lemma 7.4.6. If $\xi > \eta$ are chosen from Γ in this order, then combining the equalities (7.4.12) and (7.4.15) we have

$$\rho(\alpha_\xi, \beta_\eta) = \rho(\alpha_\xi, \alpha) = \rho(\alpha_\xi, \gamma),$$

which is again sufficient to give us the conclusion of Lemma 7.4.6. This finishes the proof. $\square$

Lemma 7.4.7. *Suppose κ is λ-inaccessible*[3] *for some $\lambda < \kappa$ and that A is a family of size κ of subsets of κ^+, all of size $< \lambda$. Then for every ordinal $\nu < \kappa$ there is a subfamily B of A of size κ such that for all a and b in B:*

(a) *$\rho\{\alpha, \beta\} > \nu$ for all $\alpha \in a \setminus b$ and $\beta \in b \setminus a$.*
(b) *$\rho\{\alpha, \beta\} \geq \min\{\rho\{\alpha, \gamma\}, \rho\{\beta, \gamma\}\}$ for all $\alpha \in a \setminus b$, $\beta \in b \setminus a$ and $\gamma \in a \cap b$.*

Proof. Consider first the case $\lambda = \omega$. Going to a subfamily, assume A forms a Δ-system with root r and that for some fixed integer $n \geq 1$ and all $a \in A$ we have $|a \setminus r| = n$. By Lemma 7.4.5, going to a subfamily we may assume that A satisfies (a). For $a \in A$ let $a(0), \ldots, a(n-1)$ be the increasing enumeration of $a \setminus r$. Going to a subfamily, we may assume that A can be enumerated as a_ξ $(\xi < \kappa)$ in such a way that $a_\xi(i)$ $(\xi < \kappa)$ is strictly increasing for all $i < n$. By $n^2 \cdot |r|$ successive

[3] I.e., $\nu^\tau < \kappa$ for all $\nu < \kappa$ and $\tau < \lambda$.

applications of Lemma 7.4.6 we find a single unbounded set $\Gamma \subseteq \kappa$ such that for all $\gamma \in r$, $(i,j) \in n \times n$ and $\xi \neq \eta$ in Γ:

$$\rho\{a_\xi(i), a_\eta(j)\} \geq \min\{\rho\{a_\xi(i), \gamma\}, \rho\{a_\eta(j), \gamma\}\}. \tag{7.4.16}$$

This gives us the conclusion (b) of the lemma.

Let us now consider the general case. Using the λ-inaccessibility of κ we may assume that A forms a Δ-system with root r and that members of A have some fixed order-type, i.e., that for some fixed ordinal $\mu < \lambda$ we can enumerate each $a \setminus r$ for $a \in A$ in increasing order as $a(i)$ $(i < \mu)$. Using λ-inaccessibility again and refining A we may assume that there is an enumeration a_ξ $(\xi < \kappa)$ of A such that $a_\xi(i)$ $(\xi < \kappa)$ is a strictly increasing sequence for all $i < \mu$. For $i < \mu$, set

$$\delta(i) = \sup_\xi a_\xi(i).$$

Then $\mathrm{cf}(\delta(i)) = \kappa$ and therefore $\mathrm{tp}\, C_{\delta(i)} = \kappa$ for all $i < \mu$. Going to a smaller cardinal we may assume that λ is a regular cardinal $< \kappa$. Applying Lemma 7.3.10 simultaneously μ times gives us a regressive map $f : \kappa \longrightarrow \kappa$ such that for every $\eta < \kappa$ of cofinality λ, all $\xi \in [c_{\delta(i)}(f(\eta)), c_{\delta(i)}(\eta))$ and $i < \mu$:

$$\rho(\xi, a_\eta(i)) \geq \rho(\xi, c_{\delta(i)}(\eta)) = \rho(\xi, \delta(i)). \tag{7.4.17}$$

Here $c_{\delta(i)}(\xi)$ $(\xi < \kappa)$ refers to the increasing enumeration of the set $C_{\delta(i)}$ for each $i < \mu$. Pick a stationary set Γ of cofinality λ ordinals $< \kappa$ such that f is constant on Γ. It follows that for all $i, j < \mu$, if $\delta(i) = \delta(j)$, then

$$\rho\{a_\xi(i), a_\eta(j)\} \geq \rho(a_\xi(i), \delta(i)) \text{ for } \xi < \eta \text{ in } \Gamma, \tag{7.4.18}$$

$$\rho\{a_\eta(j), a_\xi(i)\} \geq \rho(a_\eta(j), \delta(i)) \text{ for } \xi > \eta \text{ in } \Gamma. \tag{7.4.19}$$

On the other hand, if $i, j < \mu$ such that $\delta(i) < \delta(j)$ we have

$$\rho\{a_\xi(i), a_\eta(j)\} \geq \rho(a_\xi(i), \delta(i)) \text{ for all } \xi < \eta \text{ in } \Gamma. \tag{7.4.20}$$

Intersecting Γ with a closed and unbounded subset of κ we also assume that

$$a_\xi(i) \geq c_{\delta(i)}(\xi) \text{ for all } i < \mu \text{ and } \xi \in \Gamma. \tag{7.4.21}$$

$$a_\xi(i) \leq a_\xi(j) \text{ for all } \xi \in \Gamma \text{ and } i, j < \mu \text{ such that } \delta(i) \leq \delta(j). \tag{7.4.22}$$

Since $\rho(\alpha, \delta(i)) \geq \mathrm{tp}(C_{\delta(i)} \cap \alpha)$ for every $i < \mu$ and $\alpha < \delta(i)$, combining (7.4.18)–(7.4.22) we see that

$$\rho\{a_\xi(i), a_\eta(j)\} \geq \min\{\xi, \eta\} \text{ for all } i, j < \mu \text{ and } \xi \neq \eta \text{ in } \Gamma. \tag{7.4.23}$$

So replacing Γ with $\Gamma \setminus (\nu + 1)$ gives us the conclusion (a) of Lemma 7.4.7.

Suppose now we are given two coordinates $i, j < \mu$ and an ordinal γ in the root r. Let

$$\alpha_\xi = a_\xi(i) \text{ and } \beta_\xi = a_\xi(j)$$

for $\xi \in \Gamma$. Note that (7.4.18) and (7.4.19) give us the conditions (7.4.5) and (7.4.6) of Case 1 (i.e., $\delta(i) = \alpha = \beta = \delta(j)$) in the proof of Lemma 7.4.6 while (7.4.20) gives us the condition (7.4.11) of Case 2 (i.e., $\delta(i) = \alpha < \beta = \delta(j)$) in the proof of Lemma 7.4.6. Observe that in this proof of Lemma 7.4.6, once the set Γ was obtained by a single application of the Pressing Down Lemma to finally give us (7.4.5) and (7.4.6) of Case 1 or (7.4.11) of Case 2, the rest of the refinements of Γ all lie in the restriction of the closed and unbounded filter to Γ. In other words, we can now proceed as in the case $\lambda = \omega$ and successively apply this refining procedure for each $\gamma \in r$ and $(i, j) \in \mu \times \mu$ obtaining a set $\Gamma_0 \subseteq \Gamma$ with the property that $\Gamma \setminus \Gamma_0$ is nonstationary and

$$\begin{aligned} &\rho\{a_\xi(i), a_\eta(j)\} \geq \min\{\rho\{a_\xi(i), \gamma\}, \rho\{a_\eta(j), \gamma\}\} \text{ for all } \gamma \in r, \\ &i, j < \mu, \text{ and } \xi \neq \eta \text{ in } \Gamma_0. \end{aligned} \tag{7.4.24}$$

Since this is clearly equivalent to the conclusion (b) of Lemma 7.4.7, the proof is completed. □

Definition 7.4.8. The function $D : [\kappa^+]^2 \longrightarrow [\kappa^+]^{<\kappa}$ is defined by

$$D(\alpha, \beta) = \{\xi \leq \alpha : \rho(\xi, \alpha) \leq \rho(\alpha, \beta)\}.^4$$

Note that $D(\alpha, \beta) = \{\xi \leq \alpha : \rho(\xi, \beta) \leq \rho(\alpha, \beta)\}$, so we could take the formula

$$D\{\alpha, \beta\} = \{\xi \leq \min\{\alpha, \beta\} : \rho(\xi, \alpha) \leq \rho\{\alpha, \beta\}\}$$

as our definition of $D\{\alpha, \beta\}$ which could serve when there is no implicit assumption about the ordering between α and β, contrary to the case whenever we write $D(\alpha, \beta)$ where we implicitly assume that $\alpha < \beta$.

Lemma 7.4.9. *If κ is λ-inaccessible for some $\lambda < \kappa$, then for every family A of size κ of subsets of κ^+, all of size $< \lambda$, there exists $B \subseteq A$ of size κ such that for all a and b in B and all $\alpha \in a \setminus b$, $\beta \in b \setminus a$ and $\gamma \in a \cap b$:*

(a) $\alpha, \beta > \gamma$ *implies* $D\{\alpha, \gamma\} \cup D\{\beta, \gamma\} \subseteq D\{\alpha, \beta\}$,
(b) $\beta > \gamma$ *implies* $D\{\alpha, \gamma\} \subseteq D\{\alpha, \beta\}$,
(c) $\alpha > \gamma$ *implies* $D\{\beta, \gamma\} \subseteq D\{\alpha, \beta\}$,
(d) $\gamma > \alpha, \beta$ *implies* $D\{\alpha, \gamma\} \subseteq D\{\alpha, \beta\}$ *or* $D\{\beta, \gamma\} \subseteq D\{\alpha, \beta\}$.

Proof. Choose $B \subseteq A$ of size κ satisfying the conclusion (b) of Lemma 7.4.7. Pick $a \neq b$ in B and consider $\alpha \in a \setminus b$, $\beta \in b \setminus a$ and $\gamma \in a \cap b$. By the conclusion of Lemma 7.4.7(b), we have

$$\rho\{\alpha, \beta\} \geq \min\{\rho\{\alpha, \gamma\}, \rho\{\beta, \gamma\}\}. \tag{7.4.25}$$

a. Suppose $\alpha, \beta > \gamma$. Note that in this case a single inequality $\rho(\gamma, \alpha) \leq \rho\{\alpha, \beta\}$ or $\rho(\gamma, \beta) \leq \rho\{\alpha, \beta\}$ given to us by (7.4.25) implies that we actually have both

[4] Note that by our boundary requirements $\rho(\alpha, \alpha) = 0$, we have that in particular $\alpha \in D(\alpha, \beta)$.

inequalities simultaneously holding. The subadditivity of ρ gives us $\rho(\xi, \alpha) \leq \rho\{\alpha, \beta\}$, or equivalently $\rho(\xi, \beta) \leq \rho\{\alpha, \beta\}$ for any $\xi \leq \gamma$ with $\rho(\xi, \gamma) \leq \rho(\gamma, \alpha)$ or $\rho(\xi, \gamma) \leq \rho(\gamma, \beta)$. This is exactly the conclusion of 7.4.9(a).

b. Suppose that $\beta > \gamma > \alpha$. Using the subadditivity of ρ we see that in both cases given to us by (7.4.25), we have that $\rho(\alpha, \gamma) \leq \rho\{\alpha, \beta\}$. So the inclusion $D\{\alpha, \gamma\} \subseteq D\{\alpha, \beta\}$ follows immediately.

c. Suppose that $\alpha > \gamma > \beta$. The conclusion $D\{\beta, \gamma\} \subseteq D\{\alpha, \beta\}$ follows from the previous case by symmetry.

d. Suppose that $\gamma > \alpha, \beta$. Then $\rho(\alpha, \gamma) \leq \rho\{\alpha, \beta\}$ gives $D\{\alpha, \gamma\} \subseteq D\{\alpha, \beta\}$ while $\rho(\beta, \gamma) \leq \rho\{\alpha, \beta\}$ gives us $D\{\beta, \gamma\} \subseteq D\{\alpha, \beta\}$.

This completes the proof. □

Remark 7.4.10. Note that $\min\{x, y\} \in D\{x, y\}$ for every $\{x, y\} \in [\kappa^+]^2$, so the conclusion (a) of Lemma 7.4.9 in particular means that $\gamma < \min\{\alpha, \beta\}$ implies $\gamma \in D\{\alpha, \beta\}$. In applications, one usually needs this consequence of 7.4.9(a) rather than 7.4.9(a) itself. We shall also need the convention that $D\{x, y\} = \{x\}$ whenever $x = y$.

We finish this section with a remark about the behavior of D on families of sets of size κ^+ rather than κ.

Lemma 7.4.11. *Suppose that $\kappa^{<\lambda} = \kappa$ and that A is a family of subsets of κ^+ of size $< \lambda$. Then there is a subfamily $B \subseteq A$ of size κ^+ forming an increasing Δ-system with root r*[5] *such that for all $a, b \in B$ such that $a \setminus r < b \setminus r$, we have that $a \cap (\alpha + 1) \subseteq D(\alpha, \beta)$ for all $\alpha \in a \setminus r$ and $\beta \in b \setminus b$.*

Proof. By our assumption $\kappa^{<\lambda} = \kappa$ and the Δ-system lemma at that level, and by going to a subfamily of A of size κ^+, we may assume that A already forms an increasing Δ-system with root r. Going to a further subfamily of A of equal size, we may assume that for some ordinal $\nu < \kappa$ and all $a \in A$,

$$\rho(\xi, \alpha) < \nu \text{ for all } \xi \leq \alpha \text{ in } a. \tag{7.4.26}$$

Since κ is assumed to be a regular cardinal, since ρ is based on a $\square_\kappa$ sequence which obviously avoids cofinality κ ordinals, and since $\rho = \rho^\kappa$, Theorem 7.2.13 applies giving us a subfamily $B \subseteq A$ of size κ^+ such that for all $a, b \in B$ with $a \setminus r < b \setminus r$,

$$\rho(\alpha, \beta) > \nu \text{ for all } \alpha \in a \setminus r \text{ and } \beta \in b \setminus r. \tag{7.4.27}$$

Combining these two last equations, we see that for $a, b \in B$ with $a \setminus r < b \setminus r$, for $\alpha \in a \setminus r$ and $\beta \in b \setminus r$, and for $\xi \in a \cap (\alpha + 1)$, we have

$$\rho(\xi, \alpha) < \nu < \rho(\alpha, \beta). \tag{7.4.28}$$

Since $D(\alpha, \beta) = \{\xi \leq \alpha : \rho(\xi, \alpha) \leq \rho(\alpha, \beta)\}$, this gives us the conclusion of the lemma. □

[5]We say that B forms an *increasing* Δ-system with root r if for all $a \neq b$ in B, we have that $a \cap b = r$ and either $a \setminus r < b \setminus r$, or else $b \setminus r < a \setminus r$.

7.5 Forcing constructions based on ρ

We start with the most typical application of Lemma 7.4.9 in forcing constructions.

Theorem 7.5.1. *If $\square_\kappa$ holds for some regular cardinal κ that is λ-inaccessible, then there is a λ-closed κ-cc forcing notion $\mathcal{P}$ that introduces a Souslin tree of height κ^+.*

Proof. We let $\mathcal{P}$ be the collection of all pairs $p = (T_p, \leq_p)$, where T_p is a subset of κ^+ of size $< \lambda$ and where $\leq_p$ is a tree-ordering on T_p compatible with the usual ordering between the ordinals and with the property that

$$\begin{aligned}&\delta \leq_p \alpha \text{ and } \delta \leq_p \beta \text{ implies that there is } \gamma \in D\{\alpha, \beta\} \cap T_p \\ &\text{such that } \delta \leq_p \gamma \text{ and } \gamma \leq_p \alpha \text{ and } \gamma \leq_p \beta.\end{aligned} \tag{7.5.1}$$

We let p be a *stronger condition* than q, and write $p \leq q$, whenever $T_p \supseteq T_q$ and $\leq_p\restriction T_q^2 = \leq_q$. Clearly, $\mathcal{P}$ is a λ-closed forcing notion, so let us check that it satisfies the κ-chain condition. Consider a family $\mathcal{A} \subseteq \mathcal{P}$ of size κ. Applying Lemma 7.4.9, we obtain a subfamily $\mathcal{B} \subseteq \mathcal{A}$ of size κ such that the family of sets T_p $(p \in \mathcal{B})$ satisfies the conditions (a), (b), (c), and (d) of the lemma. By a further thinning, we may assume that T_p $(p \in \mathcal{B})$ forms a Δ-system with root T. We may also assume that T_p's are isomorphic trees via the order-isomorphisms between their domains as sets of ordinals and that fixes T. So in particular $\leq_p$ $(p \in \mathcal{B})$ all agree on T^2. Moreover, we may assume that

$$D\{\alpha, \beta\} \cap (T_p \setminus T) = \emptyset \text{ for all } \alpha, \beta \in T \text{ and } p \in \mathcal{B}. \tag{7.5.2}$$

Consider $p \neq q$ in $\mathcal{B}$. Let $r = (T_r, \leq_r)$, where $T_r = T_p \cup T_q$, and where $\leq_r$ is the minimal tree-ordering on T_r containing $\leq_p$ and $\leq_q$. We finish the proof of κ-cc for $\mathcal{P}$, by checking that r has the property (7.5.1).

Suppose that $\delta \leq_r \alpha$ and $\delta \leq_r \beta$. We need to find $\gamma \in D\{\alpha, \beta\} \cap T_r$ such that $\delta \leq_r \gamma$ and $\gamma \leq_r \alpha$ and $\gamma \leq_r \beta$.

Case 1. $\delta \in T$. Assuming that one of the α of β is in T would reduce our problem to the property (7.5.1) of p of q. So we may assume that $\alpha \in T_p \setminus T$ and $\beta \in T_q \setminus T$. Then Lemma 7.4.9 (a) gives us that $\delta \in D\{\alpha, \beta\}$, so we could take $\gamma = \delta$.

Case 2. $\delta \notin T$. By symmetry, we may consider only the case $\delta \in T_p \setminus T_q$. Since p has the property (7.5.1), we may assume that one of the α or β does not belong to T_p. By symmetry, let us assume that $\beta \in T_q \setminus T$.

Subcase 2.1. $\alpha \in T_q \setminus T$. By the minimality of the ordering $\leq_r$, we can find $\alpha_p, \beta_p \in T$ such that $\delta \leq_p \alpha_p \leq_q \alpha$ and $\delta \leq_p \beta_p \leq_q \beta$. By the property (7.5.1) for p there is $\gamma_p \in D\{\alpha_p, \beta_p\} \cap T_p$ such that $\delta \leq_p \gamma_p$ and $\gamma_p \leq_p \alpha_p$ and $\gamma_p \leq_p \beta_p$. By (7.5.2), γ_p must belong to T, and therefore to T_q. Applying (7.5.1) for q, we can find $\gamma \in D\{\alpha, \beta\} \cap T_q$ such that $\gamma_p \leq_q \gamma$ and $\gamma \leq_q \alpha$ and $\gamma \leq_q \beta$. Then γ is as required.

Subcase 2.2. $\alpha \in T$. By our assumption $\delta <_r \beta$ and by the minimality of $\leq_r$ there must be $\beta_p \in T$ such that $\delta \leq_p \beta_p \leq_q \beta$. By the property (7.5.1) of p

there is $\gamma' \in D\{\alpha, \beta_p\}$ such that $\delta \leq_p \gamma'$, $\gamma' \leq_p \alpha$ and $\gamma' \leq_p \beta_p$. By (7.5.2), γ' must belong to T and therefore to T_q, so by the property (7.5.1) of q there must be $\gamma \in D\{\alpha, \beta\} \cap T_q$ such that $\gamma' \leq_q \gamma$ and $\gamma \leq_q \alpha$ and $\gamma \leq_q \beta$. Then this γ is a witness that r has the property (7.5.1) for the triple α, β and δ and so we are done in this subcase as well.

Subcase 2.3. $\alpha \in T_p \setminus T$. Take again $\beta_p \in T$ such that $\delta \leq_p \beta_p \leq_q \beta$. Applying Lemma 7.4.9 (b) for the triple α, β, and β_p in place of α, β, and γ, respectively, we conclude that $D\{\alpha, \beta_p\} \subseteq D\{\alpha, \beta\}$. Applying the property (7.5.1) for p in the case of the inequalities $\delta \leq_p \alpha$ and $\delta \leq_p \beta_p$, we can find $\gamma \in D\{\alpha, \beta_p\} \cap T_p$ such that $\delta \leq_p \gamma$ and $\gamma \leq_p \alpha$ and $\gamma \leq_p \beta_p$. Since $D\{\alpha, \beta_p\} \subseteq D\{\alpha, \beta\}$ the ordinal γ belongs also to $D\{\alpha, \beta\} \cap T_r$, as required.

Since $\mathcal{P}$ obviously has cardinality κ^+, one condition from $\mathcal{P}$ must force that the generic tree $\dot{T}$ has cardinality κ^+. So the proof is finished, once we show that $\mathcal{P}$ forces that $\dot{T}$ has no chains nor antichains of cardinality κ^+. This reduces to showing that, given any family $\mathcal{A} \subseteq \mathcal{P}$ of size κ^+ and given an assignment $\xi_p \in T_p$ ($p \in \mathcal{A}$) such that $\xi_p \neq \xi_q$ whenever $p \neq q$, we can find distinct $p, q \in \mathcal{A}$ and $r \leq p, q$ such that ξ_p and ξ_q are $\leq_r$-incomparable, as well as distinct $p, q \in \mathcal{A}$ and $r \leq p, q$ such that ξ_p and ξ_q are $\leq_r$-comparable. Let us concentrate on the first of these goals. Going to a subfamily of $\mathcal{A}$ of size κ^+, we may assume that T_p ($p \in \mathcal{A}$) forms a Δ-system with root T which forms an initial segment of all the T_p's which we assume as isomorphic as trees via order-isomorphisms between their domains as sets of ordinals, the isomorphisms that fix T_p. Moreover, we may assume that the order-isomorphisms serve also as isomorphisms between the structures $(T_p, \rho \restriction T_p^2)$ ($p \in \mathcal{A}$). Note that this in particular means that for some $\nu_0 < \kappa$,

$$\sup\{\rho(\alpha, \beta) : \alpha, \beta \in T_p\} = \nu_0 \text{ for all } p \in \mathcal{A}. \tag{7.5.3}$$

By Theorem 7.3.12, we can find $\mathcal{B} \subseteq \mathcal{A}$ of cardinality κ^+ such that

$$\rho\{\alpha, \beta\} > \nu_0 \text{ for } \alpha \in T_p \setminus T, \beta \in T_q \setminus T, \text{ where } p \neq q \text{ in } \mathcal{B}. \tag{7.5.4}$$

Fix two different members p and q of $\mathcal{B}$. Let $T_r = T_p \cup T_q$ and let $\leq_r$ be the minimal tree-ordering on T_r extending $\leq_p$ and $\leq_q$. Since $\xi_p \in T_p \setminus T$ and $\xi_q \in T_q \setminus T$, and since T is a common initial segment of T_p and T_q, we conclude that ξ_p and ξ_q are incomparable relative to $\leq_r$. So it remains only to show that r is a member of $\mathcal{P}$, i.e., that it satisfies (7.5.1). However, note that this has already been done during the course of the proof that $\mathcal{P}$ satisfies the λ-chain condition because (7.5.3) and (7.5.4) are giving us the necessary input for that argument.

Choose now p and q in $\mathcal{B}$ such that every ordinal of $T_p \setminus T$ is smaller than every ordinal of $T_q \setminus T$. Let $T_r = T_p \cup T_q$ and let $\leq_r$ be the minimal tree-ordering on T_r extending $\leq_p$ and $\leq_q$ such that $\xi_p <_r \xi_q$. It remains to show that r satisfies (7.5.1). Let $\delta \leq_r \alpha$ and $\delta \leq_r \beta$. Note that α and β belong to the set $T_q \setminus T$ and they do not have common $\leq_q$-predecessors in that sets they do not have common $\leq_r$-predecessors either. This takes care of the case $\alpha, \beta \in T_p \setminus T_q$. The case $\alpha, \beta \in T_p \setminus T$

is also immediate. So it remains to consider the case $\alpha \in T_p \setminus T$ and $\beta \in T_q \setminus T$. Combining (7.5.3) and (7.5.4) we get that

$$D\{\alpha, \beta\} \cap T_q = T_p \cap (\alpha + 1). \tag{7.5.5}$$

It follows that $\delta \in D\{\alpha, \beta\} \cap T_q$, so we can take $\gamma = \delta$ to obtain the conclusion of (7.5.1). This finishes the proof. $\square$

Corollary 7.5.2. *If $\square_{\omega_1}$ holds, there is a property K forcing notion that adds a Souslin tree on ω_2.* $\square$

Theorem 7.5.3. *If $\square_\kappa$ holds for some regular cardinal κ that is λ-inaccessible, then there is a λ-closed κ-cc forcing notion $\mathcal{P}$ that introduces a mapping $g : [\kappa^+]^2 \longrightarrow \lambda$ with property that $E_g(\alpha, \beta) = \{\xi < \alpha : g(\xi, \alpha) = g(\xi, \beta)\}$ has cardinality $< \lambda$ for all $\alpha < \beta < \kappa^+$.*

Proof. Let $\mathcal{P}$ be the collection of all mappings of the form $p : [D_p]^2 \longrightarrow \lambda$, where D_p is a subset of κ^+ of cardinality $< \lambda$ and where the following condition is satisfied for every triple $\xi < \alpha < \beta$ of elements of D_p,

$$\xi \notin D(\alpha, \beta) \text{ implies } p(\xi, \alpha) \neq p(\xi, \beta) \tag{7.5.6}$$

For p and q in $\mathcal{P}$, let $p \leq q$ if and only if $D_p \supseteq D_q$, $p \restriction [D_q]^2 = q$, and

$$p(\xi, \alpha) \neq p(\xi, \beta) \text{ for every } \alpha < \beta \text{ in } D_q \text{ and } \xi \in (D_p \setminus D_q) \cap \alpha. \tag{7.5.7}$$

It is clear that $\mathcal{P}$ is a λ-closed forcing notion, so let us show that it satisfies the κ-chain condition. Let $\mathcal{A}$ be a given subset of $\mathcal{P}$ of cardinality κ. By our assumption about λ-inaccessibility of κ, we can find $\mathcal{B} \subseteq \mathcal{A}$ of cardinality κ forming a Δ-system with root r. By Lemma 7.4.9 and by going to a subfamily of $\mathcal{B}$ of size κ, we may assume that the family of sets D_p $(p \in \mathcal{B})$ satisfies the conditions (a), (b), (c), and (d) of the lemma. Choose two different members p and q from $\mathcal{B}$ and find a mapping $s : [D_p \cup D_q]^2 \longrightarrow \lambda$ extending p and q such that s is one-to-one on the set

$$[D_p \cup D_q]^2 \setminus ([D_p]^2 \cup [D_q]^2$$

and such that the s-image of this set is disjoint from the ranges of p and q. It remains to check that s satisfies the condition (7.5.6) for every triple $\xi < \alpha < \beta$ of elements of $D_s = D_p \cup D_q$. Fix such a triple $D_s = D_p \cup D_q$ and assume that $p(\xi, \alpha) = p(\xi, \beta)$. We need to show that $\xi \in D(\alpha, \beta)$. From our definition of s it follows that

$$\{\xi, \alpha\}, \{\xi, \beta\} \in [D_p]^2 \cup [D_q]^2.$$

If $\{\xi, \alpha\}, \{\xi, \beta\} \in [D_p]^2$ or if $\{\xi, \alpha\}, \{\xi, \beta\} \in [D_p]^2$ the required conclusion $\xi \in D(\alpha, \beta)$ follows from the fact that p and q satisfy the condition (7.5.6). So it remains to consider the case when, say, $\{\xi, \alpha\} \in [D_p]^2 \setminus [D_q]^2$ and $\{\xi, \beta\} \in [D_q]^2 \setminus [D_p]^2$. Note that this means that

$$\xi \in D_r = D_p \cap D_q, \ \alpha \in D_p \setminus D_r \text{ and } \beta \in D_q \setminus D_r. \tag{7.5.8}$$

This in turn means that the conclusion (a) of Lemma 7.4.9 applies, and so we have that

$$\xi \in (D(\xi,\alpha) \cup D(\xi,\beta)) \subseteq D(\alpha,\beta), \tag{7.5.9}$$

as required. Having checked that $\mathcal{P}$ satisfies the κ-chain condition, note that the generic filter of $\mathcal{P}$ gives us the mapping $g : [\kappa^+]^2 \longrightarrow \lambda$ with the property that $E_g(\alpha,\beta)$ has cardinality $< \lambda$ for all $\alpha < \beta < \kappa^+$. □

Corollary 7.5.4. *Suppose that $\square_\kappa$ holds for some regular cardinal κ that is λ-inaccessible and that $\lambda = \nu^+$ for some cardinal ν. Then there is a λ-closed κ-cc forcing notion $\mathcal{P}$ that introduces a mapping $f : (\kappa^+)^3 \longrightarrow \nu$ which is not constant on any product $A \times B \times C$ of three subsets of κ^+ of cardinality λ.*

Proof. It suffices to construct such a map $f : (\kappa^+)^3 \longrightarrow \nu$ on the basis of map $g : [\kappa^+]^2 \longrightarrow \lambda$ satisfying the conclusion of Theorem 7.5.3. For this, we first choose a mapping $e : [\lambda]^2 \longrightarrow \nu$ with the property that

$$e(\alpha,\gamma) \neq e(\beta,\gamma) \text{ whenever } \alpha < \beta < \gamma. \tag{7.5.10}$$

Define $f : (\kappa^+)^3 \longrightarrow \nu^4$ as follows:

$$f(\alpha,\beta,\gamma) = (p(\alpha,\beta,\gamma), e(g(\alpha,\beta), g(\alpha,\gamma)), e(g(\alpha,\beta), g(\beta,\gamma)), e(g(\alpha,\gamma), g(\beta,\gamma))),$$

where $p(\alpha,\beta,\gamma)$ codes the order and equality relations between the ordinals α, β and γ. Consider three subsets A, B and C of κ^+ of cardinality λ. We need to show that f is not constant on the product $A \times B \times C$. We may assume that A, B and C have order-types equal to λ. Since we have incorporated the mapping p into f the three sets must have different suprema. So we may, in fact, assume that

$$\sup A < \sup B < \sup C.$$

Applying the basic property of g, we can find subsets $B_0 \subseteq B$ and $C_0 \subseteq C$ of cardinality ν and $\alpha \in A$ such that

$$g(\alpha,\beta) \neq g(\alpha,\gamma) \text{ for } \beta \in B_0 \text{ and } \gamma \in C_0. \tag{7.5.11}$$

Going to subsets of B_0 and C_0, we may also assume that

$$g(\alpha,\beta) < g(\alpha,\beta') \text{ for } \beta < \beta' \text{ in } B_0, \text{ and} \tag{7.5.12}$$

$$g(\alpha,\gamma) < g(\alpha,\gamma') \text{ for } \gamma < \gamma' \text{ in } C_0. \tag{7.5.13}$$

Let $\delta = \sup\{g(\alpha,\beta) : \beta \in B_0\}$ and $\varepsilon = \sup\{g(\alpha,\gamma) : \gamma \in C_0\}$. We may assume that $\delta \leq \varepsilon$. Let $\beta < \beta'$ be the first two elements of B_0. Choose $\gamma \in C_0$ such that

$$g(\alpha,\gamma) > g(\alpha,\beta') > g(\alpha,\beta).$$

By the choice of the mapping e, we have that

$$e(g(\alpha,\beta), g(\alpha,\gamma)) \neq e(g(\alpha,\beta'), g(\alpha,\gamma)).$$

It follows that the second term of $f(\alpha,\beta,\gamma)$ is not equal to the second term of $f(\alpha,\beta',\gamma)$, and so in particular $f(\alpha,\beta,\gamma) \neq f(\alpha,\beta',\gamma)$, as required. □

Corollary 7.5.5. *If $\square_{\mathfrak{c}^+}$ holds, then there is a σ-closed $\mathfrak{c}^+$-cc forcing notion that introduces a mapping $f : (\mathfrak{c}^{++})^3 \longrightarrow \omega$ which is not constant on any product $A \times B \times C$ of three uncountable subsets of $\mathfrak{c}^{++}$.* □

Definition 7.5.6. The Δ-*function* of some family $\mathcal{F}$ of subsets of some ordinal κ (respectively, a family of functions with domain κ) is the function $\Delta : [\mathcal{F}]^2 \longrightarrow \kappa$ defined by

$$\Delta(f, g) = \min(f \bigtriangleup g),$$

(respectively, $\Delta(f, g) = \min\{\xi : f(\xi) \neq g(\xi)\}$).

Note the following property of Δ:

Lemma 7.5.7. $\Delta(f, g) \geq \min\{\Delta(f, h), \Delta(g, h)\}$ *for all* $\{f, g, h\} \in [\mathcal{F}]^3$. □

Remark 7.5.8. This property can be very useful when transferring objects that live on κ to objects on $\mathcal{F}$. This is especially interesting when $\mathcal{F}$ is of size larger than κ while all of its restrictions $\mathcal{F} \restriction \nu = \{f \cap \nu : f \in \mathcal{F}\}$ $(\nu < \kappa)$ have size $< \kappa$, i.e., when $\mathcal{F}$ is a *Kurepa family* (see for example [18]). We shall now see that it is possible to have a Kurepa family $\mathcal{F} = \{f_\alpha : \alpha < \kappa^+\}$ whose Δ-function is dominated by ρ, i.e., $\Delta(f_\alpha, f_\beta) \leq \rho(\alpha, \beta)$ for all $\alpha < \beta < \kappa^+$.

Theorem 7.5.9. *If $\square_\kappa$ holds and κ is λ-inaccessible, then there is a λ-closed κ-cc forcing notion $\mathcal{P}$ that introduces a Kurepa family on κ.*

Proof. Going to a larger cardinal, we may assume that λ is regular. Put p in $\mathcal{P}$, if p is a one-to-one function from a subset of κ^+ of size $< \lambda$ into the family of all subsets of κ of size $< \lambda$ such that for all α and β in $\mathrm{dom}(p)$:

$$p(\alpha) \cap p(\beta) \text{ is an initial part of } p(\alpha) \text{ and of } p(\beta), \tag{7.5.14}$$

$$\Delta(p(\alpha), p(\beta)) \leq \rho(\alpha, \beta) \text{ provided that } \alpha \neq \beta. \tag{7.5.15}$$

Let $p \leq q$ whenever $\mathrm{dom}(p) \supseteq \mathrm{dom}(q)$ and $p(\alpha) \supseteq q(\alpha)$ for all $\alpha \in \mathrm{dom}(q)$. Clearly $\mathcal{P}$ is a λ-closed forcing notion. To see that it satisfies the κ-chain condition, let A be a given subset of $\mathcal{P}$ of size κ. By the assumption that κ is λ-inaccessible, going to a subfamily of A of size κ, we may assume that A forms a Δ-system, or more precisely, that $\mathrm{dom}(p)$ $(p \in A)$ forms a Δ-system with root d, that $\bigcup \mathrm{rng}(p)$ $(p \in A)$ forms a Δ-system with root c and that any two members of A generate isomorphic structures via the natural isomorphism that fixes the roots. By Lemma 7.4.7 there exists $B \subseteq A$ of size κ such that for all $p, q \in B$, all $\alpha \in \mathrm{dom}(p) \setminus \mathrm{dom}(q)$, all $\beta \in \mathrm{dom}(q) \setminus \mathrm{dom}(p)$ and all $\gamma \in \mathrm{dom}(p) \cap \mathrm{dom}(q) (= d)$:

$$\rho\{\alpha, \beta\} > \max(c), \tag{7.5.16}$$

$$\rho\{\alpha, \beta\} \geq \min\{\rho\{\alpha, \gamma\}, \rho\{\beta, \gamma\}\}. \tag{7.5.17}$$

Suppose $p, q \in B$ and that $a = (\bigcup \mathrm{rng}(p)) \setminus c$ lies entirely below $b = (\bigcup \mathrm{rng}(q)) \setminus c$. Define $r : \mathrm{dom}(p) \cup \mathrm{dom}(q) \longrightarrow [\kappa]^{<\lambda}$ as follows. For $\gamma \in d$, let $r(\gamma) = p(\gamma) \cup q(\gamma)$.

If $\beta \in \operatorname{dom}(q) \setminus \operatorname{dom}(p)$ and if there is $\gamma \in d$ such that $\Delta(q(\beta), q(\gamma)) \in b$, the γ must be unique. Put $r(\beta) = p(\gamma) \cup q(\beta)$ in this case. If such γ does not exist, put $r(\beta) = q(\beta) \cup \{\max(c) + 1\}$. For $\alpha \in \operatorname{dom}(p) \setminus \operatorname{dom}(q)$, put $r(\alpha) = p(\alpha)$. Note that this choice of r does not change the behavior of Δ on pairs coming from $[\operatorname{dom}(p)]^2$ or $[\operatorname{dom}(q)]^2$, so in checking that r satisfies (7.5.14) it suffices to assume $\alpha \in \operatorname{dom}(p) \setminus \operatorname{dom}(q)$ and $\beta \in \operatorname{dom}(q) \setminus \operatorname{dom}(p)$. If $r(\beta)$ was defined as $p(\gamma) \cup q(\beta)$ for some $\gamma \in d$, we have that $\Delta(r(\alpha), r(\beta)) = \Delta(p(\alpha), p(\gamma))$ and that $\Delta(q(\beta), q(\gamma)) \in b$. It follows that

$$\begin{aligned}\Delta(r(\alpha), r(\beta)) &= \min\{\Delta(p(\alpha), p(\gamma)), \Delta(q(\beta), q(\gamma))\} \\ &\leq \min\{\rho\{\alpha, \gamma\}, \rho\{\beta, \gamma\}\}.\end{aligned} \tag{7.5.18}$$

Applying (7.5.17) we get the desired inequality $\Delta(r(\alpha), r(\beta)) \leq \rho\{\alpha, \beta\}$. If $r(\beta)$ was chosen to be $q(\beta) \cup \{\max(c) + 1\}$, then $\Delta(r(\alpha), r(\beta)) \leq \max(c) + 1$ and this is $\leq \rho\{\alpha, \beta\}$ by (7.5.16).

Let $\dot{G}$ be the generic filter of $\mathcal{P}$ and let $\dot{\mathcal{F}} = \{\dot{f}_\alpha : \alpha < \kappa^+\}$, where $\dot{f}_\alpha$ is the union of $p(\alpha)$ $(p \in \dot{G})$. A simple genericity argument shows that each $\dot{f}_\alpha$ intersects every interval of the form $[\nu, \nu + \lambda)$ $(\nu < \kappa)$. It follows (see (7.5.14)) that $\dot{\mathcal{F}} \restriction \nu$ has size at most λ for all $\nu < \kappa$. This finishes the proof. □

Corollary 7.5.10. *If $\square_{\omega_1}$ holds, and so in particular if ω_2 is not a Mahlo cardinal in the constructible universe, then there is a property K poset, forcing the Kurepa hypothesis.* □

Remark 7.5.11. This is a slight strengthening of a result of Jensen who proved that under $\square_{\omega_1}$ there is a ccc poset forcing the Kurepa hypothesis. It should be noted that Jensen also proved (see [53]) that in the Levy collapse of a Mahlo cardinal to ω_2 there is no ccc poset forcing the Kurepa hypothesis. Veličković [125] was the first to use ρ to re-prove Jensen's original result, though his proof worked only in case $\kappa = \omega_1$ and produced ccc rather than property K forcing. He has also shown (see [125]) that ρ easily yields an example of a function with the 'Δ-property' in the sense of Baumgartner and Shelah [11]. Recall that a function with the Δ-property in the sense of [11] is a function that satisfies some of the properties of the function D of Lemma 7.4.9 and a function that forms a ground for the well-known forcing construction of a locally compact scattered topology on ω_2, all of whose Cantor–Bendixson ranks are countable. We now use the deeper analysis of ρ to improve the chain-condition of the Baumgartner–Shelah forcing.

Theorem 7.5.12. *If $\square_{\omega_1}$ holds, then there is a property K forcing notion that introduces a locally compact scattered topology on ω_2, all of whose Cantor–Bendixson ranks are countable.*

Proof. Let $\mathcal{P}$ be the set of all $p = \langle D_p, \leq_p, M_p \rangle$ where D_p is a finite subset of ω_2, where $\leq_p$ is a partial ordering of D_p compatible with the well-ordering and $M_p : [D_p]^2 \longrightarrow [\omega_2]^{<\omega}$ has the following properties:

$$M_p\{\alpha, \beta\} \subseteq D\{\alpha, \beta\} \cap D_p, \tag{7.5.19}$$

$$M_p\{\alpha,\beta\} = \{\alpha\} \text{ if } \alpha \leq_p \beta \text{ and } M_p\{\alpha,\beta\} = \{\beta\} \text{ if } \beta \leq_p \alpha, \tag{7.5.20}$$

$$\gamma \leq_p \alpha,\beta \text{ for all } \gamma \in M_p\{\alpha,\beta\}, \tag{7.5.21}$$

$$\text{for every } \delta \leq_p \alpha,\beta \text{ there is } \gamma \in M_p\{\alpha,\beta\} \text{ such that } \delta \leq_p \gamma. \tag{7.5.22}$$

We let $p \leq q$ if and only if $D_p \supseteq D_q$, $\leq_p\restriction D_q = \leq_q$ and $M_p \restriction [D_q]^2 = M_q$. To verify that $\mathcal{P}$ satisfies Knaster's chain condition, let A be a given uncountable subset of $\mathcal{P}$. Shrinking A we may assume that A forms a Δ-system with root R and that the conditions of A generate isomorphic structures over R. By Lemma 7.4.9 there is an uncountable subset B of A such that for all p and q in B and all $\alpha \in D_p \setminus D_q$, $\beta \in D_q \setminus D_p$ and $\gamma \in R = D_p \cap D_q$:

$$\alpha,\beta > \gamma \quad \text{implies} \quad D\{\alpha,\gamma\} \cup D\{\beta,\gamma\} \subseteq D\{\alpha,\beta\}, \tag{7.5.23}$$

$$\beta > \gamma \quad \text{implies} \quad D\{\alpha,\gamma\} \subseteq D\{\alpha,\beta\}, \tag{7.5.24}$$

$$\alpha > \gamma \quad \text{implies} \quad D\{\beta,\gamma\} \subseteq D\{\alpha,\beta\}. \tag{7.5.25}$$

Consider two conditions p and q from B. Let $r = \langle D_r, \leq_r, M_r\rangle$ be defined as follows: $D_r = D_p \cup D_q$ and $\alpha \leq_r \beta$ if and only if $\alpha \leq_p \beta$, or $\alpha \leq_q \beta$, or there is $\gamma \in R$ such that $\alpha \leq_p \gamma \leq_q \beta$ or $\alpha \leq_q \gamma \leq_p \beta$. It is easily checked that $\leq_r$ is a partial ordering on D_r compatible with the well-ordering such that $\leq_r\restriction D_p = \leq_p$ and $\leq_r\restriction D_q = \leq_q$. Define $M_r : [D_r]^2 \longrightarrow [\omega_2]^{<\omega}$ by letting $M_r \restriction [D_p]^2 = M_p$, $M_r \restriction [D_q]^2 = M_q$, $M_r\{\alpha,\beta\} = \{\alpha\}$ if $\alpha \leq_r \beta$, $M_r\{\alpha,\beta\} = \{\beta\}$ if $\beta \leq_r \alpha$ and if $\alpha \in D_p \setminus R$ and $\beta \in D_q \setminus R$ are $\leq_r$-incomparable, set

$$M_r\{\alpha,\beta\} = \{\gamma \in D\{\alpha,\beta\} \cap D_r : \gamma \leq_r \alpha,\beta\}. \tag{7.5.26}$$

It is nontrivial to check that a so-defined r satisfies (7.5.22). So let $\delta \leq_r \alpha,\beta$ be given. If $\alpha,\beta \in D_p \setminus D_q$ and $\delta \in D_q \setminus D_p$, then there exist $\gamma_\alpha,\gamma_\beta \in R$ such that $\delta \leq_q \gamma_\alpha \leq_p \alpha$ and $\delta \leq_q \gamma_\beta \leq_p \beta$. By (7.5.22) for q there exists $\gamma \in M_q\{\gamma_\alpha,\gamma_\beta\}$ such that $\delta \leq_q \gamma$. Since $M_q\{\gamma_\alpha,\gamma_\beta\} \subseteq R$ we conclude that $\gamma \in R$. Thus $\gamma \leq_p \alpha,\beta$, so by (7.5.22) there is $\bar{\gamma} \in M_p\{\alpha,\beta\}$ such that $\gamma \leq_p \bar{\gamma}$ and therefore $\delta \leq_r \bar{\gamma}$. The case $\alpha \in R$, $\beta \in D_p \setminus D$ and $\delta \in D_q \setminus D_p$ is quite similar.

Consider now the case when $\alpha \in D_p \setminus D_q$, $\beta \in D_q \setminus D_p$ and they are $\leq_r$-incomparable, i.e., $M_r\{\alpha,\beta\}$ is given by (7.5.26). For symmetry we may consider only the subcase $\delta \leq_r \alpha,\beta$ with $\delta \in D_p$. Then $\delta \leq_p \alpha$ and $\delta \leq_p \gamma \leq_q \beta$ for some $\gamma \in R$. By (7.5.22) for p there is $\bar{\gamma} \in M_p\{\alpha,\gamma\}$ such that $\delta \leq_p \bar{\gamma}$. It remains to show that $\bar{\gamma}$ was put in $M_r\{\alpha,\beta\}$ in (7.5.26), i.e., that $\bar{\gamma}$ belongs to $D\{\alpha,\beta\}$. Note that since $\leq_r$ agrees with the well-ordering, we have that $\beta > \gamma$, so by (7.5.24) we have that $D\{\alpha,\gamma\} \subseteq D\{\alpha,\beta\}$. By (7.5.19) for p we have that $\bar{\gamma} \in M_p\{\alpha,\gamma\} \subseteq D\{\alpha,\gamma\}$ so we can conclude that $\bar{\gamma} \in D\{\alpha,\beta\}$. The rest of the cases are symmetric to the ones above.

Let $\dot{G}$ be a generic filter of $\mathcal{P}$ and let $\leq_{\dot{G}}$ be the union of $\leq_p$ $(p \in \dot{G})$. For $\alpha < \omega_2$ set $\dot{B}_\alpha = \{\xi : \xi \leq_{\dot{G}} \alpha\}$. Let τ be the topology on ω_2 generated by $\dot{B}_\alpha$ $(\alpha < \omega_2)$ as a clopen subbasis. It is easily checked that τ is a Hausdorff locally compact scattered topology on ω_2. A simple density argument and induction on α shows that the interval $[\omega\alpha, \omega\alpha + \omega)$ is the αth Cantor-Bendixson rank of the space (ω_2, τ). □

Definition 7.5.13. A function $f : [\omega_2]^2 \longrightarrow [\omega_2]^{\leq\omega}$ has *property* Δ if for every uncountable set A of finite subsets of ω_2 there exist a and b in A such that for all $\alpha \in a \setminus b$, $\beta \in b \setminus a$ and $\gamma \in a \cap b$:

$$\begin{array}{llll} (a) & \alpha, \beta > \gamma & \text{implies} & \gamma \in f\{\alpha, \beta\}, \\ (b) & \beta > \gamma & \text{implies} & f\{\alpha, \gamma\} \subseteq f\{\alpha, \beta\}, \\ (c) & \alpha > \gamma & \text{implies} & f\{\beta, \gamma\} \subseteq f\{\alpha, \beta\}. \end{array}$$

Remark 7.5.14. This definition is due to Baumgartner and Shelah [11] who have used it in their well-known forcing construction. It should be noted that they were also able to force a function with the property Δ using a σ-closed ω_2-cc poset. As shown above (a fact first checked in [11, p. 129]), the function

$$D\{\alpha, \beta\} = \{\xi \leq \min\{\alpha, \beta\} : \rho(\xi, \alpha) \leq \rho\{\alpha, \beta\}\}$$

has property Δ. However, Lemma 7.4.9 shows that the function D has many other properties that are of independent interest and that are likely to be needed in other forcing constructions of this sort. The reader is referred to papers of Koszmider [60] and Rabus [84] for further work in this area.

7.6 The function ρ on successors of singular cardinals

In the previous section we have shown that the function $\rho : [\kappa^+]^2 \longrightarrow \kappa$ based on a $\square_\kappa$-sequence C_α $(\alpha < \kappa^+)$ can be a quite useful tool in stepping-up objects from κ to κ^+. In this section we analyse the stepping-up power of ρ under the assumption that κ is a singular cardinal of cofinality ω. So let κ_n $(n < \omega)$ be a strictly increasing sequence of regular cardinals converging to κ fixed from now on. This immediately gives rise to a rather striking tree decomposition $<_n$ $(n < \omega)$ of the $\in$-relation on κ^+:

$$\alpha <_n \beta \text{ if and only if } \alpha < \beta \text{ and } \rho(\alpha, \beta) \leq \kappa_n. \tag{7.6.1}$$

Lemma 7.6.1.

(1) $\in\restriction \kappa^+ = \bigcup_{n<\omega} <_n$,

(2) $<_n \subseteq <_{n+1}$,

(3) $(\kappa^+, <_n)$ *is a tree of height* $\leq \kappa_n^+$. □

The following information about the tree-orderings $<_n$ could sometimes be useful.

Lemma 7.6.2. *For every $n < \omega$ the tree-ordering $<_n$ is a closed relation in the sense that $\{\xi < \alpha : \xi <_n \alpha\}$ is a closed subset of α for every $\alpha < \kappa^+$. In, particular, the tree $(\kappa^+, <_n)$ has no chains of size κ_n^+.*

Proof. The first conclusion follows from Lemma 7.4.1. To see the second conclusion of the lemma, let b be a maximal chain of the tree $(\kappa^+, <_n)$ and suppose that $\delta = \sup(b)$ is a limit ordinal of uncountable cofinality. Then by the first conclusion, the set b is a closed and unbounded subset of δ. It follows that if C'_δ denotes the set of limit points of C_δ, then the intersection $c = b \cap C'_\delta$ is also closed and unbounded in δ. Consider two ordinals $\alpha < \beta$ from c. Then $\alpha <_n \beta$, and therefore,

$$\kappa_n \geq \rho(\alpha, \beta) \geq \mathrm{tp}(C_\beta \cap \alpha).$$

Since $C_\delta \cap \alpha = C_\alpha$ and $C_\delta \cap \beta = C_\beta$, we conclude that $\mathrm{tp}(C_\delta \cap \alpha) \leq \kappa_n$. Since this is true for all $\alpha \in c$, it follows that C_δ has order-type at most κ_n. So the tree $(\kappa^+, <_n)$ cannot have chains of order-type κ_n^+. □

For example, one can use Lemma 7.6.2 to prove the following unboundedness property of ρ.

Theorem 7.6.3. *Suppose that for some $n < \omega$ we have a family A of κ_n^+ pairwise-disjoint finite subsets of κ^+. Then there is a subfamily $B \subseteq A$ of size κ_n^+ such that for every $a \neq b$ in B, we have that $\rho(\alpha, \beta) > \kappa_n$ for all $\alpha \in a$ and $\beta \in b$.*

Proof. By Lemma 7.6.2 the tree $(\kappa^+, <_n)$ cannot contain a Souslin subtree of height κ_n^+. So in particular, every subset of this tree of cardinality κ_n^+ contains an antichain of equal size. Using this, and arguing as in the proof of Theorem 7.4.5, one sees that an application of the Dushnik–Miller theorem gives us the conclusion. □

Definition 7.6.4. For $\alpha < \kappa^+$ and $n < \omega$, set

$$f_\alpha(n) = \mathrm{tp}(P_n(\alpha)),$$

where $P_n(\alpha) = \{\xi \leq \alpha : \rho(\xi, \alpha) \leq \kappa_n\}$. Let

$$L(\rho) = \{f_\alpha : \alpha < \kappa^+\},$$

considered as a linearly ordered set with the lexicographical ordering.

Lemma 7.6.5. *The linearly ordered set $L(\rho)$ has an order-dense subset of size κ, and so in particular it contains no well-ordered subset of size κ^+.*

Proof. For each finite sequence t of ordinals $< \kappa$ let $\alpha(t) < \kappa^+$ be chosen in the following manner. If there is $\alpha < \kappa^+$ such that f_α is the lexicographically minimal member of $L(\rho)$ end-extending t we let $\alpha(t)$ be equal to this α. Otherwise, let $\alpha(t)$ be the minimal $\alpha < \kappa^+$ such that f_α end-extends t, and if such an α cannot be found, let $\alpha(t) = 0$. Then $\{f_{\alpha(t)} : t \in \kappa^{<\omega}\}$ is an order-dense subset of $L(\rho)$ of size κ. □

The following result shows that, while $L(\rho)$ contains no well-ordered subset of size κ^+, every subset of $L(\rho)$ of size smaller than κ^+ is the union of countably many well-ordered subsets.

Lemma 7.6.6. *For every $\beta < \kappa^+$, the set $L_\beta(\rho) = \{f_\alpha : \alpha < \beta\}$ can be decomposed into countably many well-ordered subsets.*

Proof. Let $L_{\beta n}(\rho) = \{f_\alpha : \alpha \in P_n(\beta)\}$ for $n < \omega$. Note that the projection $f \longmapsto f \restriction (n+1)$ is one-to-one and lexicographically order preserving on $L_{\beta n}(\rho)$. It follows that each $L_{\beta n}(\rho)$ is a lexicographically well-ordered set. □

Example 7.6.7. Note that

$$\mathcal{K}(\rho) = \{\{(n, f_\alpha(n)) : n < \omega\} : \alpha < \kappa^+\}$$

is a family of countable subsets of $\omega \times \kappa$ which has the property that

$$\mathcal{K}(\rho) \restriction X = \{K \cap X : K \in \mathcal{K}\}$$

has size $\leq |X| + \aleph_0$ for every $X \subseteq \omega \times \kappa$ of size $< \kappa$. In other words, when $\mathrm{cf}(\kappa) = \omega$, the function $\rho : [\kappa^+] \to \kappa$ is giving us a particular example of a *Kurepa family* on κ. We now give an application of the existence of such a family of countable sets. This requires a few standard notions from measure theory and real analysis.

Definition 7.6.8. A *Radon measure space* is a Hausdorff topological space X with a σ-finite measure μ defined on a field of subsets of X containing open sets with the property that the measure of every set is equal to the supremum of measures of its compact subsets. When $\mu(X) = 1$, then (X, μ) is called *Radon probability space*. Finally recall that the *Maharam type* of (X, μ) is the least cardinality of a subset of the measure algebra of μ which completely generates the algebra.

Definition 7.6.9. A directed set D is *Tukey reducible* to a directed set E, written as $D \leq E$, if there is $f : D \longrightarrow E$ which maps unbounded sets to unbounded sets. In case D and E are ideals of subsets, this is equivalent to saying that there is $q : E \longrightarrow D$ which is monotone and has cofinal range. We say that D and E are *Tukey equivalent*, $D \equiv E$, whenever $D \leq E$ and $E \leq D$.

We shall study Tukey reducibility in a class of directed sets that includes the particular example of a directed set, the directed set $[\kappa]^\omega$ of all countable subsets of κ. The following turns out to be an important notion in this study.

Definition 7.6.10. A subset $\mathcal{K}$ of $[\kappa]^\omega$ is *locally countable* if $\{x \in \mathcal{K} : x \subseteq a\}$ is countable for all $a \in [\kappa]^\omega$.

As a closely related notion, we have the following well-known classical concept of local smallness introduced by Kurepa in [66].

Definition 7.6.11. A subset $\mathcal{K}$ of $[\kappa]^\omega$ is called a *Kurepa family*, or in short *K-family* if the restriction

$$\mathcal{K} \restriction a = \{x \cap a : x \in \mathcal{K}\}$$

is countable for every countable subset a of κ.

Clearly, every Kurepa family is locally countable but not vice versa. But these two notions are closely related as the following fact shows.

Lemma 7.6.12. *If $[\kappa]^\omega$ admits a locally countable family of some size θ, then it also admits a Kurepa family of size θ.*

Proof. Let $\mathcal{K}$ be a fixed locally countable family of countable and infinite subsets of some infinite cardinal κ and let $\theta = |\mathcal{K}|$. Let S be the set $\kappa^{<\omega}$ of finite sequences of ordinals $< \kappa$. Since S is equinumerous with κ, it suffices to produce a Kurepa family $\mathcal{K}^* \subseteq [S]^\omega$ of cardinality θ. For $x \in \mathcal{K}$, fix an onto map $f_x : \omega \longrightarrow x$, and set $x^* = \{f_x \restriction n : n < \omega\}$. Finally, set

$$\mathcal{K}^* = \{x^* : x \in K\}.$$

Since $x \mapsto x^*$ is a bijection between $\mathcal{K}$ and $\mathcal{K}^*$, the two families are equinumerous. To see that $\mathcal{K}^*$ satisfies Kurepa's condition consider a countable subset a^* of S. Replacing a^* by a bigger countable subset of S, we may assume that a^* is closed under maps $t \mapsto t \restriction n$ $(n < \omega)$. Let a be the countable subset of κ consisting of all ordinals mentioned in a finite sequence t appearing in a^*. Let

$$\mathcal{K}_a = \{x \in \mathcal{K} : x \subseteq a\}.$$

Then by our assumption about $\mathcal{K}$, the family $\mathcal{K}_a$ is countable. On the other hand, if an element x^* of $\mathcal{K}^*$ has an infinite intersection with a^*, then it is actually a subset of a^*. It follows that

$$\{x^* \cap a^* : x^* \in \mathcal{K}^*\} \subseteq \{x^* : x \in \mathcal{K}_a\} \cup [a^*]^{<\omega}.$$

It follows that $\{x^* \cap a^* : x^* \in \mathcal{K}^*\}$ is a countable family, which was to be shown. □

Here is a typical property that distinguishes between these two closely related notions.

Lemma 7.6.13. *If there is a locally countable subset of $[\kappa]^\omega$ of size* $\mathrm{cf}[\kappa]^\omega$, *then there is a locally countable family $\mathcal{K} \subseteq [\kappa]^\omega$ that is moreover cofinal*[6] *in $[\kappa]^\omega$.*

Proof. Fix a locally countable $\mathcal{H} \subseteq [\kappa]^\omega$ of cardinality $\mathrm{cf}[\kappa]^\omega$. It follows that there is a mapping $\varphi : \mathcal{H} \longrightarrow [\kappa]^\omega$ such that

(1) $a \subseteq \varphi(a)$ for every $a \in \mathcal{H}$,

(2) $\mathcal{K} = \{\varphi(a) : a \in \mathcal{H}\}$ is cofinal in $[\kappa]^\omega$.

[6]$\mathcal{K} \subseteq [\kappa]^\omega$ is *cofinal* in $[\kappa]^\omega$ if for every $a \in [\kappa]^\omega$ there is $x \in \mathcal{K}$ such that $a \subseteq x$.

Then for every countable subset b of κ,

$$\{x \in \mathcal{K} : x \subseteq b\} \subseteq \{\varphi(a) : a \in \mathcal{H},\ a \subseteq b\},$$

so by our assumption that $\mathcal{H}$ is locally countable, we conclude that $\{x \in \mathcal{K} : x \subseteq b\}$ is also countable. Thus $\mathcal{K}$ is a cofinal locally countable subset of $[\kappa]^\omega$. □

Corollary 7.6.14. *If for some infinite cardinal κ we have that $\mathrm{cf}[\kappa]^\omega = \kappa$, then there is a cofinal locally countable subset of $[\kappa]^\omega$.*

Proof. Any family of κ pairwise-disjoint countable subsets of κ satisfies the hypothesis of Lemma 7.6.13. □

Corollary 7.6.15. *For every positive integer n there is a cofinal locally countable subset of $[\omega_n]^\omega$.* □

Here is a typical application of the concept of cofinal locally countable families

Theorem 7.6.16. *If there is a locally countable subset of $[\kappa]^\omega$ of size $\mathrm{cf}[\kappa]^\omega$, or equivalently, if there is a locally countable cofinal subset of $[\kappa]^\omega$, then the null ideal of every atomless Radon probability space of Maharam type κ is Tukey equivalent to the product $\mathcal{N} \times [\kappa]^\omega$ where $\mathcal{N}$ is the ideal of measure zero subsets of the unit interval.*

Proof. We shall give an argument for the case $X = \{0,1\}^\kappa$ and μ the Haar measure of X, referring the reader to the general result from [37] that the null ideal of any atomless probability space of Maharam type κ is Tukey equivalent to the null ideal $\mathcal{N}_\kappa$ of $\{0,1\}^\kappa$. The only nontrivial direction is to produce a Tukey map from $\mathcal{N}_\kappa$ into $\mathcal{N}_\omega \times [\kappa]^\omega$. Fix a locally countable cofinal family $\mathcal{C} \subseteq [\kappa]^\omega$. For $c \in \mathcal{C}$ fix a bijection $e_c : c \longrightarrow \omega$. This gives us a way to define a function π_c mapping cylinders of $\{0,1\}^\kappa$ with supports included in c into 2^ω in the natural way. Let $\mathcal{N}_\kappa^{cl}$ be the family of all null subsets of $\{0,1\}^\kappa$ that are countably supported cylinders of $\{0,1\}^\kappa$. Clearly, $\mathcal{N}_\kappa^{cl}$ is cofinal in $\mathcal{N}_\kappa$, so it suffices to produce a Tukey map

$$f : \mathcal{N}_\kappa^{cl} \longrightarrow \mathcal{N}_\omega \times [\kappa]^\omega.$$

Given $N \in \mathcal{N}_\kappa^{cl}$, find $c(N) \in \mathcal{C}$ such that $\mathrm{supp}(N) \subseteq c(N)$ and put

$$f(N) = (\pi_{c(N)}{}''N, c(N)).$$

It is straightforward to check that f maps unbounded subsets of $\mathcal{N}_\kappa^{cl}$ to unbounded subsets of the product $\mathcal{N}_\omega \times [\kappa]^\omega$. □

Remark 7.6.17. Tukey theory of this kind originated in the two papers of Isbell [50] and [49] though Theorem 7.6.16 is due to Fremlin [37]. Note that the hypothesis of Fremlin's theorem is true when $[\kappa]^\omega$ has cofinality κ and so it is always satisfied for $\kappa < \aleph_\omega$. By Example 7.6.7 above, the hypothesis of Theorem 7.6.16 is satisfied for all κ if one assumes $\square_\kappa$ and $\mathrm{cf}[\kappa]^\omega = \kappa^+$ for every κ of countable cofinality. In fact, under these assumptions, we get a Kurepa family of cardinality $\mathrm{cf}[\kappa]^\omega = \kappa^+$. Note, however, that even if $\mathcal{K} \subseteq [\kappa]^\omega$ is assumed to be a Kurepa family of size $\mathrm{cf}[\kappa]^\omega$, the transfer to a cofinal locally countable $\mathcal{C} = \{a \cup h(a) : a \in \mathcal{K}\}$ seen above does not necessarily give us that $\mathcal{C}$ is a Kurepa family. To obtain a Kurepa family that is cofinal in $[\kappa]^\omega$, one needs to work a bit harder and use stronger assumptions.

Definition 7.6.18. If a family $\mathcal{K} \subseteq [S]^\omega$ is at the same time a Kurepa family and cofinal in $[S]^\omega$, then we call it a *cofinal Kurepa family* (cofinal K-family in short). Two cofinal K-families $\mathcal{H}$ and $\mathcal{K}$ are *compatible* if $H \cap K \in \mathcal{H} \cap \mathcal{K}$ for all $H \in \mathcal{H}$ and $K \in \mathcal{K}$. We say that $\mathcal{K}$ *extends* $\mathcal{H}$ if they are compatible and if $\mathcal{H} \subseteq \mathcal{K}$.

Remark 7.6.19. Note that the size of any cofinal K-family $\mathcal{K}$ on a set S is equal to the cofinality of $[S]^\omega$. Note also that for every $X \subseteq S$ there is $Y \supseteq X$ of size $\mathrm{cf}[X]^\omega$ such that $K \cap Y \in \mathcal{K}$ for all $K \in \mathcal{K}$.

Definition 7.6.20. Define $\mathrm{CK}(\theta)$ to be the statement that every sequence $\mathcal{K}_n$ ($n < \omega$) of comparable cofinal K-families with domains included in θ which are closed under $\cup$, $\cap$ and $\setminus$ can be extended to a single cofinal K-family on θ, which is also closed under these three operations.

Lemma 7.6.21. *$\mathrm{CK}(\omega_1)$ is true, and if $\mathrm{CK}(\theta)$ is true for some θ such that $\mathrm{cf}[\theta]^\omega = \theta$, then $\mathrm{CK}(\theta^+)$ is also true.*

Proof. The easy proof of $\mathrm{CK}(\omega_1)$ is left to the reader, so we concentrate on the rest of the conclusion of the lemma.

Suppose $\mathrm{CK}(\theta)$ and let $\mathcal{K}_n$ ($n < \omega$) be a given sequence of compatible cofinal K-families as in the hypothesis of $\mathrm{CK}(\theta^+)$. By Remark 7.6.19 there is a strictly increasing sequence δ_ξ ($\xi < \theta^+$) of ordinals $< \theta$ such that $\mathcal{K}_n \restriction \delta_\xi \subseteq \mathcal{K}_n$ for all $\xi < \theta^+$ and $n < \omega$. Recursively on $\xi < \theta^+$ we construct a chain $\mathcal{H}_\xi$ ($\xi < \theta^+$) of cofinal K-families as follows. If $\xi = 0$ or $\xi = \eta + 1$ for some η, using $\mathrm{CK}(\delta_\xi)$ we can find a cofinal K-family $\mathcal{H}_\xi$ on δ_ξ extending $\mathcal{H}_{\xi-1}$ and $\mathcal{K}_n \restriction \delta_\xi$ ($n < \omega$). If ξ has uncountable cofinality, then the union of $\bar{\mathcal{H}}_\xi = \bigcup_{\eta<\xi} \mathcal{H}_\eta$ is a cofinal K-family with domain included in δ_ξ, so using $\mathrm{CK}(\delta_\xi)$ we can find a cofinal K-family $\mathcal{H}_\xi$ on δ_ξ extending $\bar{\mathcal{H}}_\xi$ and $\mathcal{K}_n \restriction \delta_\xi$ ($n < \omega$). If ξ has countable cofinality, pick a sequence $\{\xi_n\}$ converging to ξ and use $\mathrm{CK}(\delta_\xi)$ to find a cofinal K-family $\mathcal{H}_\xi$ on δ_ξ extending $\mathcal{H}_{\xi_n}$ ($n < \omega$) and $\mathcal{K}_n$ ($n < \omega$). When the recursion is done, set $\mathcal{H} = \bigcup_{\xi<\theta^+} \mathcal{H}_\xi$. Then $\mathcal{H}$ is a cofinal K-family on θ^+ extending $\mathcal{K}_n$ ($n < \omega$). □

Corollary 7.6.22. *For each $n < \omega$ there is a cofinal Kurepa family on ω_n.* □

Definition 7.6.23. Let κ be a cardinal of cofinality ω. A *Jensen matrix* on κ^+ is a matrix $\mathrm{J}_{\alpha n}$ ($\alpha < \kappa^+, n < \omega$) of subsets of κ^+ with the following properties, where κ_n ($n < \omega$) is some fixed increasing sequence of cardinals converging to κ:

(1) $|\mathrm{J}_{\alpha n}| \leq \kappa_n$ for all $\alpha < \kappa^+$ and $n < \omega$,

(2) for all $\alpha < \beta$ and $n < \omega$ there is $m < \omega$ such that $\mathrm{J}_{\alpha n} \subseteq \mathrm{J}_{\beta m}$,

(3) $\bigcup_{n<\omega}[\mathrm{J}_{\beta n}]^\omega = \bigcup_{\alpha<\beta}\bigcup_{n<\omega}[\mathrm{J}_{\alpha n}]^\omega$ whenever $\mathrm{cf}(\beta) > \omega$,

(4) $[\kappa^+]^\omega = \bigcup_{\alpha<\kappa^+}\bigcup_{n<\omega}[\mathrm{J}_{\alpha n}]^\omega$.

Remark 7.6.24. The notion of a Jensen matrix is the combinatorial essence behind Silver's proof of Jensen's model-theoretic two-cardinal transfer theorem in the constructible universe (see [52, appendix]), so the matrix could equally well be called 'Silver matrix'. It has been implicitly or explicitly used in several places in the literature. The reader is referred to the paper of Foreman and Magidor [35] which gives a quite complete discussion of this notion and its occurrence in the literature.

Lemma 7.6.25. *Suppose some cardinal κ of countable cofinality carries a Jensen matrix $\mathrm{J}_{\alpha n}$ $(\alpha < \kappa^+, n < \omega)$ relative to some sequence of cardinals κ_n $(n < \omega)$ that converge to κ. If $\mathrm{CK}(\kappa_n)$ holds for all $n < \omega$, then $\mathrm{CK}(\kappa^+)$ is also true.*

Proof. Let $\mathcal{K}_n$ $(n < \omega)$ be a given sequence of compatible cofinal K-families with domains included in κ^+. Given $\mathrm{J}_{\alpha n}$, there is a natural continuous chain $\mathrm{J}^{\xi}_{\alpha n}$ $(\xi < \omega_1)$ of subsets of κ^+ of size $\leq \kappa_n$ such that $\mathrm{J}^0_{\alpha n} = \mathrm{J}_{\alpha n}$ and $\mathrm{J}^{\xi+1}_{\alpha n}$ equal to the union of all $K \in \bigcup_{n<\omega} \mathcal{K}_n$ which intersect $\mathrm{J}^{\xi}_{\alpha n}$. Let $\mathrm{J}^*_{\alpha n} = \bigcup_{\xi<\omega_1} \mathrm{J}^{\xi}_{\alpha n}$. It is easily seen that $\mathrm{J}^*_{\alpha n}$ $(\alpha < \kappa^+, n < \omega)$ is also a Jensen matrix. By recursion on α and n we define a sequence $\mathcal{H}_{\alpha n}$ $(\alpha < \kappa^+, n < \omega)$ of compatible cofinal K-families as follows. If $\alpha = \beta + 1$ or $\alpha = 0$ and $n < \omega$, using $\mathrm{CK}(\kappa_n)$ we can find a cofinal K-family $\mathcal{H}_{\alpha n}$ with domain $\mathrm{J}^*_{\alpha n}$ compatible with $\mathcal{H}_{\alpha m}$ $(m < n)$, $\mathcal{H}_{(\alpha-1)m}$ $(m < \omega)$ and $\mathcal{K}_m \restriction \mathrm{J}^*_{\alpha n}$ $(m < \omega)$. If $\mathrm{cf}(\alpha) = \omega$ let α_n $(n < \omega)$ be an increasing sequence of ordinals converging to α. Using $\mathrm{CK}(\kappa_n)$ we can find a cofinal K-family $\mathcal{H}_{\alpha n}$ which extends $\mathcal{H}_{\alpha m}$ $(m < n)$, $\mathcal{K}_m \restriction \mathrm{J}^*_{\alpha n}$ $(m < \omega)$ and each of the families $\mathcal{H}_{\alpha_i k}$ $(i < \omega, k < \omega$ and $\mathrm{J}^*_{\alpha_i k} \subseteq \mathrm{J}^*_{\alpha n})$. Finally suppose that $\mathrm{cf}(\alpha) > \omega$. For $n < \omega$, set

$$\mathcal{H}_{\alpha n} = [\mathrm{J}^*_{\alpha n}]^{\omega} \cap \Big(\bigcup_{\xi<\alpha} \bigcup_{m<\omega} \mathcal{H}_{\xi m}\Big).$$

Using the properties of the Jensen matrix (especially (3)) as well as the compatibility of $\mathcal{H}_{\xi m}$ $(\xi < \alpha, m < \omega)$ one easily checks that $\mathcal{H}_{\alpha n}$ is a cofinal K-family with domain $\mathrm{J}^*_{\alpha n}$ which extends each member of $\mathcal{H}_{\alpha m}$ $(m < n)$ and $\mathcal{K}_m \restriction \mathrm{J}^*_{\alpha n}$ $(m < \omega)$ and which is compatible with all of the previously constructed families $\mathcal{H}_{\xi m}$ $(\xi < \alpha, m < \omega)$. When the recursion is done we set

$$\mathcal{H} = \bigcup_{\alpha<\kappa^+} \bigcup_{n<\omega} \mathcal{H}_{\alpha n}.$$

Using the property (4) of $\mathrm{J}^*_{\alpha n}$ $(\alpha < \kappa^+, n < \omega)$, it follows easily that $\mathcal{H}$ is a cofinal K-family on κ^+ extending $\mathcal{K}_n$ $(n < \omega)$. □

Theorem 7.6.26. *If a Jensen matrix exists on any successor of a cardinal of cofinality ω, then a cofinal Kurepa family exists on any domain.* □

It should be noted that the ρ-function $\rho : [\kappa^+]^2 \longrightarrow \kappa$ associated with a $\square_\kappa$-sequence C_α $(\alpha < \kappa^+)$ for some singular cardinal κ of cofinality ω leads to the matrix

$$F_n(\alpha) = \{\xi < \alpha : \rho(\xi, \alpha) \leq n\} (\alpha < \kappa^+, n < \omega) \tag{7.6.2}$$

which has all the properties (1)–(3) of Definition 7.6.23 as well as some other properties not captured by the definition of a Jensen matrix, so let us expose some of them. The additional power of the ρ-matrix is most easily seen if one additionally has a sequence a_α $(\alpha < \kappa^+)$ of countable subsets of κ^+ that is cofinal in $[\kappa^+]^\omega$. Then one can extend the matrix (7.6.2) as follows:

$$M_{\beta n} = \bigcup_{\alpha <_n \beta} (a_\alpha \cup \{\alpha\}) \; (\beta < \kappa^+, n < \omega). \tag{7.6.3}$$

(Recall that $<_n$ is the tree-ordering on κ^+ defined by the formula $\alpha <_n \beta$ iff $\rho(\alpha, \beta) \leq \kappa_n$ where κ_n is a fixed increasing sequence of cardinals converging to κ.) The matrix

$$M_{\beta n} \; (\beta < \kappa^+, n < \omega)$$

has properties not captured by Definition 7.6.23 that are of independent interest, so let us state them in a separate lemma.

Lemma 7.6.27.

(1) $\alpha <_n \beta$ *implies* $M_{\alpha n} \subseteq M_{\beta n}$.

(2) $M_{\alpha m} \subseteq M_{\alpha n}$ *whenever* $m < n$.

(3) *If* $\beta = \sup\{\alpha : \alpha <_n \beta\}$ *then* $M_{\beta n} = \bigcup_{\alpha <_n \beta} M_{\alpha n}$.

(4) *Every countable subset of* κ^+ *is covered by some* $M_{\beta n}$.

(5) $\mathcal{M} = \{M_{\beta n} : \beta < \kappa^+, n < \omega\}$ *is a locally countable family if we have started with a locally countable* $\mathcal{K} = \{a_\alpha : \alpha < \kappa^+\}$. □

Remark 7.6.28. One can think of the matrix $\mathcal{M} = \{M_{\beta n} : \beta < \kappa^+, n < \omega\}$ as a version of a 'morass' for the singular cardinal κ (see [127]). It would be interesting to see how far one can go in this analogy. We give a few applications just to illustrate the usefulness of the families we have constructed so far.

Definition 7.6.29. A *Bernstein decomposition* of a topological space X is a function $f : X \longrightarrow 2^{\mathbb{N}}$ with the property that f takes all the values from $2^{\mathbb{N}}$ on any subset of X homeomorphic to the Cantor set.

Remark 7.6.30. The classical construction of Bernstein [14] can be interpreted by saying that every space of size at most continuum admits a Bernstein decomposition. For larger spaces one must assume Hausdorff's separation axiom, a result of Nešetril and Rődl (see [80]). In this context Malykhin was able to extend Bernstein's result to all spaces of size $< \mathfrak{c}^{+\omega}$ (see [73]). To extend this to all Hausdorff spaces, some use of square sequences seems natural. In fact, the first Bernstein decompositions of an arbitrary Hausdorff space were constructed using $\square_\kappa$ and $\kappa^\omega = \kappa^+$ for every $\kappa > \mathfrak{c}$ of cofinality ω by Weiss [132] and Wolfsdorf [134]. We shall now see that cofinal K-families are quite natural tools in constructions of Bernstein decompositions.

Theorem 7.6.31. *Suppose every regular cardinal $\theta > \mathfrak{c}$ supports a cofinal Kurepa family of size θ. Then every Hausdorff space admits a Bernstein decomposition.*

Proof. Let us say that a subset S of a topological space X is *sequentially closed in X* if it contains all limit points of sequences $\{x_n\} \subseteq S$ that converge in X. We say S is *σ-sequentially closed in X* if it can be decomposed into countably many sequentially closed sets. For a given cardinal θ, let $B(\theta)$ denote the statement that for every Hausdorff space X of size $\leq \theta$, every σ-sequentially closed subset S of X and every Bernstein decomposition $f : S \longrightarrow 2^{\mathbb{N}}$, there is a Bernstein decomposition $g : X \longrightarrow 2^{\mathbb{N}}$ extending f. Using the fact that a Hausdorff space of size $\leq \mathfrak{c}$ has at most $\mathfrak{c}$ copies of the Cantor set and following the idea of Bernstein's original proof, one sees that $B(\mathfrak{c})$ holds. Note also that by our assumption $\theta^{\aleph_0} = \theta$ for every cardinal $\theta \geq \mathfrak{c}$ of uncountable cofinality. So if we have $B(\theta)$ for such θ, one proves $B(\theta^+)$ by constructing a chain f_ξ $(\xi < \theta^+)$ of partial Bernstein decompositions on any Hausdorff topology τ on θ^+, using the fact that the set C of all $\delta < \theta^+$ which are σ-sequentially closed in (θ^+, τ) is closed and unbounded in θ^+. Thus if δ_ξ $(\xi < \theta^+)$ enumerates C increasingly, we will have $g_\xi : \delta_\xi \longrightarrow 2^{\mathbb{N}}$ and g_ξ compatible with the given partial Bernstein decomposition f of a given σ-sequentially closed set $S \subseteq \theta^+$. Note that $S \cap \delta_\xi$ will be σ-sequentially closed, so the use of $B(\delta_\xi)$ to get f_ξ will be possible. This inductive procedure encounters a crucial difficulty only when $\theta = \kappa^+$ for some $\kappa > \mathfrak{c}$ of cofinality ω. This is where a cofinal K-family $\mathcal{K}$ on κ^+ is useful. Let τ be a given Hausdorff topology on θ^+ and for $K \in \mathcal{K}$ let K' be the collection of all limits of τ-convergent sequences of elements of K. Note that if a set $\Gamma \subseteq \kappa^+$ is $\mathcal{K}$-closed in the sense that $\mathcal{K} \restriction \Gamma \subseteq \mathcal{K}$ (i.e., $K \cap \Gamma \in \mathcal{K}$ for all $K \in \mathcal{K}$), then Γ' will be equal to the union of all K' $(K \in \mathcal{K} \restriction \Gamma)$. So if $|\Gamma| > \mathfrak{c}$ is a cardinal of uncountable cofinality, then $|\Gamma'| = |\Gamma|$. It follows that every $\delta < \theta^+$ with the property that $\delta = \lim \delta_n$ for some sequence δ_n $(n < \omega)$ of ordinals $\leq \delta$ with the property that $\mathcal{K} \restriction \delta_n \subseteq \mathcal{K}$ for all n can be decomposed into countably many sets $\Gamma_{\delta n}$ $(n < \omega)$ such that for each n, the set $\Gamma_{\delta n}$ is σ-sequentially closed in τ and $|\Gamma_{\delta n}|$ is a cardinal $> \mathfrak{c}$ of uncountable cofinality. However, note that the set C of all such $\delta < \theta^+$ is closed and unbounded in θ^+ so as above we use our inductive hypothesis $B(\theta)$ $(\theta < \kappa)$ to construct a chain $g_\xi : \delta_\xi \longrightarrow 2^{\mathbb{N}}$ $(\xi < \theta^+)$ of Bernstein decompositions. This completes the proof. □

It is interesting that various less pathological classes of spaces admit a local version of Theorem 7.6.31.

Theorem 7.6.32. *A sigma-product*[7] *of the unit interval admits a Bernstein decomposition if its index set carries a cofinal Kurepa family. So in particular, any metric space that carries a cofinal Kurepa family admits a Bernstein decomposition.*

[7] Recall that a *sigma-product* is the subspace of some Tychonoff cube $[0,1]^\Gamma$ consisting of all mappings x such that $\mathrm{supp}(x) = \{\xi \in \Gamma : x(\xi) \neq 0\}$ is countable. The set Γ is its *index-set* .

Proof. So let $\mathcal{K}$ be a fixed well-founded cofinal K-family on Γ and let $<_w$ be a well-ordering of $\mathcal{K}$ compatible with $\subseteq$. For $K \in \mathcal{K}$ fix a Bernstein decomposition $f_K : X_K \longrightarrow 2^{\mathbb{N}}$ where $X_K = \{x \in X : \mathrm{supp}(x) \subseteq K\}$. Define $f : X \longrightarrow 2^{\mathbb{N}}$ by letting $f(x) = f_K(x)$ where K is the $<_w$-minimal $K \in \mathcal{K}$ such that $\mathrm{supp}(x) \subseteq K$. It is easily checked that f is a Bernstein decomposition of the sigma-product. □

Definition 7.6.33. Recall the notion of a *coherent family* of partial functions indexed by some ideal $\mathfrak{I}$, a family of the form $f_a : a \longrightarrow \omega$ $(a \in \mathfrak{I})$ with the property that $\{x \in a \cap b : f_a(x) \neq f_b(x)\}$ is finite for all $a, b \in \mathfrak{I}$.

It can be seen (see [122]) that the P-ideal dichotomy (see Definition 2.4.11) has a strong influence on such families provided $\mathfrak{I}$ is a P-ideal of countable subsets of some set Γ. The following result is one way to express this influence.

Theorem 7.6.34. *Assuming the P-ideal dichotomy, for every coherent family of functions $f_a : a \longrightarrow \omega$ $(a \in \mathfrak{I})$ indexed by some P-ideal $\mathfrak{I}$ of countable subsets of some set Γ, either*

(1) *there is an uncountable $\Delta \subseteq \Gamma$ such that $f_a \restriction \Delta$ is finite-to-one for all $a \in \mathfrak{I}$, or*

(2) *there is $g : \Gamma \longrightarrow \omega$ such that $g \restriction a =^* f_a$ for all $a \in \mathfrak{I}$.*

Proof. Let $\mathfrak{L}$ be the family of all countable subsets b of Γ for which one can find an a in $\mathfrak{I}$ such that $b \setminus a$ is finite and f_a is finite-to-one on b. To see that $\mathfrak{L}$ is a P-ideal, let $\{b_n\}$ be a given sequence of members of $\mathfrak{L}$ and for each n fix a member a_n of $\mathfrak{I}$ such that f_{a_n} is finite-to-one on b_n. Since $\mathfrak{I}$ is a P-ideal, we can find $a \in \mathfrak{I}$ such that $a_n \setminus a$ is finite for all n. Note that for all n, $b_n \setminus a$ is finite and that f_a is finite-to-one on b_n. For $n < \omega$, let

$$b_n^* = \{\xi \in b_n \cap a : f_a(\xi) > n\}.$$

Then b_n^* is a cofinite subset of b_n for each n, so if we set b to be equal to the union of the b_n^*'s, we get a subset of a which almost includes each b_n and on which f_a is finite-to-one. It follows that b belongs to $\mathfrak{L}$. This finishes the proof that $\mathfrak{L}$ is a P-ideal. Applying the P-ideal dichotomy to $\mathfrak{L}$, we get the two alternatives that translate into the alternatives (1) and (2) of Theorem 7.6.34. □

This leads to the natural question whether for any set Γ one can construct a family $\{f_a : a \longrightarrow \omega\}$ of finite-to-one mappings indexed by $[\Gamma]^\omega$. This question was answered by Koszmider [59] using the notion of a Jensen matrix discussed above. We shall present Koszmider's result using the notion of a cofinal Kurepa family instead.

Theorem 7.6.35. *If Γ carries a cofinal Kurepa family, then there is a coherent family $f_a : a \longrightarrow \omega$ $(a \in [\Gamma]^\omega)$ of finite-to-one mappings.*

Proof. Let $\mathcal{K}$ be a fixed well-founded cofinal K-family on Γ and let $<_w$ be a well-ordering of $\mathcal{K}$ compatible with $\subseteq$. It suffices to produce a coherent family of finite-to-one mappings indexed by $\mathcal{K}$. This is done by induction on $<_w$. Suppose $K \in \mathcal{K}$ and $f_H : H \longrightarrow \omega$ is determined for all $H \in \mathcal{K}$ with $H <_w K$. Let H_n $(n < \omega)$ be a sequence of elements of $\mathcal{K}$ that are $<_w K$ and have the property that for every $H \in \mathcal{K}$ with $H <_w K$ there is $n < \omega$ such that $H \cap K =^* H_n \cap K$. So it suffices to construct a finite-to-one $f_K : K \longrightarrow \omega$ which coheres with each f_{H_n} $(n < \omega)$, a straightforward task. □

Corollary 7.6.36. *For every non-negative integer n there is a coherent family*

$$f_a : a \longrightarrow \omega \quad (a \in [\omega_n]^\omega)$$

of finite-to-one mappings. □

Remark 7.6.37. It is interesting that 'finite-to-one' cannot be replaced by 'one-to-one' in these results. For example, there is no coherent family of one-to-one mappings $f_a : a \longrightarrow \omega$ $(a \in [\mathfrak{c}^+]^\omega)$. We finish this section with a typical application of coherent families of finite-to-one mappings discovered by Scheepers [87].

Theorem 7.6.38. *If there is a coherent family $f_a : a \longrightarrow \omega$ $(a \in [\Gamma]^\omega)$ of finite-to-one mappings, then there is a mapping*

$$S : [[\Gamma]^\omega]^2 \longrightarrow [\Gamma]^{<\omega}$$

with the property that $\bigcup_{n<\omega} a_n \subseteq \bigcup_{n<\omega} S(a_n, a_{n+1})$ for every strictly $\subseteq$-increasing sequence a_n $(n < \omega)$ of countable subsets of Γ.

Proof. For $a \in [\Gamma]^\omega$ let $x_a : \omega \longrightarrow \omega$ be defined by letting

$$x_a(n) = |\{\xi \in a : f_a(\xi) \leq n\}|.$$

Note that x_a is eventually dominated by x_b whenever a is a proper subset of b. Choose $\Phi : \omega^\omega \longrightarrow \omega^\omega$ with the property that $x <^* y$ implies $\Phi(y) <^* \Phi(x)$, where $<^*$ is the ordering of eventual dominance[8] on ω^ω. Define another family of functions $g_a : a \longrightarrow \omega$ $(a \in [\Gamma]^\omega)$ by letting

$$g_a(\xi) = \Phi(x_a)(f_a(\xi)). \tag{7.6.4}$$

Note the following interesting property of g_a $(a \in [\Gamma]^\omega)$:

$$S(a,b) = \{\xi \in a : g_b(\xi) \geq g_a(\xi)\} \text{ is finite for all } a \subsetneq b \text{ in } [\Gamma]^\omega. \tag{7.6.5}$$

Let a_n $(n < \omega)$ be a given strictly $\subseteq$-increasing sequence of countable subsets of Γ. Fix an ordinal $\bar{\xi}$ belonging to some $a_{\bar{n}}$. Then the sequence of integers

$$g_{a_n}(\bar{\xi}) \ (\bar{n} \leq n < \omega)$$

[8] Defined by, $x <^* y$ if and only if $x(n) < y(n)$ for all but finitely many n's.

cannot be strictly decreasing, so there must be some $n \geq \bar{n}$ with the property that $g_{a_n}(\bar{\xi}) \leq g_{a_{n+1}}(\bar{\xi})$. In other words, for every $\bar{n} < \omega$ and $\bar{\xi} \in a_{\bar{n}}$ there exists $n \geq \bar{n}$ such that $\bar{\xi} \in S(a_n, a_{n+1})$, as required. $\square$

Remark 7.6.39. Note that if κ is a singular cardinal of cofinality ω with the property that $\mathrm{cf}[\theta]^\omega < \kappa$ for all $\theta < \kappa$, then the existence of a cofinal Kurepa family on κ^+ implies the existence of a Jensen matrix on κ^+. So these two notions appear to be quite close to each other. The three basic properties of the function $\rho : [\kappa^+] \longrightarrow \kappa$ (Lemmas 7.3.8 and 7.3.7(a),(b)) seem much stronger in view of the fact that the linear ordering as in Lemma 7.6.6 cannot exist for κ above a supercompact cardinal and the fact that Foreman and Magidor [35] have produced a model with a supercompact cardinal that carries a Jensen matrix on any successor of a singular cardinal of cofinality ω. Note that Chang's conjecture of the form $(\kappa^+, \kappa) \twoheadrightarrow (\omega_1, \omega)$ for some singular κ of cofinality ω implies that every locally countable family $\mathcal{K} \subseteq [\kappa]^\omega$ must have size $\leq \kappa$. So, one of the models of set theory that has no cofinal K-family on, say $\aleph_{\omega+1}$, is the model of Levinski, Magidor and Shelah [70]. It seems still unknown whether the conclusion of Theorem 7.6.38 can be proved without additional set-theoretic assumptions.

Chapter 8

The Oscillation Mapping and the Square-bracket Operation

8.1 The oscillation mapping

In what follows, θ will be a fixed regular infinite cardinal.

$$\mathrm{osc} : \mathcal{P}(\theta)^2 \longrightarrow \mathrm{Card} \tag{8.1.1}$$

is defined by

$$\mathrm{osc}(x, y) = |x \setminus (\sup(x \cap y) + 1)/ \sim |, \tag{8.1.2}$$

where $\sim$ is the equivalence relation on $x \setminus (\sup(x \cap y) + 1)$ defined by letting $\alpha \sim \beta$ iff the closed interval determined by α and β contains no point from y. Hence, $\mathrm{osc}(x, y)$ is simply the number of convex pieces the set $x \setminus (\sup(x \cap y) + 1)$ is split by the set y (see Figure 8.1). Note that this is slightly different from the way we have defined the oscillation between two subsets x and y of ω_1 in Section 2.3 above, where $\mathrm{osc}(x, y)$ was the number of convex pieces the set x is split by into the set $y \setminus x$. Since the variation is rather minor, we keep the same old notation as there is no danger of confusion. The oscillation mapping has proven to be a useful device in various schemes for coding information. Its usefulness in a given context depends very much on the corresponding 'oscillation theory', a set of definitions and lemmas that disclose when it is possible to achieve a given number as oscillation between two sets x and y in a given family $\mathcal{X}$. The following definition reveals the notion of largeness relevant to the oscillation theory that we develop in this section.

Definition 8.1.1. A family $\mathcal{X} \subseteq \mathcal{P}(\theta)$ is *unbounded* if for every closed and unbounded subset C of θ there exist $x \in \mathcal{X}$ and an increasing sequence $\{\delta_n : n < \omega\} \subseteq C$ such that $\sup(x \cap \delta_n) < \delta_n$ and $[\delta_n, \delta_{n+1}) \cap x \neq \emptyset$ for all $n < \omega$.

This notion of unboundedness has proven to be the key behind a number of results asserting the complex behavior of the oscillation mapping on $\mathcal{X}^2$. The case $\theta = \omega$ seems to contain the deeper part of the oscillation theory known so far (see [110], [111, §1] and [121]), though in this section we shall only consider the case $\theta > \omega$. We shall also restrict ourselves to the family $\mathcal{K}(\theta)$ of all closed bounded subsets of θ rather than the whole power-set of θ. The following is the basic result about the behavior of the oscillation mapping in this context.

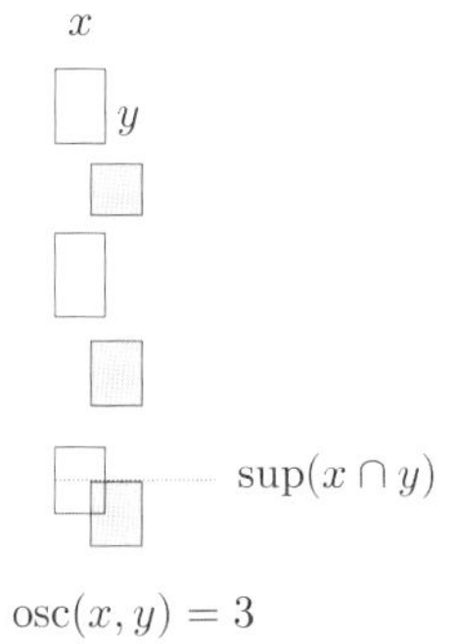

Figure 8.1: The oscillation mapping on the power-set of θ.

Lemma 8.1.2. *If $\mathcal{X}$ is an unbounded subfamily of $\mathcal{K}(\theta)$, then for every positive integer n there exist x and y in $\mathcal{X}$ such that* $\operatorname{osc}(x, y) = n$.

Proof. Choose an $\in$-chain $\mathcal{M}$ of length θ of elementary submodels M of some large enough structure of the form H_λ such that $\mathcal{X} \in M$ and $M \cap \theta \in \theta$. Applying the fact that $\mathcal{X}$ is unbounded with respect to the closed and unbounded set $\{M \cap \theta : M \in \mathcal{M}\}$, we can find $y \in \mathcal{X}$ such that:

$$\delta_i = M_i \cap \theta \notin y \text{ for all } i < n, \tag{8.1.3}$$

$$\delta_0 \cap y \neq \emptyset \text{ and } y \setminus \delta_{n-1} \neq \emptyset, \tag{8.1.4}$$

$$[\delta_{i-1}, \delta_i) \cap y \neq \emptyset \text{ for all } 0 < i < n. \tag{8.1.5}$$

Let

$$J_0 = [0, \sup(y \cap \delta_0)], \tag{8.1.6}$$

$$J_i = [\delta_{i-1}, \max(y \cap \delta_i)] \text{ for } 0 < i < n, \tag{8.1.7}$$

$$J_n = [\delta_{n-1}, \max(y)]. \tag{8.1.8}$$

Let $\mathcal{F}_n$ be the collection of all increasing sequences $I_0 < I_1 < \cdots < I_n$ of closed intervals of θ such that $I_0 = J_0$ and such that there is $x = x_I$ in $\mathcal{X}$ such that

$$x \subseteq I_0 \cup I_1 \cup \cdots \cup I_n, \tag{8.1.9}$$

$$\max(I_i) \in x \text{ for all } i \leq n. \tag{8.1.10}$$

Clearly, $\mathcal{F}_n \in M_0$ and $\langle J_0, J_1, \ldots, J_n\rangle \in \mathcal{F}_n$. Let $\mathcal{F}_{n-1}$ be the collection of all $\langle I_0, \ldots, I_{n-1}\rangle$ such that for every $\alpha < \theta$ there exists an interval I of θ above α such that $\langle I_0, \ldots, I_{n-1}, I\rangle \in \mathcal{F}_n$. Note that $\mathcal{F}_{n-1} \in M_0$ and that, using the elementarity of M_{n-1}, one proves that $\langle J_0, \ldots, J_{n-1}\rangle \in \mathcal{F}_{n-1}$. Let $\mathcal{F}_{n-2}$ be the collection of all $\langle I_0, \ldots, I_{n-2}\rangle$ such that for every $\alpha < \theta$ there exists interval I above α such that $\langle I_0, \ldots, I_{n-2}, I\rangle \in \mathcal{F}_{n-1}$. Then again $\mathcal{F}_{n-2} \in M_0$ and, using the elementarity of M_{n-2}, one shows that the restriction $\langle J_0, \ldots, J_{n-2}\rangle$ belongs to $\mathcal{F}_{n-2}$. Continuing in this way we construct families $\mathcal{F}_n, \mathcal{F}_{n-1}, \ldots, \mathcal{F}_0$ such that:

$$\mathcal{F}_i \in M_0 \text{ for all } i \leq n, \tag{8.1.11}$$

$$\mathcal{F}_0 = \{J_0\}. \tag{8.1.12}$$

$$\begin{aligned}&\langle I_0, \ldots, I_{n-i}\rangle \in \mathcal{F}_{n-i} \text{ implies that for all } \alpha < \theta \text{ there exists}\\ &\text{an interval } I \text{ above } \alpha \text{ such that } \langle I_0, \ldots, I_{n-i}, I\rangle \in \mathcal{F}_{n-i+1}.\end{aligned} \tag{8.1.13}$$

Recursively in $i \leq n$ we choose intervals $I_0 < I_1 < \cdots < I_n$ such that:

$$\langle I_0, \ldots, I_i\rangle \in \mathcal{F}_i \text{ for all } i \leq n, \tag{8.1.14}$$

$$J_{i-1} < I_i \in M_{i-1} \text{ for } 0 < i \leq n. \tag{8.1.15}$$

Clearly, there is no problem in choosing the sequence using the elementarity of the submodels M_{i-1} and the condition (8.1.13). By the definition $\langle I_0, \ldots, I_n\rangle \in \mathcal{F}_n$ means that there is $x \in \mathcal{X}$ satisfying (8.1.9) and (8.1.10). It follows that $\max(x \cap y) = \max(I_0)$ and that

$$x \cap I_i \neq \emptyset \quad (0 < i \leq n) \tag{8.1.16}$$

are the convex pieces of $x \setminus (\max(x \cap y) + 1)$ into which this set is split by y. It follows that $\operatorname{osc}(x, y) = n$. This finishes the proof. □

Lemma 8.1.2 also has a rectangular form.

Lemma 8.1.3. *If $\mathcal{X}$ and $\mathcal{Y}$ are two unbounded subfamilies of $\mathcal{K}(\theta)$, then for all but finitely many positive integers n there exist $x \in \mathcal{X}$ and $y \in \mathcal{Y}$ such that $\operatorname{osc}(x, y) = n$.* □

Recall the notion of a nontrivial C-sequence C_α $(\alpha < \theta)$ on θ from Section 6.3, a C-sequence with the property that, for every closed and unbounded subset C of θ, there is a limit point δ of C such that $C \cap \delta \nsubseteq C_\alpha$ for all $\alpha < \theta$.

Definition 8.1.4. For a subset D of θ let $\lim(D)$ denote the set of all $\alpha < \theta$ with the property that $\alpha = \sup(D \cap \alpha)$. A subsequence C_α $(\alpha \in \Gamma)$ of some C-sequence C_α $(\alpha < \theta)$ is *stationary* if the union of all $\lim(C_\alpha)$ $(\alpha \in \Gamma)$ is a stationary subset of θ.

Lemma 8.1.5. *A stationary subsequence of a nontrivial C-sequence on θ is an unbounded family of subsets of θ.*

Proof. Let C_α $(\alpha \in \Gamma)$ be a given stationary subsequence of a nontrivial C-sequence on θ. Let C be a given closed and unbounded subset of θ. Let Δ be the union of all $\lim(C_\alpha)$ $(\alpha \in \Gamma)$. Then Δ is a stationary subset of θ. For $\xi \in \Delta$ choose $\alpha_\xi \in \Gamma$ such that $\xi \in \lim(C_{\alpha_\xi})$. Applying the assumption that C_α $(\alpha \in \Gamma)$ is a nontrivial C-subsequence, we can find $\xi \in \Delta \cap \lim(C)$ such that

$$C \cap [\eta, \xi) \not\subseteq C_{\alpha_\xi} \text{ for all } \eta < \xi. \tag{8.1.17}$$

If such ξ cannot be found using the stationarity of the set $\Delta \cap \lim(C)$, we would be able to use the Pressing Down Lemma on the regressive mapping that would give us an $\eta < \xi$ violating (8.1.17) and get that a tail of C trivializes C_α $(\alpha \in \Gamma)$. Using (8.1.17) and the fact that $C_{\alpha_\xi} \cap \xi$ is unbounded in ξ, we can find a strictly increasing sequence δ_n $(n < \omega)$ of elements of $(C \cap \xi) \setminus C_{\alpha_\xi}$ such that $[\delta_n, \delta_{n+1}) \cap C_{\alpha_\xi} \neq \emptyset$ for all n. So the set C_{α_ξ} satisfies the conclusion of Definition 8.1.1 for the given closed and unbounded set C. □

Notation 8.1.6. Recall that $\mathbb{Q}_\theta$ denotes the set of all finite sequences of ordinals $< \theta$ and that we consider it totally ordered by the relation of right lexicographical ordering.[1] We need the notation for the following two further orderings on $\mathbb{Q}_\theta$:

$$s \sqsubseteq t \text{ if and only if } s \text{ is an initial segment of } t. \tag{8.1.18}$$

$$s \sqsubset t \text{ if and only if } s \text{ is a proper initial segment of } t. \tag{8.1.19}$$

Definition 8.1.7. Given a C-sequence C_α $(\alpha < \theta)$ we can define a *partial action* $(\alpha, t) \longmapsto \alpha_t$ of $\mathbb{Q}_\theta$ on θ recursively on the ordering $\sqsubseteq$ of $\mathbb{Q}_\theta$ as follows:

$$\alpha_\emptyset = \alpha, \tag{8.1.20}$$

$$\alpha_{\langle\xi\rangle} = \text{ the } \xi\text{th member of } C_\alpha \text{ if } \xi < \operatorname{tp}(C_\alpha), \tag{8.1.21}$$

$$\alpha_{\langle\xi\rangle} \text{ is undefined if } \xi \geq \operatorname{tp}(C_\alpha), \tag{8.1.22}$$

$$\alpha_{t^\frown\langle\xi\rangle} = (\alpha_t)_{\langle\xi\rangle}. \tag{8.1.23}$$

Whenever we write α_t, we implicitly assume that this ordinal is defined.

Remark 8.1.8. Note that if $\rho_0(\alpha, \beta) = t$ for some $\alpha < \beta < \theta$, then $\beta_t = \alpha$. In fact, if $\beta = \beta_0 > \cdots > \beta_n = \alpha$ is the walk from β to α along the C-sequence, each member of the trace $\operatorname{Tr}(\alpha, \beta) = \{\beta_0, \beta_1, \ldots, \beta_n\}$ has the form β_s where s is the uniquely determined initial part of t. Note, however, that in general $\beta_t = \alpha$ does not imply that $\rho_0(\alpha, \beta) = t$.

Notation 8.1.9. Given a C-sequence C_α $(\alpha < \theta)$ on θ we shall use the shorter notation

$$\operatorname{osc}(\alpha, \beta) = \operatorname{osc}(C_\alpha, C_\beta)$$

for the oscillation between two members of the C-sequence.

[1] Recall the notion of right lexicographical ordering: $s <_r t$ iff either t is an initial segment of s or else $s(\alpha) < t(\alpha)$ for $\alpha = \min\{\xi : s(\xi) \neq t(\xi)\}$.

We are now ready to define the higher cardinal analogue of the basic oscillation mapping on ω_1 exposed above in Section 2.3.

Definition 8.1.10. To every C-sequence C_α $(\alpha < \theta)$ we associate an oscillation mapping

$$o : [\theta]^2 \longrightarrow \omega$$

as follows. Given a pair of ordinals $\alpha < \beta < \theta$, if there is $t \sqsubseteq \rho_0(\alpha, \beta)$ such that α_t is defined and

(i) $\mathrm{osc}(\alpha_t, \beta_t) > 1$, but
(ii) $\mathrm{osc}(\alpha_s, \beta_s) = 1$ for all $s \sqsubset t$,

let $o(\alpha, \beta) = \mathrm{osc}(\alpha_t, \beta_t) - 2$; otherwise, let $o(\alpha, \beta) = 0$.

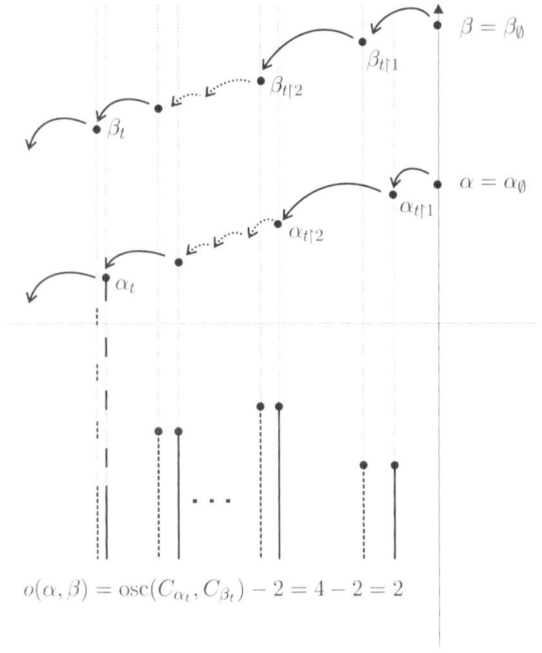

Figure 8.2: The oscillation mapping o.

Theorem 8.1.11. *If C_α $(\alpha < \theta)$ is a nontrivial C-sequence on θ and if*

$$o : [\theta]^2 \longrightarrow \omega$$

is the corresponding oscillation mapping, then for every unbounded set $\Gamma \subseteq \theta$ and every non-negative integer n there exist $\alpha < \beta$ in Γ such that $o(\alpha, \beta) = n$.

Proof. Choose an elementary submodel M of some large enough structure of the form H_λ such that $\bar{\delta} = M \cap \theta \in \theta$ and M contains all the relevant objects. Choose $\bar{\beta} \in \Gamma$ above $\bar{\delta}$ and let

$$\bar{\beta} = \bar{\beta}_0 > \bar{\beta}_1 > \cdots > \bar{\beta}_k = \bar{\delta}$$

be the minimal walk from $\bar{\beta}$ to $\bar{\delta}$ along the C-sequence. Let $\bar{k} = k-1$ if $C_{\bar{\beta}_{k-1}} \cap \bar{\delta}$ is unbounded in $\bar{\delta}$; otherwise $\bar{k} = k$. Recall our implicit assumption about essentially all C-sequences that we consider here:

(c) If γ is a limit ordinal, then an ordinal $\xi \in C_\gamma$ occupying a successor place in C_γ must be a successor ordinal.

In this case, since δ is a limit ordinal and since $\bar{\delta} \in C_{\bar{\beta}_{k-1}}$, either

(1) $\bar{\beta}_{k-1}$ is a limit ordinal and $\bar{\delta} = \sup(C_{\bar{\beta}_{k-1}} \cap \bar{\delta})$, or

(2) $\bar{\beta}_{k-1} = \bar{\delta} + 1$.

In any of the two cases, it follows that $\bar{\beta}_{\bar{k}}$ is a limit ordinal. Set $t = \rho_0(\bar{\beta}_{\bar{k}}, \bar{\beta})$. Let

$$\emptyset = t \restriction 0, t \restriction 1, \dots, t \restriction \bar{k} = t$$

be our notation for all the restrictions of the sequence t. Note that $\bar{\beta}_i = \bar{\beta}_{t \restriction i}$ for $i \leq k$. Let Γ_0 be the set of all $\beta \in \Gamma$ such that:

$$\beta = \beta_\emptyset > \beta_{t\restriction 1} > \cdots > \beta_t, \tag{8.1.24}$$

$$\sup(C_{\beta_{t\restriction i}} \cap \beta_t) = \sup(C_{\bar{\beta}_i} \cap \bar{\beta}_{\bar{k}}) \text{ for all } i < \bar{k}. \tag{8.1.25}$$

Note that $\Gamma_0 \in M$ and $\bar{\beta} \in \Gamma_0$. Let $\Delta = \{\beta_t : \beta \in \Gamma_0\}$. Then $\Delta \in M$ and $\bar{\beta}_{\bar{k}} \in \Delta$. By the very choice of $\bar{k}$, the ordinal $\bar{\delta}$ belongs to $\lim(C_{\bar{\beta}_{\bar{k}}})$, and so, using the elementarity of M, one concludes that C_δ $(\delta \in \Delta)$ is a stationary subsequence of our original sequence. Shrinking Δ a bit if necessary, we may assume that for every $\gamma \in \Delta$ an ordinal $\beta \in \Gamma_0$ such that $\beta_t = \gamma$ can be found below the next member of Δ above γ. By Lemma 8.1.5, the sequence C_δ $(\delta \in \Delta)$ forms an unbounded family of subsets of θ, and so we can apply the proof of our basic oscillation Lemma 8.1.2 and find an ordinal $\gamma < \delta$ of the form β_t for some $\beta \in \Gamma_0 \cap \delta$ such that

$$\operatorname{osc}(\beta_t, \bar{\beta}_t) = n + 2.$$

Using the fact that $\beta < \delta$ and the fact that both ordinals β and $\bar{\beta}$ satisfy (8.1.25), we conclude that $\operatorname{osc}(\beta_{t\restriction i}, \bar{\beta}_{t\restriction i}) = 1$ for all $i < \bar{k}$. It follows that $o(\beta, \bar{\beta}) = n$. This finishes the proof. □

Remark 8.1.12. The class of all regular cardinals θ that carry a nontrivial C-sequence is quite extensive. It includes not only all successor cardinals but also some inaccessible as well as hyperinaccessible cardinals such as for example, first inaccessible or first Mahlo cardinals. In other words, we would like to point out that Theorem 8.1.11 has corollaries of the following kind.

Corollary 8.1.13. *If θ is the first weakly inaccessible uncountable cardinal, then there is a mapping*

$$o : [\theta]^2 \longrightarrow \omega$$

which takes all the values from ω on any set of the form $[\Gamma]^2$ for Γ an unbounded subset of θ. □

In view of the well-known Ramsey-theoretic characterization of weak compactness, the statement of Theorem 8.1.11 leads us now to the following natural question.

Question 8.1.14. Can the weak compactness of a strong limit regular uncountable cardinal be characterized by the fact that for every mapping $f : [\theta]^2 \longrightarrow \omega$ there exists an unbounded set $\Gamma \subseteq \theta$ such that $f''[\Gamma]^2 \neq \omega$? This is true when ω is replaced by 2, but can any other number besides 2 be used in this characterization?

We shall now see how to adapt the basic idea of the proof of Theorem 8.1.11 in order to establish the following more informative and potentially more useful property of the oscillation mapping o.

Theorem 8.1.15. *If a regular uncountable cardinal θ carries a nontrivial C-sequence, then the corresponding oscillation mapping $o : [\theta]^2 \longrightarrow \omega$ has the property that for every family A of θ pairwise-disjoint finite subsets of θ, all of some fixed size n, there exist a subfamily $B \subseteq A$ of size θ, a mapping $h : n \times n \rightarrow \omega$, and an equivalence relation $\sim$ on the index-set $n = \{0, 1, \ldots, n-1\}$ such that*

$$o(a(i), b(j)) = h(i,j) \text{ for all } a < b \text{ in } B \text{ and } i \nsim j.$$

Moreover, for every subfamily $C \subseteq B$ of size θ there is $g : n \times n \rightarrow \omega$ which agrees with h on non-equivalent pairs such that every integer m there exist $a < b$ in C such that

$$o(a(i), b(j)) = g(i,j) + m \text{ for all } i, j < n \text{ such that } i \sim j.$$

Proof. As indicated above, we will build on the proof of Theorem 8.1.11 so a familiarity with that proof will be assumed below. As before we choose an elementary submodel M of some large enough structure of the form H_λ such that $\delta = M \cap \theta \in \theta$ and M contains all the relevant objects. Let $b \in A$ be minimal such that all of its ordinals are strictly above δ. For $\beta \in b$, consider the minimal walk

$$\beta = \beta_0(\delta) > \cdots > \beta_{k(\beta)}(\delta) = \delta$$

along the fixed nontrivial C-sequence C_α $(\alpha < \theta)$. Let $\bar{k}(\beta) = k(\beta) - 1$ if

$$\sup(C_{\beta_{k(\beta)-1}(\delta)}) \cap \delta = \delta;$$

otherwise, put $\bar{k}(\beta) = k(\beta)$. Let $t_\beta = \rho_0(\beta_{\bar{k}(\delta)}, \beta)$. Finally, for $i, j < n$, set $i \sim_b j$ if and only if when we set $\beta = b(i)$ and $\gamma = b(j)$, then we have:

(1) $k(\beta) = k(\gamma)$, $\bar{k}(\beta) = \bar{k}(\gamma)$ and $t_\beta = t_\gamma$,
(2) $\sup(C_{\beta_l(\delta)} \cap \delta) = \sup(C_{\gamma_l(\delta)} \cap \delta)$ for all $l < k$.

Using the elementarity of M we can find a stationary set $\Delta \subseteq \theta$ and for each $\delta \in \Delta$ a member b_δ of A above δ, a fixed equivalence relation $\sim$ on $\{0, 1, \ldots, n-1\}$ and for each equivalence class e of $\sim$ an integer k_e, a sequence t_e of ordinals $< \theta$, and a sequence ξ_l $(l < k_e)$ such that for each $\zeta \in \Delta$ and for each $\beta, \gamma \in b_\zeta$ occupying ith and jth place, respectively, with $i \sim j$, we have:

(3) $k(\beta) = k(\gamma)$, $\bar{k}(\beta) = \bar{k}(\gamma)$ and $t_\beta = t_\gamma = t_e$,
(4) $\sup(C_{\beta_l(\delta)} \cap \delta) = \sup(C_{\gamma_l(\delta)} \cap \delta) = \xi_l$ for all $l < k_e$,

where e denotes the equivalence class of i and j. Shrinking Δ, we may assume that for $\zeta < \eta$ in Δ, we have that $b_\zeta \subseteq \eta$. Let $B = \{b_\zeta : \zeta \in \Delta\}$. By our definition of the mapping o, it follows that for $a < b$ in B,

$$o(a(i), b(j)) = 0 \text{ whenever } i \nsim j. \tag{8.1.26}$$

For $\zeta \in \Delta$ and $i < n$, let $D_\zeta(i) = C_{\beta_{\bar{k}}(\zeta)} \cap \zeta$, where $\beta \in b_\zeta$ occupies the ith place. Note that each $D_\zeta(i)$ is a closed and unbounded subset of ζ. Clearly, we may assume that all these objects belong to our submodel M, that $\delta = M \cap \theta$ belongs to Δ and that $b = b_\delta$. Using a sequence of continuous $\in$-chains $\mathcal{M}_p$ of elementary submodels of H_λ, such that $\mathcal{M}_p \in M$ and such that $\mathcal{M}_p \in \min \mathcal{M}_{p+1}$, working as in the proof of Lemma 8.1.5 and applying the assumption about the nontriviality of the C-sequence C_α $(\alpha < \theta)$, we can find an $\in$-chain M_p $(p < \omega)$ of elementary submodels of H_λ containing the relevant objects such that $\delta_p = M_p \cap \theta \in \theta$, and

$$\sup(D_\delta(i) \cap \delta_p) < \delta_p \text{ and } D_\delta(i) \cap [\delta_p, \delta_{p+1}) \neq \emptyset \text{ for all } i < n \text{ and } p < \omega. \tag{8.1.27}$$

Define $h : n \times n \longrightarrow \omega$ by letting $h(i,j) = 0$ for $i \nsim j$ and let

$$h(i,j) = \operatorname{osc}(D_\delta(i) \cap \delta_0, D_\delta(j) \cap \delta_0) \text{ for } i \sim j. \tag{8.1.28}$$

Let m be a given positive integer. Using (8.1.27) and the elementarity of the first $m+2$ of our submodels M_p $(p \leq m+1)$, we can find an $a = b_\gamma \in B \cap M$ that add $m+2$ more oscillations to the existing ones, or in other words,

$$o(a(i), b(j)) = \operatorname{osc}(D_\gamma(i) \cap \delta_0, D_\delta(j) \cap \delta_0) + m \text{ whenever } i \sim j. \tag{8.1.29}$$

It is clear that using the elementarity of M_0, we can also arrange that

$$\operatorname{osc}(D_\gamma(i) \cap \delta_0, D_\delta(j) \cap \delta_0) = \operatorname{osc}(D_\delta(i) \cap \delta_0, D_\delta(j) \cap \delta_0) \text{ whenever } i \sim j. \tag{8.1.30}$$

Referring to the definition of h, we infer that $o(a(i), b(j)) = h(i,j) + m$ for all $i, j < n$ such that $i \sim j$, as required. This proves the lemma for $C = B$. The general case reduces to this one by getting first $B \subseteq A$ of the particular form $B = \{b_\zeta : \zeta \in \Delta\}$ as above for which the corresponding equivalence relation cannot be made finer by going to such a further subset of B. □

Definition 8.1.16. The mapping o of Theorem 8.1.15 has particularly useful projections when the cardinal θ is not bigger than the continuum. In order to define this projection, we need a one-to-one sequence r_α $(\alpha < \theta)$ of elements of 2^ω, and a list h_l $(l < \omega)$ of all mappings of the form $h : 2^n \times 2^n \to \omega$, where $n = n(h) < \omega$, in such a way that for every such mapping h the set $P_h = \{l : h_l = h\}$ contains an infinite arithmetic progression. Then the projection of o is defined as

$$o^*(\alpha, \beta) = h_{o(\alpha,\beta)}(r_\alpha \restriction n(h_{o(\alpha,\beta)}),\ r_\beta \restriction n(h_{o(\alpha,\beta)})).$$

Then Theorem 8.1.15 has the following immediate consequence deduced from it in a similar manner as in the case of the oscillation mapping and the square-bracket operation on ω_1 (see Chapter 5).

Theorem 8.1.17. *Suppose that a regular uncountable cardinal θ that is not bigger than the continuum carries a nontrivial C-sequence. Then the projection*

$$o^* : [\theta]^2 \longrightarrow \omega$$

of the oscillation mapping has the property that, for every family A of θ pairwise-disjoint finite subsets of θ, all of some fixed size n, there is a subfamily $B \subseteq A$ of size θ, a mapping $h : n \times n \to \omega$, and an equivalence relation on the index set $n = \{0, 1, \ldots, n-1\}$ such that

$$o^*(a(i), b(j)) = h(i, j) \text{ for all } a < b \text{ in } B \text{ and } i \not\sim j.$$

Moreover, for every subfamily $C \subseteq B$ of size θ and $g : n \times n \to \omega$ there exist $a < b$ in C such that

$$o^*(a(i), b(j)) = g(i) \text{ for all } i, j < n \text{ such that } i \sim j. \qquad \square$$

We finish this section with some typical applications of this projection. For example, the following result is proved on the basis of Theorem 8.1.17 in a similar manner as Theorem 5.3.3 above was proven on the basis of the corresponding result about the square-bracket operation.

Theorem 8.1.18. *Let Γ be an index-set of uncountable regular cardinality θ not bigger than the continuum and supporting a nontrivial C-sequence and let $\ell_1(\Gamma)$ be the complex version of the standard Banach space. Then there is a homogeneous polynomial*

$$P : \ell_1(\Gamma) \longrightarrow \mathbb{C}$$

that takes all the complex values on any closed subspace of $\ell_1(\Gamma)$ of density θ. $\square$

Corollary 8.1.19. *Suppose that the continuum $\mathfrak{c}$ is regular and not bigger than the first weakly inaccessible cardinal. Then the complex version of the space $\ell_1(\mathfrak{c})$ admits a complex-valued homogeneous polynomial of degree 2 which takes all the values on any closed subspace of $\ell_1(\mathfrak{c})$ of density continuum.* $\square$

Let us now mention another application of Theorem 8.1.17 deduced from it in a similar manner as Theorem 5.5.4 above.

Theorem 8.1.20. *Suppose that a regular uncountable cardinal θ that is not bigger than the continuum carries a nontrivial C-sequence. Let V and W be two vector spaces over the same field K such that V has dimension θ while W is countable. Then there is a non-degenerate symmetric bilinear mapping $\vartheta : V \times V \to W$ such that $\{\vartheta(x, y) : x, y \in X,\ x \neq y\} = W$ for every subset X of V of cardinality θ.* □

It should be clear (and we will come to this point in the next section) that Theorem 8.1.20 allows also its asymmetric version with essentially the same proof. To state this version, recall that a bilinear map $\vartheta : V \times V \to W$ is *alternating* if $\vartheta(x, y) = -\vartheta(y, x)$ and $\vartheta(x, x) = 0$ for all $x, y \in V$.

Theorem 8.1.21. *Suppose that a regular uncountable cardinal θ that is not bigger than the continuum carries a nontrivial C-sequence. Let V and W be two vector spaces over the same field K such that V has dimension θ while W is countable. Then there is a non-degenerate alternating bilinear mapping $\vartheta : V \times V \to W$ such that $\{\vartheta(x, y) : x, y \in X,\ x \neq y\} = W$ for every subset X of V of cardinality θ.* □

Corollary 8.1.22. *Suppose that the continuum is regular and not bigger than the first weakly inaccessible cardinal. Let V and W be two vector spaces over the same field K such that V has dimension continuum while W is countable. Then there is a non-degenerate alternating bilinear mapping $\vartheta : V \times V \to W$ such that $\{\vartheta(x, y) : x, y \in X,\ x \neq y\} = W$ for every subset X of V of cardinality θ.* □

Recall that a vector space V over a field K equipped with a bilinear form $\vartheta : V \times V \to K$ is called *symplectic space* if the bilinear form ϑ is alternating on V. The symplectic space (V, ϑ) is *non-degenerate* if for all $x \in V$ there is $y \in V$ such that $\vartheta(x, y) \neq 0$.

Theorem 8.1.23. *Suppose that a regular uncountable cardinal θ that is not bigger than the continuum carries a nontrivial C-sequence. Then for every countable (including finite) field K and for every vector space V over K of dimension θ there is a symplectic space of the form (V, ϑ) without a null subspace of cardinality θ.* □

Corollary 8.1.24. *Suppose that the continuum is regular and not bigger than the first weakly inaccessible cardinal. Then for every countable field K and for every vector space V over K of dimension continuum, there is a symplectic space of the form (V, ϑ) without a null subspace of cardinality θ.* □

Remark 8.1.25. We remark again that the results of this section cover a large class of regular cardinals θ that are not necessarily successor cardinals. In the case of successors of regular cardinals it is more effective to use the square-bracket operation that forms the subject matter of the following section.

8.2 The trace filter and the square-bracket operation

A square-bracket operation on a regular uncountable cardinal carrying a nontrivial C-sequence is a transformation which to every pair $\alpha < \beta$ of ordinals below θ assigns a point $[\alpha\beta]$ belonging to the trace

$$\beta = \beta_0 > \beta_1 > \cdots > \beta_{\rho_2(\alpha,\beta)} = \alpha$$

of the walk from β to α. We want the operation to be chosen in such a way that the set of all $\gamma < \theta$ of the form $[\alpha\beta]$ for α and β running inside the same (and sometimes two different) unbounded subsets of θ contains a closed and unbounded set relative to some fixed stationary subset Δ of θ. The set Δ has to be chosen carefully relative to the C-sequence on which we base our walks. We have seen above that ω_1 admits this kind of operation (see Definition 5.1.3) and the purpose of this section is to show that the basic idea extends to a general setting on an arbitrary uncountable regular cardinal θ that carries a nontrivial C-sequence C_α $(\alpha < \theta)$. Not surprisingly, the idea is based again on the oscillation map defined in the previous section and, in particular, on the property of this map exposed in Theorem 8.1.11. Recall also the definition of the partial action $(\beta, t) \mapsto \beta_t$, defined recursively by $\beta_\emptyset = \beta$ and

$$\beta_t = \xi\text{th member of } C_{\beta_s}$$

for $t = s^\frown\xi$, if ξ is smaller than the order type of C_{β_s}; otherwise, we leave β_t undefined.

Definition 8.2.1. For $\alpha < \beta < \theta$ we set $[\alpha\beta] = \beta_t$, where $t \sqsubseteq \rho_0(\alpha, \beta)$ is minimal for which α_t is defined and

$$\xi^{\text{th}} \text{ element of } C_{\alpha_t} \neq \xi^{\text{th}} \text{ element of } C_{\beta_t}, \tag{8.2.1}$$

where $\xi = \text{tp}(C_{\beta_t} \cap \alpha)$; if such t does not exist, we let $[\alpha\beta] = \alpha$.

Remark 8.2.2. Thus, $[\alpha\beta]$ is the first place β_m visited by β on its walk down to α such that a nontrivial oscillation occurs between C_{β_m} and C_{α_m}, where α_m is the corresponding place of a walk that starts with α and has the same full code up to that point. Note that Theorem 8.1.11 is telling us that, in particular, the nontrivial oscillation indeed happens most of the time. Results that would say that the set of values $\{[\alpha\beta] : \{\alpha, \beta\} \in [\Gamma]^2\}$ is in some sense large no matter how small the unbounded set $\Gamma \subseteq \theta$ is, would correspond to the results of Lemmas 5.1.4–5.1.5 about the square-bracket operation on ω_1. It turns out that this is indeed possible and to describe it we need the following definition.

Definition 8.2.3. A C-sequence C_α $(\alpha < \theta)$ on θ *avoids* a given subset Δ of θ if $C_\alpha \cap \Delta = \emptyset$ for all limit ordinals $\alpha < \theta$.

Lemma 8.2.4. *Suppose C_α $(\alpha < \theta)$ is a given C-sequence on θ that avoids a set $\Delta \subseteq \theta$. Then for every unbounded set $\Gamma \subseteq \theta$, the set of elements of Δ not of the form $[\alpha\beta]$ for some $\alpha < \beta$ in Γ is nonstationary in θ.*

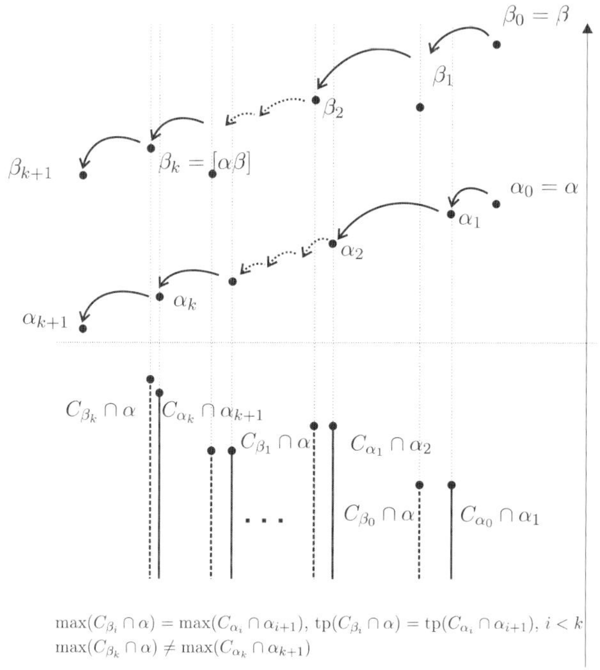

Figure 8.3: The square-bracket operation.

Proof. Let Ω be a given stationary subset of Δ. We need to find $\alpha < \beta$ in Γ such that $[\alpha\beta] \in \Omega$. For a limit $\delta \in \Omega$ fix a $\beta = \beta(\delta) \in \Gamma$ above δ and let

$$\beta = \beta_0 > \beta_1 > \cdots > \beta_{k(\delta)} = \delta$$

be the walk from β to δ along the sequence C_α $(\alpha < \theta)$. Since $\delta \notin C_\alpha$ for any limit α and since $C_{\alpha+1} = \{\alpha\}$ for all α, we must have that $\beta_{k(\delta)-1} = \delta + 1$. In particular,

$$C_{\beta_i} \cap \delta \text{ is bounded in } \delta \text{ for all } i < k(\delta). \tag{8.2.2}$$

Applying the Pressing Down Lemma gives us a stationary set $\Omega_0 \subseteq \Omega$, an integer k, ordinals ξ_i $(i < k)$, and a sequence $t \in \mathbb{Q}_\theta$ such that for all $\delta \in \Omega_0$:

$$k(\delta) = k \text{ and } \rho_0(\delta, \beta(\delta)) = t, \tag{8.2.3}$$

$$\max(C_{\beta(\delta)_i} \cap \delta) = \xi_i \text{ for all } i < k. \tag{8.2.4}$$

Intersecting Ω_0 with a closed and unbounded subset of θ, we may assume for every $\delta \in \Omega_0$ that $\beta(\delta)$ is smaller than the next member of Ω_0 above δ. Note that any C-sequence on a regular uncountable cardinal that avoids a stationary subset of the cardinal, in particular, must be nontrivial. Thus C_δ $(\delta \in \Omega_0)$ is a stationary subsequence of a nontrivial C-sequence, and so by Lemma 8.1.5, unbounded as a family of subsets of θ. By the basic oscillation Lemma 8.1.2, we can find $\gamma < \delta$ in Ω_0 such that $\mathrm{osc}(C_\gamma, C_\delta) = 2$. Then by the choice of Ω, if $\alpha = \beta(\gamma)$ and $\beta = \beta(\delta)$, then $\gamma = \alpha_t$, $\delta = \beta_t$ and $\alpha < \delta$. Comparing this with (8.2.3) and (8.2.4) we conclude that $[\alpha\beta] = \beta_t = \delta \in \Omega$. This finishes the proof. □

It is clear that the argument gives the following more general result.

Lemma 8.2.5. *Suppose C_α ($\alpha < \theta$) avoids $\Delta \subseteq \theta$ and let A be a family of size θ consisting of pairwise-disjoint finite sets, all of some fixed size n. Then the set of all elements δ of Δ that are not of the form*

$$\delta = [a(1)b(1)] = [a(2)b(2)] = \cdots = [a(n)b(n)]$$

for some $a < b$ [2] *in A is not stationary in θ.* □

The following analogue of Lemma 5.1.7 gives us in particular an explanation about the behavior of the square-bracket operation on rectangles over pairs of unbounded subsets of θ.

Theorem 8.2.6. *For every family A of θ pairwise-disjoint subsets of θ, all of same fixed size n, there exist an equivalence relation $\sim$ on $n = \{0, \ldots, n-1\}$, a subfamily $B \subseteq A$ of size θ, and a mapping $h_b : n \times n \to \theta$ for each $b \in B$ such that,*

$$[a(i)b(i)] = h_b(i,j) \text{ for all } a < b \text{ in } B \text{ and } i,j < n \text{ such that } i \nsim j.$$

Moreover, for every $C \subseteq B$ of size θ, for all but a nonstationary set of $\delta \in \Delta$ there exist $a < b$ in C such that

$$[a(i)b(j)] = \delta \text{ for all } i,j < n \text{ such that } i \sim j.$$

Proof. For each $\delta \in \Delta$ choose $b_\delta \in A$ such that $b_\delta > \delta$. Applying the Pressing Down Lemma, we find a stationary set $\Delta_0 \subseteq \Delta$ and a sequence t_i ($i < n$) of elements of $\theta^{<\omega}$ such that

$$\rho_0(\delta, b_\delta(i)) = t_i \text{ for all } i < n \text{ and } \delta \in \Delta_0. \tag{8.2.5}$$

For $i < n$, let $l_i = |t_i|$. Since our C-sequence avoids Δ, for each $i < n$, if

$$b_\delta(i) = (b_\delta(i))_0 > (b_\delta(i))_1 > \cdots > (b_\delta(i))_{l_i} = \delta$$

is the trace of the walk from $b_\delta(i)$ down to δ, then

$$\lambda_{ik}(\delta) = \max(C_{(b_\delta(i))_k} \cap \delta) < \delta \text{ for all } k < l_i. \tag{8.2.6}$$

[2]Recall that if a and b are two sets of ordinals, then the notation $a < b$ means that $\max(a) < \min(b)$.

So, going to a further stationary subset of Δ_0, we may assume that for some fixed sequence λ_{ik} $(i < n, k < l_i)$ of ordinals $< \delta$, we have

$$\lambda_{ik}(\delta) = \lambda_{ik} \text{ for all } i < n, k < l_i, \text{ and } \delta \in \Delta_0. \tag{8.2.7}$$

For $\delta \in \Delta_0$, set

$$T_\delta = \{t \in \theta^{<\omega} : t_i \sqsubset t \sqsubseteq t_j \text{ for some } i, j < n \text{ such that } (b_\delta(i))_t \text{ is defined}\}.$$

Going to a stationary subset of Δ_0, we may assume that for some T and all $\delta \in \Delta_0$, we have $T_\delta = T$. For $\delta \in \Delta_0$ and $i < n$, let

$$T_\delta(i) = \{t \in T : t \sqsupset t_i \text{ and } (b_\delta(i))_t \text{ is defined}\}.$$

Going again to a stationary subset of Δ_0, we find a sequence $T(i)$ $(i < n)$, such that $T_\delta(i) = T(i)$ for all $\delta \in \Delta_0$. Applying the Pressing Down Lemma, and going to a stationary subset of Δ_0, we may assume to have a fixed sequence ν_t $(t \in T(i), i < n)$ such that

$$(b_\delta(i))_t = \nu_t \text{ for all } t \in T(i) \text{ and } i < n. \tag{8.2.8}$$

Moreover, we may assume that $\gamma < \delta$ in Δ_0 implies $b_\gamma < \delta$ and that for some ordinal $\nu < \theta$,

$$\lambda_{ik}, \nu_t < \nu \in C_\delta \text{ for all } t \in T(i), i < n, k < l_i, \text{ and } \delta \in \Delta_0. \tag{8.2.9}$$

For $i, j < n$, set $i \sim j$ if and only if $t_i = t_j$ and $\lambda_{ik} = \lambda_{jk}$ for all $k < l_i = l_j$. Clearly, $\sim$ is an equivalence relation. Let us check that $B = \{b_\delta : \delta \in \Delta_0\}$ satisfies the second conclusion of the lemma. Consider $\gamma < \delta$ in Δ_0 and $i, j < n$ such that $i \sim j$. Then, applying 8.2.6 and 8.2.7, we conclude that for $t = t_i = t_j$,

(1) $\rho_0(b_\gamma(i), b_\delta(j)) = t^\frown \rho_0(b_\gamma(i), \delta)$,
(2) $(b_\gamma(i))_t = \gamma$, $(b_\delta(j))_t = \delta$,
(3) $(b_\gamma(i))_s = b_\gamma(i)_{|s|}$ and $(b_\delta(j))_s = b_\delta(j)_{|s|}$ for $s \sqsubset t$.

Note that by (8.2.6) and (8.2.7), we can also conclude that if for some $k < l_i = l_j$, we let $\alpha = b_\gamma(i)_k$ and $\beta = b_\delta(j)_k$, then $\mathrm{osc}(C_\alpha, C_\beta) = 1$. Combining all this, we conclude that

$$[b_\gamma(i) b_\delta(j)] = [\gamma\delta] \text{ for all } i \sim j \text{ and } \gamma < \delta \text{ in } \Delta_0, \tag{8.2.10}$$

so the second conclusion of the lemma follows from Lemma 8.2.4.

Consider now $i, j < n$ such that $i \nsim j$. If $t_i = t_j$, then there must be $k < l_i = l_j$ such that $\lambda_{ik} \neq \lambda_{jk}$. For $\delta \in \Delta_0$, put $h_{b_\delta}(i, j) = b_\delta(j)_k$ for $k < l_i = l_j$ minimal with this property. Then in this case, $[b_\gamma(i) b_\delta(j)] = h_{b_\delta}(i, j)$ for all $\gamma < \delta$ in Δ_0.

Assume now that $t_i \neq t_j$ and that there is $k < \min\{l_i, l_j\}$ such that $\lambda_{ik} \neq \lambda_{jk}$. For $\delta \in \Delta_0$, put $h_{b_\delta}(i,j) = b_\delta(j)_k$ for $k < \min\{l_i, l_j\}$ minimal with this property. Going back to the definition of the square-bracket operation, we conclude that, in this case as well, $[b_\gamma(i) b_\delta(j)] = h_{b_\delta}(i,j)$ for all $\gamma < \delta$ in Δ_0.

Assume now that t_i is a strict end-extension of t_j but there is $k < l_j$ such that $\lambda_{ik} \neq \lambda_{jk}$. For $\delta \in \Delta_0$, put $h_{b_\delta}(i,j) = b_\delta(j)_k$ for $k < l_i = l_j$ minimal with this property. Then again, $[b_\gamma(i) b_\delta(j)] = h_{b_\delta}(i,j)$ for all $\gamma < \delta$ in Δ_0. Similar definition and conclusion holds if t_j strictly end-extends t_i but there is $k < l_i$ such that $\lambda_{ik} \neq \lambda_{jk}$.

Assume now that t_i is a strict end-extension of t_j and $\lambda_{ik} = \lambda_{jk}$ for all $i < l_j$. For $\delta \in \Delta_0$, put $h_{b_\delta}(i,j) = \delta$. Let us check that $[b_\gamma(i) b_\delta(j)] = h_{b_\delta}(i,j)$ for all $\gamma < \delta$ in Δ_0. First of all note that $\delta = (b_\delta(j))_{t_j}$. Let $\xi = \mathrm{tp}(C_\delta \cap (b_\gamma(i))_{t_j})$. Then either,

$$\xi^{\text{th}} \text{ element of } C_{(b_\gamma(i))_{t_j}} \leq \lambda_{il_j}, \text{ or else } \xi^{\text{th}} \text{ element of } C_{(b_\gamma(i))_{t_j}} \geq \gamma.$$

So, by (8.2.9), in both cases

$$\xi^{\text{th}} \text{ element of } C_{(b_\gamma(i))_{t_j}} \neq \xi^{\text{th}} \text{ element of } C_\delta.$$

On the other hand, for $s \sqsubset t_j$,

$$\xi^{\text{th}} \text{ element of } C_{(b_\gamma(i))_s} = \lambda_{i|s|} = \lambda_{j|s|} = \xi^{\text{th}} \text{ element of } C_{(b_\beta(j))_s},$$

for $\xi = \mathrm{tp}(C_{(b_\delta(j))_s} \cap (b_\gamma(i))_s)$. It follows that $[b_\gamma(i) b_\delta(j)] = \delta$, as required.

Consider now the final case that t_j is a strict end-extension of t_i and $\lambda_{ik} = \lambda_{jk}$ for all $i < l_i$. Fix a δ in Δ_0. If there is no $\gamma < \delta$ in Δ_0, set $h_{b_\delta}(i,j) = 0$. Otherwise, fix $\gamma' < \delta$ in Δ_0. Note that by (8.2.8) and (8.2.9), it must be that

$$[b_{\gamma'}(i) b_\delta(j)] = (b_\delta(j))_t \text{ for some } t_i \sqsubseteq t \sqsubseteq t_j. \tag{8.2.11}$$

Moreover, the choice of $t_i \sqsubseteq t \sqsubseteq t_j$ is independent of γ'. Namely, the fact that there is a nontrivial oscillation between $C_{(b_\delta(j))_{t_j}}$ and $C_{(b_\delta(i))_{t_j}}$, if defined, is clear. Other than that, this equation depends on the relationships between the ξth point of $C_{(b_{\gamma'}(i))_s}$ and the ξth point of $C_{(b_\delta(j))_s}$, for $s \sqsubset t_j$ and

$$\xi = \mathrm{tp}(C_{(b_\delta(j))_s} \cap (b_{\gamma'}(i))_s) = \mathrm{tp}(C_{(b_\delta(j))_s} \cap \delta) = t_j(|s|),$$

which according to (8.2.7), (8.2.8), and (8.2.9) are all independent on γ'. Thus, if we put $h_{b_\delta}(i,j) = (b_\delta(j))_t$ for t satisfying (8.2.11), we have that $[b_\gamma(i) b_\delta(j)] = h_{b_\delta}(i,j)$ for *all* $\gamma < \delta$ in Δ_0. □

For sufficiently large cardinals θ, and for some special choice of the set Γ, we have the following variation on a theme first encountered above in Theorem 6.3.2 that could be used in defining a square-bracket operation with an even more complex behavior on families of θ pairwise-disjoint finite subsets of θ.

Lemma 8.2.7. *Suppose $\theta > \mathfrak{c}$ and that $\mathrm{cf}(\gamma) > \omega$ for all $\gamma \in \Gamma$. Let A be a family of θ pairwise-disjoint finite subsets of θ, all of some fixed size n. Then for every stationary $\Gamma_0 \subseteq \Gamma$ there exist $s, t \in \omega^n$ and a positive integer k such that for every $l < \omega$ there exist $a < b$ in A and $\delta_0 > \delta_1 > \cdots > \delta_l$ in $\Gamma_0 \cap (\max(a), \min(b))$ such that:*

(1) $\rho_2(\delta_{i+1}, \delta_i) = k$ *for all* $i < l$,

(2) $\rho_0(a(i), b(j)) = \rho_0(\delta_0, b(j))^\frown \rho_0(\delta_1, \delta_0)^\frown \cdots ^\frown \rho_0(\delta_l, \delta_{l-1})^\frown \rho_0(a(i), \delta_l)$ *for all* $i, j < n$,

(3) $\rho_2(\delta_0, b(j)) = t_j$ *and* $\rho_2(a(i), \delta_l) = s_i$ *for all* $i, j < n$.

Proof. For $\Delta \subseteq \Gamma$ and $\delta < \theta$ set

$$S_\delta(A) = \bigcap_{\gamma<\delta} \{\langle \rho_2(a(i), \delta) : i < n\rangle : a \in A \restriction [\gamma, \delta)\}, \tag{8.2.12}$$

$$T_\delta(\Delta) = \bigcap_{\gamma<\delta} \{\rho_2(\alpha, \delta) : \alpha \in \Delta \cap [\gamma, \delta)\}. \tag{8.2.13}$$

Note that $S_\delta(A)$ and $T_\delta(\Delta)$ range over subsets of ω^n and ω, respectively, so if δ has uncountable cofinality and if A and Δ are unbounded in δ, then both of these sets, $S_\delta(A)$ and $T_\delta(\Delta)$, are nonempty. Let Γ_0 be a given stationary subset of Γ. Since θ has size bigger than the continuum, starting from Γ_0 one can find an infinite sequence $\Gamma_0 \supseteq \Gamma_1 \supseteq \cdots$ of subsets of Γ such that $\Gamma_\omega = \bigcap_{i<\omega} \Gamma_i$ is stationary in θ and

$$S_\delta(A) = S_\varepsilon(A) \neq \emptyset \text{ for all } \delta, \varepsilon \in \Gamma_1, \tag{8.2.14}$$

$$T_\delta(\Gamma_i) = T_\varepsilon(\Gamma_i) \neq \emptyset \text{ for all } \delta, \varepsilon \in \Gamma_{i+1} \text{ and } i < \omega. \tag{8.2.15}$$

Pick $\delta \in \Gamma_\omega$ such that $\Gamma_\omega \cap \delta$ is unbounded in δ and choose $b \in A$ above δ. Let $t_j = \rho_2(\delta, b_j)$ for $j < n$. Now let s be an arbitrary member of $S_\delta(A)$ and k an arbitrary member of $T_\delta(\Gamma_\omega)$. To check that these objects satisfy the conclusion of the theorem, let $l < \omega$ be given. Let $\delta_0 = \delta$ and let $\gamma_0 < \delta_0$ be an upper bound for all sets of the form

$$C_\xi \cap \delta_0 \ (\xi \in \mathrm{Tr}(\delta_0, b(j)), \xi \neq \delta_0, j < n).$$

Then the walk from any $b(j)$ to any $\alpha \in [\gamma_0, \delta_0)$ must pass through δ_0. Since $k \in T_\delta(\Gamma_\omega) \subseteq T_\delta(\Gamma_l)$ we can find $\delta_1 \in \Gamma_l \cap (\gamma_0, \delta_0)$ such that $\rho_2(\delta_1, \delta_0) = k$. Choose $\gamma_1 \in (\gamma_0, \delta_1)$ such that the walk from δ_0 to any $\alpha \in [\gamma_1, \delta_1)$ must pass through δ_1. Since $k \in T_\delta(\Gamma_\omega) \subseteq T_\delta(\Gamma_{l-1}) = T_{\delta_1}(\Gamma_{l-1})$ there exists $\delta_2 \in \Gamma_{l-1} \cap (\gamma_1, \delta_1)$ such that $\rho_2(\delta_2, \delta_1) = k$. Choose $\gamma_2 \in (\gamma_1, \delta_2)$ such that the walk from δ_1 to any $\alpha \in [\gamma_2, \delta_2)$ must pass through δ_2, and so on. Proceeding in this way, we find $\delta = \delta_0 > \delta_1 > \cdots > \delta_l > \gamma_l > \gamma_{l-1} > \cdots > \gamma_0$ such that:

$$\delta_i \in \Gamma_{l-i+1} \text{ for all } 0 < i \leq l, \tag{8.2.16}$$

$$\rho_2(\delta_{i+1}, \delta_i) = k \text{ for all } i < l, \tag{8.2.17}$$

for every $i < l$ the walk from δ_i to some $\alpha \in [\gamma_{i+1}, \delta_{i+1})$ passes through δ_{i+1}. (8.2.18)

Since $s \in S_\delta(A) = S_{\delta_l}(A)$, we can find $a \in A \restriction [\gamma_l, \delta_l)$ such that $\rho_2(a(i), \delta_l) = s(i)$ for all $i < n$. Now, given $i, j < n$, the walk from $b(j)$ to $a(i)$ first passes through all the δ_i ($i \leq l$), so $\rho_0(a(i), b(j))$ is simply equal to the concatenation

$$\rho_0(\delta_0, b(j))^\frown \rho_0(\delta_1, \delta_0)^\frown \cdots {}^\frown \rho_0(\delta_l, \delta_{l-1})^\frown \rho_0(a(i), \delta_l).$$

This together with (8.2.17) gives us the three conclusions of Lemma 8.2.7. □

It turns out that Lemma 8.2.7 gives us a way to define another square-bracket operation which has complex behavior not only on squares of unbounded subsets of θ but also on rectangles formed by two unbounded subsets of θ. To define this variation, we need a mapping $n \mapsto n^\circ$ from ω into ω such that:

for every $k, m, n, p < \omega$ and $s \in \omega^n$ there exist $l < \omega$ such that $(m + l \cdot k + s(i))^\circ = m + p$ for all $i < n$. (8.2.19)

Definition 8.2.8. Define a square-bracket operation $[\cdot\cdot]_\circ : [\theta]^2 \longrightarrow \theta$ by letting,

$$[\alpha\beta]_\circ = \beta_t \text{ where } t = \rho_0(\alpha, \beta) \restriction (\rho_2(\alpha, \beta))^\circ.$$

Thus, if $\beta = \beta_0 > \beta_1 > \cdots > \beta_{\rho_2(\alpha,\beta)} = \alpha$ is the decreasing enumeration of the upper trace of the walk from β to α, then

$$[\alpha\beta]_\circ = \beta_k, \text{ where } k = (\rho_2(\alpha, \beta))^\circ.$$

It is clear that Lemma 8.2.7 and the choice of $n \mapsto n^\circ$ in (8.2.19) give us the following conclusion.

Lemma 8.2.9. *Suppose $\theta > \mathfrak{c}$ and that $\mathrm{cf}(\gamma) > \omega$ for all $\gamma \in \Gamma$. Let A be a family of θ pairwise-disjoint finite subsets of θ, all of some fixed size n, and let Ω be an unbounded subset of θ. Then almost every*[3] *$\delta \in \Gamma$ has the form $[a(0)\beta]_\circ = [a(1)\beta]_\circ = \cdots = [a(n-1)\beta]_\circ$ for some $a \in A$, $\beta \in \Omega$, $a < \beta$.* □

Let us now turn back to the case of an arbitrary C-sequence on a regular uncountable cardinal θ avoiding an arbitrary stationary set $\Gamma \subseteq \theta$. Since $[\alpha\beta]$ belongs to the trace $\mathrm{Tr}(\alpha, \beta)$ of the walk from β to α it is not surprising that $[\cdot\cdot]$ strongly depends on the behavior of Tr. For example the following should be clear from the proof of Lemma 8.2.4.

[3] Here 'almost every' is to be interpreted by 'all except a nonstationary set'.

Lemma 8.2.10. *The following are equivalent for a given C-sequence and a set Ω:*

(a) $\Omega \setminus \{[\alpha\beta] : \{\alpha, \beta\} \in [\Gamma]^2\}$ *is nonstationary in* θ,

(b) $\Omega \setminus \bigcup\{\mathrm{Tr}(\alpha, \beta) : \{\alpha, \beta\} \in [\Gamma]^2\}$ *is nonstationary in* θ. □

This fact suggests the following definition.

Definition 8.2.11. The *trace filter* of a given C-sequence C_α $(\alpha < \theta)$ is the normal filter on θ generated by sets of the form $\bigcup\{\mathrm{Tr}(\alpha, \beta) : \{\alpha, \beta\} \in [\Gamma]^2\}$ where Γ is an unbounded subset of θ.

Remark 8.2.12. Having a proper trace filter (i.e., a trace filter that does not contain all subsets of θ) is a strengthening of the nontriviality requirement on a given C-sequence C_α $(\alpha < \theta)$. For example, if a C-sequence avoids a stationary set $\Omega \subseteq \theta$, then its trace filter is nontrivial and in fact no stationary subset of Ω is a member of it, and in fact, any member of the trace filter contains a closed and unbounded subset relative to the set Ω. Note the following analogue of Lemma 8.2.10: The trace filter of a given C-sequence is the normal filter generated by sets of the form $\{[\alpha\beta] : \{\alpha, \beta\} \in [\Gamma]^2\}$ where Γ is an unbounded subset of θ. So to obtain the analogues of the results of Section 5.1 about the square-bracket operation on ω_1, one needs a C-sequence C_α $(\alpha < \theta)$ on θ whose trace filter is not only nontrivial but also not θ-saturated, i.e., it allows a family of θ pairwise-disjoint positive sets. It turns out that the hypothesis of Lemma 8.2.4 is sufficient for both of these conclusions.

Lemma 8.2.13. *If a C-sequence on θ avoids a stationary subset of θ, then there exist θ pairwise-disjoint subsets of θ that are positive with respect to its trace filter.*[4]

Proof. This follows from the well-known fact (see [55]) that if there is a normal, nontrivial and θ-saturated filter on θ, then for every stationary $\Omega \subseteq \theta$ there exists $\lambda < \theta$ such that $\Omega \cap \lambda$ is stationary in λ (and the fact that the stationary set which is avoided by the C-sequence does not reflect in this way). □

Corollary 8.2.14. *If a regular cardinal θ admits a nonreflecting stationary subset, then there is $c : [\theta]^2 \longrightarrow \theta$ which takes all the values from θ on any set of the form $[\Gamma]^2$ for some unbounded set $\Gamma \subseteq \theta$.* □

To get such a c, one composes the square-bracket operation of some C-sequence, that avoids a stationary subset of θ, with a mapping $* : \theta \longrightarrow \theta$ with the property that the $*$-preimage of each point from θ is positive with respect to the trace filter of the square sequence. In other words c is equal to the composition of $[\cdot\cdot]$ and $*$, i.e., $c(\alpha, \beta) = [\alpha\beta]^*$. Note that, as in Section 5.1, the property of the square-bracket operation from Lemma 8.2.5 leads to the following rigidity result which corresponds to Lemma 5.1.9.

---

[4] A subset A of the domain of some filter $\mathcal{F}$ is *positive* with respect to $\mathcal{F}$ if $A \cap F \neq \emptyset$ for every $F \in \mathcal{F}$.

Lemma 8.2.15. *The algebraic structure* $(\theta, [\cdot\cdot], *)$ *has no nontrivial automorphisms.* □

Remark 8.2.16. Note that every θ which is a successor of a regular cardinal κ admits a nonreflecting stationary set. For example $\Omega = \{\delta < \theta : \mathrm{cf}\ \delta = \kappa\}$ is such a set. Thus any C-sequence on θ that avoids Ω leads to a square-bracket operation which allows analogues of all the results from Section 5.1 about the square-bracket operation on ω_1. We shall now explore this by examining properties of a particular projection of the square-bracket operation that are analogous to those of Lemma 5.1.7 and Theorem 5.2.5.

8.3 Projections of the square-bracket operation on accessible cardinals

Suppose now that θ is not a strong limit cardinal and that it carries a C-sequence C_α $(\alpha < \theta)$ avoiding a stationary set $\Omega \subseteq \theta$. As indicated above, any successor of a regular cardinal has these properties. Let $[\cdot\cdot]$ be the corresponding square-bracket operation. Let λ be the minimal cardinal such that $2^\lambda \geq \theta$. Choose a sequence r_ξ $(\xi < \theta)$ of distinct elements of 2^λ. Let $\mathcal{G}$ be the collection of all finitely supported[5] maps $g : 2^\xi \times 2^\xi \to \theta$ where $\xi = \xi(g) < \lambda$. Clearly, $\mathcal{G}$ has cardinality θ, so we can find a mapping $\pi : \theta \to \mathcal{G}$ such that $\pi^{-1}(g) \cap \Omega$ is stationary for all $g \in \mathcal{G}$. Finally, define a binary operation $[\![\cdot\cdot]\!]$ on θ as follows:

$$[\![\alpha\beta]\!] = \pi([\alpha\beta])(r_\alpha \restriction \xi(\pi([\alpha\beta])), r_\beta \restriction \xi(\pi([\alpha\beta]))). \tag{8.3.1}$$

The following is a simple consequence of the property of the square-bracket operation stated in Lemma 8.2.5 above and its proof is very similar to the proof of Lemma 5.2.2 above (see also the proof of Theorem 8.3.2 below).

Lemma 8.3.1. *Suppose that θ is an uncountable regular and not strong limit cardinal carrying a C-sequence that avoids a stationary subset Γ of θ. Then the projection $[\![\cdot\cdot]\!]$ of the square-bracket operation $[\cdot\cdot]$ has the property that for every family A of size θ consisting of pairwise-disjoint finite subsets of θ, all of some fixed size n, and every sequence $\xi_0, \ldots, \xi_{n-1}$ of ordinals $< \theta$, there exist $a \neq b$ in A such that $[\![a(i)b(i)]\!] = \xi_i$ for all $i < n$.* □

As noted before, a typical input to this result is a successor of a regular cardinal $\theta = \kappa^+$ and $\Omega = \{\gamma < \kappa^+ : \mathrm{cf}(\gamma) = \kappa.\}$ The additional information given in Theorem 8.2.6 allows us to obtain in this case a particularly interesting projection of the square-bracket operation.

[5] i.e., maps g with $\mathrm{supp}(g) = \{(s,t) : g(s,t) \neq 0\}$ finite.

Theorem 8.3.2. *Suppose that θ is a successor of a regular cardinal κ. Then the square-bracket operation $[\cdot\cdot]$ admits a projection*

$$[\![\cdot\cdot]\!]_\kappa : [\theta]^2 \to \kappa$$

with the property that for every family A of size θ of pairwise-disjoint finite subsets of θ, all of some fixed size n, there exist $B \subseteq A$ of size θ, a mapping $h : n \times n \to \kappa$, and an equivalence relation $\sim$ on $n = \{0, 1, \ldots, n-1\}$ such that for all $a < b$ in B, we have that

$$[\![a(i)b(j)]\!]_\kappa = h(i,j) \text{ whenever } i \nsim j.$$

Moreover, for every subfamily C of B of size θ and for every mapping $g : n \times n \to \kappa$ there exist $a < b$ in C such that

$$[\![a(i)b(j)]\!]_\kappa = g(i,j) \text{ for all } i < n \text{ such that } i \sim j.$$

Proof. Let $\lambda \leq \kappa$ be minimal such that $2^\lambda \geq \theta$. As before, choose a sequence r_ξ $(\xi < \theta)$ of distinct elements of 2^λ. Let $\mathcal{G}_\kappa$ be the collection of all finitely supported maps $g : 2^\xi \times 2^\xi \to \kappa$ where $\xi = \xi(g) < \lambda$. Clearly, $\mathcal{G}_\kappa$ has cardinality κ. Let $\pi : \theta \to \mathcal{G}_\kappa$ be a map with the property that $\pi^{-1}(g) \cap \Omega$ is stationary for all $g \in \mathcal{G}_\kappa$. Finally, define a binary operation $[\![\cdot\cdot]\!]_\kappa : [\theta]^2 \to \kappa$ as follows:

$$[\![\alpha\beta]\!]_\kappa = \pi([\alpha\beta])(r_\alpha \restriction \xi(\pi([\alpha\beta])), r_\beta \restriction \xi(\pi([\alpha\beta]))). \tag{8.3.2}$$

Let A be a given family of θ pairwise-disjoint finite subsets of θ, all of some fixed size n. Applying Lemma 8.2.6, we get a subfamily B of size θ, an equivalence relation $\sim$ on n, and a mapping $h_b : n \times n \to \theta$ such that, on one hand, for all $a < b$ in B, we have $[a(i)b(j)] = h_b(i,j)$ for $i \nsim j$, while on the other hand, for almost all $\gamma < \theta$ of cofinality κ, we can find $a < b$ in B such that $[a(i)b(j)] = \gamma$ whenever $i \sim j$. Note that for each $b \in B$, the range of the composition mapping $\pi \circ h_b : n \times n \to \mathcal{G}_\kappa$ has size κ, so there are at most κ such mappings. So going to a subset of B of size θ, we may assume that for some fixed map $f : n \times n \to \mathcal{G}_\kappa$, we have that

$$\pi \circ h_b = f \text{ for all } b \in B.$$

For $i, j < n$, let $f(i,j) : 2^{\xi(i,j)} \to \kappa$. Since $|2^\eta| \leq \kappa$ for all $\eta < \kappa$, shrinking the family B, we can assume to have an ordinal $\xi < \kappa$ and a sequence t_i $(i < n)$ of elements of 2^ξ such that:

(1) $\xi > \xi(i,j)$ for all $i, j < n$,

(2) $r_{b(i)} \restriction \xi = t_i$ for all $i < n$ and $b \in B$,

(3) $t_i \neq t_j$ whenever $i \neq j$.

Define $h : n \times n \to \kappa$, by

$$h(i,j) = f(i,j)(t_i \restriction \xi(i,j), t_j \restriction \xi(i,j)).$$

Going back through the definitions, we see that for all $a < b$ in B,

$$[\![a(i)b(j)\beta]\!]_\kappa = f(i,j)(t_i \restriction \xi(i,j), t_j \restriction \xi(i,j)) = h(i,j) \text{ for all } i \not\sim j,$$

as required.

To check the second conclusion of the lemma, suppose that $g : n \times n \to \kappa$ is a given map. Choose a finitely supported map $\overline{g} : 2^\xi \times 2^\xi \to \kappa$, such that

$$\overline{g}(t_i, t_j) = g(i,j) \text{ for all } i, j < n.$$

By our choice of π the set $\Gamma_{\overline{g}} = \{\gamma < \theta : \mathrm{cf}(\gamma) = \kappa \text{ and } \pi(\gamma) = \overline{g}\}$ is stationary, so by the conclusion of Theorem 8.2.6, we can find $\gamma \in \Gamma_{\overline{g}}$ and $a < b$ in B such that

$$[a(i)b(j)] = \gamma \text{ for all } i, j < n, \ i \sim j.$$

Going back through the definitions, we conclude that,

$$[\![a(i)b(j)]\!]_\kappa = \overline{g}(t_i, t_j) = g(i,j) \text{ for all } i, j < n \text{ such that } i \sim j,$$

as required. □

Working as in the case of the corresponding projection of the square-bracket operation on ω_1, we get the following kind of application.

Theorem 8.3.3. *Suppose that θ is a successor of a regular cardinal κ. Let V and W be vector spaces over the same field K such that W has cardinality at most κ while the dimension, and therefore cardinality, of V is equal to θ. Then there is a non-degenerate alternating bilinear mapping $\vartheta : V \times V \to W$ such that $\{\vartheta(x,y) : x, y \in X, \ x \neq y\} = W$ for all $X \subseteq V$ of cardinality θ.*

Proof. Let $B = \{v_\alpha : \alpha < \theta\}$ be a fixed linear basis of V and in order to simplify the notation, we assume that the range of $[\![\cdot\cdot]\!]_\kappa$ is equal to W rather than κ. Define $\vartheta \restriction B \times B$ by letting $\vartheta(v_\alpha, v_\alpha) = 0$ for all $\alpha < \theta$, while when $\alpha, \beta < \theta$ are different, we set

$$\vartheta(v_\alpha, v_\beta) = [\![\alpha\beta]\!]_\kappa \text{ for } \alpha < \beta \text{ and } \vartheta(v_\alpha, v_\beta) = -[\![\beta\alpha]\!]_\kappa \text{ for } \alpha > \beta.$$

Then $\vartheta \restriction B \times B$ extend to a unique bilinear map $\vartheta : V \times V \to W$. It is easily checked that the extension is alternating, i.e., that $\vartheta(x,x) = 0$ for all x, or equivalently that $\vartheta(y,x) = -\vartheta(x,y)$ for all x and y.

To check the basic property of ϑ, consider an arbitrary subset X of V of cardinality θ. Going to a subset of X of size θ, we may assume that for some fixed integer n and some fixed sequence λ_i $(i < n)$ of elements of K, every vector $x \in X$ has a representation

$$x = \Sigma_{i<n} \lambda_i v_{a_x(i)},$$

where a_x is some subset of θ of cardinality n. Going to a further subset of X of size θ, we may assume that the a_x's form a Δ-system with root a. Let us consider first the case $a = \emptyset$, i.e., the case when the sets a_x are pairwise-disjoint. Applying Theorem 8.3.2, we find a $Y \subseteq X$ of size θ, a mapping $h : n \times n \to W$, and an equivalence relation $\sim$ on $n = \{0, 1, \ldots, n-1\}$ such that $[\![a_x(i)a_y(j)]\!]_\kappa = h(i,j)$ for all $x, y \in Y$ such that $a_x < a_y$ and all i, j with $i \nsim j$. Changing h, we may assume that $h(i,j) = 0$ whenever $i \sim j$. Let

$$w = \Sigma_{(i,j)\neq(0,0)}\lambda_i\lambda_j h(i,j).$$

Let $u = t - w$ and let $v = (1/\lambda_0^2)u$. Define $g : n \times n \to W$ by letting $g(0,0) = v$ and $g(i,j) = 0$ for $(i,j) \neq (0,0)$. By the second conclusion of Theorem 8.3.2, we can find $x, y \in Y$ such that $a_x < a_y$ and $[\![a_x(i)a_y(j)]\!]_\kappa = g(i,j)$ for all $i < n$ such that $i \sim j$. Then

$$\vartheta(x,y) = \Sigma_{i\nsim j}\lambda_i\lambda_j[\![a_x(i)a_y(j)]\!]_\kappa + \Sigma_{i\sim j}\lambda_i\lambda_j[\![a_x(i)a_y(j)]\!]_\kappa = w + \lambda_0^2(1/\lambda_0^2)u,$$

which gives us the desired conclusion $\vartheta(x,y) = t$.

Assume now that the root a is not empty. Fix x_0 in X and find a subset Y of $X \setminus \{x_0\}$ of size θ such that for some w_0 in W and all $y \in Y$, we have that $\vartheta(x_0, y) = w_0$. Since the supports of vectors from the family $\{y - x_0 : y \in Y\}$ are pairwise-disjoint, by the first part of the proof, we can find $y, z \in Y$ such that $a_x \setminus a < a_z \setminus a$ and

$$\vartheta(y - x_0, z - x_0) = t + \vartheta(x_0, x_0).$$

It follows that $\vartheta(y,z) = t$, as required. □

Recall that vector space V over a field K equipped with a bilinear form $\vartheta : V \times V \to K$ is called *symplectic space* if the bilinear form is alternating, i.e., it has the properties $\vartheta(y,x) = -\vartheta(x,y)$ and $\vartheta(x,x) = 0$ for all $x, y \in V$. The symplectic space (V, ϑ) is *non-degenerate* if for all $x \in V$ there is $y \in V$ such that $\vartheta(x,y) \neq 0$. A subspace W of a symplectic space (V, ϑ) is called *isotropic* if $\vartheta(x,y) = 0$ for all $x, y \in W$.

Corollary 8.3.4. *For every regular cardinal κ, every field K of size at most κ, and every vector space V over K of dimension κ^+, there is a non-degenerate symplectic space over K of the form (V, ϑ) which contains no isotropic subspace of dimension κ^+.* □

Adjusting a construction from [8], Ehrenfeucht and Faber [27] have found the following general application of bilinear mappings.

Lemma 8.3.5. *Suppose $\varphi : V \times V \to W$ is a bilinear form, where V and W are vector spaces over the same field K. Then the product $V \times W$ can be given a structure of a 2-nilpotent group, denoted by $V\varphi W$, in which the commutators*[6] *have the form $[(x,a),(y,b)] = (0,\ \varphi(x,y) - \varphi(y,x))$.*

[6] i.e., products of the form $[x,y] = xyx^{-1}y^{-1}$.

Proof. The multiplication operation of $V\varphi W$ is defined by

$$(x, a)(y, b) = (x + y,\ a + b + \varphi(x, y)). \tag{8.3.3}$$

This is easily seen to be a group operation whose identity is $(0, 0)$ and where $(x, a)^{-1} = (-x, -a + \varphi(x, x))$. Checking the equation

$$(x, a)(y, b)(x, a)^{-1}(y, b)^{-1} = (0,\ \varphi(x, y) - \varphi(y, x)) \tag{8.3.4}$$

is also straightforward. □

Applying this construction to the bilinear mappings of Theorem 8.3.3, we get the following consequence.

Corollary 8.3.6. *For every regular cardinal κ there is a 2-nilpotent group G of cardinality κ^+ such that every abelian subgroup of G has cardinality at most κ.* □

In order to investigate the homomorphisms between groups of the form $V\varphi W$ we need to fix some notation. Fixing a linear basis $B = \{v_\alpha : \alpha < \theta\}$, a vector x from V has its *support* defined to be equal to the finite set

$$\mathrm{supp}(x) = \{\alpha < \theta : \lambda_\alpha \neq 0\},$$

where $x = \Sigma_{\alpha<\theta}\lambda_\alpha v_\alpha$ is the unique representation of x in terms of B. For an arbitrary element (x, a) of the group $V\varphi W$, we let

$$\mathrm{supp}(x, a) = \mathrm{supp}(x).$$

The following result shows that if the bilinear mapping is based on the projection of the square-bracket operation as in Theorem 8.3.3, then the corresponding group $V\vartheta W$ holds a considerable amount of rigidity.

Theorem 8.3.7. *Let θ be the successor of some regular cardinal κ, let V and W be vector spaces over the same field K such that W has cardinality at most κ while the dimension of V is equal to θ, and let $\vartheta : V \times V \to W$ be the bilinear mapping of Theorem 8.3.3. Suppose Φ is an isomorphism between two subgroups of $V\vartheta W$. Then there is $\gamma < \theta$ such that for all $(x, a) \in V\vartheta W$,*

$$\mathrm{supp}(\Phi(x, a)) \setminus \gamma = \mathrm{supp}(x, a) \setminus \gamma.$$

Proof. Suppose that the conclusion is false. Then for every $\gamma < \theta$ we can select (x_γ, a_γ) in $V\vartheta W$ such that

$$\mathrm{supp}(\Phi(x)) \setminus \gamma \neq \mathrm{supp}(x) \setminus \gamma.$$

By the Pressing Down Lemma, we can find a stationary set $\Gamma \subseteq \theta$, vectors $\bar{x}$, $\bar{y}$, y_γ, z_γ, x^γ, y^γ $(\gamma \in \Gamma)$ from V, and vectors $\bar{a}$ and $\bar{b}$ from W such that for all $\gamma \in \Gamma$,

(1) $a_\gamma = \bar{a}$ and $\Phi(x_\gamma, \bar{a}) = (y_\gamma, \bar{b})$,

(2) $x_\gamma \upharpoonright \gamma = \bar{x}$ and $y_\gamma \upharpoonright \gamma = \bar{y}$,

(3) $x_\gamma \upharpoonright [\gamma, \theta) = z_\gamma + x^\gamma$ and $y_\gamma \upharpoonright [\gamma, \theta) = z_\gamma + y^\gamma$,

(4) x^γ, y^γ, and z_γ have pairwise-disjoint supports.

By symmetry, we may assume that $\mathrm{supp}(\Phi(x)) \setminus \gamma \nsubseteq \mathrm{supp}(x) \setminus \gamma$, or equivalently that $\mathrm{supp}(y^\gamma) \neq \emptyset$ for all $\gamma \in \Gamma$. Our intention is to try to find $\gamma < \delta$ in Γ such that $\vartheta(x_\gamma, x_\delta) = 0$ but $\vartheta(y_\gamma, y_\delta) \neq 0$ showing that $(x_\gamma, \bar{a})$ and $(x_\delta, \bar{a})$ commute in $V\vartheta W$ though their images $\Phi(x_\gamma, \bar{a}) = (y_\gamma, \bar{b})$ and $\Phi(x_\delta, \bar{a}) = (y_\delta, \bar{b})$ do not commute, giving us the desired contradiction. Since the mapping ϑ is alternating, shrinking Γ, we may assume that $\vartheta(\bar{x}, x_\gamma)$ $(\gamma \in \Gamma)$ and $\vartheta(\bar{y}, y_\gamma)$ $(\gamma \in \Gamma)$ are constant sequences. It follows that for all $\gamma < \delta$ in Γ,

$$\vartheta(x_\gamma, x_\delta) = \vartheta(z_\gamma + x^\gamma, z_\delta + x^\delta) \text{ and } \vartheta(y_\gamma, y_\delta) = \vartheta(z_\gamma + y^\gamma, z_\delta + y^\delta). \tag{8.3.5}$$

For $\gamma \in \Gamma$ let $s_\gamma = \mathrm{supp}(x^\gamma) \cup \mathrm{supp}(x^\gamma) \cup \mathrm{supp}(z_\gamma)$. Shrinking Γ further, we may assume that for some integer n and all $\gamma \in \Gamma$, we have that $|s_\gamma| = n$. Moreover, we may assume that for some fixed sequence λ_i $(i < n)$ of scalars and some fixed decomposition $\{0, 1, \ldots, n-1\} = H \cup I \cup J$, we have that for all $\gamma \in \Gamma$

$$x^\gamma = \Sigma_{i \in H} \lambda_i v_{s_\gamma(i)}, \; y^\gamma = \Sigma_{i \in I} \lambda_i v_{s_\gamma(i)}, \text{ and } z_\gamma = \Sigma_{i \in J} \lambda_i v_{s_\gamma(i)}. \tag{8.3.6}$$

By Theorem 8.3.2, we can find an unbounded subset Ξ of Γ, an equivalence relation $\sim$ on n, and $h : n \times n \to W$ such that for all $\gamma < \delta$ in Ξ,

$$[\![s_\gamma(i) s_\delta(j)]\!]_\kappa = h(i, j) \text{ whenever } i \nsim j. \tag{8.3.7}$$

Moreover, for every unbounded subset Ω of Ξ and for every mapping $g : n \times n \to W$ there exist $\gamma < \delta$ in Ω such that

$$[\![s_\gamma(i) s_\delta(j)]\!]_\kappa = g(i, j) \text{ for all } i < n \text{ such that } i \sim j. \tag{8.3.8}$$

Let $i_0 = \min(I)$ and let $j_0 = \min(H \cup J)$. We may assume that $H \cup J$ is indeed nonempty, since in this case our job reduces to a straightforward application of Theorem 8.3.3. Note that $i_0 \neq j_0$. Let $h_x : (H \cup J) \times (H \cup J) \to W$ be a mapping that extends the restriction of h to this domain $D_x = (H \cup J) \times (H \cup J)$ and which takes the value 0 otherwise. Let

$$u_x = \Sigma_{(i,j) \in D_x \setminus \{(j_0, j_0)\}} \lambda_i \lambda_j h_x(i, j)$$

and let $h_y : (I \cup J) \times (I \cup J) \to W$ be a mapping that extends the restriction of h to this domain $D_y = (I \cup J) \times (I \cup J)$ and takes the value 0 on all new pairs except perhaps on the pair (j_0, j_0) in case $j_0 \in J$ in which case we put $h_y(j_0, j_0) = -(1/\lambda_{j_0}^2 u_x)$. Let

$$u_y = \Sigma_{(i,j) \in D_y \setminus \{(i_0, i_0)\}} \lambda_i \lambda_j h_y(i, j).$$

Choose a mapping $g : n \times n \to W$ that extend h_x and h_y on all pairs from their respective domains except (j_0, j_0) and (i_0, i_0) for which we put $g(j_0, j_0) = -(1/\lambda_{j_0}^2 u_x)$ and $g(i_0, i_0) = u_y$ if $u_y \neq 0$ and put $g(i_0, i_0)$ to be an arbitrary nonzero element of W, otherwise. Find $\gamma < \delta$ satisfying the equation (8.3.8) for this choice of g. Then on one hand

$$\vartheta(x_\gamma, x_\delta) = \vartheta(z_\gamma + x^\gamma, z_\delta + x^\delta) = u_x + \lambda_{j_0}^2(-1/\lambda_{j_0}^2 u_x) = 0,$$

and on the other hand

$$\vartheta(y_\gamma, y_\delta) = \vartheta(z_\gamma + y^\gamma, z_\delta + y^\delta) = u_y + \lambda_{i_0}^2 g(i_0, i_0) \neq 0,$$

as required. □

Corollary 8.3.8. *Suppose that θ is the successor of some regular cardinal κ, and that V and W are vector spaces over the same field K such that W has cardinality at most κ while the dimension of V is equal to θ. Let $\vartheta : V \times V \to W$ be the bilinear mapping of Theorem* 8.3.3. *Then the corresponding group $V\varphi W$ has a family of 2^θ pairwise non-isomorphic subgroups.*

Proof. For a subset Γ of θ, let

$$G_\Gamma = \{(x, a) \in V\vartheta W : \operatorname{supp}(x) \subseteq \Gamma\}.$$

Clearly each G_Γ forms a subgroup of $V\vartheta W$. By Theorem 8.3.7, G_Γ is not isomorphic to a subgroup of G_Σ whenever the difference $\Gamma \setminus \Sigma$ is unbounded in θ. Since there is a family $\mathcal{F}$ of size 2^θ of subsets of θ such that $\Gamma \setminus \Sigma$ whenever $\Gamma \neq \Sigma \in \mathcal{F}$, the conclusion of the corollary follows. □

Remark 8.3.9. The special case of Theorem 8.3.3, where $W = K = \mathbb{Z}_p$ for some prime p and where V is the vector space of finitely supported maps from θ into $\mathbb{Z}_p$, produces a bilinear form with a particularly interesting group $V\varphi\mathbb{Z}_p$. Namely, in this case the group $G = V\varphi\mathbb{Z}_p$ is an example of what is called in [124] an *extraspecial p-group* G whose derived group G' is equal to its center $Z(G)$ and is in turn isomorphic to $\mathbb{Z}_p$, while the corresponding quotient G/G' is isomorphic to a direct product of copies of $\mathbb{Z}_p$. The first use of the square-bracket operation to construct an extraspecial p-group of cardinality θ without an abelian subgroup of cardinality θ was done in [98].

8.4 Two more variations on the square-bracket operation

In this section we present two variations of the square-bracket operation which allow projections with quite different properties from those encountered in the previous section. In fact these properties can be realized only at cardinals $\theta > \omega_1$

and as we will see this is reflected in the fact that these square-bracket operations can be defined only on the basis of C-sequences that avoid stationary subsets Γ of θ consisting of ordinals of high cofinalities. For example, while our first variation works on the basis that $\mathrm{cf}(\gamma) > \omega$ for all $\gamma \in \Gamma$, we must assume that θ is bigger than the continuum $\mathfrak{c}$. The second variation does not need the assumption $\theta > \mathfrak{c}$ though it works only for C-sequences avoiding a stationary set Γ with the property $\mathrm{cf}(\gamma) > \omega_1$ for all $\gamma \in \Gamma$.

We start this section with a variation of the square-bracket operation that works in the case $\theta > \mathfrak{c}$ and under the assumption that the C-sequence on θ avoids some fixed stationary set $\Gamma \subseteq \theta$ with the property that $\mathrm{cf}(\gamma) > \omega$ for all $\gamma \in \Gamma$. In order to define it, we need to fix a function $\xi \longmapsto \xi^\star$ from θ to ω such that

$$\Gamma_n = \{\xi \in \Gamma : \xi^\star = n\}$$

is stationary in θ for all n. The mapping $\xi \longmapsto \xi^*$ leads us naturally to the variation

$$\rho_\star : [\theta]^2 \rightarrow \omega^{<\omega}$$

of the mapping ρ_0 which we define recursively by the formula

$$\rho_\star(\alpha, \beta) = \langle \beta^\star \rangle^\frown \rho_\star(\alpha, \min(C_\beta \setminus \alpha)), \tag{8.4.1}$$

with the boundary value $\rho_\star(\alpha, \alpha) = \langle \alpha^\star \rangle$ for all α. The variation $\rho_\star$ of ρ_0 shares many properties with ρ_0 such as the following that will be frequently and implicitly used below and where we are using the already standard notation

$$\lambda(\alpha, \beta) = \max\{\max(C_\xi \cap \alpha) : \xi \in \mathrm{Tr}(\alpha, \beta) \text{ and } \xi \neq \alpha\}$$

for the upper bound on the lower trace.

Lemma 8.4.1. *If $\lambda(\beta, \gamma) < \alpha < \beta < \gamma$, then $\rho_\star(\alpha, \gamma) = \rho_\star(\beta, \gamma)^\frown \rho_\star(\alpha, \beta)$.* □

It should also be clear that the proof of Lemma 8.2.7 allows adding the following additional conclusions.

Lemma 8.4.2. *Under the hypothesis of Lemma* 8.2.7, *its conclusion can be extended by adding the following two new statements:*

(4) $\rho_\star(\delta_1, \delta_0) = \cdots = \rho_\star(\delta_l, \delta_{l-1})$.
(5) *The maximal term of the sequence $\rho_\star(\delta_1, \delta_0) = \cdots = \rho_\star(\delta_l, \delta_{l-1})$ is bigger than the maximal term of any of the sequences $\rho_\star^*(\delta_0, b(j))$ or $\rho_\star(a(i), \delta_l)$ for $i, j < n$.* □

Having this in mind, the following variation on the square-bracket operation suggests itself.

Definition 8.4.3. For $\alpha < \beta < \theta$, let $[\alpha\beta]^\star = \beta_t$ for t the minimal initial part of $\rho_0(\alpha, \beta)$ such that $(\beta_t)^\star = \max(\rho_\star(\alpha, \beta))$.

In other words, if $\beta = \beta_0 > \beta_1 > \cdots > \beta_{\rho_2(\alpha,\beta)} = \alpha$ is the decreasing enumeration of the upper trace of the walk from β to α, then

$$[\alpha\beta]^\star = \beta_k, \text{ where } k = \min\{j : (\beta_j)^\star = \max\{(\beta_i)^\star : i \leq \rho_2(\alpha,\beta)\}\}.$$

So $[\alpha\beta]^\star$ is the first place in the walk from β to α where the function $\xi \to \xi^\star$ reaches its maximum among all other places visited during the walk. Note that combining the conclusions Lemmas 8.2.7 and 8.4.2, we arrive at the following additional information.

Lemma 8.4.4. *Under the hypothesis of Lemma* 8.2.7, *its conclusion can be extended by adding the following:*

(6) $[a(i)b(j)]^\star = [\delta_1\delta_0]^\star$ *for all* $i, j < n$. □

Having in mind the property of $[\cdot\cdot]_\circ$ stated in Lemma 8.2.9, the following variation is now quite natural.

Definition 8.4.5. $[\alpha\beta]^\star_\circ = [\alpha[\alpha\beta]^\star]_\circ$ for $\alpha < \beta < \theta$.

Using Lemmas 8.2.7, 8.4.2, and 8.4.4 one gets the following conclusion.

Lemma 8.4.6. *Let A be a family of θ pairwise-disjoint finite subsets of θ, all of some fixed size n. Then for all but a nonstationary set of $\delta \in \Gamma$ one can find $a < b$ in A such that $[a(i)b(j)]^\star_\circ = \delta$ for all $i, j < n$.* □

Composing $[\cdot\cdot]^\star_\circ$ with a mapping $\pi : \theta \longrightarrow \theta$ with the property that $\pi^{-1}(\xi) \cap \Gamma$ is stationary for all $\xi < \theta$, one gets a projection of $[\cdot\cdot]^\star_\circ$ for which the conclusion of Lemma 8.4.6 is true for *all* $\delta < \kappa$. Assuming that θ is moreover a non-inaccessible regular cardinal bigger than the continuum (such as, for example, a successor of a regular cardinal $\kappa \geq \mathfrak{c}$, in which case Γ can be taken to be the set $\{\delta < \kappa^+ : \mathrm{cf}\,\delta = \kappa\}$), and proceeding as in Lemma 8.3.1 above we get a projection $[\![\cdot\cdot]\!]^\star_\circ$ with the following property.

Theorem 8.4.7. *Suppose that a regular non-strong limit cardinal $\theta > \mathfrak{c}$ carries a C-sequence avoiding a stationary set of ordinals $< \theta$ of uncountable cofinality. Then the square-bracket operation $[\cdot\cdot]^\star_\circ$ has a projection $[\![\cdot\cdot]\!]^\star_\circ$ with the property that for every family A of pairwise-disjoint finite subsets of θ, all of some fixed size n, and for every $q : n \times n \longrightarrow \theta$, there exist $a < b$ in A such that $[\![a(i)b(j)]\!]^\star_\circ = q(i, j)$ for all $i, j < n$.* □

Remark 8.4.8. The first example of a cardinal with such a complex binary operation was given by the author [110] using the oscillation mapping described above in Section 8.1. It was the cardinal $\mathfrak{b}$, the minimal cardinality of an unbounded subset of ω^ω under the ordering of eventual dominance. The oscillation mapping restricted to some well-ordered unbounded subset W of ω^ω is perhaps still the most interesting example of this kind, due to the fact that its properties are preserved in forcing extensions that do not change the unboundedness of W (although they can collapse cardinals and therefore destroy the properties of the square-bracket operations on them). This absoluteness of osc is the key feature behind its applications in various coding procedures (see, e.g., [118]).

We are now ready for another variation of the square-bracket operation. It works for a regular cardinal θ that carries a C-sequence C_α $(\alpha < \theta)$ avoiding a stationary subset Γ of θ with the property that $\mathrm{cf}(\gamma) > \omega_1$ for all $\gamma \in \Gamma$. The additional parameters in the definition are a mapping $\gamma \mapsto \gamma^*$ from θ to ω_1 such that

$$\Gamma_\xi = \{\gamma \in \Gamma : \gamma^* = \xi\}$$

is stationary in θ for all $\xi < \omega_1$, and a mapping $\xi \mapsto \xi^\bullet$ from ω_1 into ω with the property that

$$\Xi_n = \{\xi < \omega_1 : \xi^\bullet = n\}$$

is stationary in ω_1 for all $n < \omega$. For $\alpha < \beta < \theta$, let

$$\beta = \beta_0 > \beta_1 > \cdots > \beta_{\rho_2(\alpha,\beta)} = \alpha$$

be the trace of the walk from β to α, and set,

$$u(\alpha,\beta) = \{k > \rho_2(\alpha,\beta)/2 : \beta_k^* > \beta_l^* \text{ for all } l < \rho_2(\alpha,\beta)/2\}, \tag{8.4.2}$$

$$v(\alpha,\beta) = \{l < \rho_2(\alpha,\beta)/2 : \beta_l^* > \beta_k^* \text{ for all } k > \rho_2(\alpha,\beta)/2,\ k \notin u(\alpha,\beta)\}. \tag{8.4.3}$$

Let

$$k(\alpha,\beta) = \min u(\alpha,\beta) \text{ and } l(\alpha,\beta) = (\min\{\gamma_l^* : l \in v(\alpha,\beta)\})^\bullet. \tag{8.4.4}$$

If $k(\alpha,\beta) - l(\alpha,\beta) \leq \rho_2(\alpha,\beta,)$ set

$$[\alpha\beta]^* = \beta_{k(\alpha,\beta)-l(\alpha,\beta)}. \tag{8.4.5}$$

Otherwise, $[\alpha\beta]^* = \beta$.

We extend the mapping $\gamma \to \gamma^*$, to a mapping $\tau \to \tau^*$ from $\theta^{<\omega}$ into $\omega_1^{<\omega}$ in the natural way, by letting τ^* be a finite sequence of the same length as τ and $\tau^*(i) = \tau(i)^*$ for all $i < |\tau|$. The mapping $\gamma \to \gamma^*$ also leads us to a natural variation $\rho_* : [\theta]^2 \to \omega_1^{<\omega}$ of the mapping ρ_0 defined by the recursive formula

$$\rho_*(\alpha,\beta) = \langle\beta^*\rangle^\frown \rho_*(\alpha, \min(C_\beta \setminus \alpha)),$$

with the boundary value $\rho_*(\alpha,\alpha) = \langle\alpha^*\rangle$. Thus, if

$$\beta = \beta_0 > \cdots > \beta_{\rho_2(\alpha,\beta)} = \alpha$$

is the decreasing enumeration of the trace of the minimal walk from β to α along our C-sequence, then

$$\rho_*(\alpha,\beta) = \tau^*, \text{ where } \tau = \langle\beta_0, \beta_1, \ldots, \beta_{\rho_2(\alpha,\beta)}\rangle.$$

As before, the variation ρ_* shares many of the common properties with the original function ρ_0 including the following.

Lemma 8.4.9. *If $\lambda(\beta,\gamma) < \alpha < \beta < \gamma$, then $\rho_*(\alpha,\gamma) = \rho_*(\beta,\gamma)^\frown\rho_*(\alpha,\beta)$.* □

We shall need another piece of notation, where for a given $\sigma \in (\omega_1)^{\omega_1}$ and $\xi < \omega_1$, we let σ^ξ be the sequence in $(\omega_1+1)^{<\omega}$ which has the length $|\sigma^\xi|$ equal to the length $|\sigma|$ of σ, and

$$\sigma^\xi(i) = \sigma(i) \text{ if } \sigma(i) < \xi;\quad \sigma^\xi(i) = \omega_1 \text{ if } \sigma(i) \geq \xi.$$

The analysis of the square-bracket operation $[\cdot\cdot]^*$ is greatly facilitated by the following notion.

Definition 8.4.10. Given an unbounded subset Σ of θ and $\delta < \theta$, a sequence $\tau \in (\omega_1+1)^{<\omega}$ is called a (Σ,δ)-*lower limit point* if for every $\xi < \omega_1$ and $\alpha < \delta$ there is $\gamma \in (\alpha,\delta) \cap \Sigma$ such that $\rho_*(\gamma,\delta)^\xi = \tau$.

Lemma 8.4.11. *For every unbounded $\Sigma \subseteq \theta$ and almost all ordinals $\delta < \theta$ of cofinality $> \omega$ there is a (Σ,δ)-lower limit point in $(\omega_1+1)^{<\omega}$.*

Proof. Choose an arbitrary $\delta < \theta$ of uncountable cofinality belonging to the set $\bigcap_{\xi<\omega_1} \lim(\Gamma_\xi \cap \lim(\Sigma))$. Consider first the case $\mathrm{cf}(\delta) = \omega_1$. Fix a strictly increasing sequence δ_ξ $(\xi < \omega_1)$ converging to δ. For $\xi < \omega_1$, choose $\gamma_\xi \in \Gamma_\xi \cap \lim(\Sigma) \cap (\delta_\xi,\delta)$. Then, since our C-sequence avoids the set Γ, for each ξ,

$$\lambda(\gamma_\xi,\delta) = \max\{\max(C_\alpha \cap \gamma_\xi) : \alpha \in \mathrm{Tr}(\gamma_\xi,\delta) \text{ and } \alpha \neq \gamma_\xi\} < \gamma_\xi,$$

so we can find $\beta_\xi \in (\delta_\xi,\gamma_\xi) \cap \Sigma$ such that $\beta_\xi > \lambda(\gamma_\xi,\delta)$. It follows that $\gamma_\xi \in \mathrm{Tr}(\beta_\xi,\delta)$ for all $\xi < \omega_1$, and therefore $\xi = \gamma_\xi^*$ is a term of the sequence $\rho_*(\beta_\xi,\delta)$ for all $\xi < \omega_1$. Find a stationary set $\Xi_0 \subseteq \omega_1$ and $n < \omega$ such that $|\rho_*(\beta_\xi,\delta)| = n$, and such that for all $i < n$, either $\rho_*(\beta_\xi,\delta)(\xi) < \xi$ for all $\xi \in \Xi_0$, or else $\rho_*(\beta_\xi,\delta)(\xi) \geq \xi$ for all $\xi \in \Xi_0$. So, by a repeated application of the Pressing Down Lemma, we can find $\tau \in (\omega_1+1)^{<\omega}$ and a stationary set $\Xi_1 \subseteq \Xi_0$ such that

$$\rho_*(\beta_\xi,\delta)^\xi = \tau \text{ for all } \xi \in \Xi_1.$$

Then τ is one of the (Σ,δ)-lower limit points.

Let us consider now the case $\kappa = \mathrm{cf}(\delta) > \omega_1$. Choose a strictly increasing sequence δ_η $(\eta < \kappa)$ converging to δ. Working as in the first case, for each $\eta < \kappa$, we find a stationary set $\Xi \subseteq \omega_1$, a sequence

$$\beta_{\eta\xi} \in (\delta_\eta,\delta) \quad (\xi \in \Xi_\eta),$$

and $\tau_\eta \in (\omega_1+1)^{<\omega}$ such that $\rho_*(\beta_{\eta\xi},\delta)^\xi = \tau_\eta$ for all $\xi \in \Xi_\eta$. Choose $\tau \in (\omega_1+1)^{<\omega}$ such that $\{\eta < \kappa : \tau_\eta = \tau\}$ has cardinality κ. Then τ is a (Σ,δ)-lower limit point. □

Definition 8.4.12. Given an unbounded subset Σ of θ and $\delta < \theta$, a sequence $\tau \in (\omega_1+1)^{<\omega}$ is called a (Σ,δ)-*upper limit point* if for every $\xi < \omega_1$ there is $\gamma \in \Sigma \setminus (\delta+1)$ such that $\rho_*(\delta,\gamma)^\xi = \tau$. If in addition we are given a subset Ξ of ω_1, we say that such a sequence τ is a (Σ, δ, Ξ)-*upper limit point* if for every $\xi < \omega_1$, we can find $\gamma \in \Sigma \setminus (\delta + 1)$ such that

$$\rho_*(\delta, \gamma)^\xi = \tau \text{ and } \min\{\rho_*(\gamma, \delta)(i) : \tau(i) = \omega_1\} \in \Xi.$$

Lemma 8.4.13. *For every unbounded $\Sigma \subseteq \theta$ and stationary $\Xi \subseteq \omega_1$, for all but a bounded set of $\delta < \theta$, there exists a (Σ, δ, Ξ)-upper limit point.*

Proof. For $\xi < \omega_1$, let

$$\Omega_\xi = \{\delta < \theta : \xi \text{ is not a term of } \rho_*(\delta, \gamma) \text{ for all } \gamma \in \Sigma \setminus (\delta + 1)\}.$$

Since for each $\xi < \omega_1$, the set Γ_ξ is stationary in θ and is avoided by our C-sequence, one easily concludes that $|\Omega_\xi| < \theta$ for all $\xi < \omega_1$. So it suffices to show that if $\delta < \theta$ dominates all the Ω_ξ's, then there is always a (Σ, δ, Ξ)-upper limit point. So fix such δ and for each $\xi < \omega_1$ a $\gamma_\xi \in \Sigma \setminus (\delta + 1)$ such that ξ appears as a term of $\rho_*(\delta, \gamma_\xi)$. Then as before, we can choose a stationary $\Xi_0 \subseteq \Xi$ and $n < \omega$ such that $|\rho_*(\delta, \gamma_\xi)| = n$ for all $\xi \in \Xi_0$, and such that for all $i < n$, either $\rho_*(\delta, \gamma_\xi)(i) < \xi$ for all $\xi \in \Xi_0$, or else $\rho_*(\delta, \gamma_\xi)(i) \geq \xi$ for all $\xi \in \Xi_0$. So by a repeated application of the Pressing Down Lemma, we can find a stationary set $\Xi_1 \subseteq \Xi_0$ and $\tau \in (\omega_1)^{<\omega}$ such that

$$\rho_*(\delta, \gamma_\xi)^\xi = \tau \text{ for all } \xi \in \Xi_1.$$

Since for $\xi \in \Xi_1$,

$$\min\{\rho_*(\delta, \gamma_\xi)(i) : i < n,\ \tau(i) = \omega_1\} = \xi \in \Xi_1 \subseteq \Xi,$$

we conclude that τ is a (Σ, δ, Ξ)-upper limit point. □

Remark 8.4.14. Note that if $\mathrm{cf}(\delta) > \omega_1$ in the above proof, then the sequence $\lambda(\delta, \gamma_\xi)$ $(\xi < \omega)$ is bounded in δ, so if $\lambda < \delta$ is one of its upper bounds, then we say that the resulting sequence τ is a $(\Sigma, \delta, \Xi, \lambda)$-upper limit point.

We are now ready to state and prove the first important property of the square-bracket operation $[\cdot\cdot]^*$.

Lemma 8.4.15. *For every pair Σ and Ω of unbounded subsets θ, almost all $\gamma \in \Gamma$ are of the form $[\alpha\beta]^*$ for some $\alpha \in \Sigma$ and $\beta \in \Omega$ such that $\alpha < \beta$.*

Proof. Let Γ' be a given stationary subset of Γ. We need to find $\alpha \in \Sigma$ and $\beta \in \Omega$ with $\alpha < \beta$ such that $[\alpha\beta]^\circ \in \Gamma'$. Consider $\delta \in \Gamma' \cap \lim(\Sigma)$. Since $\mathrm{cf}(\delta) > \omega_1$, we can find $\sigma_\delta^- \in (\omega_1)^{<\omega}$ such that

$$\sup\{\alpha \in \Sigma \cap \delta : \rho_*(\alpha, \delta) = \sigma_\delta^-\} = \delta. \tag{8.4.6}$$

For the same $\delta \in \Gamma'$, we fix also $\beta_\delta \in \Omega \setminus (\delta + 1)$ and set $\sigma_\delta^+ = \rho_*(\delta, \beta_\delta)$. Since $\lambda(\delta, \beta_\delta) < \delta$, we can find a stationary set $\Gamma'' \subseteq \Gamma'$, sequences $\sigma^-, \sigma^+ \in (\omega_1)^{<\omega}$, and $\lambda' < \theta$ such that

$$\lambda(\delta, \beta_\delta) = \lambda',\ \sigma_\delta^- = \sigma^-, \text{ and } \sigma_\delta^+ = \sigma^+ \text{ for all } \delta \in \Gamma''. \tag{8.4.7}$$

By Lemma 8.4.11, there is a stationary set $\Upsilon \subseteq \Gamma''$ and $\varsigma \in (\omega_1+1)^{<\omega}$ such that

$$\varsigma \text{ is a } (\Gamma'', \delta)\text{-lower limit point for all } \delta \in \Upsilon. \tag{8.4.8}$$

Let

$$i_0 = \min\{i < |\varsigma| : \varsigma(i) = \omega_1.\} \tag{8.4.9}$$

By Lemma 8.4.11, there is a stationary set $\Upsilon' \subseteq \Upsilon$ and $\tau \in (\omega_1+1)^{<\omega}$ such that

$$\tau \text{ is a } (\Upsilon, \delta)\text{-lower limit point for all } \delta \in \Upsilon'. \tag{8.4.10}$$

By Lemma 8.4.13 and Remark 8.4.14, for every $\delta \in \Upsilon'$ there is $\upsilon_\delta \in (\omega_1+1)^{<\omega}$ and $\lambda' < \lambda_\delta < \delta$ such that υ_δ is a $(\Upsilon', \delta, \Xi_{i_0+|\tau|}, \lambda_\delta)$-upper limit point. So applying the Pressing Down Lemma, we can find ordinal $\lambda'' < \theta$, a sequence $\upsilon \in (\omega_1+1)^{<\omega}$, and stationary $\Upsilon'' \subseteq \Upsilon'$ such that

$$\upsilon \text{ is a } (\Upsilon', \delta, \Xi_{i_0+|\tau|}, \lambda'')\text{-upper limit point for all } \delta \in \Upsilon''. \tag{8.4.11}$$

By Theorem 6.3.2, we can find $\gamma' < \gamma''$ in Υ'' such that

$$\rho_2(\gamma', \gamma'') > |\sigma^-| + |\sigma^+| + |\tau| + |\upsilon|. \tag{8.4.12}$$

Choose $\xi_0 < \omega_1$ such that

$$\xi_0 > (\operatorname{rang}(\rho_*(\gamma', \gamma'')) \cup \operatorname{rang}(\upsilon) \cup \operatorname{rang}(\sigma^+)) \cap \omega_1. \tag{8.4.13}$$

Since τ is a (Υ, γ')-limit point there is $\beta'' \in (\lambda'', \gamma')$ such that

$$\rho_*(\beta'', \gamma')^{\xi_0} = \tau. \tag{8.4.14}$$

Pick $\xi_1 < \omega_1$ such that

$$\xi_1 > \operatorname{rang}(\rho_*(\beta'', \gamma')). \tag{8.4.15}$$

By (8.4.11) and the fact that γ'' belongs to Υ'', we can find $\delta \in \Upsilon' \setminus (\gamma''+1)$ such that

$$\rho_*(\gamma'', \delta)^{\xi_1} = \upsilon \text{ and } \lambda(\gamma'', \delta) < \lambda''. \tag{8.4.16}$$

Since $\delta \in \Gamma''$, by (8.4.7), we conclude that

$$\rho_*(\delta, \beta_\delta) = \sigma^+ \text{ and } \lambda(\delta, \beta_\delta) = \lambda'. \tag{8.4.17}$$

Choose $\xi_2 < \omega_1$ such that

$$\xi_2 > \operatorname{rang}(\rho_*(\gamma'', \delta)). \tag{8.4.18}$$

Since $\beta'' \in \Upsilon$, by (8.4.8), there is $\beta' \in (\lambda'', \beta'') \cap \Gamma''$ such that

$$\rho_*(\beta', \beta'')^{\xi_2} = \varsigma. \tag{8.4.19}$$

Since $\beta'' \in \Gamma''$, by (8.4.7), we can find $\alpha \in \Sigma$ such that

$$\alpha > \max\{\lambda'', \lambda(\beta', \beta'')\} \text{ and } \rho_*(\alpha, \beta') = \sigma^-. \tag{8.4.20}$$

Let $\beta = \beta_\delta$. Then $\alpha \in \Sigma$, $\beta \in \Omega$, and

$$\rho_0(\alpha,\beta) = \rho_0(\delta,\beta)^\frown\rho_0(\gamma'',\delta)^\frown\rho_0(\gamma',\gamma'')^\frown\rho_0(\beta'',\gamma')^\frown\rho_0(\beta',\beta'')^\frown\rho_0(\alpha,\beta').$$

Note that the middle point(s) of the trace $\beta = \beta_0 > \beta_1 > \cdots > \beta_{\rho_2(\alpha,\beta)} = \alpha$ is (are) positioned way inside the trace of the walk from γ'' to γ'. The set $u(\alpha,\beta)$ consists of the positions k belonging to the segment $\rho_0(\beta',\beta'')$ of the walk such that $\rho_*(\beta',\beta'')(k) \geq \xi_2$. It follows that

$$p(\alpha,\beta) = m + |\rho_0(\beta'',\gamma')| + i_0 = m + |\tau| + i_0,$$

where m is determined by the equation $\beta_m = \gamma'$. The set $v(\alpha,\beta)$ consists of the positions k belonging to the segment $\rho_0(\gamma'',\delta)$ of the walk such that $\rho_*(\gamma'',\delta)(k) \geq \xi_1$, since any such value is bigger than $\rho_*(\alpha,\beta)(l)$ for every $l > \rho_2(\alpha,\beta)/2$ with $l \notin u(\alpha,\beta)$. The portion corresponding to $\rho_0(\beta'',\gamma')$ and the sequence τ with values $\geq \xi_0$ was inserted in order to provide that no other $k < \rho_2(\alpha,\beta)/2$ have this property. Note also that by the choice of δ and γ'' and the sequence v, we conclude that

$$\min\{\beta_k : k \in v(\alpha,\beta)\} \in \Xi_{i_0+|\tau|}.$$

It follows that $q(\alpha,\beta) = i_0 + |\tau|$, and therefore,

$$[\alpha\beta]^* = \beta_{p(\alpha,\beta)-q(\alpha,\beta)} = \beta_m = \gamma' \in \Gamma',$$

as required. □

In fact, we have here the following slightly stronger property of $[..]^\circ$.

Theorem 8.4.16. *Suppose that a regular cardinal θ carries a C-sequence avoiding a stationary set Γ of cofinality $> \omega_1$ ordinals. Then the corresponding square-bracket operation $[\cdot\cdot]^*$ has the property that for every family A of θ pairwise-disjoint finite subsets of θ, all of some fixed size n, for almost all $\gamma \in \Gamma$ there exist $a < b$ in A such that $[a(i)b(j)]^* = \gamma$ for all $i, j < n$.*

Proof. This has been really established during the course of the proof of Lemma 8.4.15, so let us only point out the adjustment needed for the formally stronger result. As before we fix a stationary set $\Gamma' \subseteq \Gamma$. Consider $\delta \in \Gamma' \cap \lim(\Sigma)$. Since $\mathrm{cf}(\delta) > \omega_1$, we can find $\sigma_\delta^-(i) \in (\omega_1)^{<\omega}$ for $i < n$ such that for all $\alpha < \delta$ there is $a \in A$ that is included in the interval (α,δ) such that

$$\rho_*(a(i),\delta) = \sigma_\delta^-(i) \text{ for all } i < n. \tag{8.4.21}$$

For the same $\delta \in \Gamma'$, we fix also $b_\delta \in A$ such that $b > \delta$ and set $\sigma_\delta^+(i) = \rho_*(\delta, b_\delta(i))$. Since $\lambda(\delta, b_\delta(i)) < \delta$ for all $i < n$, we can find a stationary set $\Gamma'' \subseteq \Gamma'$, sequences $\sigma^-(i), \sigma^+(i) \in (\omega_1)^{<\omega}$ $(i < n)$ and $\lambda'(i) < \theta$ $(i < n)$ such that for all $i < n$,

$$\lambda(\delta, b_\delta(i)) = \lambda'(i),\ \sigma_\delta^-(i) = \sigma^-(i),\ \text{and } \sigma_\delta^+(i) = \sigma^+(i) \text{ for all } \delta \in \Gamma''.$$

We now proceed as in the previous proof, choosing of course $\gamma' < \gamma''$ such that

$$\rho_2(\gamma',\gamma'') > |\tau| + |v| + \sum_{i<n}(|\sigma^-(i)| + |\sigma^+(i)|),$$

and choosing ξ_0 such that

$$\xi_0 > (\mathrm{rang}(\rho_*(\gamma',\gamma'')) \cup \mathrm{rang}(v) \cup \bigcup_{i<n} \mathrm{rang}(\sigma^+(i))) \cap \omega_1.$$

This will result in finding $b = b_\delta > \delta$ and $a \subseteq (\max\{\lambda'', \lambda(\beta',\beta'')\}, \beta')$, both in A, such that

$$[a(i)b(j)]^* = \gamma' \in \Gamma' \text{ for all } i, j < n,$$

as required. □

Working as in the case of the original square-bracket operation $[\cdot\cdot]$, one obtains the following result.

Theorem 8.4.17. *Suppose that a regular non-strong limit cardinal θ carries a C-sequence avoiding a stationary set of cofinality $> \omega_1$ ordinals. Then the corresponding square-bracket operation $[\cdot\cdot]^*$ has a projection $[\![\cdot\cdot]\!]^*$ with the property that for every family A of pairwise-disjoint finite subsets of θ, all of some fixed size n, and for every $q : n \times n \to \theta$, there exist $a < b$ in A such that $[\![a(i)b(j)]\!]^* = q(i,j)$ for all $i, j < n$.* □

Remark 8.4.18. A typical cardinal θ satisfying the hypothesis of the previous result is a successor of a regular cardinal $\kappa > \omega_1$. Thus, $\theta = \omega_3$ is the minimal instance to which this result applies. The case $\theta = \omega_2$ requires a special treatment as we will obtain it, not through projecting a square-bracket operation on ω_2, but rather through stepping up the square-bracket operation of ω_1.

To describe a coloring that would cover the case $\theta = \omega_2$, we choose a mapping $\delta \mapsto \delta^*$ from ω_2 onto ω_1 such that

$$\Delta_\xi = \{\delta < \omega_2 : \mathrm{cf}(\delta) = \omega_1 \text{ and } \delta^* = \xi\}$$

is stationary in ω_2 for all $\xi < \omega_1$. Our C-sequence C_α $(\alpha < \omega_2)$ will have the usual properties including the property that $\mathrm{tp}(C_\alpha) \leq \omega_1$ for all α. This in particular gives that the C-sequence avoids all the sets Δ_ξ $(\xi < \omega_1)$. As before, the mapping $\delta \mapsto \delta^*$ leads us to the variation $\rho_* : [\omega_2]^2 \to \omega_1$ of ρ_0 defined by the recursive formula

$$\rho_*(\alpha,\beta) = \langle\beta^*\rangle^\frown \rho_*(\alpha, \min(C_\beta \setminus \alpha)),$$

with the boundary value $\rho_*(\alpha,\alpha) = \langle\alpha^*\rangle$ and which has the property that,

$$\lambda(\beta,\gamma) < \alpha < \beta < \gamma \text{ implies } \rho_*(\alpha,\gamma) = \rho_*(\beta,\gamma)^\frown\rho_*(\alpha,\beta). \tag{8.4.22}$$

Finally, define $g : [\omega_2] \to \omega_1$, by

$$g(\alpha,\beta) = f(\rho_*(\alpha,\beta)), \tag{8.4.23}$$

where f is the mapping of Lemma 5.2.6.

Lemma 8.4.19. *For every family A of ω_2 pairwise disjoint finite subsets of ω_2, all of some fixed size n, and for every $\xi < \omega_1$, there exist $a < b$ in A such that $g(a(i), b(j)) = \xi$ for all $i, j < n$.*

Proof. Let

$$C = \{\delta < \omega_2 : \text{ for all } \gamma < \delta \text{ there is } a \in A,\ a \subseteq (\gamma, \delta)\}.$$

Clearly, C is a closed and unbounded subset of ω_2. For $\delta < \omega_2$, we fix $b_\delta \in A$ such that $b_\delta > \delta$. Then, since our C-sequence avoids Γ, for every $\delta \in \Gamma$,

$$\lambda(\delta, \beta) = \max\{\max(C_\alpha \cap \delta) : \alpha \in \mathrm{Tr}(\delta, \beta) \text{ and } \alpha \neq \delta\} < \delta.$$

So, we can find a single ordinal $\lambda_0 < \omega_2$, and for each $\xi < \omega_1$ a stationary set $\Sigma_\xi \subseteq \Gamma_\xi \cap C$ such that for all $\xi < \omega_1$,

$$\lambda(\delta, \beta) < \lambda_0 \text{ for all } \delta \in \Sigma_\xi \text{ and } \beta \in b_\delta. \tag{8.4.24}$$

Choose an ordinal $\varepsilon < \omega$ of cofinality ω_1 that is a common limit point of all sets Σ_ξ ($\xi < \omega_1$). For $\xi < \omega_1$ choose $\delta_\xi \in \Sigma_\xi \setminus (\varepsilon + 1)$. Then as before, for all $\xi < \omega_1$ and $\beta \in b_{\delta_\xi}$, we have $\lambda(\varepsilon, \beta) < \varepsilon$. So we can choose a strictly increasing sequence $\gamma_\xi \in \Sigma_\xi \cap \varepsilon$ ($\xi < \omega_1$) converging to ε such that for all $\xi < \omega_1$,

$$\gamma_\xi > \lambda(\varepsilon, \beta) \text{ for all } \beta \in b_{\delta_\eta} \text{ and } \eta < \xi. \tag{8.4.25}$$

Note that for the same reason for each $\xi < \omega_1$, we have $\lambda(\gamma_\xi, \varepsilon) < \gamma_\xi$, so since this ordinal has cofinality ω_1 and since it belongs to C, we can find $a_\xi < \gamma_\xi$ in A such that

$$a_\xi > \gamma_\eta \text{ for all } \eta < \xi, \text{ and} \tag{8.4.26}$$

$$a_\xi > \lambda(\gamma_\xi, \varepsilon) \text{ and } a_\xi > \lambda(\varepsilon, \beta) \text{ for all } \beta \in b_{\delta_\eta} \text{ and } \eta < \xi. \tag{8.4.27}$$

Combining all this we conclude that for $\eta < \xi < \omega_1$,

$$\rho_*(\alpha, \varepsilon) = \rho_*(\gamma_\xi, \varepsilon)^\frown \rho_*(\alpha, \gamma_\xi) \text{ for } \alpha \in a_\xi, \text{ and} \tag{8.4.28}$$

$$\rho_*(\alpha, \beta) = \rho_*(\beta, \varepsilon)^\frown \rho_*(\gamma_\xi, \varepsilon)^\frown \rho_*(\alpha, \gamma_\xi) \text{ for } \alpha \in a_\xi \text{ and } \beta \in b_{\delta_\eta}. \tag{8.4.29}$$

For $\xi < \omega_1$, let τ_ξ be concatenation of all sequences of the form $\rho_*(\alpha, \varepsilon)$ for $\alpha \in a_\xi$ of the form $\rho(\varepsilon, \beta)$ for $\beta \in b_{\delta_\xi}$ in some order arbitrarily chosen. Clearly ξ is a term of τ_ξ for all $\xi < \omega_1$. Given $\vartheta < \omega_1$, by Lemma 5.2.6, we can find $\eta < \xi$ such that $f(\sigma^\frown \upsilon) = \vartheta$ whenever σ is a subsequence of τ_η and υ is a subsequence of τ_ξ and ξ appears as a term in both of them. Given $\alpha \in a_\xi$ and $\beta \in b_{\delta_\eta}$, letting $\sigma = \rho_*(\varepsilon, \beta)$ and $\upsilon = \rho_*(\alpha, \varepsilon)$, we see that they do have ξ as one of their terms, and therefore, by (8.4.28) and (8.4.29),

$$g(\alpha, \beta) = f(\rho_*(\alpha, \beta)) = f(\rho_*(\varepsilon, \beta)^\frown \rho_*(\alpha, \varepsilon)) = \vartheta,$$

as required. □

Since ω_2 is not a strong limit cardinal, the already standard argument gives us the following consequence of Lemma 8.4.19.

Lemma 8.4.20. *The mapping g of Lemma 8.4.19 has a projection $[g]$ with the property that for every family A of ω_2 pairwise-disjoint finite subsets of ω_2, all of some fixed size n, and every mapping $h : n \times n \to \omega_1$, there exist $a < b$ in A such that $[g](s(i), b(j)) = h(i,j)$ for all $i, j < n$.* □

We finish this section by giving some typical applications of these results.

Theorem 8.4.21. *Suppose that $\theta = \omega_2$, or θ is a regular cardinal carrying a C-sequence avoiding a stationary set γ of cofinality $> \omega_1$ ordinals, or $\theta > \mathfrak{c}$ and θ carries a C-sequence avoiding a stationary set Γ of cofinality $> \omega$ ordinals. Then the θ-chain condition is not productive, i.e., there exist two partially-ordered sets $\mathcal{P}_0$ and $\mathcal{P}_1$ satisfying the θ-chain condition but their cartesian product $\mathcal{P}_0 \times \mathcal{P}_1$ fails to have this property.*

Proof. Since in all three cases the proofs are rather similar let us show how it works in the second case. Fix two disjoint stationary subsets Γ_0 and Γ_1 of Γ. For $i < 2$, let

$$\mathcal{P}_i = \{p \in [\theta]^{<\omega} : [\alpha\beta]^* \in \Gamma_i \text{ for all } \alpha < \beta \text{ in } p\}.$$

By Lemma 8.4.6, $\mathcal{P}_0$ and $\mathcal{P}_1$ are θ-cc posets. The cartesian product $\mathcal{P}_0 \times \mathcal{P}_1$ fails the θ-chain condition since it has the sequence $\langle \{\alpha\}, \{\alpha\} \rangle$ $(\alpha < \theta)$ of pairwise incomparable conditions. □

Corollary 8.4.22. *For every regular uncountable cardinal κ, the κ^+-chain condition is not productive, i.e., there exist two partially-ordered sets $\mathcal{P}_0$ and $\mathcal{P}_1$ satisfying the κ^+-chain condition but their cartesian product $\mathcal{P}_0 \times \mathcal{P}_1$ fails to have this property.*
□

Corollary 8.4.23. *The ω_2-chain condition is not productive.* □

Much finer examples of this sort can be obtained by relying on the projection $[\![\cdot\cdot]\!]^*$ of the square-bracket operation $[\cdot\cdot]^*$ rather than the square-bracket operation itself. A typical application of this sort is given in the following result.

Theorem 8.4.24. *For every regular uncountable cardinal κ there is a topological group G of cellularity[7] κ, though the cellularity of its square G^2 is bigger than κ.*

Proof. All that we will need in constructing the group G is a mapping $f : [\kappa^+]^3 \to 3$ with the property that for every family A of κ^+ pairwise-disjoint finite subsets of κ^+, all of some fixed size n, and for every $h : n \times n \to 3$, there exist $a < b$ in A such that $f(a(i), b(j)) = h(i,j)$ for all $i, j < n$. By going to an interval of the form $[\lambda, \kappa^+)$, we may assume that

$$\{\beta < \kappa^+ : f(\alpha, \beta) = i\} \text{ has size } \kappa^+ \text{ for all } \alpha < \kappa^+ \text{ and } i < 3. \tag{8.4 30}$$

[7] Recall that *cellularity* of a topological space X is defined to be the supremum of cardinalities of families of pairwise-disjoint open subsets of X.

Let $S = \kappa^+ \times 2$. An element of S of the form (α, i) will be denoted by α^i. Let G be the collection of all finite subsets of S considered as a group with the symmetric difference $+_2$ as the group operation.For $a \in G$, let

$$\mathrm{supp}(a) = \{\alpha < \kappa^+ : (\alpha, i) \in a \text{ for some } i < 2\}.$$

For $\beta < \kappa^+$, let G_β be the collection of all finite subsets a of S such that:

(1) $\beta^i \notin a$ for all $i < 2$,
(2) $\alpha < \beta$ and $f(\alpha, \beta) = i$ for some $i < 2$ imply that $a^{1-i} \notin a$,
(3) $\alpha < \beta$ and $f(\alpha, \beta) = 2$ imply that either $\alpha^i \in a$ for all $i < 2$, or else, $\alpha^i \notin a$ for all $i < 2$.

Clearly, each G_β is a subgroup of G and by (8.4.30), $\{\emptyset\} = \bigcap_{\beta<\kappa^+} G_\beta$. Thus, if for subset b of κ^+, we let $G_b = \bigcap_{\beta \in b} G_\beta$, we get that G_b $(b \in [\kappa^+]^{<\omega})$ forms a neighborhood base of $\emptyset$ in the topological group structure that we decide to put on G. We shall first show that the topological group G contains no family of more than κ pairwise-disjoint open sets.

Suppose $a_\xi +_2 G_{b_\xi}$ $(\xi < \kappa^+)$ is a given sequence of basic open sets of G. We need to find $\xi < \eta$ such that

$$(a_\xi +_2 G_{b_\xi}) \cap (a_\eta +_2 G_{b_\eta}) \neq \emptyset.$$

We may assume that a_ξ's form a Δ-system with root a and that b_ξ's form a Δ-system with root b. For $\xi < \kappa^+$, let $\bar{a}_\xi = a_\xi \setminus a$, let $\bar{b}_\xi = b_\xi \setminus b$, and let

$$c_\xi = \bar{b}_\xi \cup \mathrm{supp}(\bar{a}_\xi).$$

We may assume that c_ξ's have some fixed size n and that for some fixed subset I of $n = \{0, 1, \dots, n-1\}$, a fixed mapping $\psi : I \to 3$, and all $\xi < \kappa^+$, we have:

(4) $\bar{a}_\xi = \{c_\xi(i) : i \in I\}$,
(5) $\psi(i) = 0$ if $c_\xi(i)^0 \in \bar{a}_\xi$ and $c_\xi(i)^1 \notin \bar{a}_\xi$,
(6) $\psi(i) = 1$ if $c_\xi(i)^1 \in \bar{a}_\xi$ and $c_\xi(i)^0 \notin \bar{a}_\xi$,
(7) $\psi(i) = 2$ if $c_\xi(i)^0 \in \bar{a}_\xi$ and $c_\xi(i)^1 \in \bar{a}_\xi$.

Choose a mapping $h : n \times n \to 3$ such that $h(i, j) = \psi(i)$ for $i \in I$ and arbitrary $j < n$. By the basic property of f, we can find $\xi < \eta$ such that

$$f(c_\xi(i), c_\eta(j)) = h(i, j) \text{ for all } i, j < n.$$

Moreover, we may assume that $a \cup b < c_\xi < c_\eta$. Going back through the definitions, one easily verifies that $\bar{a}_\xi \in G_{b_\eta}$, and this gives us that

$$a_\xi \cup a_\eta = a_\eta +_2 \bar{a}_\xi \in a_\eta +_2 G_{b_\eta}.$$

Since the definition of G_β gives no restriction on sets whose supports lie above β, we conclude that $\bar{a}_\eta \in G_{b_\xi}$. It follows that

$$a_\xi \cup a_\eta = a_\xi +_2 \bar{a}_\eta \in a_\xi +_2 G_{b_\xi},$$

and therefore $a_\xi \cup a_\eta \in (a_\xi +_2 G_{b_\xi}) \cap (a_\eta +_2 G_{b_\eta})$, as required.

We prove the second conclusion by verifying that the members of the family

$$\{\beta^0\} +_2 G_\beta) \times (\{\beta^1\} +_2 G_\beta \quad (\beta < \kappa^+)$$

of open subsets of G^2 are pairwise-disjoint. Consider $\alpha < \beta < \kappa^+$, and suppose that some pair (x, y) of finite subsets of S belongs to the intersection

$$(\{\alpha^0\} +_2 G_\alpha) \times (\{\alpha^1\} +_2 G_\alpha) \cap (\{\beta^0\} +_2 G_\beta) \times (\{\beta^1\} +_2 G_\beta).$$

Note that since no set from G_α can contain α^0 or α^1, the set x must contain α^0. Let $x = \{\beta^0\} +_2 \bar{x}$ for some $\bar{x} \in G_\beta$. Then $\alpha^0 \in \bar{x} \in G_\beta$, so by (2), we conclude that $f(\alpha, \beta) \neq 0$. Working with the set y and the second coordinate, we similarly conclude that $f(\alpha, \beta) \neq 1$. It follows that $f(\alpha, \beta) = 2$. Applying (3) and the fact that $\alpha^0 \in \bar{x} \in G_\beta$, we conclude that $\alpha^1 \in \bar{x}$, and therefore $\alpha^1 \in \{\beta^0\} +_2 \bar{x} = x$. It follows that $\alpha^1 \in x \in \{\alpha^0\} +_2 G_\alpha$, and so α^1 must be a member of some finite set belonging to G_α, contradicting (1). This finishes the proof. □

Corollary 8.4.25. *There is a topological group G of cellularity $\aleph_1$ whose square has cellularity $> \aleph_1$.* □

Remark 8.4.26. Theorem 8.4.21, Corollary 8.4.22 and Corollary 8.4.23 are due to Shelah [93], [92], and [95] who also used the method of minimal walks for proving these results. The example of topological group witnessing of the non-productiveness of cellularity comes from p. 12 of [111] and from [115]. An exposition of the general problem of productiveness in the context of topological groups can be found in [72]. Recall that the countable chain condition can be productive. For example MA_{ω_1} implies that the cartesian product of two posets satisfying the countable chain condition also satisfies this chain condition(see, [36]). First examples of cardinals θ with θ-cc non-productive without using additional set-theoretical assumptions were given by the author in [107] (see also [111]). For example, $\theta = \mathfrak{b}$ or $\theta = \mathrm{cf}\ \mathfrak{c}$ are some of these cardinals. The first examples of non-productiveness of the κ^+-chain condition that do not use additional axioms of set theory were given by the author in [108] using what is today known under the name 'pcf theory'. For example, it has been proved in [108] that if $\kappa = \mathfrak{c}^{+\omega}$, then the κ^+-chain condition is not productive. After the full development of pcf theory it became apparent that the basic construction from [108] applies to every successor of a singular cardinal [94]. For an overview of recent advances in this area, the reader is referred to [76]. The following problem, however, still seems open.

Question 8.4.27. Suppose that θ is a regular strong limit cardinal and the θ-chain condition is productive. Is θ necessarily a weakly compact cardinal?

Chapter 9

Unbounded Functions

9.1 Partial square-sequences

In this section κ is a regular cardinal and C_α $(\alpha < \kappa^+)$ is a fixed C-sequence with the property that $\mathrm{tp}(C_\alpha) \leq \kappa$ for all $\alpha < \kappa^+$. When the C-sequence is not necessarily coherent, then it is natural to define the corresponding mapping

$$\rho : [\kappa^+]^2 \longrightarrow \kappa \tag{9.1.1}$$

as follows:

$$\rho(\alpha, \beta) = \sup\{\mathrm{tp}(C_\beta \cap \alpha), \rho(\alpha, \min(C_\beta \setminus \alpha)), \rho(\xi, \alpha) : \ \xi \in C_\beta \cap \alpha\}, \tag{9.1.2}$$

with the boundary value $\rho(\alpha, \alpha) = 0$ for all $\alpha < \kappa^+$, a definition that is slightly different from the one given above in (7.3.2) above. Clearly,

$$\rho(\alpha, \beta) \geq \rho_1(\alpha, \beta) \text{ for all } \alpha < \beta < \kappa^+, \tag{9.1.3}$$

and so, using Lemma 6.2.1, we have the following fact.

Lemma 9.1.1. *For $\nu < \kappa$, $\alpha < \kappa^+$ the set $P_\nu(\alpha) = \{\xi \leq \alpha : \rho(\xi, \alpha) \leq \nu\}$ has size no more than $|\nu| + \aleph_0$.* □

The proof of the following subadditivity properties of ρ is very similar to the proof of the corresponding fact for the function ρ from Section 7.3.

Lemma 9.1.2. *For all $\alpha \leq \beta \leq \gamma < \kappa^+$,*

(a) $\rho(\alpha, \gamma) \leq \max\{\rho(\alpha, \beta), \rho(\beta, \gamma)\}$,
(b) $\rho(\alpha, \beta) \leq \max\{\rho(\alpha, \gamma), \rho(\beta, \gamma)\}$. □

We mention a typical application of this function to the problem of existence of partial square-sequences which, for example, have some applications in the pcf theory (see [15]).

Theorem 9.1.3. *For every regular uncountable cardinal $\lambda < \kappa$ and stationary $\Gamma \subseteq \{\delta < \kappa^+ : \mathrm{cf}(\delta) = \lambda\}$, there is a stationary set $\Sigma \subseteq \Gamma$ and a sequence C_α $(\alpha \in \Sigma)$ such that:*

(1) *C_α is a closed and unbounded subset of α,*

(2) *$C_\alpha \cap \xi = C_\beta \cap \xi$ for every $\xi \in C_\alpha \cap C_\beta$.*

Proof. For each $\delta \in \Gamma$, choose $\nu = \nu(\delta) < \kappa$ such that the set $P_{<\nu}(\delta) = \{\xi < \delta : \rho(\xi, \delta) < \nu\}$ is unbounded in δ and closed under taking suprema of sequences of size $< \lambda$. Then there is $\bar{\nu}, \bar{\mu} < \kappa$ and stationary $\Sigma \subseteq \Gamma$ such that $\nu(\delta) = \bar{\nu}$ and $\mathrm{tp}(P_{<\bar{\nu}}(\delta)) = \bar{\mu}$ for all $\delta \in \Sigma$. Let C be a fixed closed and unbounded subset of $\bar{\mu}$ of order-type λ. Finally, for $\delta \in \Gamma$ set

$$C_\delta = \{\alpha \in P_{<\bar{\nu}}(\delta) : \mathrm{tp}(P_{<\bar{\nu}}(\alpha)) \in C\}.$$

Using Lemma 9.1.2, one easily checks that C_α $(\alpha \in \Sigma)$ satisfies the conditions (1) and (2). $\square$

Another application concerns the fact exposed above in Section 7.6, that the inequalities in Lemma 9.1.2(a), (b) are particularly useful, when κ has cofinality ω. Also consider the well-known phenomenon first discovered by K.Prikry (see [55]), that in some cases, the cofinality of a regular cardinal κ can be changed to ω, while preserving all cardinals.

Theorem 9.1.4. *In any cardinal-preserving extension of the universe, which has no new bounded subsets of κ, but in which κ has a cofinal ω-sequence diagonalizing the filter of closed and unbounded subsets of κ restricted to the ordinals of cofinality $> \omega$, there is a sequence $C_{\alpha n}(\alpha \in \lim\ (\kappa^+), n < \omega)$ such that for all $\alpha < \beta$ in $\lim(\kappa^+)$:*

(1) *$C_{\alpha n}$ is a closed subset of α for all n,*

(2) *$C_{\alpha n} \subseteq C_{\alpha m}$, whenever $n \leq m$,*

(3) *$\alpha = \bigcup_{n<\omega} C_{\alpha n}$,*

(4) *$\alpha \in \lim(C_{\beta n})$ implies $C_{\alpha n} = C_{\beta n} \cap \alpha$.*

Proof. For $\alpha < \kappa^+$, let D_α be the collection of all $\nu < \kappa$ for which $P_{<\nu}(\alpha)$ is *σ-closed*, i.e., closed under supremums of all of its countable bounded subsets. Clearly, D_α contains a closed unbounded subset of κ, restricted to cofinality $> \omega$ ordinals. Note that

$$\nu \in D_\beta \text{ and } \rho(\alpha, \beta) < \nu \text{ imply that } \nu \in D_\alpha. \tag{9.1.4}$$

In the extended universe, pick a strictly increasing sequence ν_n $(n < \omega)$ which converges to κ and has the property that for each $\alpha < \kappa^+$, there is $n < \omega$ such that $\nu_m \in D_\alpha$ for all $m \geq n$. Let $n(\alpha)$ be the minimal such integer n.

Fix $\alpha \in \lim(\kappa^+)$ and $n < \omega$. If there is $\gamma \geq \alpha$ such that $n \geq n(\gamma)$ and

$$\alpha \in \lim(P_{<\nu_n}(\gamma)), \tag{9.1.5}$$

let $\gamma(\alpha)$ be the minimal such γ and let

$$C_{\alpha n} = \overline{P_{<\nu_n}(\gamma(\alpha))} \cap \alpha. \tag{9.1.6}$$

If such a $\gamma \geq \alpha$ does not exist, let $C_{\alpha n} = \emptyset$ for $n < n(\alpha)$ and let

$$C_{\alpha n} = \overline{P_{<\nu_n}(\alpha)} \cap \alpha \tag{9.1.7}$$

for $n \geq n(\alpha)$. Clearly, this choice of $C_{\alpha n}$'s satisfies (1),(2) and (3). To verify (4), let α be a limit point of some $C_{\beta n}$.

Suppose first that $\mathrm{cf}(\alpha) = \omega$. If $C_{\beta n}$ was defined according to the case (9.1.6) so is $C_{\alpha n}$. Since $P_{<\nu_n}(\gamma(\beta))$ is σ-closed we conclude that $\alpha \in P_{<\nu_n}(\gamma(\beta))$ which by Lemma 9.1.2(a),(b) means that $P_{<\nu}(\alpha) = P_{<\nu}(\gamma(\beta)) \cap \alpha$ for all $\nu \geq \nu_n$ and therefore $D_{\gamma(\beta)} \setminus \nu_n \subseteq D_\alpha$. It follows that $n(\alpha) \geq n$ and therefore that $\gamma(\alpha) = \alpha$. Hence, $C_{\alpha n} = C_{\beta n} \cap \alpha$. If $C_{\beta n}$ was defined according to the case (9.1.7) then our assumption $\alpha \in \lim(C_{\beta n})$ and the fact that $n \geq n(\beta)$ means that $C_{\alpha n}$ was defined according to (9.1.6). Similarly as above we conclude that actually $\alpha \in P_{<\nu_n}(\beta)$ which gives us $n \geq n(\alpha)$ and $\gamma(\alpha) = \alpha$. It follows that $C_{\alpha n} = C_{\beta n} \cap \alpha$ in this case as well.

Suppose now that $\mathrm{cf}(\alpha) > \omega$. If $C_{\beta n}$ was defined according to (9.1.6) then so is $C_{\alpha n}$ and so $P_{\nu_n}(\gamma(\alpha)) \cap \alpha$ and $P_{\nu_n}(\gamma(\beta)) \cap \alpha$ are two σ-closed and unbounded subsets of α. So their intersection is unbounded in α which by Lemma 9.1.2(a), (b) gives us that $P_{\nu_n}(\gamma(\alpha)) \cap \alpha = P_{\nu_n}(\gamma(\beta)) \cap \alpha$. It follows that $C_{\alpha n} = C_{\beta n} \cap \alpha$ in this case. Suppose now that $C_{\beta n}$ was defined according to (9.1.7). Then $\gamma = \beta$ satisfies $n \geq n(\gamma)$ and (9.1.5). So $C_{\alpha n}$ was defined according to the case (9.1.6). It follows that $P_{\nu_n}(\gamma(\alpha)) \cap \alpha$ and $P_{\nu_n}(\beta) \cap \alpha$ are two σ-closed and unbounded subsets of α so their intersection is unbounded in α. Applying Lemma 9.1.2(a),(b) we get that $P_{\nu_n}(\gamma(\alpha)) \cap \alpha = P_{\nu_n}(\beta) \cap \alpha$. It follows that $C_{\alpha n} = C_{\beta n} \cap \alpha$ in this case as well. □

Remark 9.1.5. The combinatorial principle appearing in the statement of Theorem 9.1.4 is a member of a family of square principles that has been studied systematically by Schimmerling and others (see, e.g., [88]). It is definitely a principle sufficient for all of the applications of $\square_\kappa$ appearing in Section 7.6 above.

9.2 Unbounded subadditive functions

We now turn our attention to various unboundedness properties of functions of the form $f : [\kappa^+]^2 \longrightarrow \kappa$.

Definition 9.2.1. A function $f : [\kappa^+]^2 \longrightarrow \kappa$ is *unbounded* if $f''[\Gamma]^2$ is unbounded in κ for every $\Gamma \subseteq \kappa^+$ of size κ. We shall say that such an f is *strongly unbounded* if for every family A of size κ^+, consisting of pairwise-disjoint finite subsets of κ^+, and every $\nu < \kappa$, there exists $A_0 \subseteq A$ of size κ such that $f(\alpha, \beta) > \nu$ for all $\alpha \in a$, $\beta \in b$ and $a \neq b$ in A_0.

Lemma 9.2.2. *If $f : [\kappa^+]^2 \longrightarrow \kappa$ is unbounded and subadditive (i.e., it satisfies the two inequalities in Lemma* 9.1.2(a), (b)*), then f is strongly unbounded.*

Proof. For $\alpha < \beta < \kappa^+$, set $\alpha <_\nu \beta$ if and only if $f(\alpha,\beta) \leq \nu$. Then our assumption about f satisfying Lemma 9.1.2(a) and (b) reduces to the fact that each $<_\nu$ is a tree-ordering on κ^+ compatible with the usual ordering on κ^+. Note that the unboundedness property of f is preserved by any forcing notion satisfying the κ-chain condition, so in particular no tree $(\kappa^+, <_\nu)$ can contain a Souslin subtree of height κ. In the proof of Lemma 7.4.5 above we have seen that this property of $(\kappa^+, <_\nu)$ alone is sufficient to conclude that every family A of κ many pairwise-disjoint subsets of κ^+ contains a subfamily A_0 of size κ such that for every $a \neq b$ in A_0 every $\alpha \in a$ is $<_\nu$-incomparable to every $\beta \in b$, which is exactly the conclusion of f being strongly unbounded. □

Lemma 9.2.3. *The following are equivalent for a regular uncountable cardinal κ:*

(1) *There is a structure $(\kappa^+, \kappa, <, R_n)_{n<\omega}$ with no elementary substructure B of size κ such that $B \cap \kappa$ is bounded in κ.*
(2) *There is an unbounded function $f : [\kappa^+]^2 \longrightarrow \kappa$.*
(3) *There is a strongly unbounded subadditive function $f : [\kappa^+]^2 \longrightarrow \kappa$.*

Proof. Only the implication from (2) to (3) needs some argument. So let $f : [\kappa^+]^2 \longrightarrow \kappa$ be a given unbounded function. Increasing f, we may assume that each of its sections $f_\alpha : \alpha \longrightarrow \kappa$ is one-to-one. For a set $\Gamma \subseteq \kappa^+$, let the f-closure of Γ, $\mathrm{CL}_f(\Gamma)$, be the minimal $\Delta \supseteq \Gamma$ with the property that $f''[\Delta]^2 \subseteq \Delta$ and $f_\alpha^{-1}(\xi) \in \Delta$, whenever $\alpha \in \Delta$ and $\xi < \sup(\Delta \cap \kappa)$. Define $e : [\kappa^+]^2 \longrightarrow \kappa$ by letting

$$e(\alpha,\beta) = \sup(\kappa \cap \mathrm{CL}_f(\rho(\alpha,\beta) \cup P_{\rho(\alpha,\beta)}(\alpha))). \tag{9.2.1}$$

We show the two subadditive properties for e:

$$e(\alpha,\gamma) \leq \max\{e(\alpha,\beta), e(\beta,\gamma)\} \text{ whenever } \alpha \leq \beta \leq \gamma. \tag{9.2.2}$$

By Lemma 9.1.2(a) either $\rho(\alpha,\gamma) \leq \rho(\alpha,\beta)$ in which case $P_{\rho(\alpha,\gamma)}(\alpha) \subseteq P_{\rho(\alpha,\beta)}(\alpha)$ holds, or $\rho(\alpha,\gamma) \leq \rho(\beta,\gamma)$ when we have $P_{\rho(\alpha,\gamma)}(\alpha) \subseteq P_{\rho(\beta,\gamma)}(\beta)$. By the definition of e, the first case gives us the inequality $e(\alpha,\gamma) \leq e(\alpha,\beta)$ and the second gives us $e(\alpha,\gamma) \leq e(\beta,\gamma)$. Then

$$e(\alpha,\beta) \leq \max\{e(\alpha,\gamma), e(\beta,\gamma)\} \text{ whenever } \alpha \leq \beta \leq \gamma. \tag{9.2.3}$$

By Lemma 9.1.2(b) either $\rho(\alpha,\beta) \leq \rho(\alpha,\gamma)$ in which case $P_{\rho(\alpha,\beta)}(\alpha) \subseteq P_{\rho(\alpha,\gamma)}(\alpha)$ holds, or $\rho(\alpha,\beta) \leq \rho(\beta,\gamma)$ when we have $P_{\rho(\alpha,\beta)}(\alpha) \subseteq P_{\rho(\beta,\gamma)}(\beta)$. In the first case we get $e(\alpha,\beta) \leq e(\alpha,\gamma)$ and in the second $e(\alpha,\beta) \leq e(\beta,\gamma)$.

By Lemma 9.2.2, to get the full conclusion of (3) it suffices to show that e is unbounded. So let Γ be a given subset of κ^+ of order-type κ and let $\nu < \kappa$ be a given ordinal. Since $e(\alpha,\beta) \geq \rho(\alpha,\beta)$ for all $\alpha < \beta < \kappa^+$ we may assume that

ρ is bounded on Γ, or more precisely, that for some $\bar{\nu} \leq \nu$ and all $\alpha < \beta$ in Γ, $\rho(\alpha, \beta) \leq \bar{\nu}$. For $\alpha \in \Gamma$, let $\beta(\alpha)$ be the minimal point of Γ above α. Then there is $\mu \leq \bar{\nu}$ and $\Delta \subseteq \Gamma$ of size κ such that $\rho(\alpha, \beta(\alpha)) = \mu$ for all $\alpha \in \Delta$. Note that the sequence of sets $P_\mu(\alpha)$ $(\alpha \in \Delta)$ forms a chain under end-extension. Let Γ^* be the union of this sequence of sets. Since f is unbounded, there exist $\xi < \eta$ in Γ^* such that $f(\xi, \eta) > \nu$. Pick $\alpha \in \Delta$ above η. Then $\xi, \eta \in P_\mu(\alpha)$ and therefore $e(\alpha, \beta(\alpha)) \geq f(\xi, \eta) > \nu$. This finishes the proof. □

Remark 9.2.4. Recall that *Chang's conjecture* is the model-theoretic transfer principle asserting that every structure of the form $(\omega_2, \omega_1, <, \ldots)$ with a countable signature has an uncountable elementary submodel B with the property that $B \cap \omega_1$ is countable. This principle shows up in many considerations including the first two uncountable cardinals ω_1 and ω_2. For example, it is known that it is preserved by ccc forcing extensions, that it holds in the Silver collapse of an ω_1-Erdős cardinal, and that it in turn implies that ω_2 is an ω_1-Erdős cardinal in the core model of Dodd and Jensen (see, e.g., [20],[55]).

Corollary 9.2.5. *The negation of Chang's conjecture is equivalent to the statement that there exists* $e : [\omega_2]^2 \longrightarrow \omega_1$ *such that:*

(a) $e(\alpha, \gamma) \leq \max\{e(\alpha, \beta), e(\beta, \gamma)\}$ *whenever* $\alpha \leq \beta \leq \gamma$.

(b) $e(\alpha, \beta) \leq \max\{e(\alpha, \gamma), e(\beta, \gamma)\}$ *whenever* $\alpha \leq \beta \leq \gamma$.

(c) *For every uncountable family A of pairwise-disjoint finite subsets of ω_2 and every $\nu < \omega_1$, there exists an uncountable $A_0 \subseteq A$ such that $e(\alpha, \beta) > \nu$ whenever $\alpha \in a$ and $\beta \in b$ for every $a \neq b \in A_0$.* □

Remark 9.2.6. Note that if a mapping $e : [\omega_2]^2 \longrightarrow \omega_1$ has properties (a),(b) and (c) of Corollary 9.2.5, then $D_e : [\omega_2]^2 \longrightarrow [\omega_2]^{\aleph_0}$ defined by

$$D_e\{\alpha, \beta\} = \{\xi \leq \min\{\alpha, \beta\} : e(\xi, \alpha) \leq e\{\alpha, \beta\}\}$$

satisfies the weak form of Definition 7.5.13, where only the conclusion (a) is kept. The full form of Definition 7.5.13, however, cannot be achieved assuming only the negation of Chang's conjecture. This shows that the function ρ, based on a $\square_{\omega_1}$-sequence, is a considerably deeper object than an $e : [\omega_2]^2 \longrightarrow \omega_1$ satisfying Corollary 9.2.5(a),(b),(c).

Note that by Corollary 9.2.5, Chang's conjecture is equivalent to the statement that within every decomposition of the usual ordering on ω_2 as an increasing chain of tree-orderings, one of the trees has an uncountable chain. Is it possible to have decompositions of $\in \restriction \omega_2$ into an increasing ω_1-chain of tree-orderings of countable heights? It turns out that the answer to this question is equivalent to a different well-known combinatorial statement about ω_2 rather than Chang's conjecture itself. Recall that $f : [\kappa^+]^2 \longrightarrow \kappa$ is *transitive* if

$$f(\alpha, \gamma) \leq \max\{f(\alpha, \beta), f(\beta, \gamma)\} \text{ whenever } \alpha \leq \beta \leq \gamma. \tag{9.2.4}$$

Definition 9.2.7. Given a transitive map $f : [\kappa^+]^2 \to \kappa$, one defines

$$\rho_f : [\kappa^+]^2 \longrightarrow \kappa$$

recursively on $\alpha \leq \beta < \kappa$ using the formula

$$\rho_f(\alpha, \beta) = \sup\{f(\min(C_\beta \setminus \alpha), \beta),\ \mathrm{tp}(C_\beta \cap \alpha),\ \rho_f(\alpha, \min(C_\beta \setminus \alpha)),\ \rho_f(\xi, \alpha) : \xi \in C_\beta \cap \alpha\},$$

with the boundary condition $\rho_f(\alpha, \alpha) = 0$ for all α.

Lemma 9.2.8. *For every transitive map $f : [\kappa^+]^2 \longrightarrow \kappa$ the corresponding function $\rho_f : [\kappa^+]^2 \longrightarrow \kappa$ has the following properties:*

(a) $\rho_f(\alpha, \gamma) \leq \max\{\rho_f(\alpha, \beta), \rho_f(\beta, \gamma)\}$ *whenever* $\alpha \leq \beta \leq \gamma$,
(b) $\rho_f(\alpha, \beta) \leq \max\{\rho_f(\alpha, \gamma), \rho_f(\beta, \gamma)\}$ *whenever* $\alpha \leq \beta \leq \gamma$,
(c) $|\{\xi \leq \alpha : \rho_f(\xi, \alpha) \leq \nu\}| \leq |\nu| + \aleph_0$ *for* $\nu < \kappa$ *and* $\alpha < \kappa^+$,
(d) $\rho_f(\alpha, \beta) \geq f(\alpha, \beta)$ *for all* $\alpha < \beta < \kappa^+$.

Proof. Proofs of (a)–(c) are almost identical to the corresponding proofs for the function ρ itself and so we avoid repetition. The inequality (d) is deduced from the definition of ρ_f using the transitivity of f as follows:

$$\rho_f(\alpha, \beta) \geq \max_{i<n} f(\beta_{i+1}, \beta_i) \geq f(\alpha, \beta),$$

where $\beta = \beta_0 > \beta_1 > \cdots > \beta_{n-1} > \beta_n = \beta$ is the walk from β to α along the given C-sequence. □

Remark 9.2.9. Transitive maps are frequently used combinatorial objects, especially when one works with quotient structures. Adding the extra subadditivity condition Lemma 9.2.8(b), one obtains a considerably more subtle object which is much less understood.

Definition 9.2.10. Let $f_\alpha : \kappa \longrightarrow \kappa$ $(\alpha < \kappa^+)$ be a given sequence of functions such that $f_\alpha <^* f_\beta$ whenever $\alpha < \beta$ (i.e., $\{\nu < \kappa : f_\alpha(\nu) \geq f_\beta(\nu)\}$ is bounded in κ whenever $\alpha < \beta$). Then the *corresponding transitive map* $f : [\kappa^+]^2 \longrightarrow \kappa$ is defined by

$$f(\alpha, \beta) = \min\{\mu < \kappa : f_\alpha(\nu) < f_\beta(\nu) \text{ for all } \nu \geq \mu\}. \tag{9.2.5}$$

Let ρ_f be the corresponding ρ-function that dominates this particular f and for $\nu < \kappa$ let $<_\nu^f$ be the corresponding tree-ordering of κ^+, i.e.,

$$\alpha <_\nu^f \beta \text{ if and only if } \rho_f(\alpha, \beta) \leq \nu. \tag{9.2.6}$$

Lemma 9.2.11. *Suppose $f_\alpha \leq g$ for all $\alpha < \kappa^+$ where $\leq$ is the ordering of everywhere dominance. Then for every $\nu < \kappa$ the tree $(\kappa^+, <_\nu^f)$ has height $\leq g(\nu)$.*

Proof. Let P be a maximal chain of $(\kappa^+, <^f_\nu)$. $f(\alpha,\beta) \leq \rho_f(\alpha,\beta) \leq \nu$ for every $\alpha < \beta$ in P. It follows that $f_\alpha(\nu) < f_\beta(\nu) \leq g(\nu)$ for all $\alpha < \beta$ in P. So P has order-type $\leq g(\nu)$. □

Note that if we have a function $g : \kappa \longrightarrow \kappa$ which bounds the sequence f_α $(\alpha < \kappa^+)$ in the ordering $<^*$ of eventual dominance, then the new sequence $\bar{f}_\alpha = \min\{f_\alpha, g\}$ $(\alpha < \kappa^+)$ is still strictly $<^*$-increasing but now bounded by g even in the ordering of everywhere dominance. So this proves the following result of Galvin and Keisler (see [57], [48], [85]).

Corollary 9.2.12. *The following two conditions are equivalent for every regular cardinal κ.*

(1) *There is a sequence $f_\alpha : \kappa \longrightarrow \kappa$ $(\alpha < \kappa^+)$ which is strictly increasing and bounded in the ordering of eventual dominance.*

(2) *The usual order-relation of κ^+ can be decomposed into an increasing κ-sequence of tree-orderings of heights $< \kappa$.* □

Remark 9.2.13. The assertion that every strictly $<^*$-increasing κ^+-sequence of functions from κ to κ is $<^*$-unbounded is strictly weaker than Chang's conjecture and in the literature it is usually referred to as *weak Chang's conjecture*. This statement still has considerable large cardinal strength (see [21]). Also note the following consequence of Corollary 9.2.12 which can be deduced from Lemmas 9.1.1 and 9.1.2 above as well.

Corollary 9.2.14. *If κ is a regular limit cardinal (e.g., $\kappa = \omega$), then the usual order-relation of κ^+ can be decomposed into an increasing κ-sequence of tree-orderings of heights $< \kappa$.* □

9.3 Chang's conjecture and Θ_2

Recall the Erdős–Rado arrow notation

$$\theta \longrightarrow (\omega)^2_\omega$$

which succinctly expresses the statement that for every mapping $f : [\theta]^2 \to \omega$ there is some infinite subset X of θ such that f is constant on its symmetric square $[X]^2$. Recall also that the successor of the continuum is characterized as the minimal cardinal θ having this property, or in other words, that

$$\mathfrak{c}^+ = \min\{\theta : \theta \longrightarrow (\omega)^2_\omega\}.$$

We shall now see that considering an asymmetric version of this partition property, i.e., replacing $[\theta]^2$ by $\theta \times \theta$ and replacing symmetric squares $[X]^2$ by rectangles

$X \times Y$, one gets a characterization of a quite different cardinal. To this end, let us use the arrow notation

$$\binom{\theta}{\theta} \longrightarrow \binom{\omega}{\omega}^{1,1}_{\omega}$$

to succinctly express the statement that for every map $f : \theta \times \theta \longrightarrow \omega$, there exist infinite sets $A, B \subseteq \theta$ such that f is constant on their product. Let

$$\Theta_2 = \min\{\theta : \binom{\theta}{\theta} \longrightarrow \binom{\omega}{\omega}^{1,1}_{\omega}\}.$$

Note that $\omega_1 < \Theta_2 \leq \mathfrak{c}^+$. The following result shows that Θ_2 can have the minimal possible value ω_2 and that, in general, the cardinal Θ_2 can be considerably smaller than the continuum.

Theorem 9.3.1. *Chang's conjecture is equivalent to the statement that every ccc poset forces $\Theta_2 > \omega_2$.*

Proof. We have already noted that Chang's conjecture is preserved by ccc forcing extensions, so in order to prove the direct implication, it suffices to show that Chang's conjecture implies that every $f : \omega_2 \times \omega_2 \longrightarrow \omega$ is constant on the product of two infinite subsets of ω_2. Given such an f, using Chang's conjecture it is possible to find an elementary submodel M of H_{ω_3} such that $f \in M$, $\delta = M \cap \omega_1 \in \omega_1$ and $\Gamma = M \cap \omega_2$ is uncountable. Choose $i \in \omega$ and uncountable $\Delta \subseteq \Gamma$ such that $f(\delta, \beta) = i$ for all $\beta \in \Delta$. Choose $\alpha_0 < \delta$ such that

$$\Delta_0 = \{\beta \in \Delta : f(\alpha_0, \beta) = i\}$$

is uncountable. Such α_0 exists by elementarity of M. Pick $\beta_0 \in \Delta_0$ above ω_1. Note that by the elementarity of M, for every $\beta \in \Delta_0$ there exist unboundedly many $\alpha < \delta$ such that $f(\alpha, \beta_0) = f(\alpha, \beta) = i$. So we can choose $\alpha_1 < \delta$ above α_0 such that the set

$$\Delta_1 = \{\beta \in \Delta_0 : f(\alpha_1, \beta) = i\}$$

is uncountable. Then again for every $\beta \in \Delta_1$ there exist unboundedly many $\alpha < \delta$ such that $f(\alpha, \beta_0) = f(\alpha, \beta_1) = f(\alpha, \beta) = i$, and so on. Continuing in this way, we build two increasing sequences $\{\alpha_n\} \subseteq \delta$ and $\{\beta_n\} \subseteq \Gamma \backslash \omega_1$ such that $f(\alpha_n, \beta_m) = i$ for all $n, m < \omega$. This shows that f is constant on a product of two infinite sets, finishing the proof of the direct implication.

Assume now that Chang's conjecture fails and fix $e : [\omega_2]^2 \longrightarrow \omega$ as in Corollary 9.2.5. We shall describe a ccc forcing notion $\mathcal{P}$ which forces a mapping $f : \omega_2 \times \omega_2 \longrightarrow \omega$ which is not constant on the product of any two infinite subsets of ω_2. The conditions of $\mathcal{P}$ are simply maps of the form $p : [D_p]^2 \longrightarrow \omega$ where D_p is some finite subset of ω_2 such that

$p(\xi, \alpha) \neq p(\xi, \beta)$ whenever $\xi < \alpha < \beta$ belong to D_p and have the property that $e(\xi, \alpha) > e(\alpha, \beta)$. (9.3.1)

We order $\mathcal{P}$ by letting p extend q if and only if p extends q as a function and

$$p(\xi,\alpha) \neq p(\xi,\beta) \text{ whenever } \alpha < \beta \text{ belong to } D_q \text{ and } \xi \text{ belongs to } D_p \setminus D_q. \tag{9.3.2}$$

Let A be a given uncountable subset of $\mathcal{P}$. Refining A, we may assume that D_p $(p \in A)$ forms a Δ-system with root D and that $F_p = e''[D_p]^2$ $(p \in A)$ form an increasing Δ-system on ω_1 with root F. Moreover, we may assume that for every p and q in A, the finite structures they generate together with e are isomorphic via the isomorphism that fixes D and F. By Corollary 9.2.5(c) there is uncountable $A_0 \subseteq A$ such that for all $p, q \in A_0$,

$$e(\alpha,\beta) > \max(F) \text{ whenever } \alpha \in D_p \setminus D \text{ and } \beta \in D_q \setminus D. \tag{9.3.3}$$

We claim that every two conditions p and q from A_0 are compatible. To see this, extend $p \cup q$ to a mapping $r : [D_p \cup D_q]^2 \longrightarrow \omega$ which is one-to-one on new pairs, avoiding the old values. To see that r is indeed a member of $\mathcal{P}$ we need to check (9.3.1) for r. So let $\xi < \alpha < \beta$ be three given ordinals from D_r such that $e(\xi,\alpha) > e(\alpha,\beta)$. We have to prove that $r(\xi,\alpha) \neq r(\xi,\beta)$. By the choice of r we may assume that $\{\xi,\alpha\}$ and $\{\xi,\beta\}$ are old pairs, i.e., that they belong to $[D_p]^2 \cup [D_q]^2$. Since p and q both satisfy (9.3.1), we may consider only the case when (say) $\alpha \in D_p \setminus D_q$, $\beta \in D_q \setminus D_p$ and $\xi \in D$. Recall that $e(\xi,\alpha) > e(\alpha,\beta)$ and the subadditive properties of e (see Corollary 9.2.5(a),(b)) give us the equality $e(\xi,\alpha) = e(\xi,\beta)$, i.e., $e(\xi,\alpha)$ belongs to F contradicting (9.3.3). To see that r extends p and q, by symmetry it suffices to check that r extends, say, p. So suppose $\xi < \alpha < \beta$ are given such that $\alpha < \beta$ belong to D_p and ξ belongs to $D_q \setminus D_p$. We need to show that $r(\xi,\alpha) \neq r(\xi,\beta)$. By the choice of new pairs, this is automatic if one of the two pairs is new. So we may consider the case when both α and β are in D. Since $r(\xi,\alpha) = q(\xi,\alpha)$ and $r(\xi,\beta) = q(\xi,\beta)$, in this case the inequality $r(\xi,\alpha) \neq r(\xi,\beta)$ would follow from (9.3.1) for q if we show that $e(\xi,\alpha) > e(\alpha,\beta)$. Otherwise $e(\xi,\alpha)$ would belong to the root F and by the fact that all conditions are isomorphic, $e(\xi(s),\alpha) \in F$ holds for all $s \in A_0$, where $\xi(s) \in D_s$ $(s \in A_0)$ are the copies of ξ. This of course contradicts the property Corollary 9.2.5(c) of e, since $\xi(s) \neq \xi(t)$ when $s \neq t$.

Forcing with $\mathcal{P}$ gives us a mapping $g : [\omega_2]^2 \longrightarrow \omega$ such that

$$\{\xi < \alpha : g(\xi,\alpha) = g(\xi,\beta)\} \text{ is finite for all } \alpha < \beta < \omega_2. \tag{9.3.4}$$

Define $f : \omega_2 \times \omega_2 \longrightarrow \omega$ by letting $f(\alpha,\beta) = 2g\{\alpha,\beta\} + 2$ when $\alpha < \beta$, $f(\alpha,\beta) = 2g\{\alpha,\beta\} + 1$ when $\alpha > \beta$, and $f(\alpha,\beta) = 0$ when $\alpha = \beta$. Then f is not constant on any product of two infinite sets. $\square$

In view of considerable technical difficulties one encounters while trying to design a partial ordering that would force that $\Theta_2 > \omega_3$, the following question is natural.

Question 9.3.2. Can one prove a bound like $\Theta_2 \leq \omega_3$ without appealing to additional set-theoretic axioms?

Remark 9.3.3. One can of course consider also the higher-dimensional analogue Θ_n of Θ_2 defined as the minimal cardinal θ with property that every mapping $f : \theta^n \to \omega$ is constant on a product of n infinite subsets of θ. The problem of comparison of the two sequences of cardinals Θ_2, Θ_3, Θ_4,... and $\omega_2, \omega_3, \omega_4, \ldots$ is of considerable interest, both in set theory and model theory (see, e.g., [90], [114], [120]). In particular, the following natural questions are still unanswered.

Question 9.3.4. Can one prove any of the bounds of the form $\Theta_n \leq \omega_{n+1}$, or in general, can one determine the function $\iota : \omega \to \mathrm{Ord}$ satisfying $\Theta_n = \omega_{\iota(n)}$ without appealing to additional axioms of set theory? Is $\Theta_n \geq \omega_n$ for all n ?

It turns out that the cardinal Θ_2 is also related to the following standard concept from set theory.

Definition 9.3.5. A cardinal θ is a *Jónsson cardinal* if every algebra of the form $\langle \theta, (f_i)_{i<\omega} \rangle$ has a proper subalgebra of cardinality θ, or equivalently, if for every mapping $f : [\theta]^{<\omega} \to \theta$ there is $X \subseteq \theta$ of cardinality θ such that $f''[X]^{<\omega} \neq \theta$.

Note that our previous results about the square-bracket operation show that a Jónsson cardinal cannot be a successor of a regular cardinal and, in fact, the following simple stepping-up procedure shows that a long initial segment of successors of singular cardinals also fail to be Jónsson.

Lemma 9.3.6. *Suppose that κ_n $(n < \omega)$ is a strictly increasing sequence of regular cardinals with the property that for some non-principal ultrafilter $\mathcal{U}$ on ω the ultrapower $\prod_n \kappa_n/\mathcal{U}$ has cofinality κ^+ where $\kappa = \sup \kappa_n$. Suppose that for every $n < \omega$ there $c_n : [\kappa_n]^2 \to \kappa_n$ such that $c_n''[A]^2 = \kappa_n$ for all $A \subseteq \kappa_n$ of size κ_n. Then there is $c : [\kappa^+]^2 \to \kappa^+$ such that $c''[A]^2 = \kappa^+$ for all $A \subseteq \kappa^+$ of size κ^+.*

Proof. Let f_α $(\alpha < \kappa^+)$ be an increasing and cofinal sequence in the ultrapower $\prod_n \kappa_n/\mathcal{U}$. Define $c : [\kappa^+]^2 \to \kappa$ by

$$c(\{\alpha, \beta\}) = c_n(\{f_\alpha(n), f_\beta(n)\}) \text{ where } n = \min\{i : f_\alpha(i) \neq f_\beta(i)\}. \tag{9.3.5}$$

We claim that $c''[A]^2 = \kappa$ for all unbounded $A \subseteq \kappa^+$. To see this, fixing unbounded $A \subseteq \kappa^+$ and $\xi < \kappa$, let n be the minimal integer i such that $\kappa_n > \xi$ and $\sup\{f_\alpha(k) : \alpha \in A\} = \kappa_i$. Then we can find $B \subseteq A$ of size κ_n and $t \in \prod_{i<n} \kappa_i$ such that:

(1) $f_\beta \restriction n = t$ for all $\beta \in B$, and
(2) $f_\alpha(n) < f_\beta(n)$ whenever $\alpha < \beta$ belong to B.

Since $\{f_\beta : \beta \in B\}$ is unbounded in κ_n, by the property of c_n, we can find $\alpha < \beta$ in B such that $c_n(\{f_\alpha(n), f_\beta(n)\}) = \xi$. Since $n = \min\{k : f_\alpha(i) \neq f_\beta(i)\}$, we conclude that $c(\{\alpha, \beta\}) = c_n(\{f_\alpha(n), f_\beta(n)\}) = \xi$.

For each $\alpha < \kappa^+$, fix a one-to-one mapping $e_\alpha : \alpha \to \kappa$. Define $\bar{c} : [\kappa^+]^2 \to \kappa^+$, by

$$\bar{c}(\alpha, \beta) = e_\alpha^{-1}(c(\alpha, \beta)), \tag{9.3.6}$$

if $c(\alpha, \beta) \in \mathrm{rang}(e_\alpha)$; put $\bar{c}(\alpha, \beta) = 0$, otherwise. Then it is easily checked that $\bar{c}''[A]^2 = \kappa^+$ for all $A \subseteq \kappa^+$ of size κ^+. □

Corollary 9.3.7. *There is $c : [\omega_{\omega+1}]^2 \longrightarrow \omega_{\omega+1}$ such that $c''[A]^2 = \omega_{\omega+1}$ for all unbounded $A \subseteq \omega_{\omega+1}$.*

Proof. This follows from Lemma 9.3.6 and a standard fact from the PCF theory (see, e.g., [94]). □

Singular cardinals of larger cofinalities are handled by the following stepping-up procedure which uses a slightly different assumption.

Lemma 9.3.8. *Let κ_ξ $(\xi < \lambda)$ be a sequence of regular cardinals such that for every $\xi < \lambda$, we have that $\kappa_\xi > \sup_{\eta<\xi} \kappa_\eta$ and we can find $c_\xi : [\kappa_\xi]^2 \to \kappa_\xi$ such that $c''_\xi[A]^2 = \kappa_\xi$ for all unbounded $A \subseteq \kappa_\xi$. Suppose that there is a sequence f_α $(\alpha < \kappa^+)$ of elements of $\prod_{\xi<\lambda} \kappa_\xi$ that is increasing and cofinal in this product relative to the ordering of eventual dominance. Let $\kappa = \sup_\xi \kappa_\xi$. Then there is $c : [\kappa^+]^2 \to \kappa^+$ such that $c''[A]^2 = \kappa^+$ for all unbounded $A \subseteq \kappa^+$.*

Proof. Working as before, it suffices to construct $c : [\kappa^+]^2 \to \kappa$ such that $c''[A]^2 = \kappa$ for all unbounded $A \subseteq \kappa^+$. For $\alpha < \beta < \kappa^+$, let

$$c(\alpha, \beta) = c_\xi(f_\beta(\xi), f_\alpha(\xi)), \text{ where } \xi = \max\{\eta < \lambda : f_\alpha(\eta) \geq f_\beta(\eta)\}, \tag{9.3.7}$$

if such maximum $\xi < \lambda$ exists and if $f_\alpha(\xi) \neq f_\beta(\xi)$; otherwise, let $c(\alpha, \beta) = 0$.

Consider an unbounded $A \subseteq \kappa^+$ and $\zeta < \kappa$. For each $\xi < \lambda$ and $\delta < \kappa_\xi$, pick, if possible, an ordinal $\alpha = \alpha(\xi, \delta) \in A$ with property that $f_\alpha(\xi) = \delta$. Define $g \in \prod_\xi \kappa_\xi$, by

$$g(\xi) = \sup\{f_\alpha(\xi) : \alpha = \alpha(\eta, \delta) \text{ for some } \delta < \eta < \xi\}. \tag{9.3.8}$$

Since f_α $(\alpha < \kappa^+)$ is cofinal in $\prod_\xi \kappa_\xi$ relative to the ordering of eventual dominance, we can find cofinal $B \subseteq A$ and $\xi_0 < \lambda$ such that

$$g(\xi) < f_\beta(\xi) \text{ for all } \xi \geq \xi_0 \text{ and } \beta \in B. \tag{9.3.9}$$

Then, in particular, we have that

$$\alpha(\xi, \delta) < \beta \text{ for all } \xi < \lambda, \delta < \kappa_\xi \text{ and } \beta \in B. \tag{9.3.10}$$

Since f_β $(\beta \in B)$ is cofinal in $\prod_\xi \kappa_\xi$, we can find $\xi_1 \geq \xi_0$ such that the set $X = \{f_\beta(\xi_1) : \beta \in B\}$ is unbounded in κ_{ξ_1}. So by the property of c_{ξ_1}, we can find $\gamma < \delta$ in X such that $c_{\xi_1}(\gamma, \delta) = \zeta$. Then $\alpha = \alpha(\xi_1, \delta)$ is defined. Pick $\beta \in B$ such that $\gamma = f_\beta(\xi_1)$. Then by (9.3.9) and (9.3.10), we have that $\alpha < \beta$, $f_\alpha(\xi_1) > f_\beta(\xi_1)$, and $f_\alpha(\xi) \leq g(\xi) < f_\beta(\xi)$ for all $\xi > \xi_1$. It follows that

$$c(\alpha, \beta) = c_{\xi_1}(f_\beta(\xi_1), f_\alpha(\xi_1)) = c_{\xi_1}(\gamma, \delta) = \zeta,$$

as required. □

Corollary 9.3.9. *Suppose that κ is a singular cardinal and that for some $\nu < \kappa$, for every regular cardinal $\theta \in [\nu, \kappa)$ there is $c_\theta : [\theta]^2 \to \theta$ such that $c''_\theta[A]^2 = \theta$ for all unbounded $A \subseteq \theta$. Then there is $c : [\kappa^+]^2 \to \kappa^+$ such that $c''[A]^2 = \kappa^+$ for all unbounded $A \subseteq \kappa^+$.*

Proof. This follows again from Lemma 9.3.8 and a standard fact from the PCF theory (see, e.g., [94]). □

Corollary 9.3.10. *There is $c : [\omega_{\omega_1+1}]^2 \longrightarrow \omega_{\omega_1+1}$ such that $c''[A]^2 = \omega_{\omega_1+1}$ for all unbounded $A \subseteq \omega_{\omega_1+1}$.* □

The following fact explains the relationship between the class of Jónsson cardinals and the cardinal Θ_2.

Theorem 9.3.11. *There are no Jónsson cardinals $< \Theta_2$.*

Proof. Suppose that θ is a Jónsson cardinal and let $f : \theta^2 \to \omega$ be a given mapping. Then we can find an elementary submodel M of $(H_{\theta^+}, \in)$ such that $f \in M$, such that $M \cap \theta$ has cardinality θ, but $\theta \not\subseteq M$. Let $\xi = \min(\theta \setminus M)$ and let ξ^+ be the minimal cardinal $> \xi$. Choose $B \subseteq M \cap \theta$ of size ξ^+ and $n < \omega$ such that $f(\xi, \beta) = n$ for all $\beta \in B$. Then by the elementarity of M, for every finite subset $F \subseteq B$, the set

$$A_F = \{\alpha < \xi : f(\alpha, \beta) = n\}$$

is unbounded in ξ. Since B has bigger cardinality than ξ, there must be $\alpha_0 < \xi$ such that

$$B_0 = \{\beta \in B : f(\alpha_0, \beta) = n\}$$

has cardinality ξ^+. Let $\beta_0 = \min B_0$. Then $A_{\{\beta_0,\beta\}}$ is unbounded in ξ for all $\beta \in B_0$. So there must be $\alpha_1 < \xi$, such that $\alpha_1 > \alpha_0$ and such that

$$B_1 = \{\beta \in B_0 : \beta > \beta_0 \text{ and } f(\alpha_1, \beta) = n\}$$

has cardinality ξ^+, and so on. Proceeding in this way we constrict an increasing sequence α_i $(i < \omega)$ of ordinals $< \xi$ and an increasing sequence β_i $(i < \omega)$ such that

$$f(\alpha_i, \beta_j) = n \text{ for all } i, j < \omega.$$

Since f was an arbitrary mapping this shows that $\theta \geq \Theta_2$, as required. □

It turns out also that cardinals $< \Theta_2$ share also the following strong unboundedness property.

Theorem 9.3.12. *For every $\theta < \Theta_2$ there is $f : \theta^2 \to \omega$ such that, for every uncountable family A of pairwise-disjoint finite subsets of θ, all of some fixed size n, and for every $k < \omega$, there exist $a \neq b$ in A such that $f(a(i), b(j)) > k$ for all $i, j < n$.*

Proof. From the definition of the cardinal Θ_2, we infer that there exist a mapping $f : \theta^2 \to \omega$ which is not constant on any product of two infinite subsets of θ. Suppose that for some $k < \omega$ such pair $a \neq b$ in A cannot be found. Let A_0 and A_1 be two disjoint subsets of A such that A_0 is countable and infinite and such that A_1 is uncountable. Choose a uniform ultrafilter $\mathcal{U}_0$ on A_0 and a uniform ultrafilter $\mathcal{U}_1$ on A_1. Then there exist $\bar{k} \leq k$ and $i_0, j_0 < n$ such that

$$(\mathcal{U}_1 b)(\mathcal{U}_0 a)\ f(a(i_0), b(j_0)) = \bar{k}. \tag{9.3.11}$$

Let $X = \{a(i_0) : a \in A_0\}$ and let

$$Y = \{b(j_0) : b \in A_1 \text{ and } (\mathcal{U}_0 a)\ f(a(i_0), b(j_0)) = \bar{k}\}.$$

Then X is countable and infinite while the set Y is uncountable as it belongs to $\mathcal{U}_1$. Moreover, by (9.3.11), for every finite $F \subseteq Y$, the set

$$X_F = \{\alpha \in X : f(\alpha, \beta) = \bar{k} \text{ for all } \beta \in F\}$$

is infinite as it belongs to $\mathcal{U}_0$. So working as in the previous proof, we can select a one-to-one sequence $\alpha_l (l < \omega)$ of elements of X and a one-to-one sequence β_l $(l < \omega)$ of elements of Y such that

$$f(\alpha_l, \beta_m) = \bar{k} \text{ for all } l, m < \omega.$$

This contradicts our initial assumption that the mapping f is not constant on any product of two infinite subsets of θ. □

9.4 Higher dimensions and the continuum hypothesis

Recall that the minimal cardinal θ for which there is a ccc poset of size θ that is not σ-centered[1] is equal to the Baire category number $\mathfrak{m}$, the minimal cardinal θ for which one can find a compact ccc space without isolated points that can be decomposed into θ nowhere-dense subsets (see [104]). What is the minimal cardinality Σ of a ccc poset that cannot be decomposed into $\aleph_1$ centered subsets? In fact, it is natural to consider this question by restricting ourselves to a particular finite dimension n.

Definition 9.4.1. For an integer $n \geq 2$, an n-hypergraph is simply a pair of the form (V, H), where H is a collection of n-element subsets of V, i.e., $H \subseteq [V]^n$. Thus, a graph is a 2-hypergraph and a triple-system is a 3-hypergraph. An n-hypergraph (V, H) satisfies the countable chain condition if for every uncountable family A of finite subsets of V, either there is $a \in A$ such that $[a]^n \cap H \neq \emptyset$, or else there exist $a \neq b$ in A such that $[a \cup b]^n \cap H = \emptyset$.

[1] A subset X of a partially ordered set $\mathcal{P}$ is *centered* if for every finite subset F of X there is $q \in \mathcal{P}$ such that $q \leq p$ for all $p \in F$. A poset $\mathcal{P}$ is κ-centered if it can be decomposed into at most κ centered subsets. Of course, σ-centered signifies $\aleph_0$-centered.

Let Σ_n be the minimal cardinal θ for which one can find a ccc n-hypergraph on a vertex set of size θ that is not $\aleph_1$-chromatic[2]. First of all, we should mention the following immediate inequalities.

Lemma 9.4.2. $\Sigma_n \geq \Sigma_{n+1} \geq \Sigma > \omega_1$ *for every integer* $n \geq 2$. □

It turns out that these cardinals do have dramatic influence on standard cardinal characteristics of the continuum. For example, we have the following facts proved in [114].

Theorem 9.4.3. $\Sigma_2 > \mathfrak{b}$ *implies* $\mathfrak{b} = \omega_1$, *while* $\Sigma_3 > \mathrm{cf}(\mathfrak{c})$ *implies* $\mathrm{cf}(\mathfrak{c}) = \omega_1$ *and* $\Sigma_3 > \mathfrak{d}$ *implies* $\mathfrak{d} = \omega_1$. □

It follows that cardinal inequalities like $\Sigma_n > \mathfrak{c}$ can be considered as special forms of the Continuum Hypothesis, one for each finite dimension n. In fact, it is still unknown if the inequality $\Sigma > \omega_1$ is equivalent to CH. So, in some sense, it is not surprising that these inequalities do bear some relationship to the cardinal Θ_2 which was meant to be the Ramsey-theoretic analogue of the successor of the continuum.

Theorem 9.4.4. *If* $\Sigma_3 > \mathfrak{c}$, *then* $\Theta_2 = \omega_2$.

Proof. Suppose towards a contradiction that $\Theta_2 > \omega_2$ and fix $p : \omega_2 \times \omega_2 \to \omega$ that is not constant on any product of two infinite subsets of ω_2. By Theorem 9.4.3, we can also fix a sequence f_ξ $(\xi < \omega_1)$ of increasing functions from ω into ω which is unbounded in ω^ω relative to the ordering of eventual dominance. We shall also consider ω^ω as a partially ordered set with the ordering $\leq$ of everywhere dominance:

$$f \leq g \text{ iff } f(n) \leq g(n) \text{for all } n < \omega$$

and so we will say that $f, g \in \omega^\omega$ are *incomparable* if $f \nleq g$ and $f \ngeq g$. Thus, an *antichain* of ω^ω is any subset A of ω^ω with the property that every pair of distinct members of A are incomparable. Similar terminology will be adopted when working with the finite powers ω^m and their coordinatewise orderings $\leq$ rather than the infinite power ω^ω and its coordinatewise ordering $\leq$. Since the Continuum Hypothesis must fail, we may also assume to have a one-to-one sequence r_ξ $(\xi < \omega_2)$ of elements of 2^ω. Fix also $e : [\omega_2]^2 \to \omega_1$ such that $e(\cdot, \alpha) : \alpha \to \omega_1$ is one-to-one for all $\alpha < \omega_2$. Now we are ready to define our triple-system (ω_2, H). In fact, it will be more convenient to define the complement $H^c = [\omega_2]^3 \setminus H$. So, let H^c be the collection of all triples $\alpha < \beta < \gamma$ such that

$$r_\beta \upharpoonright p(\alpha, \beta) = r_\gamma \upharpoonright p(\alpha, \gamma) \text{ implies } \{f_{e(\alpha,\beta)}, f_{e(\alpha,\gamma)}\} \text{ is an antichain.} \tag{9.4.1}$$

The proof that the triple-system (ω_2, H) satisfies the countable chain condition and the final step towards a contradiction depends on the oscillation theory of ω^ω,

[2]Recall that an n-hypergraph (V, H) is κ-chromatic if the set V can be decomposed into κ subsets X with the property that $[X]^n \cap H = \emptyset$.

so let us reproduce some facts from this theory. More information can be found in Chapter 1 of [111].

For f and g in ω^ω, let

$$\operatorname{osc}(f,g) = |\{n < \omega : f(n) \leq g(n) \text{ and } f(n+1) > g(n+1)\}|. \tag{9.4.2}$$

Note that if $\operatorname{osc}(f,g) = 0$ and if $f(0) \leq g(0)$, then $f(n) \leq g(n)$ for all $n < \omega$, i.e., $f \leq g$. Conversely, if $f \leq g$, then $\operatorname{osc}(f,g) = 0$. The analysis of the oscillation mapping requires its natural extension to $(\omega^{\leq\omega})^2$ as follows. Given s and t in $\omega^{\leq\omega}$, let

$$\operatorname{osc}(s,t) = |\{n < \min(|s|,|t|) - 1 : s(n) \leq t(n) \text{ and } s(n+1) > t(n+1)\}|. \tag{9.4.3}$$

For f and g in ω^ω, and $m < \omega$, set

$$f <_m g \text{ iff } f(n) < g(n) \text{ for all } n \geq m. \tag{9.4.4}$$

We are now ready to state explicitly a consequence of Lemma 1.0 of [111] that will be useful in the rest of the proof of Theorem 9.4.4.

Lemma 9.4.5. *Suppose X and Y are two subsets of $\omega^{\uparrow\omega}$ that are totally ordered and unbounded relative to the ordering $<^*$ of eventual dominance. Suppose that there is a topologically dense subset D of X such that for some $m < \omega$, $h \in \omega^{\uparrow\omega}$, and $s, t \in \omega^{<\omega}$ with $|s| = |t| > m$, we have:*

(a) $f \restriction |s| = s$ *for all* $f \in X$,
(b) $g \restriction |t| = t$ *for all* $g \in Y$,
(c) $f <^* h <_m g$ *for all* $f \in D$ *and* $g \in Y$,
(d) $\sup\{g(|t|) : g \in Y\} = \infty$.

Then there exist increasing sequences s_l $(-1 \leq l < \omega)$ and t_l $(-1 \leq l < \omega)$ of elements of $\omega^{<\omega}$ and sequences f_l $(l < \omega)$ and g_l $(l < \omega)$ of elements of D and Y, respectively, such that for all $l \geq 0$:

(e) $s_{-1} = s$, $t_{-1} = t$, *and* $|t_{l-1}| < |s_l| < |t_l|$,
(f) $f_l \restriction |s_l| = s_l$, *and* $g_l \restriction |t_l| = t_l$,
(g) $f_l <_{|s_l|} h$ *and* $g_{2l}(|t_{2l-1}|) > h(|s_{2l}|)$,
(h) $\sup\{f(|s_l|) : f \restriction |s_l| = s_l \text{ and } f \in X\} = \infty$,
(i) $\sup\{g(|t_l|) : g \restriction |t_l| = t_l \text{ and } g \in Y\} = \infty$.

So, in particular $\operatorname{osc}(f_l, g_l) = \operatorname{osc}(s,t) + l$ *for all* $0 \leq l < \omega$. □

Corollary 9.4.6. *For every subset X of $\omega^{\uparrow\omega}$ that is totally ordered and unbounded in the ordering of eventual dominance and for every $l < \omega$, there exist f and g in X such that* $\operatorname{osc}(f,g) = l$. □

We shall also need the following fact that is an immediate consequence of the Claim on p. 62 of [111].

Lemma 9.4.7. *Let X be a subset of $\omega^{\uparrow\omega}$ that is totally ordered in the ordering of eventual dominance. Suppose that A is a family of pairwise-disjoint finite subsets of X, all of some fixed size n, such that for every $a \neq b$ in A either $a <^* b$ or $b <^* a$.*[3] *Let $\mathcal{P}(A)$ be the poset of all finite subsets F of A such that $a(i) \not\leq b(j)$ for all $a \neq b$ in F and $i, j < n$. Then $\mathcal{P}(A)$ satisfies the productive countable chain condition.*[4] □

We are now ready to continue the proof of Theorem 9.4.4. We first prove that the triple system (ω_2, H) satisfies the countable chain condition. So let A be a given uncountable family of finite subsets of ω_2 such that $[a]^3 \cap H = \emptyset$ for all a in A. We need to find $a \neq b$ in A such that $[a \cup b]^3 \cap H = \emptyset$. We may assume that elements of A are all of some fixed size n and that A forms a Δ-system with root w. Shrinking A further, we may assume that for some fixed integer m and all $a \in A$,

$$p(\alpha, \beta) < m \text{ and } \Delta(r_\alpha, r_\beta) < m \text{ for all } \alpha \neq \beta \text{ in } a. \tag{9.4.5}$$

Moreover, we may assume that for all $a \in A$,

$$\{f_{e(\alpha,\beta)} \restriction m, f_{e(\alpha,\gamma)} \restriction m\} \text{ is an antichain for all } \{\alpha, \beta, \gamma\} \in [a]^3 \cap H^c. \tag{9.4.6}$$

Applying the Δ-system lemma again, we may assume that for each $\alpha \in w$ and $l < m$, the family

$$a_\alpha(l) = \{e(\alpha, \beta) : \beta \in a \text{ and } p(\alpha, \beta) = l\} \quad (a \in A) \tag{9.4.7}$$

of finite subsets of ω_1 forms a Δ-system with root $v_\alpha(l)$. Applying Lemma 9.4.7, and going perhaps to some uncountable subset of A in some ccc forcing extension, we may assume that for all $a \neq b$ in A, and $\alpha \in w$ and $l < m$,

$$f_\xi \not\leq f_\eta \text{ for all } \xi \in a_\alpha(l) \setminus v_\alpha(l) \text{ and } \eta \in b_\alpha(l) \setminus v_\alpha(l). \tag{9.4.8}$$

Moreover, we may assume that the structures of the form

$$(a, <, w, e \restriction [a]^2, p \restriction [a]^2, r_\alpha \restriction m, v_\alpha(l))_{\alpha \in a, l < m} \tag{9.4.9}$$

are all isomorphic. Split A into two uncountable disjoint subsets A_0 and A_1 such that for some $\bar{m} \geq m$ and all $a \in A_0$ and $b \in A_1$,

$$\Delta(r_\alpha, r_\beta) < \bar{m} \text{ for all } \alpha \in a \text{ and } \beta \in b. \tag{9.4.10}$$

Applying Theorem 9.3.12, we can find $a \in A_0$ and $b \in A_1$ such that

$$p(\alpha, \beta) > \bar{m} \text{ for all } \alpha \in a \setminus w \text{ and } \beta \in b \setminus w. \tag{9.4.11}$$

[3] Here $a <^* b$ signifies the fact that $a(i) <^* b(j)$ for all $i, j < n$.

[4] I.e., the product of $\mathcal{P}(A)$ with every ccc poset $\mathcal{Q}$ satisfies the ccc.

We claim that $[a \cup b]^3 \cap H = \emptyset$. Consider a triple $\alpha < \beta < \gamma$ from $a \cup b$. We need to check that it belongs to H^c, i.e., that it satisfies the condition (9.4.1). So let us assume that

$$r_\beta \restriction p(\alpha, \beta) = r_\gamma \restriction p(\alpha, \gamma) \tag{9.4.12}$$

and let $l = p(\alpha, \beta) = p(\alpha, \gamma)$. Then either $l < m$ or $l > \bar{m}$. Consider first the case $l < m$. Then the pairs $\{\alpha, \beta\}$ and the pairs $\{\alpha, \gamma\}$ must be included in one of the sets a or b. It follows that in particular $\alpha \in w$. Only the subcase $\beta \in a \setminus w$ and $\gamma \in b \setminus w$ requires some argument. Let $\gamma' \in a$ be the copy of γ with respect to the isomorphism of structures of the form 9.4.9 associated with a and b. If $\beta \neq \gamma'$ then $\{f_{e(\alpha,\beta)}, f_{e(\alpha,\gamma')}\}$ forms an antichain since $[a]^2 \cap H = \emptyset$. Then by (9.4.6), the pair

$$\{f_{e(\alpha,\beta)} \restriction m, f_{e(\alpha,\gamma')} \restriction m\}$$

forms an antichain of the poset $(\omega^m, \leq)$, and since by the isomorphism

$$f_{e(\alpha,\gamma)} \restriction m = f_{e(\alpha,\gamma')} \restriction m,$$

we get the desired conclusion that $\{f_{e(\alpha,\beta)}, f_{e(\alpha,\gamma)}\}$ forms an antichain.

Consider now the subcase $\beta = \gamma'$, i.e., the case when β and γ correspond to each other in the isomorphism of the structures (9.4.9). Then the ordinal $e(\alpha, \beta)$ belongs to the root $v_\alpha(l)$ if and only if the ordinal $e(\alpha, \gamma)$ belongs to $v_\alpha(l)$ in which case they are equal, and therefore, $\{f_{e(\alpha,\beta)}, f_{e(\alpha,\gamma)}\}$ forms the trivial one-element antichain. If these two ordinals do not belong to $v_\alpha(l)$, then by (9.4.8), the pair $\{f_{e(\alpha,\beta)}, f_{e(\alpha,\gamma)}\}$ forms an antichain as well.

It remains to consider the case $l > \bar{m}$. This can only happen if α belongs to one of the set $a \setminus w$ or $b \setminus w$ and β and γ to the other. So, in particular β and γ belong to the same set a or b. But then by (9.4.5), $r_\beta \restriction m \neq r_\gamma \restriction m$, and therefore, $r_\beta \restriction l \neq r_\gamma \restriction l$, contradicting (9.4.12) and the fact that $l = p(\alpha, \beta) = p(\alpha, \gamma)$. This shows that this case is in fact not possible and finishes the proof that the triple system (ω_2, H) satisfies the countable chain condition.

Since the triple system (ω_2, H) satisfies the countable chain condition, our assumption $\Sigma_3 > \mathfrak{c}$ applies to it, and so we have a decomposition of ω_2 into $\aleph_1$ subsets X with the property that $[X]^3 \cap H = \emptyset$. So, in particular, there is an unbounded subset X of ω_2 such that $[X]^3 \cap H = \emptyset$. Since the mappings $e(\cdot, \alpha)$'s are one-to-one there must be $\alpha \in X$, $l < \omega$ and $t \in \omega^l$ such that

$$Y = \{e(\alpha, \beta) : \beta \in X, \beta > \alpha,\ p(\alpha, \beta) = l \text{ and } r_\beta \restriction l = t\}$$

is unbounded in ω_1. By Corollary 9.4.6, there exist $\xi < \eta$ in Y such that f_ξ and f_η are comparable, and in fact $f_\xi \leq f_\eta$. Choose β and γ in X above α such that

$$p(\alpha, \beta) = l = p(\alpha, \gamma) \text{ and } r_\beta \restriction l = t = r_\gamma \restriction l,$$

and $\xi = e(\alpha, \beta)$ and $\eta = e(\alpha, \gamma)$. Going back to the definition (9.4.1) of H^c, we conclude that the triple $\{\alpha, \beta, \gamma\}$ does not belong to H^c, since on one hand, the hypothesis of (9.4.1) holds but $f_\xi = f_{e(\alpha,\beta)}$ and $f_\eta = f_{e(\alpha,\gamma)}$ are comparable. This shows that $[X]^3 \cap H \neq \emptyset$, a contradiction. The proof of Theorem 9.4.4 is now completed. □

We finish this section with the following natural question asking if the dimension in Theorem 9.4.4 can be lowered.

Question 9.4.8. Does $\Sigma_2 > \mathfrak{c}$ imply that $\Theta_2 = \omega_2$?

Of course, one is also interested in determining how much of the Continuum Hypothesis is captured by inequalities between the continuum and the cardinal numbers like Σ or Σ_n.

Question 9.4.9. Is any of the inequalities like $\Sigma > \mathfrak{c}$ or $\Sigma_n > \mathfrak{c}$ equivalent to the Continuum Hypothesis?

Chapter 10

Higher Dimensions

10.1 Stepping-up to higher dimensions

The reader must have already noticed that in this book so far, we have only considered functions of the form $f : [\theta]^2 \longrightarrow I$ or equivalently sequences

$$f_\alpha : \alpha \longrightarrow I \ (\alpha < \theta)$$

of one-place functions. To obtain analogous results about functions defined on higher-dimensional cubes $[\theta]^n$, one usually develops some form of *stepping-up procedure* that lifts a function of the form $f : [\theta]^n \longrightarrow I$ to a function of the form $g : [\theta^+]^{n+1} \longrightarrow I$. The basic idea seems quite simple. One starts with a coherent sequence $e_\alpha : \alpha \longrightarrow \theta$ $(\alpha < \theta^+)$ of one-to-one mappings and wishes to define $g : [\theta^+]^{n+1} \longrightarrow I$ as follows:

$$g(\alpha_0, \alpha_1, \ldots, \alpha_n) = f(e(\alpha_0, \alpha_n), \ldots, e(\alpha_{n-1}, \alpha_n)). \tag{10.1.1}$$

In other words, we use e_{α_n} to send $\{\alpha_0, \ldots, \alpha_{n-1}\}$ to the domain of f and then apply f to the resulting n-tuple. The problem with such a simple-minded definition is that for a typical subset Γ of θ^+, the sequence of restrictions $e_\delta \restriction (\Gamma \cap \delta)$ $(\delta \in \Gamma)$ may not cohere, so we cannot produce a subset of θ that would correspond to Γ and on which we would like to apply some property of f. It turns out that the definition (10.1.1) is basically correct except that we need to replace e_{α_n} by $e_{\tau(\alpha_{n-2}, \alpha_{n-1}, \alpha_n)}$, where

$$\tau : [\theta^+]^3 \longrightarrow \theta^+ \tag{10.1.2}$$

is defined as follows (see Definition 8.1.7):

$$\tau(\alpha, \beta, \gamma) = \gamma_t, \text{ where } t = \rho_0(\alpha, \gamma) \cap \rho_0(\beta, \gamma). \tag{10.1.3}$$

The function ρ_0 to which (10.1.3) refers is of course based on some C-sequence C_α $(\alpha < \theta^+)$ on θ^+. The following result shows that if the C-sequence is carefully chosen, the function τ will serve as a stepping-up tool.

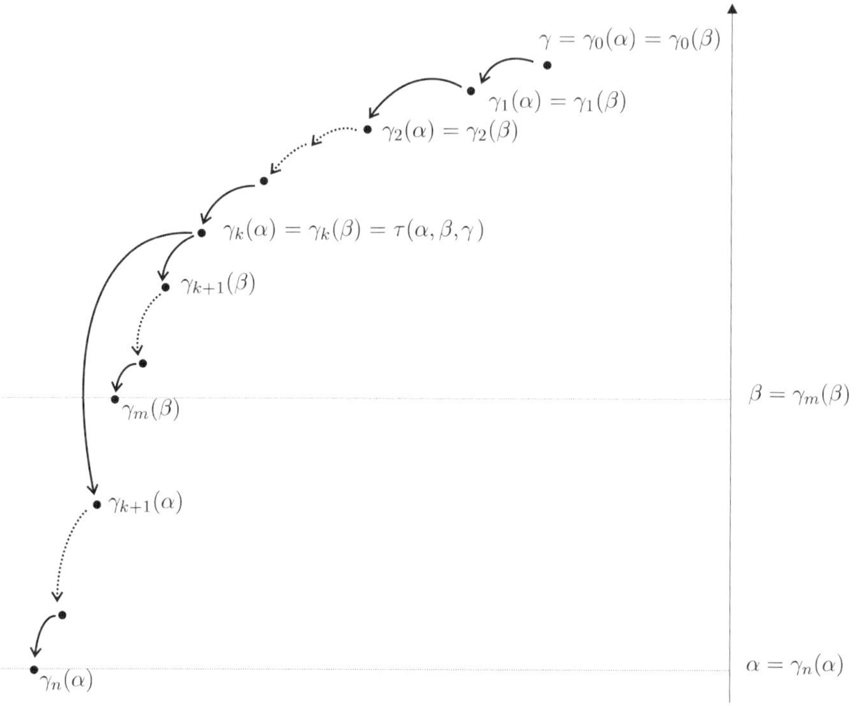

Figure 10.1: The characteristic $\tau(\alpha, \beta, \gamma)$.

Lemma 10.1.1. *Suppose ρ_0 and τ are based on some $\square_\theta$-sequence C_α ($\alpha < \theta^+$) and let κ be a regular uncountable cardinal $\leq \theta$. Then every set $\Gamma \subseteq \theta^+$ of order-type κ contains a cofinal subset Δ such that, if $\varepsilon = \sup(\Gamma) = \sup(\Delta)$, then $\rho_0(\xi, \varepsilon) = \rho_0(\xi, \tau(\alpha, \beta, \gamma))$ for all $\xi < \alpha < \beta < \gamma$ in Δ.*

Proof. For $\delta \in \lim(C_\varepsilon)$, let ξ_δ be the minimal element of Γ above δ. Let ε_δ be the maximal point of the trace $\mathrm{Tr}(\delta, \xi_\delta)$ of the walk from ξ_δ to δ along the $\square_\theta$-sequence C_α ($\alpha < \theta^+$). Thus $\varepsilon_\delta = \delta$ or $\varepsilon_\delta = \min(\mathrm{Tr}(\delta, \xi_\delta) \setminus (\delta + 1))$ and $C_\xi \cap \delta$ is bounded in δ for all $\xi \in \mathrm{Tr}(\delta, \xi_\delta)$ strictly above ε_δ. Let $f(\delta)$ be a member of $C_\varepsilon \cap \delta$ which bounds all these sets. By the Pressing Down Lemma there is a stationary set $\Sigma \subseteq \lim(C_\varepsilon)$ and $\bar{\delta} \in C_\varepsilon$ such that $f(\delta) \leq \bar{\delta}$ for all $\delta \in \Sigma$. Shrinking Σ we may assume that $\xi_\gamma < \delta$ whenever $\gamma < \delta$ are members of Σ. Note that

$$\varepsilon_\gamma = \min(\mathrm{Tr}(\xi_\alpha, \xi_\gamma) \cap \mathrm{Tr}(\xi_\beta, \xi_\gamma)) \text{ for every } \alpha < \beta < \gamma \text{ in } \Sigma. \tag{10.1.4}$$

In other words, $\tau(\xi_\alpha, \xi_\beta, \xi_\gamma) = \varepsilon_\gamma$ for all $\alpha < \beta < \gamma$ in Σ. By the definition of ε_δ for $\delta \in \lim(C_\varepsilon)$, δ belongs to C_{ε_δ} and so by the implicit assumption that nonlimit points of C_{ε_δ} are successor ordinals, we conclude that δ is a limit point of C_{ε_δ}.

Thus the section $(\rho_0)_\delta$ of the function ρ_0 is an initial part of both the sections $(\rho_0)_\varepsilon$ and $(\rho_0)_{\varepsilon_\delta}$, which is exactly what is needed for the conclusion of Lemma 10.1.1. □

Recall that for a given C-sequence C_α $(\alpha < \theta^+)$ such that $\mathrm{tp}(C_\alpha) \leq \theta$ for all $\alpha < \theta^+$, the range of ρ_0 is the collection of all finite sequences of ordinals $< \theta$. It will be convenient to identify $\mathbb{Q}_\theta$ with θ via, for example, Gödel's coding of sequences of ordinals by ordinals. This identification gives us a well-ordering $<_w$ of $\mathbb{Q}_\theta$ of length θ. We use this identification with lift-up of an arbitrary map $f : [\theta]^n \longrightarrow I$ (really, a map of the form $f : [\theta]^n \longrightarrow \mathbb{Q}_\theta$) to a map $g : [\theta^+]^{n+1} \longrightarrow I$ by the following formula:

$$g(\alpha_0, \ldots, \alpha_{n-1}, \alpha_n) = f(\rho_0(\alpha_0, \varepsilon), \ldots, \rho_0(\alpha_{n-1}, \varepsilon)), \tag{10.1.5}$$

where $\varepsilon = \tau(\alpha_{n-2}, \alpha_{n-1}, \alpha_n)$.

Let us examine how this stepping-up procedure works on a particular example motivated by a problem from [43].

Theorem 10.1.2. *Suppose θ is an arbitrary cardinal for which $\square_\theta$ holds. Suppose further that for some regular $\kappa > \omega$ and integer $n \geq 2$ there is a map $f : [\theta]^n \longrightarrow [[\theta]^{<\kappa}]^{<\kappa}$ such that:*

(1) *$A \subseteq \min(a)$ for all $a \in [\theta]^n$ and $A \in f(a)$.*
(2) *For all $\nu < \kappa$ and $\Gamma \subseteq \theta$ of size κ there exist $a \in [\Gamma]^n$ and $A \in f(a)$ such that $\mathrm{tp}(A) \geq \nu$ and $A \subseteq \Gamma$.*

Then θ^+ and κ satisfy the same combinatorial property, but with $n+1$ in place of n.

Proof. We assume that the domain of f is actually equal to the set $[\mathbb{Q}_\theta]^n$ rather than $[\theta]^n$ and define g by

$$g(\alpha_0, \ldots, \alpha_{n-1}, \alpha_n) = (\rho_0)_\varepsilon^{-1}(f(\rho_0(\alpha_0, \varepsilon), \ldots, \rho_0(\alpha_{n-1}, \varepsilon))), \tag{10.1.6}$$

where $\varepsilon = \tau(\alpha_{n-2}, \alpha_{n-1}, \alpha_n)$, and where ρ_0 and τ are based on some fixed $\square_\theta$-sequence. Note that the transformation does not necessarily preserve (1), so we intersect each member of a given $g(a)$ with $\min(a)$ in order to satisfy this condition. To check (2), let $\Gamma \subseteq \theta$ be a given set of size κ. By Lemma 10.1.1, shrinking Γ we may assume that Γ has order-type κ and that if $\varepsilon = \sup(\Gamma)$, then

$$\rho_0(\xi, \varepsilon) = \rho_0(\xi, \tau(\alpha, \beta, \gamma)) \text{ for all } \alpha < \beta < \gamma \text{ in } \Gamma. \tag{10.1.7}$$

It follows that g restricted to $[\Gamma]^{n+1}$ satisfies the formula

$$g(\alpha_0, \ldots, \alpha_{n-1}, \alpha_n) = (\rho_0)_\varepsilon^{-1}(f(\rho_0(\alpha_0, \varepsilon), \ldots, \rho_0(\alpha_{n-1}, \varepsilon))). \tag{10.1.8}$$

Shrinking Γ further we assume that the mapping $(\rho_0)_\varepsilon$, as a map from the ordered set $(\theta^+, \in)$ into the ordered set $(\mathbb{Q}_\theta, <_w)$ is strictly increasing when restricted to Γ.

Given an ordinal $\nu < \kappa$, we apply (2) for f to the set $\Delta = \{\rho_0(\alpha, \varepsilon) : \alpha \in \Gamma\}$ and find $a \in [\Delta]^n$ and $A \in f(a)$ such that $\mathrm{tp}(A) \geq \nu$ and $A \subseteq \Delta$. Let $\{\alpha_0, \ldots, \alpha_{n-1}\}$ be the increasing enumeration of the preimage $(\rho_0)_\varepsilon{}^{-1}(a)$ and pick $\alpha_n \in \Gamma$ above α_{n-1}. Let B be the preimage $(\rho_0)_\varepsilon{}^{-1}(A)$. Then $B \in g(\alpha_0, \ldots, \alpha_{n-1}, \alpha_n)$, $\mathrm{tp}(B) \geq \nu$ and $B \subseteq \Gamma$. This finishes the proof. □

Remark 10.1.3. It should be noted that Velleman [128] was the first to attempt to step-up the combinatorial property of Hajnal and Komjath [43] appearing in Theorem 10.1.2 using his version of the gap-2 morass. Unfortunately the stepping-up procedure of [128] worked only up to the value $n = 3$ so it remains unclear if the higher-gap morass could be used to reach all the other values of n.

If we apply this stepping-up procedure to the projection $[\![\cdot\cdot]\!]$ of the square-bracket operation defined in Definition 5.2.1, one obtains analogues of families $\mathcal{G}, \mathcal{H}$ and $\mathcal{K}$ defined in the course of the proof of Theorem 5.3.10 for ω_2 instead of ω_1. To make this precise, we first fix a $\square_{\omega_1}$-sequence C_α $(\alpha < \omega_2)$ and consider the corresponding mappings,

$$\rho_0 : [\omega_2]^2 \longrightarrow \omega_1^{<\omega} \text{ and } \rho : [\omega_2]^2 \longrightarrow \omega_1.$$

Let $\bar{\rho} : [\omega_2]^2 \longrightarrow \omega_1$ be defined by

$$\bar{\rho}(\alpha, \beta) = 2^{\rho(\alpha,\beta)} \cdot (2 \cdot \mathrm{tp}\{\xi \leq \alpha : \rho(\xi, \alpha) \leq \rho(\alpha, \beta)\} + 1).$$

Then as in the proof of Lemma 3.2.2 one shows that $\bar{\rho}$ has the following three useful properties.

Lemma 10.1.4. *For all $\alpha < \beta < \gamma < \omega_2$,*

(a) $\bar{\rho}(\alpha, \gamma) \neq \bar{\rho}(\beta, \gamma)$,
(b) $\bar{\rho}(\alpha, \gamma) \leq \max\{\bar{\rho}(\alpha, \beta), \bar{\rho}(\beta, \gamma)\}$,
(c) $\bar{\rho}(\alpha, \beta) \leq \max\{\bar{\rho}(\alpha, \gamma), \bar{\rho}(\beta, \gamma)\}$. □

Moreover, we have the following coherence property of $\bar{\rho}$ that is also quite useful.

Lemma 10.1.5. *$\bar{\rho}_\alpha = \bar{\rho}_\beta \restriction \alpha$ whenever $\alpha \in \lim(C_\beta)$.* □

Definition 10.1.6. The ternary operation $[\![\![\cdot\cdot]\!]\!] : [\omega_2]^3 \longrightarrow \omega_2$ is defined as follows:

$$[\![\![\alpha\beta\gamma]\!]\!] = (\bar{\rho}_{\tau(\alpha,\beta,\gamma)})^{-1}[\![\bar{\rho}_{\tau(\alpha,\beta,\gamma)}(\alpha)\ \bar{\rho}_{\tau(\alpha,\beta,\gamma)}(\beta)]\!].$$

It should be clear that the stepping-up method lifts the property of Lemma 5.2.2 of $[\![\cdot\cdot]\!]$ to the following property of the ternary operation $[\![\![\cdot\cdot]\!]\!]$:

Lemma 10.1.7. *Let A be a given uncountable family of pairwise-disjoint subsets of ω_2, all of some fixed size n. Let $\xi_1, \ldots, \xi_n$ be a sequence of ordinals $< \omega_2$ such that for all $i = 1, \ldots, n$ and all but countably many $a \in A$ we have that $\xi_i < a(i)$. Then there exist a, b and c in A such that $[\![\![a(i)b(i)c(i)]\!]\!] = \xi_i$ for $i = 1, \ldots, n$.* □

As in Theorem 5.3.10, let us identify ω_2 and $3 \times [\omega_2]^{<\omega}$ and view $[\![\cdots]\!]$ as a function

$$[\![\cdots]\!] : [\omega_2]^3 \longrightarrow 3 \times [\omega_2]^{<\omega}$$

having the range $3 \times [\omega_2]^{<\omega}$ rather than ω_2. Let

$$[\![\cdots]\!]_0 : [\omega_2]^3 \longrightarrow 3 \text{ and } [\![\cdots]\!]_1 : [\omega_2]^3 \longrightarrow [\omega_2]^{<\omega}$$

be the two projections of $[\![\cdot\cdot]\!]$. Set

$$\mathcal{G} = \{G \in [\omega_2]^{<\omega} : [\![\alpha\beta\gamma]\!]_0 = 0 \text{ for all } \{\alpha, \beta, \gamma\} \in [G]^3\},$$
$$\mathcal{H} = \{H \in [\omega_2]^{<\omega} : [\![\alpha\beta\gamma]\!]_0 = 1 \text{ for all } \{\alpha, \beta, \gamma\} \in [H]^3\}.$$

Note the following immediate consequence of Lemma 10.1.7.

Lemma 10.1.8. *For every uncountable family A of disjoint 2-element subsets of ω_2 there is an arbitrarily large finite subset B of A with $\{b(0) : b \in B\} \in \mathcal{G}$ and $\{b(1) : b \in B\} \in \mathcal{H}$.* □

Recall that the parameter in all our definitions so far is a fixed $\square_{\omega_1}$-sequence C_α $(\alpha < \omega_2)$ and the corresponding characteristics of the walk ρ_0, ρ, $\bar{\rho}$, τ, etc. We use them to define an analogue of the family $\mathcal{K}$ of Theorem 5.3.10 as follows. Let $\mathcal{K}$ now be the collection of all finite sets $\{a_i : i < n\}$ of elements of $[\omega_2]^2$ such that for all $i < j < k < n$:

$$a_i < a_j < a_k, \tag{10.1.9}$$
$$\rho(\alpha, \beta) \leq \rho(\beta, \gamma) \text{ holds if } \alpha \in a_i, \beta \in a_j \text{ and } \gamma \in a_k, \tag{10.1.10}$$
$$[\![a_i(0)a_j(0)a_k(0)]\!]_0 = [\![a_i(1)a_j(1)a_k(1)]\!]_0 = 2, \tag{10.1.11}$$
$$[\![a_i(0)a_j(0)a_k(0)]\!]_1 = [\![a_i(1)a_j(1)a_k(1)]\!]_1 = \bigcup_{l<i} a_l. \tag{10.1.12}$$

The following property of $\mathcal{K}$ is a consequence of Lemma 10.1.7 and the triangle inequalities of ρ.

Lemma 10.1.9. *For every sequence a_ξ $(\xi < \omega_1)$ of 2-element subsets of ω_2 such that $a_\xi < a_\eta$ whenever $\xi < \eta$, and for every positive integer n, there exist $\xi_0 < \xi_1 < \cdots < \xi_{n-1} < \omega_1$ such that $\{a_{\xi_i} : i < n\} \in \mathcal{K}$.* □

The particular definition of $\mathcal{K}$ is made in order to have the following property of $\mathcal{K}$ which is analogous to (5.3.15) of Section 5.1:

If K and L are two distinct members of $\mathcal{K}$, then there are no more than nine ordinals α such that $\{\alpha, \beta\} \in K$ and $\{\alpha, \gamma\} \in L$ for some $\beta \neq \gamma$. (10.1.13)

The definitions of $\mathcal{G}, \mathcal{H}$ and $\mathcal{K}$ immediately lead to the following analogues of (5.3.13) and (5.3.14):

$\mathcal{G}$ and $\mathcal{H}$ contain all singletons, are closed under subsets and are 3-orthogonal to each other. (10.1.14)

$\mathcal{G}$ and $\mathcal{H}$ are both 5-orthogonal to the family of all unions of members of $\mathcal{K}$. (10.1.15)

So we can proceed as in Theorem 5.3.10 and for a function x from ω_2 into $\mathbb{R}$, define

$$\|x\|_{\mathcal{H},2} = \sup\left\{\left(\sum_{\alpha\in\mathcal{H}} x(\alpha)^2\right)^{\frac{1}{2}} : H \in \mathcal{H}\right\},$$

$$\|x\|_{\mathcal{K},2} = \sup\left\{\left(\sum_{\{\alpha,\beta\}\in\mathcal{K}} (x(\alpha) - x(\beta))^2\right)^{\frac{1}{2}} : K \in \mathcal{K}\right\}.$$

Let $\|\cdot\| = \max\{\|\cdot\|_\infty, \|\cdot\|_{\mathcal{H},2}, \|\cdot\|_{\mathcal{K},2}\}$, let $\bar{E}_2 = \{x : \|x\| < \infty\}$, and let E_2 be the closure of the linear span of $\{1_\alpha : \alpha \in \omega_2\}$ inside $(\bar{E}_2, \|\cdot\|)$. Then as in Theorem 5.3.10 one shows, using the properties of $\mathcal{G}$, $\mathcal{H}$ and $\mathcal{K}$ listed above in Lemmas 10.1.8 and 10.1.9 and in (10.1.10–10.1.15), that E_2 is a Banach space with $\{1_\alpha : \alpha \in \omega_2\}$ as its transitive basis and with the property that every bounded linear operator $T : E_2 \longrightarrow E_2$ can be written as $S + D$, where S is an operator with separable range and D is a diagonal operator relative to the basis $\{1_\alpha : \alpha \in \omega_2\}$ with only countably many changes of the constants. The new feature here is the use of (10.1.10) in checking that the projections on countable sets of the form $\{\alpha < \beta : \rho(\alpha, \beta) < \nu\}$ $(\alpha < \omega_2, \nu < \omega_1)$ are (uniformly) bounded. This is needed among other things in showing, using Lemma 10.1.8, that T can be written as a sum of one such projection and a diagonal operator relative to $\{1_\alpha : \alpha \in \omega_2\}$.

Using the interpolation method of [17], one can turn E_2 into a reflexive example. Thus, we have the following:

Theorem 10.1.10. *Assuming $\square_{\omega_1}$, there is a reflexive Banach space E with a transitive basis of type ω_2 with the property that every bounded operator $T : E \longrightarrow E$ can be written as a sum of an operator with a separable range and a diagonal operator (relative to the basis) with only countably many changes of constants.* □

Remark 10.1.11. In [61], P. Koszmider has shown that such a space cannot be constructed on the basis of the usual axioms of set theory. We refer the reader to that paper for more details about these kinds of examples of Banach spaces.

10.2 Chang's conjecture as a 3-dimensional Ramsey-theoretic statement

In this section we shall examine the stepping-up method with less restrictions on the given C-sequence C_α $(\alpha < \theta^+)$ on which it is based. This will lead us to an interesting formulation of Chang's Conjecture as a purely Ramsey-theoretic statement.

Theorem 10.2.1. *The following are equivalent for a regular cardinal θ such that $\log\theta^+ = \theta$.* [1]

(1) *There is a substructure of the form $(\theta^{++},\theta^+,<,\ldots)$ with no substructure B of size θ^+ with $B\cap\theta^+$ of size θ.*

(2) *There is $f : [\theta^{++}]^3 \longrightarrow \theta^+$ which takes all the possible values on the cube of any subset Γ of θ^{++} of size θ^+.*

Proof. To prove the nontrivial direction from (1) to (2), we use Lemma 9.2.3 and choose a strongly unbounded and subadditive $e : [\theta^{++}]^2 \longrightarrow \theta^+$. We also choose a C-sequence C_α ($\alpha < \theta^+$) such that $\mathrm{tp}(C_\alpha) \leq \theta$ for all $\alpha < \theta^+$ and consider the corresponding function $\rho : [\theta^+]^2 \longrightarrow \theta$ defined above in (9.1.2). Finally we choose a one-to-one sequence r_α ($\alpha < \theta^{++}$) of elements of $\{0,1\}^{\theta^+}$ and consider the corresponding function $\Delta : [\theta^{++}]^2 \longrightarrow \theta^+$:

$$\Delta(\alpha,\beta) = \Delta(r_\alpha, r_\beta) = \min\{\nu : r_\alpha(\nu) \neq r_\beta(\nu)\}. \tag{10.2.1}$$

The definition of $f : [\theta^{++}]^3 \longrightarrow \theta$ is given according to the following two rules applied to a given triple $x = \{\alpha,\beta,\gamma\} \in [\theta^{++}]^3$ ($\alpha<\beta<\gamma$):

Rule 1: If $\Delta(r_\alpha, r_\beta) < \Delta(r_\beta, r_\gamma)$ and $r_\alpha <_{\mathrm{lex}} r_\beta <_{\mathrm{lex}} r_\gamma$ or $r_\alpha >_{\mathrm{lex}} r_\beta >_{\mathrm{lex}} r_\gamma$, let

$$f(\alpha,\beta,\gamma) = \min(P_\nu(\Delta(\beta,\gamma)) \setminus \Delta(\alpha,\beta)),$$

where

$$\nu = \rho(\min\{\xi \leq \Delta(\alpha,\beta) : \rho(\xi,\Delta(\alpha,\beta)) \neq \rho(\xi,\Delta(\beta,\gamma))\}, \Delta(\beta,\gamma)).$$

Rule 2: If $\alpha \in x$ is such that r_α is lexicographically between the other two r_ξ's for $\xi \in x$, if $\beta \in x\setminus\{\alpha\}$ is such that $\Delta(r_\alpha, r_\beta) > \Delta(r_\alpha, r_\gamma)$, where γ is the remaining element of x and if x does not fall under Rule 1, let

$$f(\alpha,\beta,\gamma) = \min(P_\nu(e(\beta,\gamma)) \setminus e(\alpha,\beta)),$$

where $\nu = \rho\{\Delta(\alpha,\beta), e(\beta,\gamma)\}$.

The proof of the theorem is finished once we show the following: for every stationary set Σ of cofinality θ ordinals $< \theta^+$ and every $\Gamma \subseteq \theta^{++}$ of size θ^+ there exist $\alpha<\beta<\gamma$ in Γ such that $f(\alpha,\beta,\gamma) \in \Sigma$. Clearly we may assume that Γ has order-type θ^+. Let

$$R_\Gamma = \{r_\alpha : \alpha \in \Gamma\}.$$

Case 1: R_Γ contains a subset of size θ^+ which is lexicographically well ordered or conversely well ordered. Going to a subset of Γ, we may assume that

$$\Delta(r_\alpha, r_\beta) = \Delta(r_\alpha, r_\gamma) \text{ for all } \alpha<\beta<\gamma \text{ in } \Gamma. \tag{10.2.2}$$

[1] $\log\kappa = \min\{\lambda : 2^\lambda \geq \kappa\}$.

It follows that Rule 1 applies in the definition of $f(\alpha,\beta,\gamma)$ for any triple $\alpha<\beta<\gamma$ of elements of Γ. Let $\Delta=\{\Delta(r_\alpha,r_\beta):\{\alpha,\beta\}\in[\Gamma]^2\}$. Then Δ is a subset of θ^+ of size θ^+ and the range of f on $[\Gamma]^3$ includes the range of $[\cdot\cdot]$ on $[\Delta]^2$, where $[\cdot\cdot]$ is the square-bracket operation on θ^+ defined as follows:

$$[\alpha\beta]=\min(P_\nu(\beta)\setminus\alpha), \tag{10.2.3}$$

where $\nu=\rho(\min\{\xi\leq\alpha:\rho(\xi,\alpha)\neq\rho(\xi,\beta)\},\beta)$. Note that this is the exact analogue of the square-bracket operation that has been analysed above in Section 5.1. So exactly as in Section 5.1, one establishes the following fact.

For $\Delta\subseteq\theta^+$ unbounded, the set $\{[\alpha\beta]:\{\alpha,\beta\}\in[\Delta]^2\}$ contains almost all ordinals $<\theta^+$ of cofinality θ. (10.2.4)

Case 2: R_Γ contains no subsets of size θ^+ which are lexicographically well ordered or conversely well ordered. For $t\in\{0,1\}^{<\theta^+}$, let $\Gamma_t=\{\alpha\in\Gamma:t\subseteq r_\alpha\}$. Let

$$T=\{t\in\{0,1\}^{<\theta^+}:|\Gamma_t|=\theta^+\}.$$

Shrinking Γ we may assume that T is either a downwards closed subtree of $\{0,1\}^{<\theta^+}$ of size θ and height $\lambda<\theta^+$, or a downwards closed Aronszajn subtree of $\{0,1\}^{<\theta^+}$ of height θ^+. For $\beta\in\Gamma$, let

$$e_\beta{}''\Gamma^+=\{e\{\beta,\gamma\}:\gamma\in\Gamma,r_\beta<_{\mathrm{lex}}r_\gamma\}. \tag{10.2.5}$$

Case 2.1: There exist θ^+ many $\beta\in\Gamma$ such that the set $e_\beta{}''\Gamma^+$ is bounded for all $\beta\in\Gamma$. Choose an elementary submodel M of some large enough structure of the form H_κ such that $\delta=M\cap\theta^+\in\Sigma$ and M contains all the relevant objects. By the elementarity of M, strong unboundedness of e and the fact that T contains no θ^+-chain, there exist β and γ in $\Gamma\setminus M$ such that

$$r_\gamma<_{\mathrm{lex}}r_\beta,\quad \varepsilon=e\{\beta,\gamma\}\geq\delta \text{ and } \Delta(r_\beta,r_\gamma)<\delta. \tag{10.2.6}$$

Let $\bar\nu=\rho(\delta,\varepsilon)$. Note that there exist θ many $\xi>\Delta(r_\beta,r_\gamma)$ such that $(r_\beta\restriction\xi)^\frown 0$ belongs to M and $r_\beta(\xi)=1$. This is clear in the case when T is an Aronszajn tree, by the elementarity of M and the fact that $\mathrm{cf}(\delta)=\theta$. However, this is also true in case T has height $\lambda<\theta^+$ and has size θ, since from our assumption $\log\theta^+=\theta$ the ordinal λ must have cofinality exactly θ. So by the property 9.1.1 of ρ there is one such ξ such that $\nu=\rho(\xi,\varepsilon)\geq\bar\nu$. Let $t=(r_\beta\restriction\xi)^\frown 0$. Then $t\in T\cap M$ and so Γ_t is a subset of Γ of size θ^+ which belongs to M. Using the strong unboundedness of e and the elementarity of M, there must be an α in $\Gamma_t\cap M$ such that

$$e(\alpha,\beta)>\sup(P_\nu(\varepsilon)\cap\delta). \tag{10.2.7}$$

Since $e(\alpha,\beta)$ belongs to $e_\alpha{}''\Gamma^+$, a set which, by the assumption of Case 2.1, is bounded in θ^+, we conclude that $e(\alpha,\beta)<\delta$. It follows that $f(\alpha,\beta,\gamma)$ is defined according to Rule 2, and therefore

$$f(\alpha,\beta,\gamma)=\min(P_\nu(\varepsilon)\setminus e(\alpha,\beta))=\delta. \tag{10.2.8}$$

Case 2.2: For every β in some tail of Γ the set $e_\beta''\Gamma^+$ is unbounded in θ^+. Going to a tail, we assume that this is true for all $\beta \in \Gamma$. Let M and $\delta = M \cap \theta^+ \in \Sigma$ be chosen as before. If T is Aronszajn, using elementarity of M, there must be $\beta \in \Gamma \cap M$ and $\bar{\delta} < \delta$ of cofinality θ such that there exist unboundedly many ξ below $\bar{\delta}$ such that the set

$$\Omega_\xi = \{e\{\beta, \chi\} : \chi \in \Gamma, r_\beta <_{\text{lex}} r_\chi, \Delta(r_\beta, r_\chi) = \xi\} \tag{10.2.9}$$

is unbounded in θ^+. If T is a tree of height $\lambda < \theta^+$, $\bar{\delta} = \lambda$ will satisfy this for all but θ many $\beta \in \Gamma$, so in particular, we can pick one such $\beta \in \Gamma \cap M$. Choose $\gamma \in \Gamma$ such that

$$r_\beta <_{\text{lex}} r_\gamma, \quad \varepsilon = e(\beta, \gamma) \geq \delta \text{ and } \Delta(r_\beta, r_\gamma) < \bar{\delta}. \tag{10.2.10}$$

Let $\bar{\nu} = \rho(\delta, \varepsilon)$. By the property of ρ in Lemma 9.1.1 of ρ, there is $\bar{\mu} < \bar{\delta}$ such that $\rho(\xi, \varepsilon) \geq \bar{\nu}$ for all $\xi \in [\bar{\mu}, \bar{\delta})$. Choose $\xi \in [\bar{\mu}, \bar{\delta})$ above $\Delta(r_\beta, r_\gamma)$ such that the set Ω_ξ is unbounded in θ^+. Let $\nu = \rho(\xi, \varepsilon)$. Then $\nu \geq \bar{\nu}$. The set $P_\nu(\varepsilon) \cap \delta$ being of size $< \theta$ is bounded in δ. Since $\Omega_\xi \in M$, there is $\alpha \in \Gamma \cap M$ above β such that

$$\Delta(r_\beta, r_\alpha) = \xi, \quad r_\beta <_{\text{lex}} r_\alpha \text{ and } e(\beta, \alpha) > \sup(P_\nu(\varepsilon) \cap \delta). \tag{10.2.11}$$

Note that $r_\alpha <_{\text{lex}} r_\gamma$, so the definition of $f\{\alpha, \beta, \gamma\}$ obeys Rule 2. Applying this rule, we get that

$$f\{\alpha, \beta, \gamma\} = \min(P_\nu(\varepsilon) \setminus e(\beta, \alpha)) = \delta. \tag{10.2.12}$$

This finishes the proof of Theorem 10.2.1. □

Theorem 10.2.2. *If θ is a regular strong limit cardinal carrying a nonreflecting stationary set, then there is $f : [\theta^+]^3 \longrightarrow \theta$ which takes all the values from θ on the cube of any subset of θ^+ of size θ.*

Proof. This is really a corollary of the proof of Theorem 10.2.1, so let us only indicate the adjustments. By Corollary 9.2.14 and Lemma 9.2.2, we can choose a strongly unbounded subadditive map $e : [\theta^+]^2 \longrightarrow \theta$. By the assumption about θ we can choose a C-sequence C_α ($\alpha < \theta$) avoiding a stationary set $\Sigma \subseteq \theta$ and consider the corresponding notion of a walk, trace, ρ_0-function and the square-bracket operation $[\cdot\cdot]$ as defined in (8.2.1) above. As in the proof of Theorem 10.2.1, we choose a one-to-one sequence r_α ($\alpha < \theta^+$) of elements of $\{0,1\}^\theta$ and consider the corresponding function $\Delta : [\theta^+]^2 \longrightarrow \theta$:

$$\Delta(\alpha, \beta) = \Delta(r_\alpha, r_\beta) = \min\{\nu < \theta : r_\alpha(\nu) \neq r_\beta(\nu)\}. \tag{10.2.13}$$

The definition of $f : [\theta^+]^3 \longrightarrow \theta$ is given according to the following rules, applied to a given $x \in [\theta^+]^3$.

Rule 1: If $x = \{\alpha < \beta < \gamma\}$, $\Delta(r_\alpha, r_\beta) < \Delta(r_\beta, r_\gamma)$ and $r_\alpha <_{\text{lex}} r_\beta <_{\text{lex}} r_\gamma$, or $r_\alpha >_{\text{lex}} r_\beta >_{\text{lex}} r_\gamma$, let

$$f\{\alpha, \beta, \gamma\} = [\Delta(\alpha, \beta)\Delta(\beta, \gamma)].$$

Rule 2: If $\alpha \in x$ is such that r_α is lexicographically between the other two r_ξ's for $\xi \in x$, if $\beta \in x \setminus \{\alpha\}$ is such that $\Delta(r_\alpha, r_\beta) > \Delta(r_\alpha, r_\gamma)$, where γ is the remaining element of x, and they do not satisfy the conditions of Rule 1, set

$$f\{\alpha, \beta, \gamma\} = \min(\mathrm{Tr}(\Delta(\alpha, \beta), e\{\beta, \gamma\}) \setminus e\{\alpha, \beta\}),$$

i.e., $f\{\alpha, \beta, \gamma\}$ is the minimal point on the trace of the walk from $e\{\beta, \gamma\}$ to $\Delta(\alpha, \beta)$ above the ordinal $e\{\alpha, \beta\}$; if such a point does not exist, set $f\{\alpha, \beta, \gamma\} = 0$.

The proof of the theorem is finished once we show that for every stationary $\Omega \subseteq \Sigma$ and every $\Gamma \subseteq \theta^+$ of size θ, there exist $x \in [\Gamma]^3$ such that $f(x) \in \Omega$. This is done, as in the proof of Theorem 10.2.1, by considering the two cases. Case 1 is reduced to the property in Lemma 8.2.4 of the square-bracket operation. The treatment of Case 2 is similar and a bit simpler than in the proof of Theorem 10.2.1, since the case that T is of height $< \theta$ is impossible due to our assumption that θ is a strong limit cardinal. This finishes the proof. □

Since $\log \omega_1 = \omega$, the following is an immediate consequence of Theorem 10.2.1.

Theorem 10.2.3. *Chang's conjecture is equivalent to the statement that for every* $f : [\omega_2]^3 \longrightarrow \omega_1$ *there is uncountable* $\Gamma \subseteq \omega_2$ *such that* $f''[\Gamma]^3 \neq \omega_1$. □

Remark 10.2.4. Since this same statement is stronger for functions from higher-dimensional cubes $[\omega_2]^n$ into ω_1 Theorem 10.2.3 shows that they are all equivalent to Chang's conjecture. Note also that $n = 3$ is the minimal dimension for which this equivalence holds, since the case $n = 2$ follows from the Continuum Hypothesis, which has no relationship to Chang's conjecture.

10.3 Three-dimensional oscillation mapping

In this section we examine the basic stepping-up procedure without the assumption that some form of Chang's conjecture is false. So let θ be a given regular uncountable cardinal and let C_α $(\alpha < \theta^+)$ be a fixed C-sequence such that $\mathrm{tp}(C_\alpha) \leq \kappa$ for all $\alpha < \theta^+$. Let $\rho : [\theta^+]^2 \longrightarrow \theta$ be the ρ-function defined above in (9.1.2). Recall that, in case C_α $(\alpha < \theta^+)$ is a $\square_\theta$-sequence, the key to our stepping-up procedure was the function $\tau : [\theta^+]^3 \longrightarrow \theta^+$ defined by the formula (10.1.3). Without using the assumption that C_α $(\alpha < \theta^+)$ is a $\square_\theta$-sequence, one can build a related function which turns out to be a good substitute. This is the function,

$$\chi : [\theta^+]^3 \longrightarrow \omega \tag{10.3.1}$$

defined by

$$\chi(\alpha, \beta, \gamma) = |\rho_0(\alpha, \gamma) \cap \rho_0(\beta, \gamma)|. \tag{10.3.2}$$

Thus, $\chi(\alpha, \beta, \gamma)$ is equal to the length of the common part of the walks $\gamma \to \alpha$ and $\gamma \to \beta$.

Definition 10.3.1. A subset Γ of θ^+ is *stable* if χ is bounded on $[\Gamma]^3$.

Lemma 10.3.2. *Suppose that Γ is a stable subset of θ^+ of size θ. Then*

$$\sup\{\rho(\alpha,\beta) : \{\alpha,\beta\} \in [\Omega]^2\} = \theta$$

for every $\Omega \subseteq \Gamma$ of size θ.

Proof. Suppose ρ is bounded by ν on $[\Omega]^2$ for some $\Omega \subseteq \theta^+$ of size θ. We need to show that χ is unbounded on $[\Omega]^3$. Therefore we may assume that Ω is of order-type θ and let $\lambda = \sup(\Omega)$. Construct a sequence $\Omega = \Omega_0 \supseteq \Omega_1 \supseteq \cdots$ of subsets of Ω of order-type θ such that:

$$\chi(\alpha,\beta,\gamma) \geq n \text{ for all } \{\alpha,\beta,\gamma\} \in [\Omega_n]^3. \tag{10.3.3}$$

Given Ω_n, choose an $\in$-chain $\mathcal{M}$ of length θ of elementary submodels M of some large enough structure of the form H_κ such that $M \cap \theta \in \theta$ and M contains all the relevant objects. Now choose another elementary submodel N of H_κ which contains $\mathcal{M}$ as well as all the other relevant objects such that $N \cap \theta \in \theta$. Let $\lambda_N = \sup(N \cap \lambda)$. Choose $\gamma \in \Omega_n$ above λ_N. Let $\gamma = \gamma_0 > \cdots > \gamma_k = \lambda_N$ be the walk from γ to λ_N. Let $\bar{\gamma}$ be the γ_i with the smallest index such that $C_{\gamma_i} \cap \lambda_N$ is unbounded in λ_N. Then $\bar{\gamma}$ is either equal to λ_N or γ_{k-1}. Choose $\lambda_0 \in N \cap \lambda$ such that:

$$\lambda_0 > \max(C_{\gamma_i} \cap \lambda_N) \text{ for all } i < k \text{ for which } C_{\gamma_i} \cap \lambda_N \text{ is bounded in } \lambda_N. \tag{10.3.4}$$

Then $\{\gamma_i : i < k\} \setminus \bar{\gamma} \subseteq \mathrm{Tr}(\alpha,\gamma)$ for all $\alpha \in N \cap \Omega_n$ above λ_0. So if we choose $\alpha < \beta$ in $N \cap \Omega_0$ such that $C_{\bar{\gamma}} \cap [\alpha,\beta) \neq \emptyset$, then

$$\mathrm{Tr}(\alpha,\gamma) \cap \mathrm{Tr}(\beta,\gamma) = \{\gamma_i : i \leq k\} \setminus \bar{\gamma}. \tag{10.3.5}$$

From this and our assumption $\chi(\alpha,\beta,\gamma) \geq n$ we conclude that this set has size at least n. For $M \in \mathcal{M}$, let $\lambda_M = \sup(M \cap \lambda)$ and $C = \{\lambda_M : M \in \mathcal{M}\}$. Then C is a closed and unbounded subset of λ of order-type θ. By our assumption that ρ is bounded by ν on $[\Omega]^2$, we get:

$$\begin{aligned} \mathrm{tp}(C_{\bar{\gamma}} \cap \lambda_N) &\leq \sup\{\mathrm{tp}(C_{\bar{\gamma}} \cap \alpha) : \alpha \in \Omega \cap [\lambda_0, \lambda_N)\} \\ &\leq \sup\{\rho(\alpha,\gamma) : \alpha \in \Omega \cap [\lambda_0, \lambda_N)\} \\ &\leq \nu. \end{aligned} \tag{10.3.6}$$

Since $C \cap [\lambda_0, \lambda_N)$ has order-type $> \nu$ there must be $M \in \mathcal{M} \cap N$ such that $\lambda_M \notin C_{\bar{\gamma}}$ and $\lambda_M > \lambda_0$. Thus, we have an elementary submodel M of H_κ which contains all the relevant objects and a $\gamma \in \Omega_n$ above λ_M such that the walk $\gamma \to \lambda_M$ contains all the points from the set $\{\gamma_i : i < k\} \setminus \bar{\gamma}$, but includes at least the point $\min(C_{\bar{\gamma}} \setminus \lambda_M)$ which is strictly above λ_M. Hence, if $\gamma = \gamma_0 > \gamma_1 > \cdots > \gamma_l = \lambda_M$ is the trace of $\gamma \to \lambda_M$ and if $\hat{\gamma}$ is the γ_i $(i \leq l)$ with the smallest index, subject to the requirement that $C_{\hat{\gamma}} \cap \lambda_M$ is unbounded in λ_M, then $\{\gamma_i : i < l\} \setminus \hat{\gamma}$ has size at least $n+1$. Let $\hat{\lambda}_0 \in \lambda \cap M$ be such that:

$$\hat{\lambda}_0 > \max(C_{\gamma_i} \cap \lambda_M) \text{ for all } i < l \text{ for which } C_{\gamma_i} \cap \lambda_M \text{ is bounded in } \lambda_M. \tag{10.3.7}$$

Then $\{\gamma_i : i < l\} \setminus \bar{\gamma} \subseteq \mathrm{Tr}(\alpha, \gamma)$ for all $\alpha \in M \cap \Omega_n$ above $\hat{\lambda}_0$. Using the elementarity of M we conclude that for every $\delta < \lambda$ there exist $\varepsilon \in \Omega_n \setminus \delta$ and a set $a_\varepsilon \subseteq [\delta, \varepsilon)$ of size at least $n+1$ such that $a_\varepsilon \subseteq \mathrm{Tr}(\alpha, \varepsilon)$ for all $\alpha \in \Omega_n \cap [\hat{\lambda}_0, \delta)$. Hence we can choose a closed and unbounded set D of $[\hat{\lambda}_0, \lambda)$ of order-type θ, for each $\delta \in D$ an $\varepsilon(\delta) \in \Omega_n \setminus \delta$ and an $a_{\varepsilon(\delta)} \subseteq [\delta, \varepsilon(\delta))$ such that for every $\delta \in D$:

$$a_{\varepsilon(\delta)} \subseteq \mathrm{Tr}(\alpha, \varepsilon(\delta)) \text{ for } \alpha \in [\hat{\lambda}_0, \delta), \tag{10.3.8}$$

$$\varepsilon(\delta) < \min(D \setminus (\delta + 1)). \tag{10.3.9}$$

Finally let $\Omega_{n+1} = \{\varepsilon(\delta) : \delta \in D\}$. Then

$$a_{\varepsilon(\gamma)} \subseteq \mathrm{Tr}(\varepsilon(\alpha), \varepsilon(\gamma)) \cap \mathrm{Tr}(\varepsilon(\beta), \varepsilon(\gamma)) \text{ for all } \alpha < \beta < \gamma \text{ in } \Delta. \tag{10.3.10}$$

This gives us that $\chi(\alpha, \beta, \gamma) \geq n+1$ for every triple $\alpha < \beta < \gamma$ in Ω_{n+1}, finishing the inductive step and therefore the proof of Lemma 10.3.2. □

Definition 10.3.3. The 3-*dimensional version of the oscillation mapping*,

$$\mathrm{osc} : [\theta^+]^3 \longrightarrow \omega, \tag{10.3.11}$$

is defined on the basis of the 2-dimensional version of Section 8.1 as follows:

$$\mathrm{osc}(\alpha, \beta, \gamma) = \mathrm{osc}(C_{\beta_s} \setminus \alpha, C_{\gamma_t} \setminus \alpha), \tag{10.3.12}$$

where $s = \rho_0(\alpha, \beta) \restriction \chi(\alpha, \beta, \gamma)$ and $t = \rho_0(\alpha, \gamma) \restriction \chi(\alpha, \beta, \gamma)$.

In other words, we let n be the length of the common part of the two walks $\gamma \to \alpha$ and $\gamma \to \beta$. Then we consider the walks

$$\gamma = \gamma_0(\alpha) > \cdots > \gamma_k(\alpha) = \alpha \text{ and } \beta = \beta_0(\alpha) > \cdots > \beta_l(\alpha) = \alpha \tag{10.3.13}$$

from γ to α and β to α, respectively. If both k and l are bigger than n (and note that $k \geq n$), or in other words, if $\gamma_n(\alpha)$ and $\beta_n(\alpha)$ are both defined, we let $\mathrm{osc}(\alpha, \beta, \gamma)$ be equal to the oscillation of the two sets $C_{\beta_n(\alpha)} \setminus \alpha$ and $C_{\gamma_n(\alpha)} \setminus \alpha$. If not, we let $\mathrm{osc}(\alpha, \beta, \gamma) = 0$.

Lemma 10.3.4. *Suppose that Γ is a subset of θ^+ of size κ, a regular uncountable cardinal, and that every subset of Γ of size κ is unstable. Then for every positive integer n, there exist $\alpha < \beta < \gamma$ in Γ such that* $\mathrm{osc}(\alpha, \beta, \gamma) = n$.

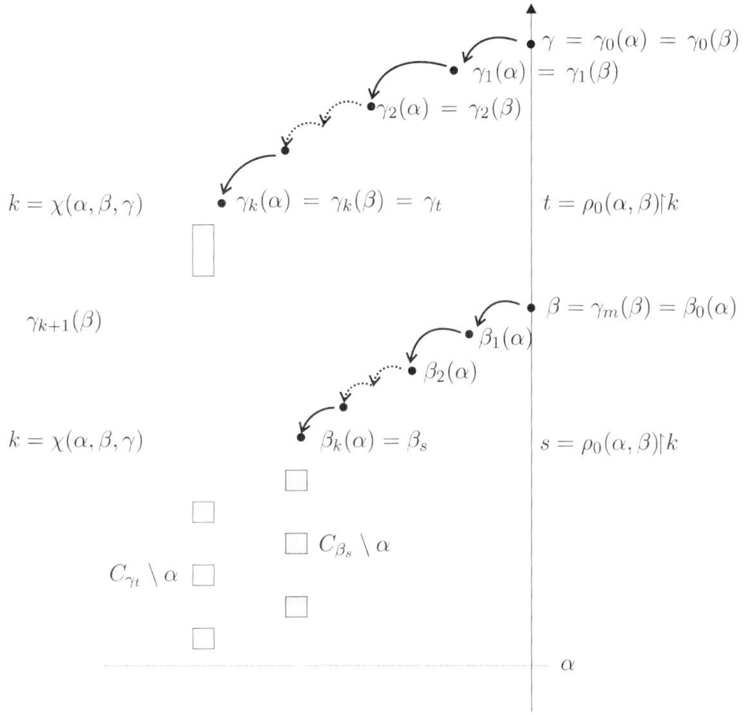

Figure 10.2: $\operatorname{osc}(\alpha, \beta, \gamma) = 3$.

Proof. We may assume that Γ is actually of order-type κ and let $\lambda = \sup(\Gamma)$. Let $\mathcal{M}$ be an $\in$-chain of length κ of elementary submodels M of $H_{\theta^{++}}$ such that $M \cap \kappa \in \kappa$ and M contains all the relevant objects. For $M \in \mathcal{M}$, let $\lambda_M = \sup(\lambda \cap M)$. Now choose another elementary submodel N of $H_{\theta^{++}}$ containing $\mathcal{M}$ as well as all the other relevant objects with $N \cap \kappa \in \kappa$. Let $\lambda_N = \sup(\lambda \cap N)$ and choose $\gamma \in \Gamma$ above λ_N and let $\gamma = \gamma_0 > \cdots > \gamma_k = \lambda_N$ be the walk from γ to λ_N. Let $\bar{\gamma}$ and λ_0 be chosen as above in (10.3.4). Then $\bar{\gamma} = \gamma_k$ or $\bar{\gamma} = \gamma_{k-1}$ and

$$\operatorname{Tr}(\alpha, \gamma) \cap \operatorname{Tr}(\beta, \gamma) = \{\gamma_i : i \leq k\} \setminus \bar{\gamma} \text{ for all } \alpha < \beta \text{ in } \Gamma \cap [\lambda_0, \lambda_N) \text{ such that } C_{\bar{\gamma}} \cap [\alpha, \beta) \neq \emptyset. \tag{10.3.14}$$

So if a tail of the set $\{\lambda_M : M \in \mathcal{M} \cap N\}$ is included in $C_{\bar{\gamma}}$, then using the elementarity of N, we would be able to construct a cofinal subset $\Gamma_0 \subseteq \Gamma$ such that χ is constantly equal to $\bar{k}$ on $[\Gamma_0]^3$, where $\bar{k}$ is the place of $\bar{\gamma}$ in the sequence $\gamma_0, \ldots, \gamma_k$. This, of course, would contradict our assumption that Γ contains no large stable subsets. It follows that the set of all λ_N for $M \in \mathcal{M} \cap N$ which do not belong to $C_{\bar{\gamma}}$ is cofinal in λ_N. So we can pick $M_1 \in M_2 \in \cdots \in M_{n+1}$ in $\mathcal{M} \cap N$

such that, if $\lambda_i = \lambda_{M_i}$ $(1 \leq i \leq n+1)$, then:

$$\lambda_i \in [\lambda_0, \lambda) \setminus C_{\bar{\gamma}} \text{ for all } 1 \leq i \leq n+1, \tag{10.3.15}$$

$$[\lambda_i, \lambda_{i+1}) \cap C_{\bar{\gamma}} \neq \emptyset \text{ for all } 1 \leq i \leq n. \tag{10.3.16}$$

Fix $\alpha \in \Gamma \cap M_1$ above $\max(C_{\bar{\gamma}} \cap \lambda_1)$. For $1 \leq i \leq n$, pick $\bar{\lambda}_i \in [\lambda_i, \lambda_{i+1}) \cap M_{i+1}$ such that $\bar{\lambda}_i \geq \max(C_{\bar{\gamma}} \cap \lambda_{i+1})$. For $1 \leq i \leq n$, let $\bar{I}_i = [\lambda_i, \bar{\lambda}_i]$, and let $\bar{I}_{n+1} = [\lambda_n, \gamma]$. Now set $\mathcal{F}$ to be the collection of all increasing sequences $I_1 < I_2 < \cdots < I_{n+1}$ of closed intervals of ordinals $< \lambda$ with the property that there is $\beta = \beta(\vec{I})$ in $\Gamma \cap I_{n+1}$ such that, if $\beta = \beta_0 > \cdots > \beta_l = \alpha$ is the walk from β to α, then:

$$l > \bar{k} \text{ and } \beta_{\bar{k}} = \min(\mathrm{Tr}(\alpha, \beta) \cap I_{n+1}), \tag{10.3.17}$$

$$C_{\beta_{\bar{k}}} \setminus \alpha \subseteq \bigcup_{i=1}^{n+1} I_i, \tag{10.3.18}$$

$$C_{\beta_{\bar{k}}} \cap I_i \neq \emptyset \text{ for all } 1 \leq n \leq n+1. \tag{10.3.19}$$

Clearly, $\mathcal{F} \in M_1$ and $\langle \bar{I}_1, \ldots, \bar{I}_{n+1} \rangle \in \mathcal{F}$, as γ is a witness of this. So working as in the proof of Lemma 8.1.2, we can find $\mathcal{F}_0 \subseteq \mathcal{F}$ in M_1 such that, if some sequence $\vec{J}$ of length $\leq n$ (including the case $\vec{J} = \emptyset$) end-extends to a member of $\mathcal{F}_0$, then for every $\xi < \lambda$ there is a closed interval K included in $[\xi, \lambda)$ such that the concatenation $\vec{J}^\frown K$ also end-extends to a sequence in $\mathcal{F}_0$. Working inductively on $1 \leq i \leq n+1$ we can select a sequence $I_1, I_2, \ldots, I_n$ such that:

$$I_1, \ldots, I_i \in M_{i+1} \text{ for } 1 \leq i \leq n, \tag{10.3.20}$$

$$I_i \subseteq [\bar{\lambda}_i, \lambda_{i+1}) \text{ for } 1 \leq i \leq n. \tag{10.3.21}$$

$$\langle I_1, \ldots, I_i \rangle \text{ end-extends to a member of } \mathcal{F}_0. \tag{10.3.22}$$

Clearly, there is no problem in choosing these objects, using the elementarity of the models M_i $(1 \leq i \leq n+1)$ and the splitting property of $\mathcal{F}_0$. Working in M_{n+1}, we find an interval I_{n+1} such that $\langle I_0, \ldots, I_n, I_{n+1} \rangle$ belongs to $\mathcal{F}_0$. Let $\beta = \beta(\langle I_0, \ldots, I_{n+1} \rangle)$. Since $C_{\bar{\gamma}}$ intersects the interval $[\lambda_1, \bar{\lambda}_1)$ it separates α from β, so by (10.3.14) we get that $\tau(\alpha, \beta, \gamma) = \bar{k}$, hence by definition

$$\mathrm{osc}(\alpha, \beta, \gamma) = \mathrm{osc}(C_{\beta_{\bar{k}}} \setminus \alpha, C_{\gamma_{\bar{k}}} \setminus \alpha). \tag{10.3.23}$$

Recall that $\gamma_{\bar{k}} = \bar{\gamma}$, so referring to (10.3.16), (10.3.18) and (10.3.19) we conclude that $I_1 \cap C_{\beta_{\bar{k}}}, \ldots, I_{n-1} \cap C_{\beta_{\bar{k}}}$ and $(I_n \cup I_{n+1}) \cap C_{\beta_{\bar{k}}}$ are the n convex pieces into which the set $C_{\bar{\gamma}} \setminus \alpha$ splits the set $C_{\beta_{\bar{k}}} \setminus \alpha$. Hence $\mathrm{osc}(C_{\beta_{\bar{k}}} \setminus \alpha, C_{\gamma_{\bar{k}}} \setminus \alpha) = n$. This completes the proof of Lemma 10.3.4. □

Applying the last two lemmas to the subsets of θ^+ of size θ, we get an interesting dichotomy:

Lemma 10.3.5. *Every $\Gamma \subseteq \theta^+$ of size θ can be refined to a subset Ω of size θ such that either:*

(1) *ρ is unbounded and therefore strongly unbounded on Ω, or*
(2) *the oscillation mapping takes all its possible values on $[\Omega]^3$.* □

The proof of the following result contains a typical application of this dichotomy where a projection of the square-bracket operation of θ^+ is lifted up to a mapping of the form $f : [\theta^{++}]^3 \longrightarrow \omega$.

Theorem 10.3.6. *Suppose θ is a regular cardinal such that $\log \theta^+ = \theta$. Then a natural projection of the square-bracket operation on θ^+ can be lifted to a mapping $f : [\theta^{++}]^3 \longrightarrow \omega$ which in particular takes all the values from ω on the cube of any subset of θ^{++} of size θ^+.*

Proof. We choose two C-sequences C_α ($\alpha < \theta^+$) and C_α^+ ($\alpha < \theta^{++}$) on θ^+ and θ^{++} respectively, such that $\mathrm{tp}(C_\alpha) \leq \theta$ for all $\alpha < \theta^+$ and $\mathrm{tp}(C_\alpha^+) \leq \theta^+$ for all $\alpha < \theta^{++}$. Let

$$\rho : [\theta^+]^2 \longrightarrow \theta \text{ and } \rho^+ : [\theta^{++}]^2 \longrightarrow \theta^+$$

be the corresponding ϱ-functions defined above in (9.1.2). Choose also a one-to-one sequence

$$\{r_\alpha : \alpha < \theta^{++}\} \subseteq \{0,1\}^{\theta^+}$$

and consider the corresponding function $\Delta : [\theta^{++}]^2 \longrightarrow \theta^+$ defined by (see (10.2.1 above),

$$\Delta(\alpha, \beta) = \Delta(r_\alpha, r_\beta) = \min\{\nu : r_\alpha(\nu) \neq r_\beta(\nu)\}. \tag{10.3.24}$$

The mapping

$$f : [\theta^{++}]^3 \longrightarrow \theta^+$$

that will satisfy the conclusion of the theorem is defined according to the following three cases where a given triple $\alpha < \beta < \gamma$ of elements of θ^{++} can fall.

Case 1: $(C_{\beta_s} \cap C_{\gamma_t}) \setminus \alpha \neq \emptyset$, where

$$s = \rho_0(\alpha, \beta) \upharpoonright \chi(\alpha, \beta, \gamma) \text{ and } t = \rho_0(\alpha, \gamma) \upharpoonright \chi(\alpha, \beta, \gamma),$$

assuming of course that $\rho_0(\alpha, \beta)$ has length at least $\chi(\alpha, \beta, \gamma)$.

Rule 1: If $\Delta(r_\alpha, r_\beta) < \Delta(r_\beta, r_\gamma)$ and $r_\alpha <_{\mathrm{lex}} r_\beta <_{\mathrm{lex}} r_\gamma$ or $r_\alpha >_{\mathrm{lex}} r_\beta >_{\mathrm{lex}} r_\gamma$, set

$$f(\alpha, \beta, \gamma) = \min(P_\nu(\Delta(\beta, \gamma)) \setminus \Delta(\alpha, \beta)),$$

where $\nu = \rho(\min\{\xi \leq \Delta(\alpha, \beta) : \rho(\xi, \Delta(\alpha, \beta)) \neq \rho(\xi, \Delta(\beta, \gamma))\}, \Delta(\beta, \gamma))$.

Rule 2: If $\bar{\alpha} \in \{\alpha, \beta, \gamma\}$ is such that $r_{\bar{\alpha}}$ is lexicographically between the other two r_ξ's for $\xi \in \{\alpha, \beta, \gamma\}$, if $\bar{\beta} \in \{\alpha, \beta, \gamma\} \setminus \{\bar{\alpha}\}$ is such that $\Delta(r_{\bar{\alpha}}, r_{\bar{\beta}}) > \Delta(r_{\bar{\alpha}}, r_{\bar{\gamma}})$, where $\bar{\gamma}$ is the remaining member of $\{\alpha, \beta, \gamma\}$, and if $\{\alpha, \beta, \gamma\}$ does not fall under Rule 1, let

$$f(\alpha, \beta, \gamma) - \min(P_\nu(\rho^+\{\bar{\beta}, \bar{\gamma}\}) \setminus \rho^+(\alpha, \beta)),$$

where $\nu = \rho\{\Delta(\alpha, \beta), \rho^+(\beta, \gamma)\}$.

Case 2: $(C_{\beta_s} \cap C_{\gamma_t}) \setminus \alpha = \emptyset$, where

$$s = \rho_0(\alpha, \beta) \upharpoonright \chi(\alpha, \beta, \gamma) \text{ and } t = \rho_0(\alpha, \gamma) \upharpoonright \chi(\alpha, \beta, \gamma),$$

assuming again that $\rho_0(\alpha, \beta)$ has length at least $\chi(\alpha, \beta, \gamma)$. Let

$$f(\alpha, \beta, \gamma) = \mathrm{osc}(\alpha, \beta, \gamma).$$

Case 3: If a given triple $\alpha < \beta < \gamma$ does not fall into one of the Cases 1 and 2, let $f(\alpha, \beta, \gamma) = 0$.

The proof of Theorem 10.3.6 is finished if we show that for every $\Gamma \subseteq \theta^{++}$ of size θ^+, the image $f''[\Gamma]^3$ either contains all positive integers or almost all ordinals $< \theta^+$ of cofinality θ.

So let Γ be a given subset of θ^{++} of order-type θ^+ and let $\lambda = \sup(\Gamma)$. By the proof of Lemma 10.3.2, if Γ is stable, then it must contain a (stable) subset Ω such that for some $n < \omega$ and all $\alpha < \beta < \gamma$ in Ω:

$$\chi(\alpha, \beta, \gamma) = n, \tag{10.3.25}$$

$$(C_{\beta_s} \cap C_{\gamma_t}) \setminus \alpha \neq \emptyset \text{ where } s = \rho_0(\alpha, \beta) \upharpoonright n \text{ and } t = \rho_0(\alpha, \gamma) \upharpoonright n. \tag{10.3.26}$$

In other words, Case 1 of the definition of $f(\alpha, \beta, \gamma)$ applies to any triple $\alpha < \beta < \gamma$ from Ω. Note that in this case, the definition follows exactly the definition of the analogous function in the proof of Theorem 10.2.1 with ρ^+ in the role of the strongly unbounded subadditive function $e : [\theta^{++}]^2 \longrightarrow \theta^+$. By Lemma 10.3.2, ρ^+ is strongly unbounded on subsets of Ω, so by the proof of Theorem 10.2.1 we conclude that in this case the image $f''[\Omega]^3$ contains almost all ordinals $< \theta^+$ of cofinality θ.

Suppose now that no subset of Γ of size θ^+ is stable. By Lemma 10.3.4 (in fact, its proof) for every integer $n \geq 1$ there exist $\alpha < \beta < \gamma$ in Γ such that for $s = \rho_0(\alpha, \beta) \upharpoonright \chi(\alpha, \beta, \gamma)$ and $t = \rho_0(\alpha, \gamma) \upharpoonright \chi(\alpha, \beta, \gamma)$, where $\rho_0(\alpha, \beta)$ has length $> \chi(\alpha, \beta, \gamma)$:

$$\mathrm{osc}(\alpha, \beta, \gamma) = n, \tag{10.3.27}$$

$$(C_{\beta_s} \cap C_{\gamma_t}) \setminus \alpha = \emptyset. \tag{10.3.28}$$

This finishes the proof. □

Corollary 10.3.7. *There is $f : [\omega_2]^3 \longrightarrow \omega$ which takes all the values on the cube of any uncountable subset of ω_2.* □

Remark 10.3.8. Note that the dimension 3 in this corollary cannot be lowered to 2 as long as one does not use some additional axioms to construct such f. Note also that the range ω cannot be replaced by a set of bigger size, as this would contradict Chang's conjecture. We have seen above that Chang's conjecture is equivalent to the statement that for every $f : [\omega_2]^3 \longrightarrow \omega_1$ there is an uncountable set $\Gamma \subseteq \omega_2$ such that $f''[\Gamma]^3 \neq \omega_1$. Is there a similar reformulation of the Continuum Hypothesis? More precisely, one can ask the following question.

Question 10.3.9. Is CH equivalent to the statement that for every $f : [\omega_2]^2 \longrightarrow \omega$ there exists an uncountable $\Gamma \subseteq \omega_2$ with $f''[\Gamma]^2 \neq \omega$?

10.4 Two-cardinal walks

The basic notion of minimal walk also makes sense in the context of structures $\mathcal{P}_\kappa(\lambda) = \{x \subseteq \lambda : |x| < \kappa\}$ and the purpose of this section is to expose some of this theory. The two-cardinal walk is based on a two-cardinal version of a C-sequence, a sequence of the form

$$C_x \quad (x \in \mathcal{P}_\kappa(\lambda)),$$

where for each x in $\mathcal{P}_\kappa(\lambda)$, the sequence $C_{x\cap\alpha}$ $(\alpha \in x \cup \{\sup(x)\})$[2] is an ordinary C-sequence modulo taking the transitive collapse of x. In other words, if θ_x is the order-type of x and if $\pi_x : x \cup \{\sup(x)\} \to \theta_x + 1$ is the order-isomorphism, then

$$C_\gamma = \pi_x'' C_{x\cap\pi^{-1}(\gamma)} \quad (\gamma \leq \theta_x)$$

is an ordinary C-sequence on $\theta_x + 1$. Given $x \in \mathcal{P}_\kappa(\lambda)$ and $\alpha \in x$, one defines *the minimal walk* from x to α to be the sequence $\beta_0 = \sup(x) > \beta_1 > \cdots > \beta_n = \alpha$, where

$$\beta_{i+1} = \min(C_{x\cap\beta_i} \setminus \alpha)$$

for $i < n$. The integer n is *the length* of the walk and is denoted by $\rho_2(\alpha, x)$, while *the upper trace* of the walk from x to $\alpha \in x$ is the finite set

$$\mathrm{Tr}(\alpha, x) = \{\beta_0, \beta_1, \ldots, \beta_n\}.$$

The sequence

$$\rho_0(\alpha, x) = \langle |C_{x\cap\beta_i}| : i < n\rangle$$

is the *the full code* of the walk from x to α. Note that for every $i \leq n$, the walk from x to β_i is the segment $\beta_0 = \sup(x) > \beta_1 > \cdots > \beta_i$ of the walk from x to α. In other words, $\mathrm{Tr}(\beta_i, x) = \{\beta_0, \ldots, \beta_i\}$ holds for every $i \leq n$. So if, in general, for $\delta \in y \in \mathcal{P}_\kappa(\lambda)$, we put

$$\lambda(\delta, y) = \max(\bigcup_{\gamma \in \mathrm{Tr}(\delta,y)\setminus\{\delta\}} C_{y\cap\gamma} \cap \delta),$$

we get the following expression for *the lower trace* of the walk from x to α,

$$\Lambda(\alpha, x) = \{\lambda(\beta_i, x) : i < n\}.$$

The theory of minimal walks in the context of $\mathcal{P}_\kappa(\lambda)$ comparable to the theory of minimal walks on ordinals is yet to be developed, but we do have some important analogues such as, for example, the *square-bracket operation*.

Definition 10.4.1. Consider $x, y \in \mathcal{P}_\kappa(\lambda)$. If $x \cup \{\sup(x)\} \not\subseteq y$ put $[xy] = 0$. If $x \cup \{\sup(x)\} \subseteq y$, let

$$\beta_0 = \sup(y) > \beta_1 > \cdots > \beta_n = \sup(x)$$

---

[2]Here, $\sup(x) = \min\{\xi \in \lambda : \alpha < \xi \text{ for all } \alpha \in x\}$. Hence, $C_{x\cap\alpha} = C_x$ for $\alpha = \sup(x)$.

be the minimal walk from y to $\sup(x)$. Let $m \leq n$ be maximal with the property that there is $\alpha \in x$ such that $\rho_0(\alpha, x) = \rho_0(\beta_m, y)$ and let $\alpha_0 = \sup(x) > \alpha_1 > \cdots > \alpha_m = \alpha$ be the minimal walk from x to α. Finally, let $[xy] = \beta_j$ where $j < m$ is determined by the following two properties:

(1) $\max(C_{x\cap\alpha_i} \cap \alpha) \leq \max(C_{y\cap\beta_i} \cap \sup(x))$ for all $i < j$.

(2) $\max(C_{x\cap\alpha_j} \cap \alpha) > \max(C_{y\cap\beta_j} \cap \sup(x))$.

If such a $j < m$ cannot be found, set $[xy] = 0$.

Recall that analysis of the square-bracket operation in the realm of walks on ordinals requires that the C-sequence avoids certain stationary sets. In particular, ordinary square-bracket operations work well on successors of regular cardinals. In the two-cardinal context this corresponds to considering $\mathcal{P}_{\kappa^+}(\lambda)$, which we denote more simply as $[\lambda]^\kappa$, assuming that κ is a regular cardinal, and choosing the C-sequence C_x $(x \in [\lambda]^\kappa)$ in the following way. One first chooses a C-sequence C_α $(\alpha < \kappa^+)$ which avoids the set $\{\delta < \kappa^+ : \mathrm{cf}(\delta) = \kappa\}$, or in other words, $C_{\alpha+1} = \{\alpha\}$ and for limit α the set C_α is a closed and unbounded in α, has order type $\mathrm{cf}(\alpha)$ and it contains no ordinals of cofinality κ. Given $x \in [\lambda]^\kappa$ one puts

$$C_x = \pi_x^{-1} C_\alpha,$$

where $\alpha = \mathrm{tp}(x)$ and where $\pi_x : x \longrightarrow \alpha$ is the unique order isomorphism.

Lemma 10.4.2. *If $\lambda > \kappa$ are regular cardinals, then for every cofinal subset U of $[\lambda]^\kappa$, the set of all ordinals $\delta < \lambda$ of cofinality κ which are not of the form $[xy]$ for some $x \subset y$ in U is not stationary in λ.*

Proof. Let S be a given stationary subset of $\{\delta < \lambda : \mathrm{cf}(\delta) = \kappa\}$. We need to find $x \subset y$ in U such that $[xy] \in S$. Choose large enough regular cardinal θ and continuous $\in$-chain $\mathcal{M}$ of elementary submodels M of $(H_\theta, \in)$ of size $< \lambda$ such that $M \cap \lambda \in \lambda$ and such that M contains S and the C-sequence C_x $(x \in [\lambda]^\kappa)$. Choose now an elementary submodel N of $(H_\theta, \in)$ of size κ such that $N \cap \kappa^+ \in \kappa^+$, $\delta = \sup(N \cap \lambda) \in S$, and such that N contains S, the C-sequence C_x $(x \in [\lambda]^\kappa)$, and the chain $\mathcal{M}$. Since U is cofinal in $[\lambda]^\kappa$, we can find $y \in U$ such that $(N \cap \lambda) \cup \{\delta\} \subseteq y$. Let

$$\beta_0 = \sup(y) > \beta_1 > \cdots > \beta_j = \delta$$

be the minimal walk from y to δ. Note that by our choice of the C-sequence C_x $(x \in [\lambda]^\kappa)$, the set $C_{y\cap\beta_i} \cap \delta$ is bounded in δ for all $i < j$. Since $\delta = \sup(N \cap \lambda)$, for each $i < j$, the minimum

$$\gamma_i = \min((N \cap \lambda) \setminus \max(C_{y\cap\beta_i} \cap \delta))$$

exists and is smaller than δ. Let

$$I = \{i < j : \gamma_i = \max(C_{y\cap\beta_i} \cap \delta)\}.$$

For $M \in \mathcal{M}$, let $\delta_M = M \cap \lambda$. Let

$$D = \{\delta_M : M \in \mathcal{M}\}.$$

Then D is a closed and unbounded subset of λ. Let D' be the set of all limit points of D of cofinality at most κ. Then $D' \in N$ and the intersection $D' \cap M$ has an indecomposable order-type bigger than κ. So in particular we can find two models $M_0, M_1 \in \mathcal{M} \cap N$ such that δ_{M_0} and δ_{M_1} belong to the set D' but $\delta_{M_0} \notin C_{y\cap\delta}$, $\delta_{M_1} \notin C_{y\cap\delta}$, and moreover,

$$\{\gamma_i : i < j\} \subseteq M_0 \in M_1, \text{ and } C_{y\cap\delta} \cap [\delta_{M_0}, \delta_{M_1}) \neq \emptyset. \tag{10.4.1}$$

Let $N_0 = N \cap M_0$ and $N_1 = N \cap M_1$. Then N_0 and N_1 are also elementary submodels of $(H_\theta, \in)$ and they have the property that

$$\delta_{M_0} = \sup(N_0 \cap \lambda) \text{ and } \delta_{M_1} = \sup(N_1 \cap \lambda).$$

Since $C_{y\cap\delta}$ is a closed subset of $y\cap\delta$ modulo taking the transitive collapse, it follows that $\max(C_{y\cap\delta} \cap \delta_{M_0})$ and $\max(C_{y\cap\delta} \cap \delta_{M_1})$ exist and that they are strictly smaller than δ_{M_0} and δ_{M_1}, respectively. By the cofinality requirement on δ_{M_0} and δ_{M_1}, the sets $N_0 \cap \lambda$ and $N_1 \cap \lambda$ are cofinal in δ_{M_0} and δ_{M_1}, respectively. So we can define ε_0 to be the minimal point of $N_0 \cap \lambda$ that is greater than or equal to $\max(C_{y\cap\delta} \cap \delta_{M_0})$. Similarly, we let ε_1 be the minimal point of $N_1 \cap \lambda$ that is greater than or equal to $\max(C_{y\cap\delta} \cap \delta_{M_1})$. Let

$$\xi_0 = \mathrm{tp}(C_{y\cap\delta} \cap \delta_{M_0}) \text{ and } \xi_1 = \mathrm{tp}(C_{y\cap\delta} \cap \delta_{M_1}).$$

Let $t = \rho_0(\delta, y)$. Note that $t \in N_0$. For $\eta < \lambda$, let U_η be the set of all $x \in U$ for which there is a minimal walk $\alpha_0 = \sup(x) > \alpha_1 > \cdots > \alpha_j$ inside x with the following properties:

(i) $\rho_0(\alpha, x) = t$,
(ii) $\max(C_{x\cap\alpha_i} \cap \alpha) = \gamma_i$ for $i \in I$,
(iii) $\max(C_{x\cap\alpha_i} \cap \alpha) < \gamma_i$ for $i < j$, $i \notin I$,
(iv) $\mathrm{tp}(C_{x\cap\alpha}) = \kappa$ and $\mathrm{tp}(C_{x\cap\alpha} \cap \varepsilon_0) = \xi_0$,
(v) $C_{x\cap\alpha} \cap [\varepsilon, \eta) = \emptyset$.

Note that for every $\eta \in N_0 \cap \lambda$, the set U_η belongs to N_0. Note also that by our choices of the objects appearing in the definition, the set y belongs to U_η for all $\eta \in N_0 \cap \lambda$. Since $y \supseteq N_0 \cap \lambda$, by the elementarity of N_0, we conclude that U_η must be cofinal in $[\lambda]^\kappa$ for all $\eta < \lambda$. Applying this in the model M_1 to the set U_η for $\eta = \varepsilon_1 + 1$, we find $x \in U_{\varepsilon_1+1} \cap M_1$ such that $\{\gamma_i : i < j\} \cup \{\varepsilon_0, \varepsilon_1\} \subseteq x$. Let $\alpha_0 = \sup(x) > \alpha_1 > \cdots > \alpha_j$ be the minimal walk with properties (i), (ii), (iii),(iv), and (v) above witnessing the membership of x in U_{ε_1+1}. Let $\alpha \in C_{x\cap\alpha_j}$ be such that $\mathrm{tp}(C_{x\cap\alpha_j} \cap \alpha) = \xi_1$. Note that ξ_1 is a successor ordinal so $C_{x\cap\alpha_j} \cap \alpha$ is bounded in $x \cap \alpha$, so the maximum $\max(C_{x\cap\alpha_j} \cap \alpha)$ exists, and since $\sup(x) < \delta_{M_1}$, we have that

$$\max(C_{x\cap\alpha_j} \cap \alpha) > \varepsilon_1 \geq \max(C_{y\cap\beta_j} \cap \sup(x)).$$

By (ii) and the definition of the set I, we conclude that

$$\max(C_{x\cap\alpha_i} \cap \alpha) = \gamma_i = \max(C_{y\cap\beta_i} \cap \sup(x)) \text{ for all } i \in I.$$

By (iii), for $i < j$ with $i \notin I$, we must have

$$\max(C_{x\cap\alpha_i} \cap \alpha) < \max(C_{y\cap\beta_i} \cap \sup(x)),$$

since otherwise, $\max(C_{x\cap\alpha_i} \cap \alpha) \in x \subseteq M_1 \subseteq N$ would be a point of $N \cap \lambda$ that is strictly smaller than γ_i and which still dominates $\max(C_{y\cap\beta_i} \cap \delta)$ contradicting the choice of γ_i. Combining all this we see that the pair $x \cup \{\sup(x)\} \subseteq y$ satisfies the requirements of Definition 10.4.1 which permits us to conclude that $[xy] = \delta \in S$. This finishes the proof. □

In order to treat the case of singular cardinals λ, we need the following relativization of the square-bracket operation.

Definition 10.4.3. Let $\kappa \leq \theta \leq \lambda$. Consider an arbitrary pair $x, y \in \mathcal{P}_\kappa(\lambda)$. If $x \cup \{\sup(x \cap \theta)\} \nsubseteq y$, put $[xy]_\theta = 0$. If $x \cup \{\sup(x \cap \theta)\} \subseteq y$, let

$$[xy]_\theta = [(x \cap \theta)(y \cap \theta)]$$

as defined in $\mathcal{P}_\kappa(\theta)$ according to the Definition 10.4.1. In other words, let

$$\beta_0 = \sup(y \cap \theta) > \beta_1 > \cdots > \beta_n = \sup(x \cap \theta)$$

be the minimal walk from $y \cap \theta$ to $\sup(x \cap \theta)$. Let $m \leq n$ be maximal with the property that there is $\alpha \in x \cap \theta$ such that $\rho_0(\alpha, x \cap \theta) = \rho_0(\beta_m, y \cap \theta)$ and let $\alpha_0 = \sup(x \cap \theta) > \alpha_1 > \cdots > \alpha_m = \alpha$ be the minimal walk from $x \cap \theta$ to α. Finally, let $[xy]_\theta = \beta_j$ where $j < m$ is determined by the following two properties:

(1) $\max(C_{x\cap\alpha_i} \cap \alpha) \leq \max(C_{y\cap\beta_i} \cap \sup(x))$ for all $i < j$.
(2) $\max(C_{x\cap\alpha_j} \cap \alpha) > \max(C_{y\cap\beta_j} \cap \sup(x))$.

If such a $j < m$ cannot be found, set $[xy]_\theta = 0$.

Then the proof of Lemma 10.4.2 adjusts easily to the proof of the following slightly more general fact.

Lemma 10.4.4. *Suppose κ is a regular uncountable cardinal and that $\lambda \geq \kappa$ is an arbitrary cardinal. Let θ_0 and θ_1 be two distinct regular uncountable cardinals such that $\kappa \leq \theta_0, \theta_1 \leq \lambda$. Let S_0 and S_1 be stationary subsets of θ_0 and θ_1, respectively, consisting only of cofinality κ ordinals. Then for every cofinal subset U of $[\lambda]^\kappa$ there exist $x \subset y$ in U such that $[xy]_{\theta_0} \in S_0$ and $[xy]_{\theta_1} \in S_1$.* □

Taking an appropriate projection of the square-bracket operation one gets the following result.

Theorem 10.4.5. *For every pair of infinite cardinals $\kappa < \lambda$ with κ regular, there is a mapping $c : [[\lambda]^\kappa]^2 \to \lambda$ such that for every cofinal subset U of $[\lambda]^\kappa$ and every $\xi \in \lambda$ there exist $x \subset y$ in U such that $c(x, y) = \xi$.*

Proof. If λ is a regular cardinal, then according to Lemma 10.4.2, composing the square-bracket operation of $[\lambda]^\kappa$ with a partition of $\{\delta < \lambda : \mathrm{cf}(\delta) = \kappa\}$ into λ disjoint stationary sets, gives us the conclusion. So let us assume that λ is a singular cardinal. Let $\theta = \max(\mathrm{cf}(\lambda), \kappa^+)$ and choose a strictly increasing sequence θ_ξ $(\xi < \mathrm{cf}(\lambda))$ of regular cardinals bigger than θ whose supremum is equal to λ. Let S_ξ $(\xi < \mathrm{cf}(\lambda))$ be a partition of $\{\delta < \theta : \mathrm{cf}(\delta) = \kappa\}$ into $\mathrm{cf}(\lambda)$ disjoint stationary sets. For each $\xi < \mathrm{cf}(\lambda)$, let T_ξ^η $(\eta < \theta_\xi)$ be a partition of $\{\delta < \theta_\xi : \mathrm{cf}(\delta) = \kappa\}$ into θ_ξ disjoint stationary sets. Finally, for $x, y \in [\lambda]^\kappa$, set

$$c(x, y) = \langle \xi, \eta \rangle \text{ if } [xy]_\theta \in S_\xi \text{ and } [xy]_{\theta_\xi} \in T_\xi^\eta,$$

and if these conditions are not satisfied for any pair from the set $P = \bigcup_{\xi<\mathrm{cf}(\lambda)} \{\xi\} \times \theta_\xi$, let $c(x, y)$ be an arbitrary member of P. Then by Lemma 10.4.4, for every cofinal $U \subseteq [\lambda]^\kappa$ and every $\langle \xi, \eta \rangle \in P$ there exist $x \subset y$ in P such that $[xy]_\theta \in S_\xi$ and $[xy]_{\theta_\xi} \in T_\xi^\eta$. Since P has cardinality λ, this finishes the proof. $\square$

Remark 10.4.6. The first attempt to generalize the square-bracket operation into the two-cardinal context was made by Velleman [129] who proved the conclusion of Theorem 10.4.5 for $\kappa = \omega$ and for cardinals λ for which $[\lambda]^\omega$ admits a stationary subset of cardinality λ. The full conclusion was obtained by the author in [113]. In certain cases when the cardinal λ is singular, one can increase the number of colors in Theorem 10.4.5. The following result of Shioya [99] gives one condition when this is possible.

Theorem 10.4.7. *Suppose κ is a regular cardinal and $\lambda > \kappa$ is such that $\kappa^{\mathrm{cf}(\lambda)} = \kappa$. Then there is a mapping $c : [[\lambda]^\kappa]^2 \to \lambda^+$ such that for every cofinal subset U of $[\lambda]^\kappa$ and every $\xi \in \lambda^+$, there exist $x \subset y$ in U such that $c(x, y) = \xi$.* $\square$

Further advances in this area are likely to involve walks that step from one element y of $\mathcal{P}_\kappa(\lambda)$ to one of its subsets x that is not necessarily its initial segment. This in turn is likely to involve two-cardinal analogues of square-sequences that would facilitate these kind of walks. The case $\kappa = \omega_1$ is of course the first to be understood. So let us present something in that direction, a result which states that the quantifier reduction analogous to the one proved before in the case $\kappa = \lambda = \omega_1$ is possible for many other cardinals λ.

Theorem 10.4.8. *Suppose S is a subset of $[\lambda]^\omega$ that is equinumerous with a locally countable subset of $[\lambda]^\omega$. Then there is a square-bracket operation*

$$[\cdot\cdot]_S : [[\lambda]^\omega]^2 \to [\lambda]^\omega$$

such that for every cofinal subset U of $[\lambda]^\omega$ the set of all $s \in S$ that are not of the form $[xy]_S$ for some $x \subset y$ in U is not stationary in $[\lambda]^\omega$. □

Recall that a *locally countable* subset of $[\lambda]^\omega$ is an arbitrary subset K of $[\lambda]^\omega$ with the property that $\{x \in K : x \subseteq y\}$ is countable for every $y \in [\lambda]^\omega$. Clearly every family of pairwise-disjoint sets is locally countable, so there is always a locally countable subset of $[\lambda]^\omega$ of cardinality λ. The interest of Theorem 10.4.8 lies in the fact that in many cases there exist locally countable subsets of $[\lambda]^\omega$ that are equinumerous with some *stationary* subsets of $[\lambda]^\omega$. In order to define the square-bracket operation of Theorem 10.4.8, we need the following simple fact.

Lemma 10.4.9. *There is $r : [\lambda]^\omega \to \{0,1\}^\omega$ such that $r_x \neq r_y$ for $x \subset y$.*

Proof. Choose a one-to-one sequence p_ξ $(\xi < \omega_1)$ of elements of $\{0,1\}^\omega$ and choose for each ordinal $\alpha < \lambda$ of cofinality ω a strictly increasing sequence (α_i) of smaller ordinals cofinal in α. It will be convenient for us to extend the domain of r to include all finite subsets of λ as well and to have range of r included in $\{0,1\}^\omega \times \{0,1\}^\omega$ rather than in $\{0,1\}^\omega$. Let $r_x = \langle p_{\mathrm{tp}(x)}, q_x \rangle$, where q_x is defined recursively on $\sup(x)$ as follows. If x has a maximal element α, set $q_x(0) = 1$ and $q_x(i+1) = q_y(i)$, where $y = x \cap \alpha$. If $\alpha = \sup(x)$ is a limit ordinal, let $q_x(0) = 0$ and $q_x(2^i(2j+1)) = r_{x_i}$, where $x_i = x \cap \alpha_1$. Then $x \mapsto r_x$ is the required mapping. □

From now on, we fix the mapping r satisfying Lemma 10.4.9. This allows us to define $\Delta : [[\lambda]^\omega]^2 \to \omega$, by

$$\Delta(x,y) = \min\{n : r_x(n) \neq r_y(n)\}.$$

We also need to fix, for every countable subset X of S, a one-to-one mapping $e_X : X \to \omega$. We start the description of the square-bracket operation satisfying Theorem 10.4.8 by fixing a one-to-one mapping $\varphi : S \to [\lambda]^\omega$ whose range is a locally countable subset of $[\lambda]^\omega$. For $y \in [\lambda]^\omega$, let

$$S_y = \{x \in S : x \cup \{\varphi(x)\} \subseteq y\}.$$

Then S_y is a countable subset of S and for it we have already fixed a one-to-one mapping $e_y = e_{S_y} : S_y \to \omega$. For an integer n and y in $[\lambda]^\omega$, set

$$S_y(n) = \{x \in S_y : e_y(x) \leq n\}.$$

Finally, for $x \subset y$ in $[\lambda]^\omega$, set

$$[xy]_S = \min\{z \in S_y(\Delta(x,y)) : z \supseteq x\},$$

if the $\subseteq$-minimal element of $S_y(\Delta(x,y))$ containing x exists; if such a minimal element does not exist, set $[xy]_S = x$.

Consider a stationary subset T of S and a cofinal subset U of $[\lambda]^\omega$. We need to find $x \subset y$ in U such that $[xy]_S$ belongs to T. Choose a sufficiently large regular cardinal θ and a countable elementary submodel M of $(H_\theta, \in)$ containing all these objects such that $z = M \cap \lambda \in T$. Choose $y \in U$ such that

$$z \cup \{\varphi(z)\} \subseteq y.$$

Then $z \in S_y$ so we can find an integer m such that $z \in S_y(m)$. For $\sigma \in \{0,1\}^{<\omega}$, set

$$U_\sigma = \{x \in U : r_x \supset \sigma\}.$$

Let $\tau = r_y \upharpoonright m$. Then $U_\tau \in M$ and $y \in U_\tau$, so by the elementarity of M, the set U_τ is cofinal in $[\lambda]^\omega$. Let

$$P = \{\sigma \in \{0,1\}^{<\omega} : \sigma \supseteq \tau \text{ and } U_\sigma \text{ is cofinal in } [\lambda]^\omega\}.$$

Applying the elementarity of M again, we conclude that $r_y \upharpoonright k \in P$ for all $k \geq |\tau|$, and so in particular, P is infinite. Note also that, by the basic property of the function $x \mapsto r_x$, an intersection of the form $\bigcap_{\sigma \in b} U_\sigma$, where b is an infinite chain of P, cannot be cofinal in $[\lambda]^\omega$ for the simple reason that it contains no two distinct elements comparable in the inclusion ordering. So in particular, P is not a chain, and therefore, we can find $\sigma \in P$ such that $r_y \not\supseteq \sigma$. Let

$$n = \min\{k : \sigma(k) \neq r_y(k)\}.$$

Then $n \geq m$ and $\Delta(x, y) = n$ for all $x \in U_\sigma$. Note that $z \in S_y(n)$. For every $w \in S_y(n)$ which does not include z, we fix $\xi_w \in z \setminus w$. Let

$$x_0 = \{\xi_w : w \in S_y(n), w \not\supseteq z\}.$$

Then x_0, being a finite subset of $z = M \cap \lambda$, is an element of M. Since U_σ is cofinal in $[\lambda]^\omega$ and since it belongs to M, by elementarity of M, we can find $x \in U_\sigma \cap M$ containing x_0. It follows that z is the $\subseteq$-minimal member of $S_y(n)$ which includes x, and therefore, $[xy] = z \in T$. This completes the proof of Theorem 10.4.8.

It turns out that the hypothesis on the set S appearing in Theorem 10.4.8, is also suitable for the following simple decomposition fact which could be used in finding an interesting projection of the corresponding square-bracket operation $[..]_S$.

Lemma 10.4.10. *Suppose S is a stationary subset of $[\lambda]^\omega$ and that K is a locally countable subset of $[\lambda]^\omega$. Then S splits into $|K|$ pairwise-disjoint stationary subsets.*

Proof. We may assume that K is infinite. In fact, we may assume that $|K|$ has an uncountable cofinality, since otherwise we would first split S into countably many disjoint stationary sets S_n and then split each S_n into sufficiently many disjoint stationary subsets so that combining all these sets we get the desired decomposition. Since K is locally countable, for every $x \in [\lambda]^\omega$, we can fix a one-to-one mapping

$$f_x : \{a \in K : a \subseteq x\} \longrightarrow \omega.$$

For n in ω and a in K, set

$$S_n^a = \{x \in S : a \subseteq x \text{ and } f_x(a) = n\}.$$

Then for each $a \in K$, the family S_n^a $(n < \omega)$ covers the set $\{x \in S : x \supseteq a\}$, so we can fix one integer $n(a)$ such that $S_{n(a)}^a$ is stationary. This gives us a countable decomposition $K_n = \{a \in K : n(a) = n\}$ $(n < \omega)$ of the family K. Note that, on the other hand, for each $n < \omega$,

$$S_n^a \cap S_n^b = \emptyset \text{ for } a \neq b \text{ in } K.$$

It follows that for each $n < \omega$, the family S_n^a $(a \in K_n)$ consists of pairwise-disjoint stationary subsets of S. So taking n such that $|K_n| = |K|$, we get the desired family of pairwise-disjoint stationary subsets of S. □

Corollary 10.4.11. *For every stationary subset S of $[\lambda]^\omega$ that is equinumerous with a locally countable subset of $[\lambda]^\omega$ there is a projection*

$$[\![..]\!]_S : [[\lambda]^\omega]^2 \to S$$

of the square-bracket operation $[..]_S$ with the property that for every cofinal $U \subseteq [\lambda]^\omega$ and every $z \in S$ there exist $x \subset y$ in U such that $[\![xy]\!]_S = z$. □

Bibliography

[1] U. Abraham and S. Shelah. Isomorphism types of Aronszajn trees. *Israel J. Math.*, 50(1-2):75–113, 1985.

[2] P.S. Alexandroff. Pages from an autobiography. *Russian Math. Surveys*, 34:267–302, 1979.

[3] P.S. Alexandroff and P. Urysohn. Mémoire sur les espaces topologique compacts. *Verh. Akad. Wetensch.*, 14:VIII+96, 1929.

[4] S.A. Argyros, J. Lopez-Abad, and S. Todorcevic. A class of Banach spaces with few non-strictly singular operators. *J. Funct. Anal.*, 222(2):306–384, 2005.

[5] R.M. Aron and P. Hajek. Odd degree polynomials on real Banach spaces. *Positivity*, 11(1):143–153, 2007.

[6] Antonio Aviles and Stevo Todorcevic. Zero subspaces of polynomials on $l_1(\gamma)$. *Journal of Mathematical Analysis and Applications*, 2007.

[7] Heinz Bachmann. *Transfinite Zahlen.* Zweite, neubearbeitete Auflage. Ergebnisse der Mathematik und ihrer Grenzgebiete, Band 1. Springer-Verlag, Berlin, 1967.

[8] Reinhold Baer. Groups with abelian central quotient group. *Trans. Amer. Math. Soc.*, 44(3):357–386, 1938.

[9] J. Bagaria. Fragments of Martin's axiom and Δ^1_3 sets of reals. *Ann. Pure Appl. Logic*, 69:1–25, 1994.

[10] James Baumgartner. All $\aleph_1$-dense sets of reals can be isomorphic. *Fund. Math.*, 79:101–106, 1973.

[11] James Baumgartner and Saharon Shelah. Remarks on superatomic Boolean algebras. *Ann. Pure Appl. Logic*, 33:109–129, 1987.

[12] James Baumgartner and Otmar Spinas. Independence and consistency proofs in quadratic form theory. *J. Symbolic Logic*, 56:1195–1211, 1991.

[13] M. Bekkali. *Topics in set theory*, volume 1476 of *Lect. notes in Math.* Springer-Verlag, 1992.

[14] Felix Bernstein. Zur Theorie der trigonometrischen Reihen. *Berichte Verh. Königl. Sächs. Gesellschaft*, 60:325–338, 1908.

[15] M. Burke and Menachem Magidor. Shelah's pcf theory and its applications. *Ann. Pure Appl. Logic*, 50:207–254, 1990.

[16] W.W. Comfort and S. Negrepontis. *Ultrafilters*. Springer-Verlag, 1974.

[17] W.J. Davis, T. Figiel, W.B. Johnson, and A. Pelczynski. Factoring weakly compact operators. *J. Functional Analysis*, 17:311–327, 1974.

[18] K.J. Devlin. *Constructibility*. Springer-Verlag, 1984.

[19] Seán Dineen. *Complex analysis in locally convex spaces*, volume 57 of *North-Holland Mathematics Studies*. North-Holland Publishing Co., Amsterdam, 1981. Notas de Matemática [Mathematical Notes], 83.

[20] H.-D. Donder, Ronald Jensen, and Koppelberg B. Some applications of the core model. In *Set theory & Model theory*, volume 872 of *Lect. notes in Math.*, pages 55–97. Springer-Verlag, 1979.

[21] H.-D. Donder and P. Koepke. On the consistency strength of 'accessible' Jonsson cardinals and of the weak Chang conjecture. *Ann. Pure Appl. Logic*, 25:233–261, 1983.

[22] H.-D. Donder and J.P. Levinski. Some principles related to Chang's conjecture. *Ann. Pure Appl. Logic*, 45:39–101, 1989.

[23] A. Dow. PFA and ω_1^*. *Topology and its Appl.*, 28:127–140, 1988.

[24] A. Dow and S. Watson. A subcategory of TOP. *Trans. Amer. Math. Soc.*, 337:825–837, 1993.

[25] B. Dushnik and E.W. Miller. Partially ordered sets. *Amer. J. Math.*, 63:600–610, 1941.

[26] H.-D. Ebbinghaus and J. Flum. Eine Maximalitätseigenschaft der Prädikatenlogik erster Stufe. *Math.-Phys. Semesterber.*, 21:182–202, 1974.

[27] A. Ehrenfeucht and V. Faber. Do infinite nilpotent groups always have equipotent abelian subgroups? *Nederl. Akad. Wetensch. Proc. Ser. A* **75**=*Indag. Math.*, 34:202–209, 1972.

[28] P. Erdős, S. Jackson, and R.D. Mauldin. On partitions of lines and spaces. *Fund. Math.*, 145:101–119, 1994.

[29] P. Erdős, S. Jackson, and R.D. Mauldin. On infinite partitions of lines and spaces. *Fund. Math.*, 152:75–95, 1997.

[30] P. Erdős and S. Kakutani. On non-denumerable graphs. *Bull. Amer. Math. Soc.*, 49:457–461, 1943.

[31] P. Erdős and A. Tarski. On families of mutually exclusive sets. *Ann. of Math.*, 44:315–329, 1943.

[32] P. Erdős and A. Tarski. On some problems involving inaccessible cardinals. In Y. Bar-Hillel, E. Poznanski, M. Rabin, and A. Robinson, editors, *Essays on the Foundations of Mathematics*, pages 50–82. Magnes Press, 1961.

[33] Paul Erdős, András Hajnal, Attila Máté, and Richard Rado. *Combinatorial set theory: partition relations for cardinals*, volume 106 of *Studies in*

Logic and the Foundations of Mathematics. North-Holland Publishing Co., Amsterdam, 1984.

[34] V. Ferenczi. Operators on subspaces of hereditarily indecomposable Banach spaces. *Bull. London Math. Soc.*, 29(3):338–344, 1997.

[35] Matthew Foreman and Menachem Magidor. A very weak square principle. *J. Symbolic Logic*, 62:175–196, 1997.

[36] D. H. Fremlin. *Consequences of Martin's axiom*, volume 84 of *Cambridge Tracts in Mathematics*. Cambridge University Press, Cambridge, 1984.

[37] D.H. Fremlin. The partially ordered sets of measure theory and Tukey's ordering. *Note di Matematica*, XI:177–214, 1991.

[38] Harvey Friedman. On the necessary use of abstract set theory. *Adv. in Math.*, 41(3):209–280, 1981.

[39] W.T. Gowers and B. Maurey. The unconditional basic sequence problem. *J. Amer. Math. Soc.*, 6(4):851–874, 1993.

[40] Ronald L. Graham, Bruce L. Rothschild, and Joel H. Spencer. *Ramsey theory*. John Wiley & Sons Inc., New York, 1980. Wiley-Interscience Series in Discrete Mathematics, A Wiley-Interscience Publication.

[41] G. Gruenhage. Irreducible restrictions of closed mappings. *Topology and its Appl.*, 85:127–135, 1998.

[42] G. Gruenhage and Y. Tanaka. Products of k-spaces. *Trans. Amer. Math. Soc.*, 273:299–308, 1982.

[43] A. Hajnal and P. Komjáth. Some higher-gap examples in combinatorial set theory. *Ann. Pure Appl. Logic*, 33(3):283–296, 1987.

[44] Andras Hajnal, Akihiro Kanamori, and Saharon Shelah. Regressive partition relations for infinite cardinals. *Trans. Amer. Math. Soc.*, 299:145–154, 1987.

[45] William Hanf. On a problem of Erdős and Tarski. *Fund. Math.*, 53:325–334, 1964.

[46] G.H. Hardy and E.M.Wright. *An introduction to the theory of numbers*. Clarendon Press, Oxford, 1979.

[47] Leo Harrington and Saharon Shelah. Some exact equiconsistency results in set theory. *Notre Dame J. Formal Logic*, 26:178–187, 1985.

[48] W. Hodges and Saharon Shelah. Infinite games and reduced products. *Annals of Mathematical Logic*, 20:77–108, 1981.

[49] John R. Isbell. Seven cofinal types. *J. London Math. Soc.* (2), 4:651–654, 1972.

[50] J.R. Isbell. The category of cofinal types. II. *Trans. Amer. Math. Soc.*, 116:394–416, 1965.

[51] Thomas Jech. *Set theory*. Springer Monographs in Mathematics. Springer-Verlag, Berlin, 2003. The third millennium edition, revised and expanded.

[52] Ronald Jensen. The fine structure of the constructible hierarchy. *Annals of Mathematical Logic*, 4:229–308, 1972.

[53] Ronald Jensen and Karl Schlechta. Results on the generic Kurepa hypothesis. *Archive for Math. Logic*, 30:13–27, 1990.

[54] Akihiro Kanamori. Regressive partitions and Borel diagonalization. *J. Symbolic Logic*, 54(2):540–552, 1989.

[55] Akihiro Kanamori. *The Higher Infinite.* Perspectives in Math. Logic. Springer-Verlag, 1997.

[56] Akihiro Kanamori and Kenneth McAloon. On Gödel incompleteness and finite combinatorics. *Ann. Pure Appl. Logic*, 33(1):23–41, 1987.

[57] H. Jerome Keisler. Logic with the quantifier "there exist uncountably many". *Ann. Math. Logic*, 1:1–93, 1970.

[58] Jussi Ketonen. Banach spaces and large cardinals. *Fund. Math.*, 81:291–303, 1974. Collection of articles dedicated to Andrzej Mostowski on the occasion of his sixtieth birthday, IV.

[59] Piotr Koszmider. On coherent families of finite-to-one functions. *J. Symbolic Logic*, 58:128–138, 1993.

[60] Piotr Koszmider. Forcing minimal extensions of Boolean algebras. *Trans. Amer. Math. Soc.*, 351:3073–3117, 1999.

[61] Piotr Koszmider. On Banach spaces of large density but few operators. preprint, 2000.

[62] Jean-Louis Krivine. *Introduction to axiomatic set theory.* Translated from the French by David Miller. D. Reidel Publishing Co., Dordrecht, 1971.

[63] Kenneth Kunen. Saturated ideals. *J. Symbolic Logic*, 43:65–76, 1978.

[64] Kenneth Kunen. *Set theory*, volume 102 of *Studies in Logic and the Foundations of Mathematics*. North-Holland Publishing Co., Amsterdam, 1980. An introduction to independence proofs.

[65] G. Kurepa. Ensembles ordonnés et ramifiés. *Publ. Math. Univ. Belgrade*, 4:1–138, 1935.

[66] Georges Kurepa. À propos d'une généralisation de la notion d'ensembles bien ordonnés. *Acta Math.*, 75:139–150, 1943.

[67] Paul Larson. The nonstationary ideal in the $\mathbb{P}_{\max}$-extension. preprint, 2005.

[68] Richard Laver. Better-quasi-orderings and a class of trees. In G.C. Rota, editor, *Studies in Foundations and Combinatorics*, volume 1 of *Advances in Math. Suppl. Stud.*, pages 31–48. Academic Press, 1978.

[69] Richard Laver and Saharon Shelah. The $\aleph_2$-Souslin hypothesis. *Trans. Amer. Math. Soc.*, 264:411–417, 1981.

[70] J.-P. Levinski, Menachem Magidor, and Saharon Shelah. Chang's conjecture for $\aleph_\omega$. *Israel J. Math.*, 69:161–172, 1990.

[71] Jordi Lopez-Abad and Stevo Todorcevic. A c_0-saturated banach space with no long unconditional basic sequences. preprint, June 2006.

[72] V. I. Malykhin. Nonpreservation of the properties of topological groups when they are squared. *Sibirsk. Mat. Zh.*, 28(4):154–161, 1987.

[73] V.I. Malykhin. On Ramsey spaces. *Sov. Math. Dokl.*, 20:894–898, 1979.

[74] B. Maurey and H. P. Rosenthal. Normalized weakly null sequence with no unconditional subsequence. *Studia Math.*, 61(1):77–98, 1977.

[75] William Mitchell. Aronszajn trees and the independence of the transfer property. *Annals of Mathematical Logic*, 5:21–46, 1972.

[76] J. Donald Monk. *Cardinal invariants on Boolean algebras.* Birkhäuser Verlag, Basel, 1996.

[77] Justin T. Moore. A five element basis for the uncountable linear orders. *Annals of Math.*, 163:669–688, 2006.

[78] Justin T. Moore. A solution to the l-space problem. *Journal of the Amer Math. Soc.*, 19:717–736, 2006.

[79] Charles Morgan. Morasses, square and forcing axioms. *Ann. Pure Appl. Logic*, 80(2):139–163, 1996.

[80] J. Nešetril and V. Rődl. Ramsey topological spaces. In J. Novak, editor, *General toplogy and its relation to modern algebra*, volume IV Part B, pages 333–337. Society of Czechoslovak Mathematicians and Physicists, Prague, 1977.

[81] T. Ohkuma. Comparability between ramified sets. *Proc. Japan Acad. Sci.*, 36:383–388, 1960.

[82] Martin Otto. EM constructions for a class of generalized quantifiers. *Archive for Math. Logic*, 31:355–371, 1992.

[83] Anatolij Plichko and Andriy Zagorodnyuk. On automatic continuity and three problems of *the scottish book* concerning the boundedness of polynomial functionals. *J. Math. Anal. Appl.*, 220(2):477–494, 1998.

[84] Mariusz Rabus. An ω_2-minimal Boolean algebra. *Trans. Amer. Math. Soc.*, 348:3235–3244, 1996.

[85] J. Roitman. A thin atomic Boolean algebra. *Algebra Universalis*, 21:137–142, 1985.

[86] M.E. Rudin. *Lectures in set-theoretic topology.* Amer. Math. Soc. Providence R.I., 1975.

[87] Marion Scheepers. Concerning n-tactics in the countable-finite game. *J. Symbolic Logic*, 56:786–794, 1991.

[88] E. Schimmerling. Combinatorial set theory and inner models. In C.A. DiPrisco et al., editors, *Set Theory: Techniques and Applications*, pages 207–212. Kluwer Academic Publishers, 1998.

[89] Thomas Schlumprecht. An arbitrarily distortable Banach space. *Israel J. Math.*, 76(1-2):81–95, 1991.

[90] J. Schmerl. Transfer theorems and their applications to logics. In J. Barwise et al., editors, *Model-theoretic logics*, pages 177–209. Springer-Verlag, 1985.

[91] James Schmerl. A partition property characterizing cardinals hyperinaccessible of finite type. *Trans. Amer. Math. Soc.*, 188:281–291, 1974.

[92] S. Shelah. Strong negative partition relations below the continuum. *Acta Math. Hungar.*, 58(1-2):95–100, 1991.

[93] Saharon Shelah. Strong negative partition above the continuum. *J. Symbolic Logic*, 55:21–31, 1990.

[94] Saharon Shelah. *Cardinal arithmetic*, volume 29 of *Oxford Logic Guides*. The Clarendon Press, Oxford University Press, New York, 1994. Oxford Science Publications.

[95] Saharon Shelah. Colouring and non-productivity of $\aleph_2$-c.c. *Ann. Pure Appl. Logic*, 84(2):153–174, 1997.

[96] Saharon Shelah and Otmar Spinas. Gross spaces. *Trans. Amer. Math. Soc.*, 348:4257–4277, 1996.

[97] Saharon Shelah and Juris Steprans. A Banach space on which there are few operators. *Proc. Amer. Math. Soc.*, 104:101–105, 1986.

[98] Saharon Shelah and Juris Steprans. Extraspecial p-groups. *Ann. Pure Appl. Logic*, 34:87–97, 1987.

[99] Masahiro Shioya. Partitioning pairs of uncountable sets. In *Logic Colloquium 2000*, volume 19 of *Lect. Notes Log.*, pages 350–364. Assoc. Symbol. Logic, Urbana, IL, 2005.

[100] I. Singer. *Bases in Banach Spaces, II.* Springer-Verlag, 1981.

[101] E. Specker. Sur un problème de Sikorski. *Colloquium Mathematicum*, 2:9–12, 1949.

[102] D.A. Talayco. Application of Cohomology to Set Theory I. *Ann. Pure Appl. Logic*, 71:69–107, 1995.

[103] D.A. Talayco. Application of Cohomology to Set Theory II. *Ann. Pure Appl. Logic*, 77:279–299, 1996.

[104] S. Todorcevic and B. Velickovic. Martin's axiom and partitions. *Compositio Math.*, 63(3):391–408, 1987.

[105] Stevo Todorcevic. Trees and linearly ordered sets. In K. Kunen and J.E. Vaughan, editors, *Handbook of Set-theoretic topology*, pages 235–293. North-Holland, 1984.

[106] Stevo Todorcevic. Aronszajn trees and partitions. *Israel J. Math.*, 52:53–58, 1985.

[107] Stevo Todorcevic. Remarks on chain conditions in products. *Compositio Math.*, 55:295–302, 1985.

[108] Stevo Todorcevic. Remarks on cellularity in products. *Compositio Math.*, 57:357–372, 1986.

[109] Stevo Todorcevic. Partitioning pairs of countable ordinals. *Acta Math.*, 159:261–294, 1987.

[110] Stevo Todorcevic. Oscillations of real numbers. In *Logic colloquium '86*, pages 325–331. North-Holland, Amsterdam, 1988.

[111] Stevo Todorcevic. *Partition problems in topology.* Amer. Math. Soc. Providence R.I., 1989.

[112] Stevo Todorcevic. Special square sequences. *Proc. Amer. Math. Soc.*, 105:199–205, 1989.

[113] Stevo Todorcevic. Partitioning pairs of countable sets. *Proc. Amer. Math. Soc.*, 111(3):841–844, 1991.

[114] Stevo Todorcevic. Remarks on Martin's axiom and the Continuum Hypothesis. *Canadian J. Math.*, 43:832–851, 1991.

[115] Stevo Todorcevic. Some applications of S and L combinatorics. volume 705 of *Ann. New York Acad. Sci.*, pages 130–167. New York Acad. Sci., New York, 1993.

[116] Stevo Todorcevic. Some partitions of three-dimensional combinatorial cubes. *J. Combin. Theory Ser. A*, 68:410–437, 1994.

[117] Stevo Todorcevic. Aronszajn orderings. *Publ. Inst. Math. Belgrade*, 57(71):29–46, 1995.

[118] Stevo Todorcevic. Comparing the continuum with the first two uncountable cardinals. In M.L. Dalla Chiara et al., editors, *Logic and Scientific Methods*, pages 145–155. Kluwer Academic Publishers, 1997.

[119] Stevo Todorcevic. Basis problems in combinatorial set theory. In *Proceedings of the International Congress of Mathematicians, Vol. II* (*Berlin*), pages 43–52, 1998.

[120] Stevo Todorcevic. Countable chain condition in partition calculus. *Discrete Math.*, 188:205–223, 1998.

[121] Stevo Todorcevic. Oscillations of sets of integers. *Adv. in Appl. Math.*, 20:220–252, 1998.

[122] Stevo Todorcevic. A dichotomy for P-ideals of countable sets. *Fund. Math.*, 166:251–267, 2000.

[123] Stevo Todorcevic. Lipschitz maps on trees. *Journal of the Inst. of Math. Jussieu*, 6:527–556, 2007.

[124] M. J. Tomkinson. FC-*groups*, volume 96 of *Research Notes in Mathematics*. Pitman (Advanced Publishing Program), Boston, MA, 1984.

[125] Boban Veličković. PhD thesis, University of Wisconsin, 1986.

[126] Dan Velleman. ω-morasses, and a weak form of Martin's axiom provable in ZFC. *Trans. Amer. Math. Soc.*, 285(2):617–627, 1984.

[127] Dan Velleman. Simplified morasses. *J. Symbolic Logic*, 49:257–271, 1984.

[128] Dan Velleman. On a combinatorial principle of Hajnal and Komjáth. *J. Symbolic Logic*, 51:1056–1060, 1986.

[129] Dan Velleman. Partitioning pairs of countable sets of ordinals. *J. Symbolic Logic*, 55(3):1019–1021, 1990.

[130] H.M. Wark. A non-separable reflexive Banach space on which there are few operators. *J. London Math. Soc.*, 64:675–689, 2001.

[131] N.M. Warren. Extending continuous functions in zero-dimensional spaces. *Proc. Amer. Math. Soc.*, 52:414–416, 1975.

[132] William Weiss. Partitioning topological spaces. In J. Nešetril and V. Rődl, editors, *Mathematics of Ramsey theory*, pages 155–171. Springer-Verlag, 1990.

[133] Neil H. Williams. *Combinatorial Set Theory*, volume 91 of *Studies in Logic and the Foundation in Mathematics*. North-Holland Publishing Company, Amsterdam, 1977.

[134] K. Wolfsdorf. Färbungen großer Würfel mit bunten Wegen. *Arch. Math.*, 40:569–576, 1983.

[135] W. Hugh Woodin. *The Axiom of Determinacy, Forcing Axioms, and the Nonstationary Ideal.* Walter de Gruyter & Co., Berlin, 1999.

Index

BIRKHÄUSER

Progress in Mathematics (PM)

Edited by
Hyman Bass, University of Michigan, USA
Joseph Oesterlé, Institut Henri Poincaré, Université Paris VI, France
Alan Weinstein, University of California, Berkeley, USA

Progress in Mathematics is a series of books intended for professional mathematicians and scientists, encompassing all areas of pure mathematics. This distinguished series, which began in 1979, includes research level monographs, polished notes arising from seminars or lecture series, graduate level textbooks, and proceedings of focused and refereed conferences. It is designed as a vehicle for reporting ongoing research as well as expositions of particular subject areas.

PM 263: Todorcevic, S.
Walks on Ordinals and Their Characteristics (2007)
ISBN 978-3-7643-8528-6

PM 262: Fargues, L. / Genestier, A. / Lafforgue, V.
L'isomorphisme entre les tours de Lubin-Tate et de Drinfield (due 2007)
ISBN 978-3-7643-8455-5

PM 261: Stout, E.L.
Polynomial Convexity (2007)
ISBN 978-0-8176-4537-3

PM 260: Holzapfel, R.-P. / Uludağ, A.M. / Yoshida, M. (Eds.)
Arithmetic and Geometry Around Hypergeometric Functions (2007)
ISBN 978-3-7643-8283-4

PM 259: Capogna, L. / Danielli, D. / Pauls, S.D. / Tyson, J.T.
An Introduction to the Heisenberg Group and the Sub-Riemannian Isoperimetric Problem (2007)
ISBN 978-3-7643-8132-5

PM 258: Gan, W.T. / Kudla, S.S. / Tschinkel, Y. (Eds.)
Eisenstein Series and Applications (due 2008). ISBN 978-0-8176-4496-3

PM 257: Lakshmibai, V. / Littelmann, P. / Seshadri, C.S. (Eds.)
Schubert Varieties (due 2007)
ISBN 978-0-8176-4153-5

PM 256: Borthwick, D.
Spectral Theory of Infinite-Volume Hyperbolic Surfaces (due 2007)
ISBN 978-0-8176-4524-3

PM 255: Kobayashi, T. / Schmid, W. / Yang, J.-H. (Eds.)
Representation Theory and Automorphic Forms (due 2007)
ISBN 978-0-8176-4505-2

PM 254: Ma, X. / Marinescu, G.
Holomorphic Morse Inequalities and Bergman Kernels (2007)
Winner of the Ferran Sunyer i Balaguer Prize 2006
ISBN 978-3-7643-8096-0

PM 253: Ginzburg, V. (Ed.)
Algebraic Geometry and Number Theory (2006). ISBN 978-0-8176-4471-7

PM 252: Maeda, Y. / Michor, P. / Ochiai, T. / Yoshioka, A. (Eds.)
From Geometry to Quantum Mechanics (2007). ISBN 978-0-8176-4512-8

PM 251: Boutet de Monvel, A. / Buchholz, D. / Iagolnitzer, D. / Moschella, U. (Eds.)
Rigorous Quantum Field Theory (2006)
ISBN 978-3-7643-7433-4

PM 250: Unterberger, A.
The Fourfold Way in Real Analysis (2006)
ISBN 978-3-7643-7544-7

Progress in Mathematics (PM)

Edited by

Hyman Bass, University of Michigan, USA
Joseph Oesterlé, Institut Henri Poincaré, Université Paris VI, France
Alan Weinstein, University of California, Berkeley, USA

BIRKHÄUSER

Progress in Mathematics is a series of books intended for professional mathematicians and scientists, encompassing all areas of pure mathematics. This distinguished series, which began in 1979, includes research level monographs, polished notes arising from seminars or lecture series, graduate level textbooks, and proceedings of focused and refereed conferences. It is designed as a vehicle for reporting ongoing research as well as expositions of particular subject areas.

PM 249: Kock, J. / Vainsencher, I.
An Invitation to Quantum Cohomology. Kontsevich's Formula for Rational Plane Curves (2006).
ISBN 978-0-8176-4456-7

PM 248: Bartholdi, L. / Ceccherini-Silberstein, T. / Smirnova-Nagnibeda, T. / Zuk, A.
(Eds.) Infinite Groups: Geometric, Combinatorial and Dynamical Aspects (2006)
ISBN 978-3-7643-7446-4

PM 247: Baues, H.-J.
The Algebra of Secondary Cohomology Operations (2006)
ISBN 978-3-7643-7448-8

PM 246: Dragomir, S. / Tomassini, G., Differential Geometry and Analysis on CR Manifolds (2006).
ISBN 978-0-8176-4388-1

PM 245: Fels, G. / Huckleberry, A.T. / J.A. Wolf
Cycle Spaces of Flag Domains. A Complex Geometric Viewpoint (2006). ISBN 978-0-8176-4391-1

PM 244: Etingof, P. / Retakh, V. / Singer, I.M. (Eds.)
The Unity of Mathematics (2006)
ISBN 978-0-8176-4076-7

PM 243: Bernstein, J. / Hinich, V. / A. Melnikov (Eds.)
Studies in Lie Theory (2006)
ISBN 0-8176-4342-3

PM 242: Dufour, J.-P. / Zung, N.T.
Poisson Structures and their Normal Forms (2005)
ISBN 978-3-7643-7334-4

PM 241: Seade, J.
On the Topology of Isolated Singularities in Analytic Spaces (2005).
Winner of the Ferran Sunyer i Balaguer Prize 2005
ISBN 978-3-7643-7322-1

PM 240: Ambrosetti, A. / Malchiodi, A.
Perturbation Methods and Semilinear Elliptic Problems in R^n (2005).
Winner of the Ferran Sunyer i Balaguer Prize 2005
ISBN 978-3-7643-7321-4

PM 239: van der Geer, G. / Moonen, B.J.J. / Schoof, R. (Eds.)
Number Fields and Function Fields – Two Parallel Worlds (2005)
ISBN 978-0-8176-4397-3

PM 238: Sabadini, I. / Struppa, D.C. / Walnut, D.F. (Eds.)
Harmonic Analysis, Signal Processing, and Complexity (2005).
ISBN 978-0-8176-4358-4

PM 237: Kulish, P.P. / Manojlovic, N. / Samtleben, H. (Eds.)
Infinite Dimensional Algebras and Quantum Integrable Systems (2005). ISBN 978-3-7643-7215-6

PM 236: Hotta, R. / Takeuchi, K. / Tanisaki, T.
D-Modules, Perverse Sheaves, and Representation Theory (due 2007)
ISBN 0-8176-4363-X

PM 235: Bogomolov, F. / Tschinkel, Y. (Eds.)
Geometric Methods in Algebra and Number Theory (2005). ISBN 0-8176-4349-4

PM 234: Kowalski, O. / Musso, E. / Perrone, D. (Eds.)
Complex, Contact and Symmetric Manifolds. (2005)
ISBN 0-8176-3850-4

PM 233: David, G.
Singular Sets of Minimizers for the Mumford–Shah Functional (2005)
ISBN 3-7643-7182-X

PM 232: Marsden, J.E. / Ratiu, T.S. (Eds.)
The Breadth of Symplectic and Poisson Geometry (2005)
ISBN 0-8176-3565-3

PM 231: Brion, M. / Kumar, S.
Frobenius Splitting Methods in Geometry and Representation Theory (2005)
ISBN 0-8176-4191-2

PM 230: Anker, J.-P. / Orsted, B. (Eds.)
Lie Theory. Harmonic Analysis on Symmetric Spaces — General Plancherel Theorems (2005)
ISBN 0-8176-3777-X

PM 229: Anker, J.-P. / Orsted, B. (Eds.)
Lie Theory. Unitary Representations and Compactifications of Symmetric Spaces (2005)
ISBN 0-8176-3526-2

PM 228: Anker, J.-P. / Orsted, B. (Eds.)
Lie Theory. Lie Algebras and Representations (2004)
ISBN 0-8176-3373-1

PM 227: Lepowski, J. / Li, H.
Introduction to Vertex Operator Algebras and Their Representations (2004)
ISBN 0-8176-3408-8

PM 226: Poonen, B. / Tschinkel, Y. (Eds.)
Arithmetic of Higher-Dimensional Algebraic Varieties (2004)
ISBN 0-8176-3259-X